10.6C
10.6A
10.6B
10.7
PL. 3
4.19
8.9
8.22
8.23
8.29
7.9
4.29
4.30
8.18
8.19
8.21
9.2
6.38
4.25
4.26

REMOTE SENSING

A SERIES OF BOOKS IN THE EARTH SCIENCES
EDITOR: ALLAN COX

REMOTE SENSING
PRINCIPLES AND INTERPRETATION

FLOYD F. SABINS, JR.

Chevron Oil Field Research Company,
University of Southern California,
and
University of California, Los Angeles

W. H. FREEMAN AND COMPANY
SAN FRANCISCO

Cover photograph from Floyd F. Sabins, "Aerial Camera Mount for 70-mm Stereo," *Photogrammetric Engineering,* vol. 19 (1973), Figure 6.

Library of Congress Cataloging in Publication Data

Sabins, Floyd F.
Remote sensing.

Includes bibliographies and index.
1. Remote sensing. I. Title.
G70.4.S15 621.36'7 77-27595
ISBN 0-7167-0023-9

Printed in the United States of America

9 8 7 6 5 4 3 2

To Janice, Barbara, and Edward

CONTENTS

PREFACE

This book is designed for a one-semester or one-quarter course for upper-class or graduate students with no previous training in remote sensing. It should also be useful for short courses and as a reference for workers in the remote-sensing field. Courses in introductory physics, physical geography, and physical geology would provide useful, but not essential, background for users of this book. The presentation attempts to strike a balance between the physical principles that control remote-sensing processes and the practical interpretation and use of the imagery for a variety of applications.

The first chapter of the book summarizes the fundamental characteristics of electromagnetic radiation and the reactions with matter that are basic to all forms of remote sensing. The vital concepts of spatial resolution and detection are explained, using the eye as an example of a remote-sensing system. The next chapters describe the following remote-sensing systems: aerial photography, manned satellite imagery, Landsat, thermal infrared, and radar imagery. For each system the following topics are covered: (1) physical properties and electromagnetic interactions of materials that control the imaging process; (2) design and operation of the imaging system; (3) characteristics of the images, including defects and geometric distortion that may distract or confuse the interpreter; (4) guidelines and examples for interpreting images. There is a growing trend toward quantitative interpretation of images and toward the use of mathematical models to understand the interaction between electromagnetic radiation and materials. These techniques are described and illustrated together with some practical examples.

The chapter on Digital Image Processing describes computer techniques for restoration and enhancement of images and for information extraction. This is a rapidly expanding technology with great potential that should be included in every remote-sensing curriculum. The remaining chapters describe practical applications of remote-sensing to resource exploration, environmental monitoring, recognition of natural hazards, and comparison of different types of images.

Because I am a geologist by training and early experience, there is a tendency to emphasize the geologic aspects of image interpretation. Over the years many of my students have been engineers, geographers, ecologists, and other nongeology majors. There has been no significant difference in the performance of these students relative to the geologists. This is partially due to the versatility of the students and also to the fact that understanding the geologic background is essential to most other applications of remote sensing.

As shown on the index maps inside the front and back covers, I have attempted to achieve some geographic balance in the distribution of image examples. There is, however, a bias toward images from the western United States because of the prevailing good flying weather and the geographic and geologic diversity of this region. In addition, I have been able to field check the western examples. Instructors may wish to supplement these examples with images from their own areas of interest. To aid in obtaining local coverage, the appropriate chapters of the book provide directions to sources of available images. These local images can also be the basis for class field trips. Course evaluations submitted by my students rate the field trips as a valuable portion of the course. The students interpret the local images before going into the field, where they then check and modify the interpretations for inclusion in their reports.

My introduction to remote sensing was through using black-and-white aerial photographs as a student and as a field geologist. Later at Chevron Oil Field Research Company in the late 1950s, I helped evaluate the exploration potential of color and infrared color aerial photographs. In the early 1960s, censored examples of side-looking airborne radar and thermal infrared imagery began to appear. These military reconnaissance techniques were classified for security reasons, but reports indicated that geologic features could be mapped on the images. Having obtained the necessary security clearances, I investigated these techniques and was convinced of their value for resource exploration. When security restraints were partially relaxed in the middle and late 1960s, Chevron began to test the new methods in exploration areas ranging from the arctic to the tropics. It was my good fortune to be assigned the responsibility for planning and executing these surveys and interpreting the results, in cooperation with Chevron geologists and pilots. My present research is directed toward the utilization of digital data acquired by the Landsat program.

For the past ten years I have conducted a graduate seminar in remote sensing at the University of Southern California. Because no textbook existed, the students relied on handout material and outside reading assignments. In 1975 I was appointed Regents' Professor of Geology for the fall quarter at the University of California at Los Angeles, where I taught a remote-sensing course for upper-class and graduate students. In preparation for this appointment I wrote a syllabus that was used by my 1975 and 1976 classes at UCLA and my 1977 class at USC. The syllabus has also been used at Stanford University by Professor R. J. P. Lyon and at Purdue University by Professor D. W. Levandowski. On the basis of these classroom experiences, in addition to comments from other colleagues, the original syllabus has been extensively revised to produce this book.

The preparation of this book would have been impossible without the generosity and cooperation of colleagues in government, industry, and universities who have provided data and images. The sources of images and data are acknowledged in the figure captions, and I would like to express my gratitude to all of these people. The complete manuscript was reviewed by G. Lennis Berlin

(Northern Arizona University), Allan V. Cox (Stanford University), C. L. Kober (Colorado State University), Donald W. Levandowski (Purdue University), Ronald J. P. Lyon (Stanford University), and Robert G. Raynolds (Stanford University). These reviews offered many useful suggestions for improving the manuscript.

The following colleagues reviewed portions of the manuscript related to their fields of interest: James C. Barnes (Environmental Research and Technology, Inc.), Laurence C. Breaker (National Oceanic and Atmospheric Administration), Pat S. Chavez (U.S. Geological Survey), Robert N. Colwell (University of California at Berkeley), Christopher S. Condit (U.S. Geological Survey), A. T. Edgerton (Aerojet Electrosystems), R. Bryan Erb (NASA), Norman L. Fritz (Eastman Kodak Research Laboratories), Seth I. Gutman (Santa Fe International), Yngvar Y. Isachsen (New York State Geological Survey), Leonard Jaffe (NASA), Hugh Kieffer (University of California at Los Angeles), Glenn H. Landis (EROS Data Center), Paul D. Lowman (NASA), Paul M. Merifield (University of California at Los Angeles), Terry W. Offield (U.S. Geological Survey), Thomas R. Ory (Daedalus Enterprises, Inc.), Donald W. Peterson (U.S. Geological Survey), A. Pressman (Environmental Protection Agency), Samuel S. Rifman (TRW Space Systems), Lawrence C. Rowan (U.S. Geological Survey), Gerald G. Schaber (U.S. Geological Survey), Robert G. Schmidt (U.S. Geological Survey), Swan A. Sie (Chevron Oil Field Research Company), Lawrence A. Soderblom (U.S. Geological Survey), Charles S. Spirakis (U.S. Geological Survey), James V. Taranik (EROS Data Center), Allen H. Watkins (EROS Data Center), Kenneth Watson (U.S. Geological Survey), and Richard S. Williams, Jr. (U.S. Geological Survey).

My interest in teaching remote sensing began in 1967 at the Geology Department of the University of Southern California and was encouraged under the successive chairmanships of William H. Easton, the late Orville N. Bandy, and Richard O. Stone. The stimulus for preparing the book came from my part-time teaching assignments in the Department of Earth and Space Sciences of the University of California at Los Angeles under the chairmanship of Clarence A. Hall. I am grateful to these universities and their students for enabling me to develop and test many of the concepts in this book.

My greatest debt of gratitude is to my employer for the past 22 years, the Chevron Oil Field Research Company, and its parent company, the Standard Oil Company of California. N. Allan Riley, James R. Baroffio, E. W. Jones, John E. McCall, and other members of the Chevron management have encouraged the outside teaching and writing activities of the staff members. I have learned much about the practical applications of remote sensing from working in the field and in the laboratory with exploration and research personnel in the Chevron organization. I am also grateful to my fellow employees in the stenographic, drafting, photographic, and reproduction sections who contributed much in the preparation of the manuscript and illustrations.

Floyd F. Sabins
La Habra, California
December 1977

REMOTE SENSING

1 FUNDAMENTAL CONSIDERATIONS

Remote sensing may be broadly defined as the collection of information about an object without being in physical contact with the object. Aircraft and satellites are the common platforms from which remote sensing observations are made. The term *remote sensing* is restricted to methods that employ electromagnetic energy as the means of detecting and measuring target characteristics. Electromagnetic energy includes light, heat, and radio waves. This definition of remote sensing excludes electrical, magnetic, and gravity surveys that measure force fields, rather than electromagnetic radiation. Magnetic and electrical surveys are commonly made from aircraft, but are considered airborne geophysical surveys rather than remote sensing.

Aerial photography is the original form of remote sensing and is widely used for topographic mapping, engineering and environmental studies, and exploration for oil and minerals. These successful applications, using only the visible portion of the electromagnetic spectrum, suggest that valuable additional information might be obtained by using other wavelength regions. In the 1960s, technologic developments enabled imagery to be acquired at other wavelengths, including thermal infrared and microwave. The development and deployment of manned and unmanned earth satellites in the 1960s provided an orbital vantage point for acquiring imagery of the earth. The description of all these methods and interpretation of the imagery are the subject of this book.

UNITS OF MEASURE

This book employs the International System of Units, which is a modernized metric system adopted in 1960. The following standard units and abbreviations will be used:

meter	m
second	sec
hertz	Hz
gram	g
radian	rad
degrees Celsius	°C

Distance is expressed in multiples and submultiples of meters, as shown in Table 1.1.

Temperature is given in degrees Celsius, °C, or in degrees Kelvin, °K, which is also known as the absolute temperature scale. A few examples are

TABLE 1.1
Metric nomenclature for distance

Unit	Symbol	Equivalent	Comment
Kilometer	km	1000 m = 10^3 m	
Meter	m	1.0 m = 10^0 m	Basic unit
Centimeter	cm	0.01 m = 10^{-2} m	
Millimeter	mm	0.001 m = 10^{-3} m	
Micrometer	μm	0.000001 m = 10^{-6} m	Formerly called micron (μ)
Nanometer	nm	10^{-9} m	
Angstrom	Å	10^{-10} m	Common unit in X-ray technology

given in degrees Fahrenheit, °F, where conversion to °C would be inconvenient. In fractional statements, units in the denominator are identified by a negative exponent. Roots are indicated by fractional exponents. For example, the property called thermal inertia, P, is expressed as

$$P = 0.53 \text{ cal} \cdot \text{cm}^{-2} \cdot \text{sec}^{-1/2} \cdot {}^{\circ}\text{C}^{-1}$$

This expression means that thermal inertia, P, equals 0.53 calories per square centimeter per second to the square root per degree Celsius.

In a few instances, where they are appropriate for clarity, the English units for distance are used. For example, American aerial photographers might not recognize their standard film size if it were given as 22.8 by 22.8 cm instead of being given in the familiar 9 by 9 in. dimensions. Where English units are used, the metric equivalents are shown in parentheses.

ELECTROMAGNETIC ENERGY

Electromagnetic energy refers to all energy that moves with the velocity of light in a harmonic wave pattern. The word *harmonic* implies that the component waves are equally and repetitively spaced in time. The wave concept explains the propagation of electromagnetic energy, but this energy is detectable only in terms of its interaction with matter. In this interaction, electromagnetic energy behaves as though it consists of many individual bodies called *photons* that have such particle-like properties as energy and momentum. The dual concept of waves and particles may be demonstrated for light. The bending (*refraction*) of light as it propagates through media of different optical densities, such as air and glass, may be analyzed in terms of waves. When the intensity of light is measured with a light meter, however, the interaction of photons with the light-sensitive photodetector produces an electric signal that varies in strength proportionally with the number of photons.

Properties of Electromagnetic Waves

Electromagnetic waves can be described in terms of their velocity, wavelength, and frequency. All electromagnetic waves travel at the same speed, c. In a vacuum $c = 299{,}793 \text{ km} \cdot \text{sec}^{-1}$, or for practical purposes $c = 3 \cdot 10^8 \text{ m} \cdot \text{sec}^{-1}$. This is commonly spoken of as the *speed of light,* although light is only one form of electromagnetic energy.

The *wavelength,* λ, of electromagnetic waves is the distance from any position in a cycle to the same position in the next cycle, measured in the standard metric system. The micrometer (Table 1.1) is a convenient unit for designating the wavelength of both visible and infrared radiation, although optical scientists commonly employ nanometers for visible light to avoid decimal numbers. Electromagnetic radiation is classified on the basis of wavelength into regions, or *bands*, such as the visible band that ranges from 0.4 to 0.7 μm in wavelength.

Frequency, v, is the number of wave crests passing a given point in a specified unit of time. Frequency was formerly expressed as "cycles per second," but today *hertz* is the unit for a frequency

TABLE 1.2
Terms used to designate frequencies

Unit	Symbol	Frequency
Hertz	Hz	1 cycle $\cdot$ sec^{-1}
Kilohertz	kHz	10^3 cycles $\cdot$ sec^{-1}
Megahertz	MHz	10^6 cycles $\cdot$ sec^{-1}
Gigahertz	GHz	10^9 cycles $\cdot$ sec^{-1}

of one cycle per second. The terms used to designate frequencies are shown in Table 1.2. Unlike velocity and wavelength, which change as electromagnetic energy is propagated through media of various densities, frequency remains constant and is therefore a more fundamental property. Frequency nomenclature is employed by electronic engineers for designating radio and radar energy regions. In this book wavelength is used (rather than frequency) to facilitate comparison of all wavelengths of electromagnetic radiation. The relationship of velocity, c, wavelength, λ, and frequency, v, is shown by the expression

$$c = \lambda v \tag{1.1}$$

Energy Distribution

Figure 1.1 shows the relationship between wavelength and the amount of energy radiated for energy sources at different temperatures expressed in degrees Kelvin (°K − 273 = °C). The sun, with a surface temperature of almost 6000°K, radiates enormous amounts of energy at wavelengths all across the ultraviolet, visible, and infrared bands with the maximum concentration of energy occurring at a wavelength of about 0.5 μm, corresponding to green light. Thus during daylight hours the maximum energy incident on the earth and reflected from the earth is in the visible band with a maximum energy peak at a wavelength of 0.5 μm. The average surface temperature of the earth is 290°K (17°C), which is also called the *ambient temperature*. The energy radiated from the earth at this temperature is distributed as a broad low curve in the infrared band with a peak concentration at 9.7 μm (Figure 1.1). This radiant energy level is very low in comparison to reflected solar energy, but is dominant at night and provides energy for remote sensing in the infrared band.

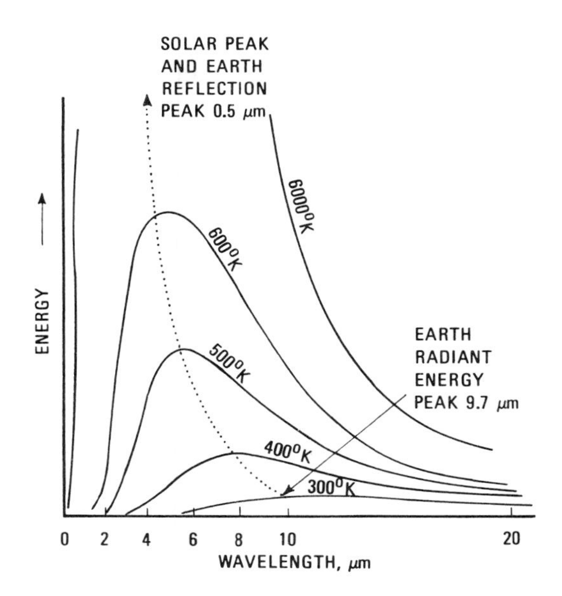

FIGURE 1.1
Spectral distribution curves of energy radiated from objects at different temperatures. From Colwell and others (1963, Figure 2).

Interaction Mechanisms

A number of interactions are possible when electromagnetic energy encounters matter, whether solid, liquid, or gas. The interactions that take place at the surface of a substance are called *surface phenomena*. Penetration of electromagnetic radiation beneath the surface of a substance results in interactions called *volume phenomena*. The surface and volume interactions with matter can produce a number of changes in the incident electromagnetic radiation; primarily changes of magnitude, direction, wavelength, polarization, and phase (Janza and others, 1975, p. 77). The science of remote sensing detects and records these changes. The resulting images and data are interpreted to identify remotely the characteristics of the matter that produced the changes in the recorded electromagnetic radiation.

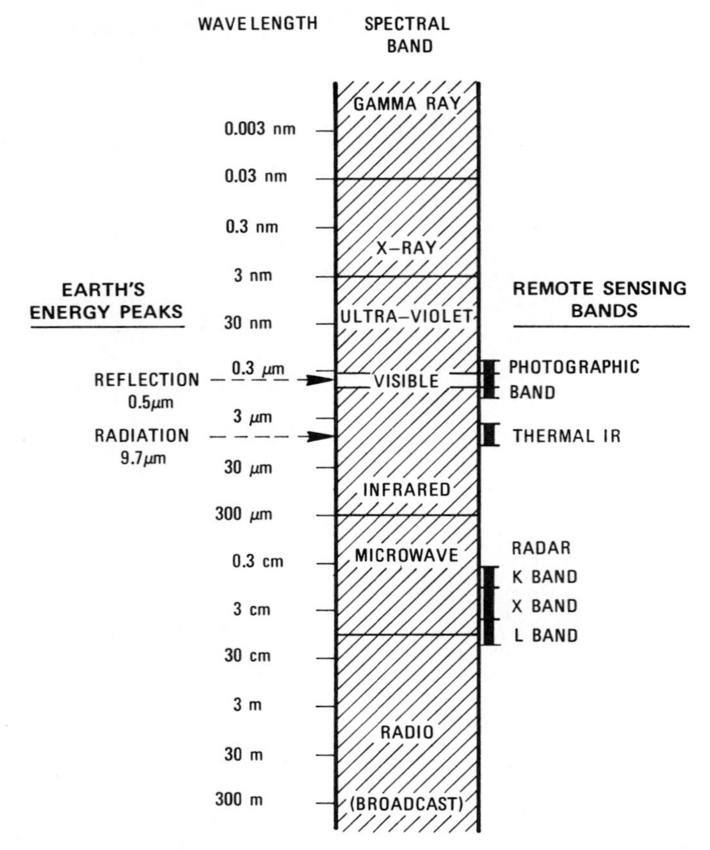

FIGURE 1.2
Electromagnetic spectrum showing bands employed in remote sensing.

During the interaction between electromagnetic radiation and matter, both mass and energy are conserved according to basic physical principles. The following interactions may occur:

1. Radiation may be *transmitted*, that is, passed through the substance. The velocity of electromagnetic radiation changes as it is transmitted from air, or a vacuum, into other substances. This velocity change is called the *index of refraction, n,* and is expressed as

$$n = \frac{c_a}{c_s} \tag{1.2}$$

   where c_a is the velocity in air and c_s is the velocity in the substance.
2. Radiation may be *absorbed* by a substance and give up its energy largely to heating the substance.
3. Radiation may be *emitted* by a substance as a function of its structure and temperature. All matter at temperatures above absolute zero, 0°K, emits energy.
4. Radiation may be *scattered*, that is deflected in all directions and lost ultimately to absorption or further scattering. The scattering of light by the atmosphere is a familiar example.
5. Radiation may be *reflected*, that is, returned unchanged from the surface of a substance with the angle of reflection equal and opposite to the angle of incidence.

These interactions with any particular form of matter are selective with regard to the wavelength of electromagnetic radiation and are specific for that form of matter, depending primarily upon its surface properties and its atomic and molecular structure (Colwell and others, 1963, p. 765). These interactions between matter and energy provide the basis for remote sensing.

TABLE 1.3
Electromagnetic spectral bands

Band	Wavelength	Remarks
Gamma ray	<0.03 nm	Incoming radiation from the sun is completely absorbed by the upper atmosphere, and is not available for remote sensing. Gamma radiation from radioactive minerals is detected by low-flying aircraft as a prospecting method.
X-ray	0.03 to 3 nm	Incoming radiation is completely absorbed by atmosphere. Not employed in remote sensing.
Ultraviolet, UV	3 nm to 0.4 μm	Incoming UV radiation at wavelengths <0.3 μm is completely absorbed by ozone in the upper atmosphere.
Photographic UV	0.3 to 0.4 μm	Transmitted through the atmosphere. Detectable with film and photodetectors, but atmospheric scattering is severe.
Visible	0.4 to 0.7 μm	Detected with film and photodetectors. Includes earth reflectance peak at about 0.5 μm.
Infrared, IR	0.7 to 300 μm	Interaction with matter varies with wavelength. Atmospheric transmission windows are separated by absorption bands.
Reflected IR	0.7 to 3 μm	This is primarily reflected solar radiation and contains no information about thermal properties of materials. Radiation from 0.7 to 0.9 μm is detectable with film and is called *photographic IR radiation.*
Thermal IR	3 to 5 μm 8 to 14 μm	These are the principal atmospheric windows in the thermal region. Imagery at these wavelengths is acquired through the use of optical-mechanical scanners, not by film.
Microwave	0.3 to 300 cm	These longer wavelengths can penetrate clouds and fog. Imagery may be acquired in the active or passive mode.
Radar	0.3 to 300 cm	Active form of microwave remote sensing.

ELECTROMAGNETIC SPECTRUM

The *electromagnetic spectrum* is the continuum of energy ranging from kilometers to nanometers in wavelength, traveling at $3 \cdot 10^8$ m $\cdot$ sec^{-1}, and capable of propagation through a vacuum such as outer space. All matter radiates a range of electromagnetic energy, with the peak intensity shifting toward progressively shorter wavelengths with increasing temperature (Figure 1.1).

Wavelength Bands

The wavelength bands of the electromagnetic spectrum are shown in Figure 1.2 and are described in Table 1.3. Boundaries between wavelength bands are gradational. Those shown in Figure 1.2 are arbitrary, and some versions of this chart show overlapping boundaries to emphasize the gradations. The right side of Figure 1.2 shows the wavelength bands commonly employed in remote sensing. In Table 1.3 note that radiation in the short wavelength portion of the IR band and the long wavelength portion of the UV band is detectable by photographic film. Therefore these wavelengths are included with the visible wavelengths and are designated as the *photographic remote-sensing band.*

Electromagnetic radiation in the different wavelength bands interacts differently with matter. The interaction mechanisms for each remote-sensing band are described in subsequent chapters of this book, together with the technology and applications.

A *passive* remote-sensing system records the energy naturally radiated or reflected from an object. An *active* remote-sensing system supplies its own source of energy, which is directed at the object in order to measure the returned energy. Flash photography is active remote sensing in contrast to available-light photography, which is passive. The other common form of active remote sensing is radar (Table 1.3), which provides its own source of electromagnetic energy in the microwave region.

Atmospheric Effects

Our eyes inform us that the atmosphere is essentially transparent to light, and we tend to assume that this condition exists for all electromagnetic

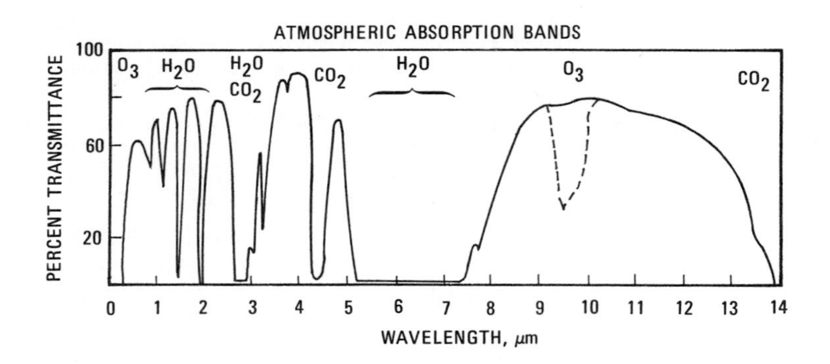

FIGURE 1.3
Transmission of energy through the atmosphere as a function of wavelength. Wavelength regions of high transmittance are atmospheric windows. Gases responsible for absorption are noted. From Santa Barbara Research Center (1975).

radiation. In fact, however, the gases of the atmosphere selectively scatter light of different wavelengths. The gases also absorb electromagnetic energy at specific wavelength intervals called *absorption bands.* The intervening regions of high energy transmittance are called *atmospheric transmission bands,* or windows. The transmission and absorption bands are shown in Figure 1.3, together with the gases responsible for the absorption bands.

Wavelengths shorter than 0.3 μm are completely absorbed by the ozone, O_3, layer in the upper atmosphere (Figure 1.3). This absorption is essential for life, because prolonged exposure to radiation of these wavelengths destroys living tissue. As an example of this effect, consider how readily sunburn occurs at high mountain elevations. Sunburn is caused by UV radiation, much of which is absorbed by the atmosphere at sea level. At higher elevations, however, there is less atmosphere to absorb the UV energy. This is the basis for concern about activities of man that might alter the ozone layer.

In clouds, water occurs as aerosol-sized particles of liquid rather than vapor. Clouds absorb and scatter electromagnetic radiation at wavelengths less than about 0.3 cm. Only radiation of microwave and longer wavelengths is capable of penetrating clouds without being scattered, reflected, or absorbed.

IMAGE CHARACTERISTICS

An *image* is the general term for any pictorial representation, irrespective of the wavelength or imaging device used to produce the image. Although image is a general term, it is commonly restricted to pictures that are detected by means other than photosensitive film. Most images are displayed on film after having been detected by a nonphotographic remote-sensing system, such as a radar antenna or thermal IR scanner. A *photograph* is an image formed by electromagnetic radiation that is detected by photosensitive chemicals on a film. For purposes of remote sensing, the electromagnetic energy detectable by film is restricted to the photographic wavelength region from approximately 0.3 to 0.9 μm.

All images can be described in terms of certain fundamental properties regardless of the wavelength at which the image is recorded. These common fundamental properties are scale, brightness, contrast, and resolution. Tone and texture of images are functions of these fundamental properties.

Scale

Scale is the ratio of the distance between two points on an image or map to the corresponding distance on the ground. A common scale on U.S.

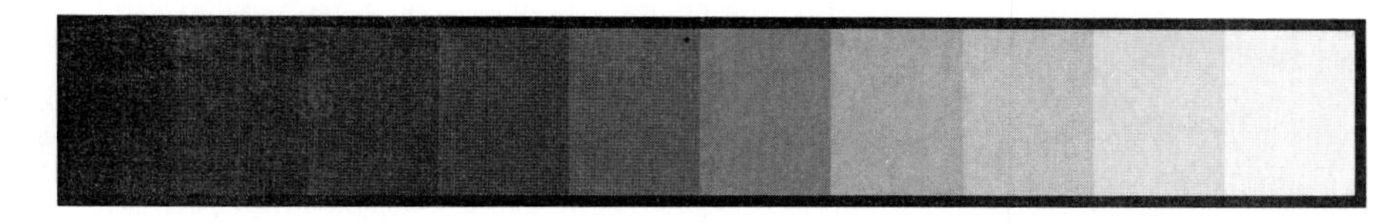

FIGURE 1.4
Gray scale.

Geological Survey maps is 1:24,000, which means that one unit on the map equals 24,000 units on the ground. Thus 1 cm on the map represents 240 m on the ground, or 1 in. represents 2,000 ft. Scales are shown graphically as bars in the maps and images of this book. Image scale is determined by:

1. The angular field of view of the remote-sensing device.
2. The altitude from which the image is acquired.
3. The magnification factor employed in reproducing the image.

The use of earth satellites to acquire images has changed the concepts of image scale. In this book, the relative scales of images are designated as follows:

Small scale < 1:500,000	1 cm = 5 km or more (1 in. = 8 mi or more)
Intermediate scale 1:50,000 to 1:500,000	1 cm = 0.5 to 5 km (1 in. = 0.8 to 8 mi)
Large scale > 1:50,000	1 cm = 0.5 km or less (1 in. = 0.8 mi or less)

This is a departure from the traditional scale designations employed for aerial photographs. Twenty years ago 1:62,500 was the smallest scale original photograph that was commercially available. Today sensor systems on high-altitude aircraft and satellites can acquire photographs and images at very small scales. The optimum scale of images is determined by the nature of the interpretation project. With the advent of satellite imagery, many investigators have been surprised at the amount and types of information that can be interpreted from very small-scale images.

Image Brightness and Tone

The electromagnetic radiation that is reflected, emitted, or scattered by an object is detected by remote-sensing systems at the wavelength bands shown in Figure 1.2. Variations in intensity of electromagnetic radiation from the terrain are commonly displayed as variations in brightness on black-and-white images. On positive images, such as those in this book, the brightness of objects is proportional to the intensity of electromagnetic radiation that is detected from that object.

Brightness is the magnitude of the response produced in the eye by light, and is a subjective sensation that can be determined only approximately. *Luminance* is a quantitative measure of the intensity of light from a source, and is measured with a device called a photometer or light meter. Image interpreters rarely make quantitative measurements of brightness variations on an image. Brightness variations may be calibrated with a *gray scale* such as the one in Figure 1.4. The term *tone* is used for each distinguishable shade from black to white. In practice, most interpreters do not use an actual gray scale the way one would use a centimeter scale; they have a mental concept of a gray scale and characterize areas on an image as light, intermediate, or dark in tone.

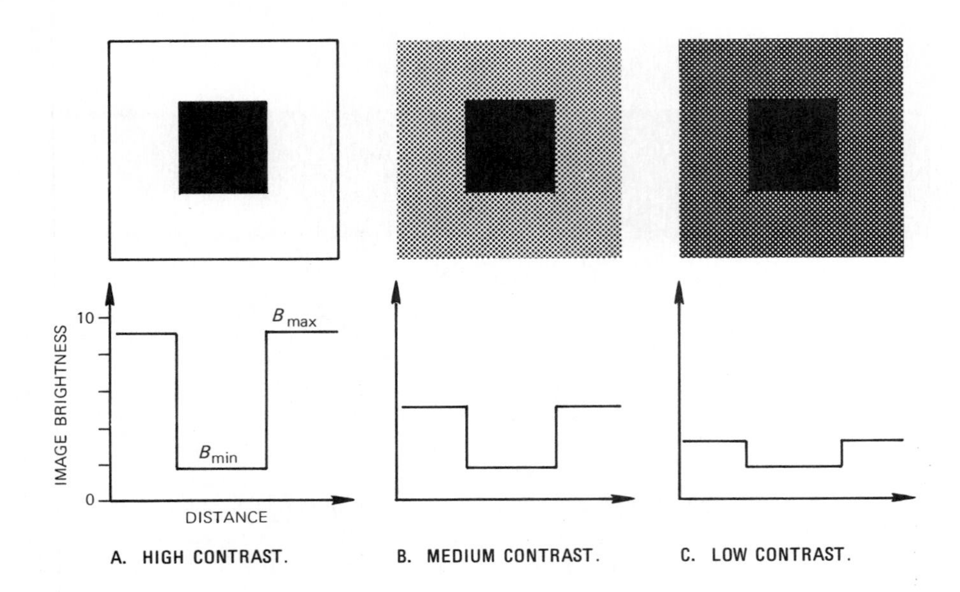

FIGURE 1.5
Images of different contrast ratio with corresponding brightness profiles.

On aerial photographs the tone of an object is primarily determined by the ability of the object to reflect incident light, although atmospheric effects and the spectral senstivity of the film are also factors. On other forms of remote sensor images, tone is determined by other physical properties of objects. On a thermal IR image the brightness of an object is proportional to the heat radiated from the object. The brightness on a radar image is determined by the extent to which the transmitted beam of radar energy is reflected back to the receiving antenna of the radar system.

Image Contrast

One definition of image contrast is the ratio between the brightest and darkest parts of the image. Images with high, medium, and low contrast are shown diagrammatically in Figure 1.5, together with plots of brightness variation across each image. Several expressions are commonly used to describe contrast, and there may be confusion in their usage. *Contrast ratio*, C_r, is widely used and is defined as:

$$C_r = \frac{B_{max}}{B_{min}} \qquad (1.3)$$

where B_{max} is maximum brightness of the scene and B_{min} is minimum brightness. On a brightness scale of 0 to 10 the scenes in Figure 1.5 have the following contrast ratios:

A. High contrast $\quad C_r = \frac{9}{2} = 4.5$

B. Medium contrast $\quad C_r = \frac{5}{2} = 2.5$

C. Low contrast $\quad C_r = \frac{3}{2} = 1.5$

Note that when $B_{min} = 0$, C_r is infinity; when $B_{min} = B_{max}$, C_r is unity. This discussion was summarized from the extensive review by Slater (1975) that describes other terms for contrast. In addition to describing an entire scene, contrast ratio is also used to describe the ratio of brightness between an object and the adjacent background on an image. Contrast ratio is a vital factor in determining the ability to resolve and detect objects.

Images with a low contrast ratio are commonly referred to as "washed out" with monotonous, nearly uniform tones of gray. Low contrast may result from the following causes:

1. The individual objects and background that make up the terrain may have a nearly uni-

form electromagnetic response at the wavelength band of energy that is recorded by the remote-sensing system. In other words, the scene itself has a low contrast ratio.

2. Scattering of electromagnetic energy by the atmosphere can reduce the contrast of a scene. This effect is most pronounced in the shorter wavelength portions of the photographic remote-sensing band and is described in Chapter 2.
3. The remote-sensing system may lack sufficient sensitivity to detect and record the contrast of the terrain. Also incorrect recording techniques can result in low-contrast imagery although the scene has a high-contrast ratio.

Images with a low contrast ratio, regardless of the cause, can have the contrast improved by photographic and digital methods that are discussed in Chapter 4 on Landsat Imagery and in Chapter 7 on Digital Image Processing.

Resolution and Resolving Power

The term resolving power applies to an imaging system or a component of the system, whereas resolution applies to the image produced by a system (Slater, 1975, p. 243). The lens and film of a camera system each have a characteristic resolving power that (together with other factors) determines the resolution of the photographs. As used in remote sensing, resolution is the ability to distinguish between two closely spaced objects on an image. More specifically, *resolution* is the minimum separation between two objects at which the objects appear distinct and separate on an image. Objects spaced closer together than the resolution limit appear as a single object on the image.

In photography, resolution and resolving power are customarily defined as the number of line-pairs per unit distance that are just discernable by the human eye looking at the photograph of a standard *resolution target* of standard contrast (Figure 1.6A) under specified conditions of illumination and magnification. The spacing of the resolution target in Figure 1.6A is expressed in line-pairs per cm, but the lines are actually black bars separated by white bars of the same width. For the target with 5 line-pairs $\cdot$ cm^{-1}, each bar is 0.1-cm wide and separated from the adjacent bar by the same distance. Human judgment and visual characteristics are critical components in this analysis, which therefore is not completely objective and reproducible. Resolution is different for objects of different shape, size, arrangement, and contrast ratio. The *modulation transfer function*, MTF, more completely describes the image-forming properties of a remote-sensing system than does resolving power. A description of MTF is not within the scope of this book, but can be found elsewhere (Perrin, 1966 and Slater, 1975).

Resolving power also may be expressed as the angle subtended between the imaging system and two targets spaced at the minimum resolvable distance. The radian system of angular measurement is commonly used to describe angular resolution. As shown in Figure 1.7 a *radian, rad,* is the angle subtended by an arc BC of a circle having a length equal to the radius AB of the circle. Because the circumference of a circle has a length equal to 2π times the radius, there are 2π rad in a circle, or 6.28 rad in a circle. A radian corresponds to 57.29° or 3,438 minutes, and a *milliradian, mrad,* is 10^{-3} radians. In the radian system of angular measurement

$$\text{angle} = \frac{L}{r} \text{ rad} \qquad (1.4)$$

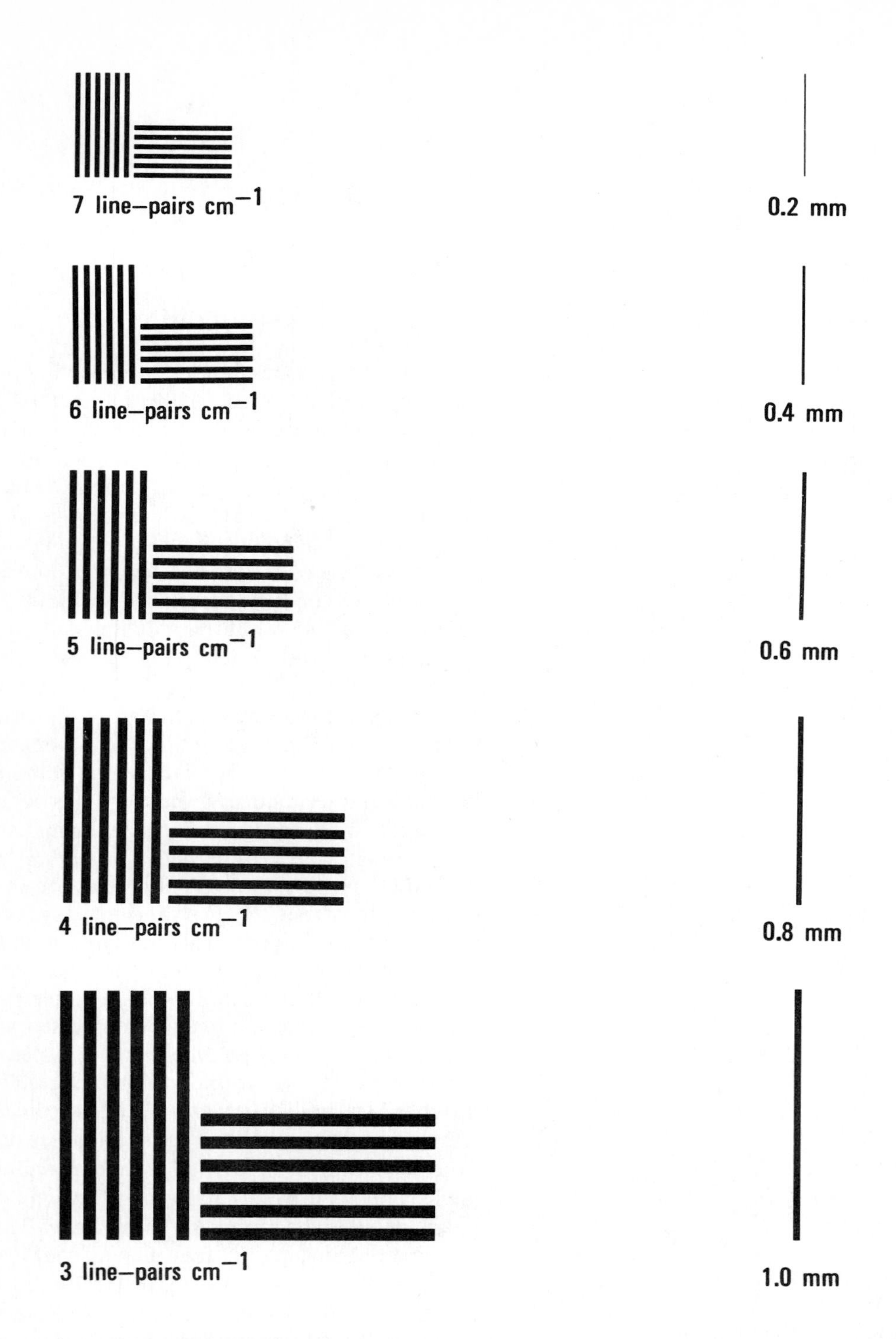

FIGURE 1.6
Resolution and detection targets with high contrast ratio. View this chart from a distance of 5 m (16.5 ft). For A, determine the most closely spaced set of bars that you can resolve. For B, determine the narrowest bar that you can detect.

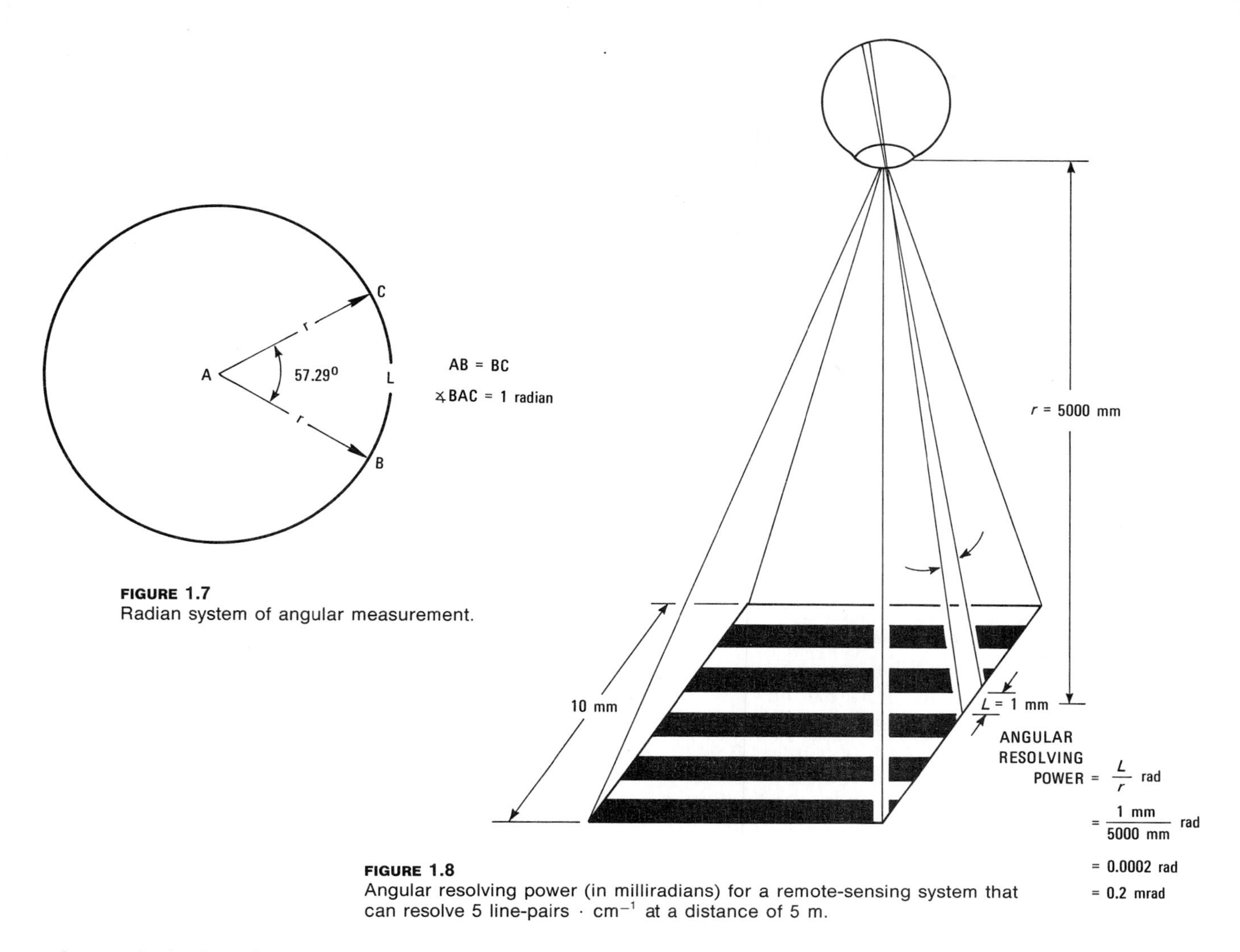

FIGURE 1.7
Radian system of angular measurement.

FIGURE 1.8
Angular resolving power (in milliradians) for a remote-sensing system that can resolve 5 line-pairs · cm^{-1} at a distance of 5 m.

where L is the length of the subtended arc and r is the radius of the circle. A convenient relationship is that at a distance r of 1,000 units, 1 mrad subtends an arc L of 1 unit. Figure 1.8 illustrates the angular resolving power of a remote-sensing system that can resolve the bar chart in the center of Figure 1.6 at a distance of 5 m. This chart has 5 line-pairs · cm^{-1}, and the bars are separated by 1 mm. For these targets with a high-contrast ratio the angular resolving power is 0.2 mrad.

Resolving power is discussed for each remote-sensing system described in this book, but the reader should remember the following points:

1. Theoretical resolving power of a system is rarely achieved in actual operation.

2. Resolution alone does not adequately describe suitability of an image for a particular application.

3. Resolution is the minimum separation between two objects for which the images appear distinct and separate; it is *not* the size of the smallest object that can be seen. By knowing the resolution and scale of an image, however, the size of the smallest detectable object can be estimated.

Other Characteristics of Images

Detectability is the ability of an imaging system to record the presence or absence of an object, although the identity of the object may be unknown. An object may be detected even though it is smaller than the resolving power of the imaging system.

Recognizability is the ability to identify an object on an image. Objects can be detected and resolved and yet not be recognizable. For example, roads on an image appear as narrow lines, but

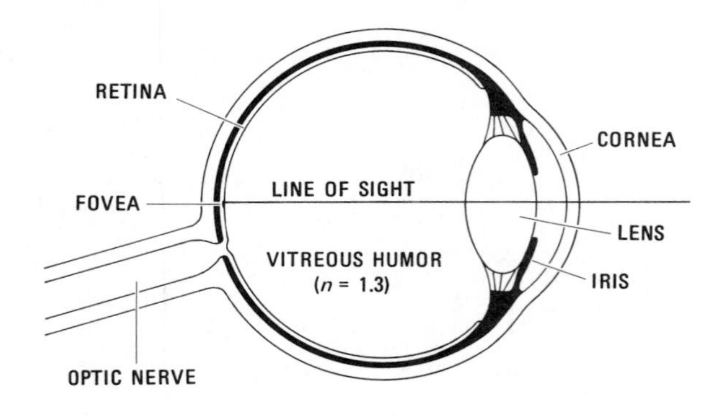

FIGURE 1.9
Structure of the human eye.

these could also be railroads or canals; therefore the lines have been detected but not recognized. Unlike resolution, there are no quantitative measures for recognizability and detectability. It is important for the interpreter to understand the significance and correct usage of these terms.

A *signature* is the expression of an object on an image that enables the object to be recognized. Signatures are determined by the characteristics of an object that determine its interaction with electromagnetic energy. Signatures of objects are generally different at different wavelengths.

Texture is the frequency of change and arrangement of tones on an image. Fine, medium, and coarse are some terms used to describe texture.

An *interpretation key* is a characteristic or combination of characteristics that enable an object to be identified on an image. Typical keys are size, shape, tone, and color. The associations of different characteristics are valuable keys. On images of cities, single-family residential areas may be recognized by the association of a dense street network, lawns, and small buildings. The associations of certain landforms and vegetation species are keys for identifying different types of rocks.

VISION

Of our five senses, two are able to detect electromagnetic radiation. The nerve endings of our sense of feel detect thermal IR radiation as heat, but do not form images. Vision is the most important sense and accounts for most of the information input to our brain. Vision is not only an important remote sensing system in its own right, but it is also the means by which we interpret the images produced by other remote-sensing systems. In the following discussion the human eye is analyzed as a remote-sensing system. Much of the information is summarized from Gregory (1966).

Structure of the Eye

The cross section of the human eye shown in Figure 1.9 is deceptively simple for such a complex structure. Light enters through the clear cornea, which is separated from the lens by fluid called *aqueous humor.* The *iris* is the pigmented part of the eye that controls the variable aperture called the *pupil.* It is commonly thought that variations in pupil size allow the eye to function over a wide range of light intensities. This is not the primary function, however, for the area of the pupil only varies over a ratio of 16:1, whereas the eye functions over a brightness range of about 100,000:1. Apparently the pupil contracts to limit the light rays to the central and optically best part of the lens, except when the full opening is needed in dim light. The pupil also contracts for near vision, increasing the depth of field for near objects.

A common misconception is that the lens refracts the incoming rays of light to form the image. The *cornea* actually forms the image because the refraction of light is determined by the difference in the refractive index, n, of the two adjacent optical media. For the eye this difference is at a maximum between the air ($n = 1.0$) and the cornea ($n = 1.3$), and this is where the maximum light refraction occurs. Although the *lens* is relatively unimportant for forming the image it is important for accommodating, or focusing, for near and far vision. In cameras, this accommodation is accomplished by changing the position of the lens relative to the film. In the eye the shape, rather than the position, of the lens is changed by varying the tension on the membrane attached to the margin of the lens. For near vision, tension is released allowing the lens to become thicker in the center and assume a more convex cross section. With age, the cells of the lens harden and the lens becomes too rigid to accommodate for different distances. This is the time in life when bifocal glasses may become necessary to provide for near and far vision.

The inverted image is focused on the *retina*, a thin sheet of interconnected nerve cells that includes the light-receptor cells called rods and cones that convert light into electrical impulses. The rods and cones are named for their longitudinal shape when viewed microscopically. The *cones* serve in daylight conditions to give color, or photopic, vision. The *rods* serve under low illumination and only give vision in tones of gray called scotopic vision. Rods and cones are not uniformly distributed throughout the retinal surface. The maximum concentration and organization of receptor cells is in the *fovea* (Figure 1.9), a small region at the center of the retina that provides maximum visual acuity. The existence and importance of the fovea can be demonstrated by concentrating on a single letter of a single word on this page. The rest of the page and even the nearby words and letters appear indistinct because they are outside the field of view of the fovea. The eye is in continual motion to bring the fovea to bear on all parts of the page or scene. Close to the fovea is the blind spot on the retina where the optic nerve joins the eye and there are no receptor cells. The electrical impulses from the receptor cells are transmitted to the brain where they are interpreted as the visual perception.

Resolving Power of the Eye

The diameter of the largest receptor cells in the fovea determines the resolving power of the eye. This maximum diameter is 3 μm, which is multiplied by the refractive index of the vitreous humor ($n = 1.3$) to determine the effective diameter of 4 μm for the receptor cells. The image distance from the retina to the lens is about 20 mm, or 20,000 μm. The effective width of the receptors is $4/20{,}000 = 1/5{,}000$ of the image distance. Image distance is proportional to object distance, which is the distance from the eye to the object. An object forms an image that fills the width of a receptor if the object width is 1/5,000 the object distance. Therefore adjacent objects must be separated by 1/5,000 the object distance in order for their images to fall on alternate receptors and be resolved by the eye.

The resolving power of the eye may be demonstrated by viewing the resolution targets of Figure 1.6A at a distance of 5 m (16.5 ft) and determining the most closely spaced set of line-pairs that can be resolved. Also, in Figure 1.6B, determine the narrowest of the bars that you can detect. The reader should make these determinations now, before proceeding further, because the following text may influence the reader's perception of the targets.

For the high-contrast resolution targets of Figure 1.6A at a distance of 5 m the normal eye should resolve the middle set that has 5 line-pairs $\cdot$ cm^{-1}. Spacing between the black bars is 1 mm, which at the distance of 5,000 mm gives the same ratio of 1/5,000 that was derived from the analysis of the visual receptors. This is also the same L/r ratio shown in Figure 1.8, which means that the eye has an angular resolving power of 0.2 mrad. This means that at a distance of 1,000 units, the eye can resolve high-contrast targets that are spaced no closer than 0.2 units.

Detection Capability of the Eye

Upon viewing the detection targets of Figure 1.6B from a distance of 5 m, most students report that they can detect the narrowest bar, which is 0.2 mm wide. Recall, however, that at this distance the minimum separation at which targets can be resolved is 1 mm. This is a graphic illustration of the difference between resolution and detection. Detection is influenced not only by the size of objects but also by their shape and orientation.

7 line–pairs cm^{-1}

0.2 mm

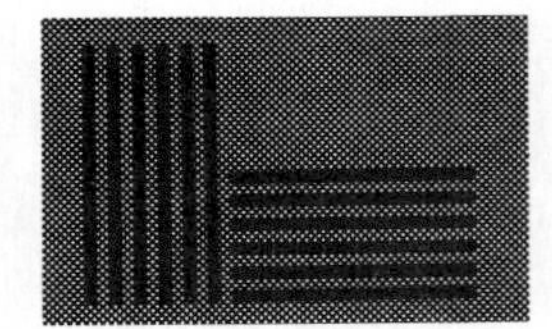

6 line–pairs cm^{-1}

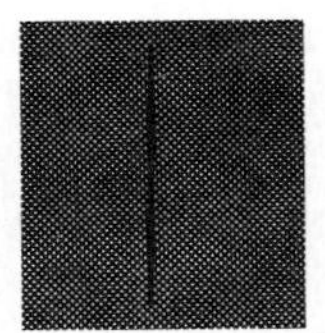

0.4 mm

5 line–pairs cm^{-1}

0.6 mm

4 line–pairs cm^{-1}

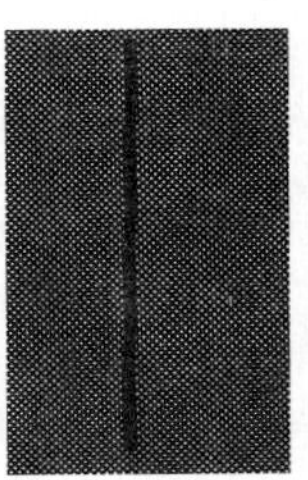

0.8 mm

3 line–pairs cm^{-1}

1.0 mm

A. RESOLUTION TARGETS.

B. DETECTION TARGETS.

FIGURE 1.10
Resolution and detection targets with medium contrast ratio. View this chart from a distance of 5 m (16.5 ft). For A, determine the most closely spaced set of bars that you can resolve. For B, determine the narrowest bar that you can detect. Compare these values with those determined from Figure 1.6.

For example, if dots were used in place of lines in Figure 1.6B, the diameter of the smallest detectable dot would be considerably larger than 0.2 mm.

Effect of Contrast Ratio on Resolution and Detection

The resolution and detection targets in Figure 1.10 have the same dimensions as those in Figure 1.6, but the contrast ratio had been lowered by the addition of a gray background. To evaluate the effect of the lower contrast ratio, the reader should view Figure 1.10 from a distance of 5 m and determine which targets can be resolved and detected. Students typically report that in Figure 1.10 they can resolve only 3 line-pairs $\cdot$ cm^{-1}, and the smallest detectable target is the 0.6 mm-wide line. These values differ from the 5 line-pairs $\cdot$ cm^{-1} and 0.2-mm line of the high-contrast target and demonstrate the importance of contrast ratio on resolution and detection.

SOURCES OF REMOTE-SENSING INFORMATION

In the United States there is no technical society or organization devoted solely to remote sensing. The American Society of Photogrammetry and Remote Sensing has broadened its original charter to include workers in all forms of remote sensing. The society publishes a monthly journal, *Photogrammetric Engineering and Remote Sensing* that includes articles on various aspects of remote sensing. In 1975 the society published the *Manual of Remote Sensing* under the editorship of R. G. Reeves of the U.S. Geological Survey. This is a valuable reference book. The journal *Remote Sensing of the Environment* is published quarterly by Elsevier Publishing Company. Several organizations sponsor regularly scheduled remote-sensing conferences and publish the results.

1. Environmental Research Institute of Michigan, Ann Arbor, Michigan: "Remote Sensing of Environment." These conferences are held every 18 months.
2. EROS Data Center of U.S. Geological Survey, Sioux Falls, South Dakota: "Pecora Symposium on Remote Sensing." This is an annual conference.
3. Laboratory for Applications of Remote Sensing, Purdue University, East Lafayette, Indiana: "Annual Symposium on Processing of Remote Sensing Data."
4. Space Institute, University of Tennessee, Tullahoma, Tennessee: "Remote Sensing of Earth Resources." This is an annual conference.

The National Aeronautics and Space Administration (NASA) convened a major remote-sensing symposium in 1975. The proceedings are available from the U.S. Government Printing Office, Washington, DC. 20402 as the following documents: *NASA Earth Resources Survey Symposium,* NASA TM X-58168, 1975: Vol. 1, "Technical Session Presentations," 2166 p. Vol. 2 "Special Session Presentations," 389 p. Vol. 3 "Summary Reports," 46 p. Remote sensing has applications to many disciplines, and papers describing specific applications appear in the technical journals of many professional and scientific organizations.

COMMENTS

Remote sensing is broadly defined as the process of collecting information about a subject without being in physical contact with the subject. In this book remote sensing is restricted to processes that (1) record information about a subject by detecting or sensing the interaction between the subject and electromagnetic radiation, and (2) produce an image of the subject. The first restriction eliminates airborne methods, such as aeromagnetic surveys, that measure force fields rather than electromagnetic radiation. The second restriction eliminates nonimaging methods, such as airborne gamma-radiation surveys. These methods are described in standard textbooks on exploration geophysics.

The electromagnetic spectrum is divided into wavelength regions, or bands, that are employed for remote sensing. These regions range from the short wavelength UV band to the long wavelength microwave and radio bands. Subsequent chapters of the book describe the interactions between matter and electromagnetic radiation of the different wavelength bands together with the technology employed in sensing the radiation.

The interpretation of images, regardless of the wavelength band at which they were sensed, depends upon the scale, brightness (or tone), texture, contrast ratio, and spatial resolution. In this book, spatial resolution, or resolving power, refers to the minimum separation at which two objects can be distinguished on an image. Resolution is not the smallest object that can be discerned on an image; this property is detectability. The human eye is used to illustrate these concepts that are vital to understanding the discussions that follow.

REFERENCES

Colwell, R. N., W. Brewer, G. Landis, P. Langley, J. Morgan, J. Rinker, J. M. Robinson, and A. L. Sorem, 1963, Basic matter and energy relationships involved in remote reconnaissance: Photogrammetric Engineering, v. 29, p. 761–799.

Gregory, R. L., 1966, Eye and brain, the psychology of seeing: World University Library, McGraw-Hill Book Co., New York.

Janza, F. J. and others, 1975, Interaction mechanisms *in* Reeves, R. G., ed., Manual of remote sensing: ch. 4, p. 75–179, American Society of Photogrammetry, Falls Church, Va.

Perrin, F. H., 1966, The structure of the developed image *in* James, T. H., ed., The theory of the photographic process: ch. 23, p. 499–551. Third Edition, Macmillan Co., New York.

Santa Barbara Research Center, 1975, The SBRC Brochure: Goleta, Calif.

Slater, P. N., 1975, Photographic systems for remote sensing *in* Reeves, R. G., ed., Manual of remote sensing: ch. 6. p. 235–323, American Society of Photogrammetry, Falls Church, Va.

ADDITIONAL READING

Colwell, R. N., 1975, Introduction *in* Reeves, R. G., ed., Manual of remote sensing: ch. 1, p. 1–26, American Society of Photogrammetry, Falls Church, Va.

Fischer, W. A. and others, 1975, History of remote sensing *in* Reeves, R. G., ed., Manual of remote sensing: ch. 2, p. 27–50, American Society of Photogrammetry, Falls Church, Va.

Neisser, U., 1974. The process of vision *in* Held, R., comp., Image, object, and illusion: p. 4–11. W. H. Freeman and Company, San Francisco, Calif.

Suits, G. H., 1975, The nature of electromagnetic radiation *in* Reeves, R. G., ed., Manual of remote sensing: ch. 3, p. 51–73. American Society of Photogrammetry, Falls Church, Va.

Woll, P. W. and W. A. Fischer, eds., 1977. Proceedings of the first annual William T. Pecord Memorial Symposium, October, 1975. Sioux Falls, South Dakota: U.S. Geological Survey Professional Paper 1015.

2
AERIAL PHOTOGRAPHY

In addition to visible radiation (0.4 to 0.7 μm), portions of the UV (0.3 to 0.4 μm) and IR (0.7 to 0.9 μm) spectral regions may be detected by photographic methods. The wavelength region from 0.3 to 0.9 μm is therefore called the *photographic* remote-sensing region. These wavelengths are also among those detectable by optical-mechanical scanning devices, such as the multispectral scanner on the Skylab satellite described in Chapter 3. Strictly speaking, the term *light* refers to wavelengths detectable by the human eye (approximately 0.4 to 0.7 μm), but it is common practice to use the term light for the entire photographic remote-sensing region. In the enthusiasm for satellite imagery and new forms of airborne remote sensing, such as thermal IR and radar, the advantages of aerial photography should not be overlooked. Topographic maps are made from aerial photographs and many engineering projects employ aerial photographs. Soil conservation studies, agricultural crop inventories, and city planning all employ aerial photographs. Geologic mapping and exploration commonly begin with an analysis of photographs. As recently as the early 1970s, aerial photographic interpretation was responsible for discovering several valuable oil fields in West Irian.

INTERACTIONS BETWEEN LIGHT AND MATTER

As with other forms of electromagnetic radiation, light may be reflected, absorbed, or transmitted by matter. Aerial photographs record the light reflected by a surface, which is determined by the property called *albedo*. Albedo designates the total radiant reflectance of a surface, and is the ratio of the reflected energy to the incident energy. Dark surfaces have a low albedo and light surfaces have a high albedo. For opaque materials, the light that is not reflected is absorbed and causes an increase in the temperature of the material. During its transmission through the atmosphere light interacts with the gases and particulate matter in a process called scattering, which may have a strong effect on aerial photographs.

Selective and Nonselective Scattering

Scattering results from multiple interactions among light waves and the molecules and particles of the atmosphere as shown in Figure 2.1. The two major processes, selective scattering and nonselective scattering, are related to the size of particles

TABLE 2.1
Atmospheric scattering processes

Scattering process	Wavelength, λ, dependence	Approximate particle size, in λ	Kind of particles
Selective			
Rayleigh	λ^{-4}	$\ll 0.1$	Gas molecules
Mie	λ^0 to λ^{-4}	0.1 to 10	Smoke, fumes, haze
Nonselective	λ^0	> 10	Dust, fog, clouds

Source: From Slater (1975, Table 6-3).

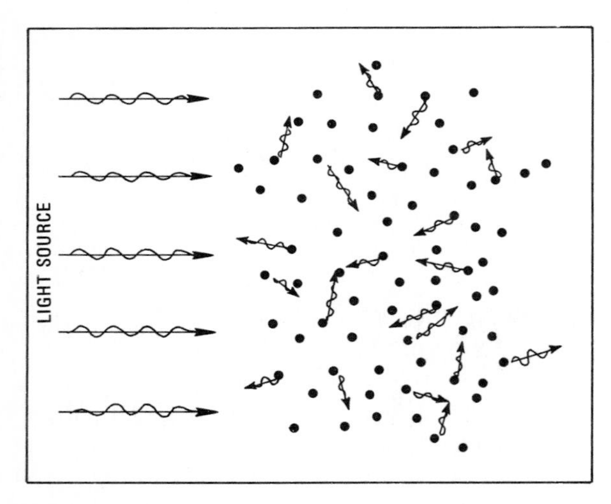

FIGURE 2.1
Scattering of light waves by particles in the atmosphere.

in the atmosphere as shown in Table 2.1. *Selective scattering* is so named because shorter wavelengths of light are selectively scattered more severely than longer wavelengths. Rayleigh scattering and Mie scattering are the two processes of selective scattering. These processes are determined by different size ranges of particles in the atmosphere, as shown in Table 2.1.

Rayleigh scattering is caused by gas molecules that are much smaller than the wavelengths of light. Rayleigh scattering is inversely proportional to the fourth power of the wavelength, λ^{-4}. This means that the shorter wavelengths of light are scattered more severely than the longer wavelengths. For every ten blue waves ($\lambda = 0.4\ \mu m$) that are scattered, there is only one scattered red wave ($\lambda = 0.7\ \mu m$) with intermediate scattering for other wavelengths. The curve for Rayleigh scattering shown in Figure 2.2 illustrates the predominance of this scattering process in the blue spectral region and particularly in the photographic UV region.

The other selective scattering process, Mie scattering (Table 2.1), is caused by particles of smoke, fumes, and haze that are of approximately the same dimension as wavelengths of light. Mie scattering is inversely proportional to wavelength at a range of exponents from 0 to 4. A general statement is that Mie scattering is inversely proportional to wavelength, λ^{-1}, as shown in Figure 2.2.

In a clear atmosphere the combination of Rayleigh and Mie processes causes the selective scattering of light shown in the shaded portion of Figure 2.2. The selective scattering of violet and blue light by the atmosphere causes the blue color of the sky. When the sun is directly overhead at noon it appears white because the light is passing through the minimum thickness of atmosphere and little selective scattering occurs. At sunrise and sunset, however, sunlight passes tangentially through a much thicker column of atmosphere in which the short wavelengths are so completely scattered that little blue light reaches the eye; hence the red color of the sun at sunrise and sunset. Selectively scattered light illuminates shadows that are never completely dark, but are bluish in color. This scattered illumination is referred to as *skylight* to distinguish it from direct sunlight. One of the striking characteristics of photographs taken by Apollo astronauts on the moon is the black tone of the shadows. The lack of atmosphere on the moon precludes any scattering of light into the shadowed areas.

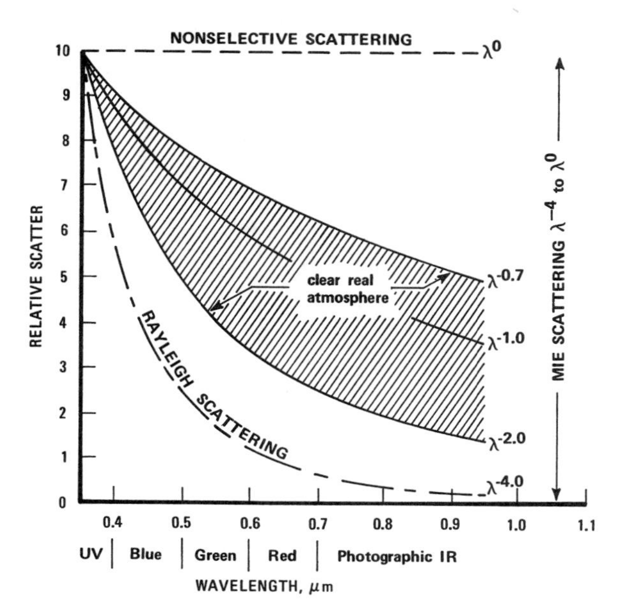

FIGURE 2.2
Selective scattering (Rayleigh and Mie) and nonselective scattering as a function of wavelength. The shaded region indicates the range of scattering caused by typical atmospheres. From Slater (1975, Figure 6-15).

Nonselective scattering is caused by dust, fog, and clouds with particle sizes more than 10 times the wavelength of light. These particles scatter all wavelengths equally, λ^0. Therefore clouds and fog appear white although their water particles are colorless.

Effects of Scattering on Aerial Photographs

Light scattered by the atmosphere that enters the camera is a source of illumination that contains no information about the terrain. This extra illumination reduces the contrast ratio of the scene, thereby reducing the spatial resolution and detectability of the photograph. Figure 2.3 illustrates diagrammatically the effect of scattered light on the contrast ratio of a scene. For the original scene with no scattered light, the contrast ratio is determined from Equation (1.3) as

$$C_r = \frac{B_{\max}}{B_{\min}}$$

$$C_r = \frac{5}{2}$$

$$C_r = 2.5$$

A. ORIGINAL SCENE.

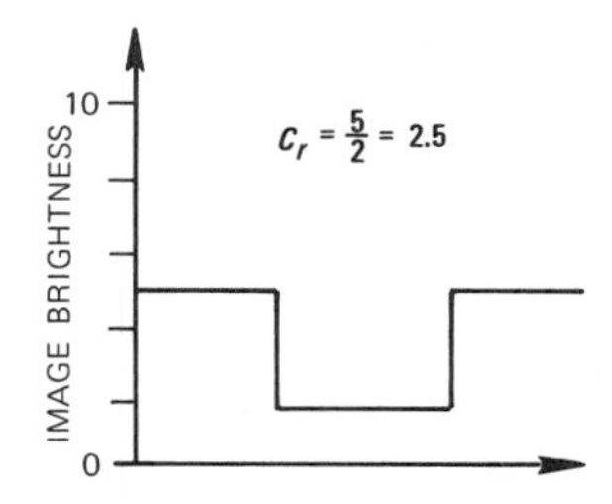

B. BRIGHTNESS PROFILE OF IMAGE WITH NO SCATTERED LIGHT.

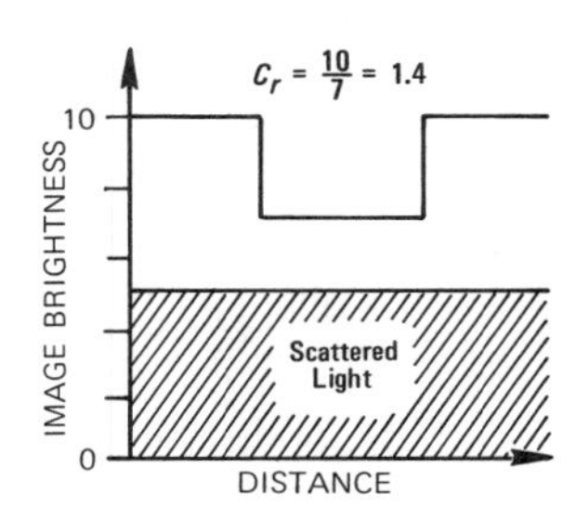

C. PROFILE OF IMAGE WITH FIVE BRIGHTNESS UNITS ADDED BY SCATTERED LIGHT.

FIGURE 2.3
Effect of scattered light on contrast ratio of an image.

Figure 2.3C shows the effect of photographing the same scene in conditions of heavy haze where five brightness units of scattered light are contributed by the atmosphere. The scattered light is added uniformly to all parts of the scene and results in a contrast ratio of

$$C_r = \frac{10}{7}$$

$$C_r = 1.4$$

Thus atmospheric scattering has reduced the contrast ratio from 2.5 to 1.4. The effect of reduced contrast ratio for a scene is to lower the spatial resolution on a photograph of that scene. This relationship between contrast ratio and resolving

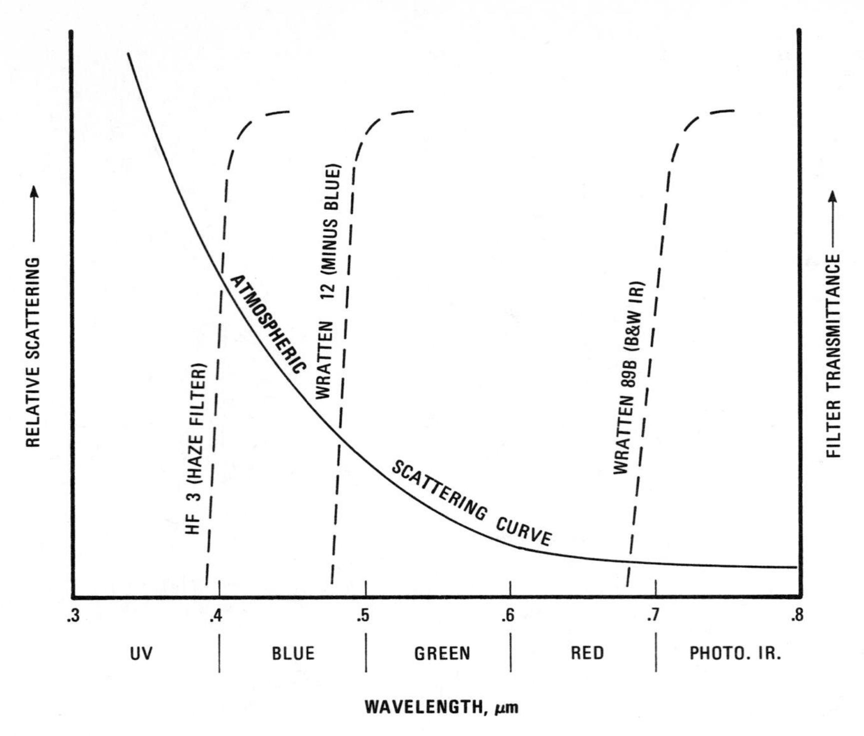

FIGURE 2.4
Atmospheric scattering diagram and transmission curves of common Kodak Wratten filters used in aerial photography. Shorter wavelengths to the left of each filter curve are absorbed. From Slater (1975, Figures 6-81, 82, 83).

power was demonstrated in Chapter 1 (Figures 1.6 and 1.10). The effect of atmospheric scattering on aerial photographs is illustrated later in this chapter with examples of IR and UV photographs.

Removing the selectively scattered shorter wavelengths with filters before they reach the film reduces the problem of atmospheric scattering. Superimposed on the scattering diagram of Figure 2.4 are the spectral transmittance curves of common optical filters used in aerial photography. The curves show the wavelengths absorbed and transmitted. The trade off with filters is that in addition to reducing haze, they remove the spectral information contained in the wavelengths that are absorbed. The characteristics of a wide range of filters are given by Kodak (1970).

FILM TECHNOLOGY

Film consists of a flexible transparent base coated with a layer of emulsion approximately 100 μm in thickness (Figure 2.5). The emulsion is a suspension in solidified gelatin of grains of silver halide salts that are a few micrometers or less in diameter. The grains are precipitated from solution rather rapidly to make them irregular with numerous points of imperfection on the surface. After the emulsion is deposited on the film base, the grains are further processed to increase their sensitivity to light. The emulsion is exposed to light when the camera shutter is triggered. When a photon strikes one of the grains, an electron from the silver halide crystal lattice is free to move about. A number of things may happen to the electron but the important thing for photography is that the electron may be trapped at an imperfection in the grain. The electron may then convert a silver ion into a silver atom (Jones, 1968, p. 116). This atom cannot last long by itself, but if two electrons are liberated within about a second, a four-atom combination of silver will form at the imperfection (Figure 2.5B). This combination is very stable and is large enough to trigger the conversion of the entire silver halide grain to metallic silver when the film is developed. At this stage the film is said to contain a *latent image*.

Photographic developing is the chemical process of changing the latent image into a real image by converting the exposed silver halide grains to

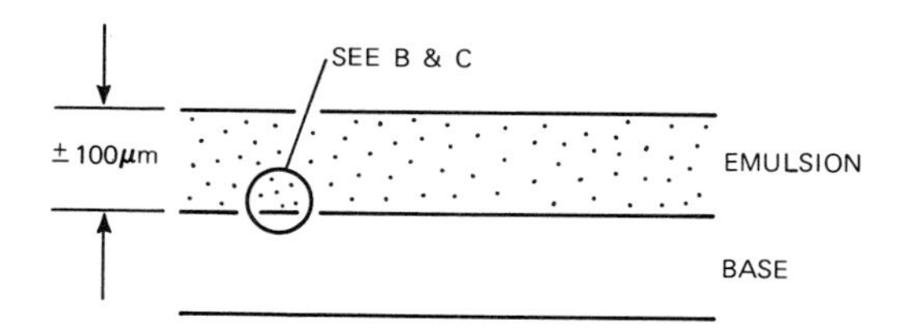

A. CROSS SECTION OF FILM.

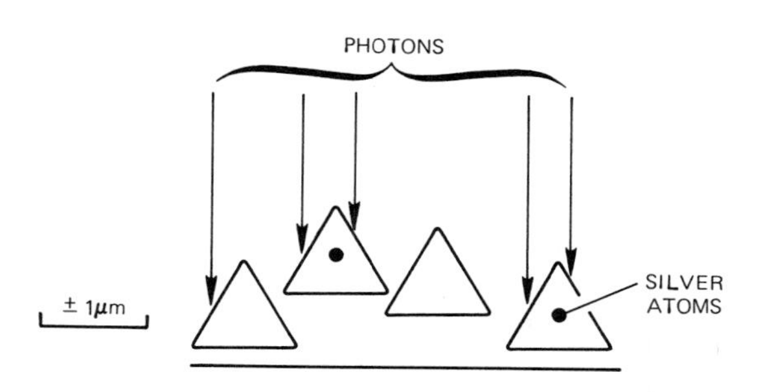

B. EXPOSURE OF SILVER HALIDE GRAINS.

FIGURE 2.5
Film technology.

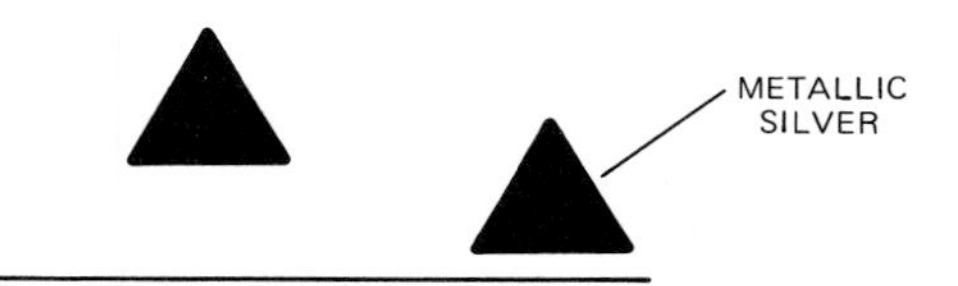

C. DEVELOPED FILM NEGATIVE.

opaque grains of silver. The film is immersed in a water solution containing a reducing agent that is unable to develop an unexposed grain. For exposed grains with silver atoms at an imperfection, however, the agent starts the chemical reduction of silver halide to silver. As shown in Figure 2.5C the entire grain is converted to silver once the reduction process begins. Unexposed grains are removed by the developing process, leaving clear areas in the emulsion. The resulting film is called a negative because bright targets form dark images on the film. When the film is printed onto photographic paper the signatures are reversed and bright targets appear bright on the print. The requirement that more than one photon be received by a silver halide grain within a short time is essential for the success of the photographic method. If only one photon were needed, photons radiated thermally would soon expose all the grains.

An enormous amplification occurs in the development process. A few photons absorbed in a grain of silver halide with a volume of 1 μm^3 produce more than 10^{10} atoms of silver. This is an amplification of more than a billion (Jones, 1968, p. 116). In addition to the amplification factor, other advantages of photographic remote sensing include the high resolving power, low cost, versatility, ease of operation, and the high information storage capacity of film. On the developed film each grain, whether exposed or unexposed, records information about the scene. There are more than 150 million ($1.5 \cdot 10^8$) such grains on a 6.5 cm^2 (1 $in.^2$) piece of film. For comparison, the same area of magnetic computer tape typically contains only 3,200 ($3.2 \cdot 10^3$) magnetic signals that record information. A standard reference to the photographic process is the book edited by James (1966). The two major disadvantages of photographic remote sensing are (1) it is restricted to the 0.3 to 0.9 μm spectral region, and (2) it is restricted by weather, lighting conditions, and atmospheric effects.

CHARACTERISTICS OF AERIAL PHOTOGRAPHS

Aerial photographs are acquired with a variety of cameras and films. Characteristics such as resolution, scale, and relief displacement are common to all aerial photographs.

Photographic Resolution

Spatial resolution on aerial photographs is largely determined by the resolving power of the camera lens and resolving power of the film. Resolving power of a lens is determined by its optical quality and size. If a lens is used to photograph a resolution target, such as the one shown in Figure 1.6, there is an upper limit to the number of line-pairs per mm that can be resolved on the resulting photograph. This maximum number of resolvable line pairs per mm is a measure of the resolving power of the lens. Atmospheric conditions and aircraft vibration and motion also affect photographic resolution.

Resolving Power of Film Resolving power of film is determined by several factors, the most important of which is granularity that is largely determined by: (1) the size distribution of silver halide grains in the emulsion, and (2) the nature of the development process. Films with higher granularity generally have lower resolving power than those with lower granularity. There is a trade off because films with high granularity are generally *faster*, meaning they are more sensitive to light.

One method for expressing resolving power of film is to photograph a resolution target and determine the maximum number of line-pairs per mm that can be distinguished on the developed film. As previously illustrated in Figures 1.6 and 1.10, targets with a high contrast ratio produce better resolution than those with low contrast ratio, which are more typical of terrain features. Film resolving power is commonly given both for targets with a high contrast ratio and a low contrast ratio. A widely used black-and-white film, Kodak Panatomic X aerial film, has a resolving power of 300 line-pairs $\cdot$ mm^{-1} for high-contrast targets and 80 line-pairs $\cdot$ mm^{-1} for low-contrast targets. *System resolution,* R_s, of a camera and film combination is a result of the resolving powers of the lens and film and typically ranges from about 25 to 100 line-pairs $\cdot$ mm^{-1}.

Ground Resolution An important term for interpreters is *ground resolution*, which expresses the ability to resolve ground features on aerial photographs and images. System resolution is converted into ground resolution by the formula

$$R_g = \frac{R_s \cdot f}{H} \tag{2.1}$$

where

R_g = ground resolution in line-pairs per m

H = camera height above ground in m. This is not to be confused with aircraft altitude above mean sea level

R_s = system resolution in line-pairs per mm

f = camera focal length in mm.

The geometric basis for this relationship is shown in Figure 2.6. For a camera lens with focal length of 152 mm producing photographs with a system resolution of 40 line-pairs $\cdot$ mm^{-1} acquired at a camera height of 6,100 m, the ground resolution may be calculated from Equation (2.1) as

$$R_g = \frac{R_s \cdot f}{H}$$

$$R_g = \frac{40 \text{ line-pairs} \cdot \text{mm}^{-1} \cdot 152 \text{ mm}}{6{,}100 \text{ m}}$$

$$R_g = 1 \text{ line-pair} \cdot \text{m}^{-1}$$

This means that under the specified conditions the most closely spaced resolution target on the ground that can be resolved on the photograph consists of 1 line-pair $\cdot$ m^{-1}. A useful expression is

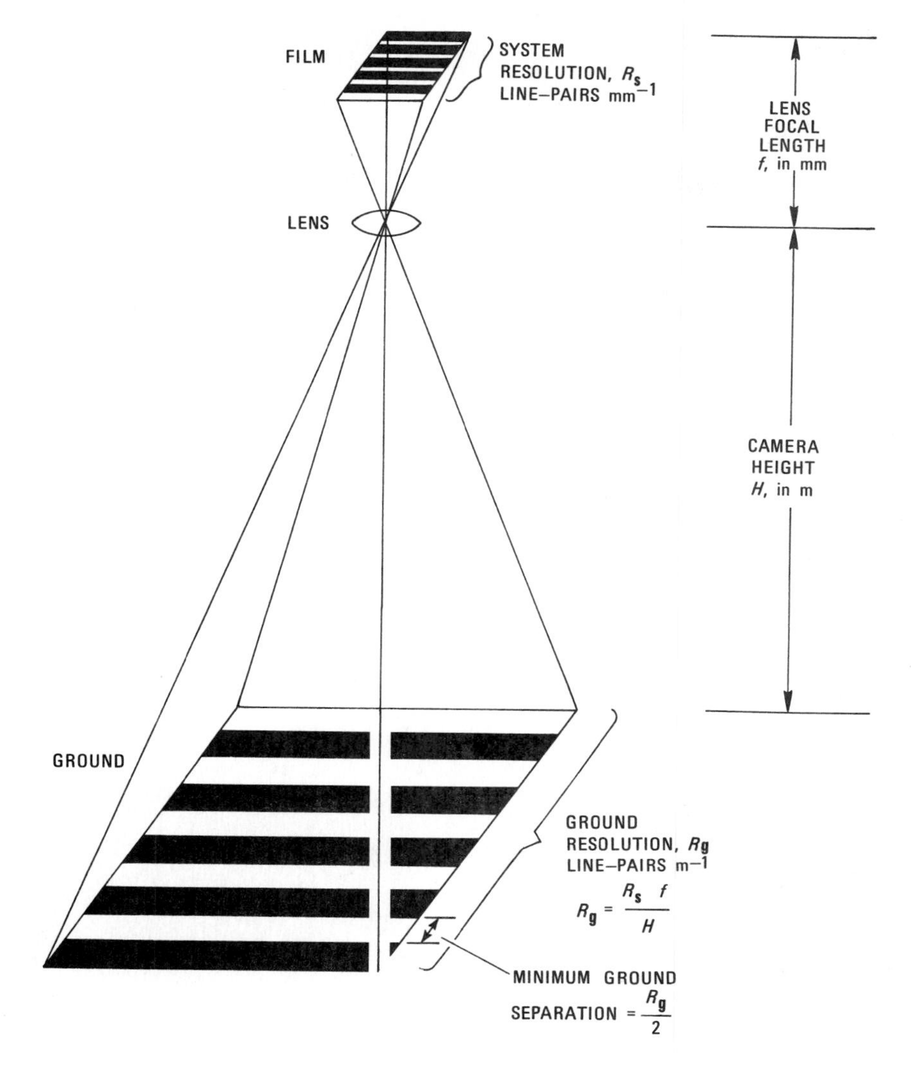

FIGURE 2.6
Ground resolution and minimum ground separation on aerial photographs.

minimum ground separation, which is $R_g/2$, and is the minimum distance between two targets on the ground at which they can be resolved on the photograph. Minimum ground separation is illustrated in Figure 2.6.

Table 2.2 lists minimum ground separation values for typical aerial photographs acquired with camera systems of medium and high system resolution. The camera heights and lens focal length of Table 2.2 correspond to the medium-resolution aerial photographs in Figure 2.7. Inspection of the photographs with a magnifier indicates that these minimum ground separation values are appropriate although the photographs lack a ground resolution target that is necessary for precise measurement. Note that prints and enlargements may have poorer resolution than the original negative from which they are made. Table 2.3 lists features that may be identified on photographs with different ground separation values. These are only guidelines that illustrate the general relationship of ground resolution to recognition.

Photographic Scale

Scale of aerial photographs is determined by the relationship

$$\text{Scale} = \frac{1}{H/(f \cdot 10^{-3})} \tag{2.2}$$

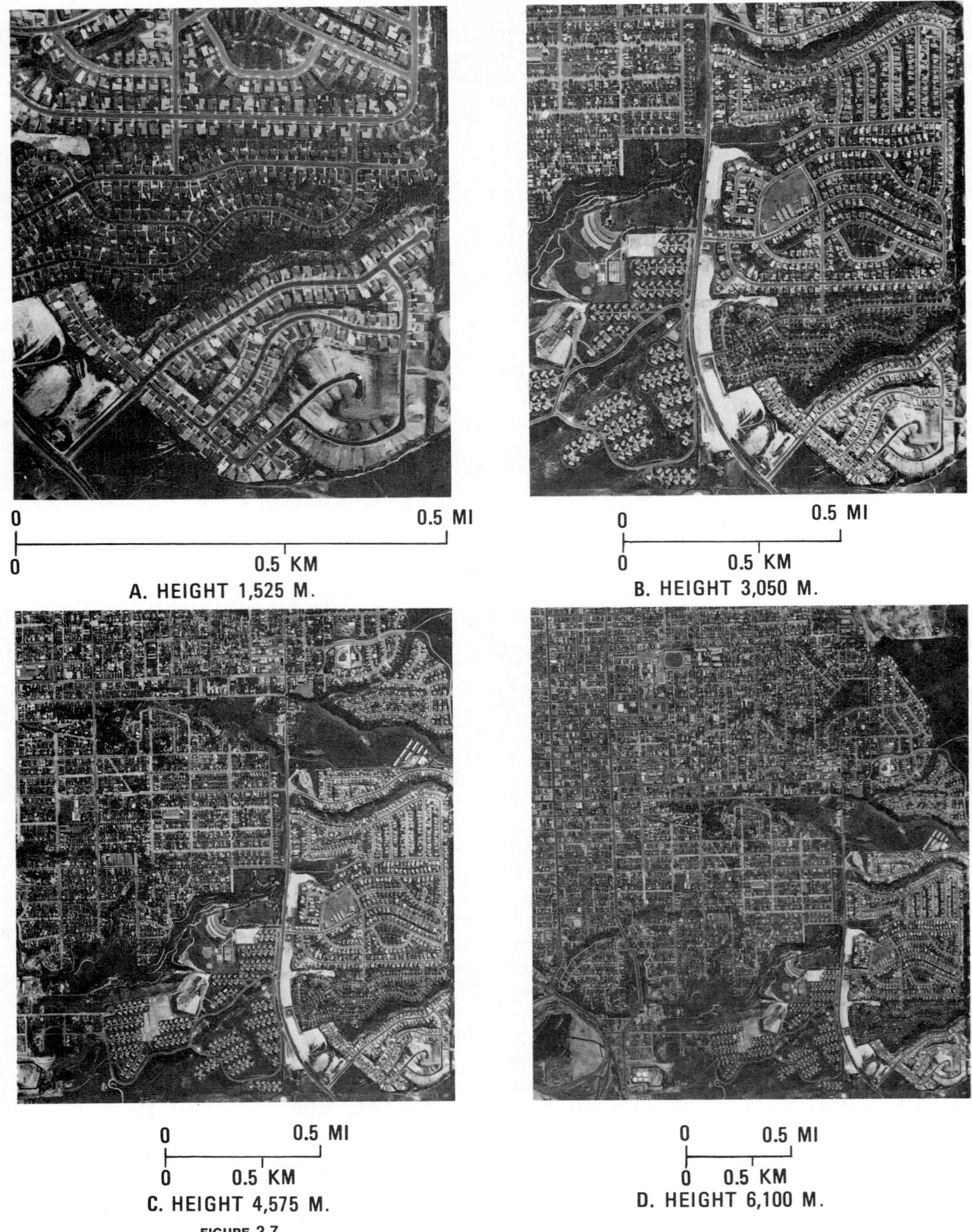

FIGURE 2.7
Aerial photographs acquired at different camera heights with a 152-mm focal length lens. Minimum ground separation values for this medium resolution system are given in Table 2.2. The southeast corner is common to all four photographs. Palos Verde Peninsula, southern California.

TABLE 2.2
Minimum ground separation on typical aerial photographs acquired at different heights (focal length, f, of camera lens = 152 mm)

Camera height, H	Scale of photographs	Minimum ground separation $R_g/2$ for system resolution R_s of:	
		40 line-pairs · mm^{-1}	100 line-pairs · mm^{-1}
6,100 m	1:40,000	0.5 m	0.2 m
4,575 m	1:30,000	0.37 m	0.15 m
3,050 m	1:20,000	0.25 m	0.1 m
1,525 m	1:10,000	0.12 m	0.05 m

Focal length (f, in mm) is multiplied by 10^{-3} to convert to the same units used for camera height (H, in m). For example, the scale of a photograph acquired at a camera height of 3,050 m with a 152-mm lens is

$$\frac{1}{3{,}050 \text{ m}/0.152 \text{ m}} = \frac{1}{20{,}000} \text{ or } 1{:}20{,}000$$

A scale of 1:20,000 means that 1 cm on the photograph represents 200 m on the ground (1 in. = 1,665 ft). Figure 2.7 illustrates the different scales that result from photographing the same area at different altitudes with the same camera.

TABLE 2.3
Features recognizable on aerial photographs at different minimum ground separation values

Minimum ground separation, $R_g/2$	Recognizable features
15 m	Geographic features, such as shore lines, rivers, mountains and water can be identified.
4.5 m	Settled areas may be differentiated from undeveloped land.
1.5 m	Roadways can be identified
0.15 m	Front of automobiles can be distinguished from the rear.
0.05 m	People can be counted, particularly if there are shadows and if the individuals are not in crowds.

Source: After Rosenblum (1968, Table 2).

Relief Displacement

Figure 2.8 illustrates the geometric distortion called *relief displacement* that is present on all vertical aerial photographs. The tops of objects are displaced from their bases in a direction radially outward from the *principal point*, or center, of the photograph. This displacement causes the buildings in Figure 2.8 to "lean" away from the center of the photograph, with the amount of displacement increasing at greater radial distances from the center. The relief displacement is greatest for buildings in the corners of the photograph. The geometry of image displacement is shown in Figure 2.9A, where light rays are traced from the terrain through the camera lens and onto the negative film. Prints made from the negative appear as though they were in the position shown by the plane of photographic print in Figure 2.9A. The vertical arrows on the terrain represent objects of various heights located at various distances from the principal point. The light ray reflected from the base of object A intersects the plane of the photographic print at position A, and the ray from the top intersects the print at A′. The distance from A to A′ is the relief displacement, d, and is shown in plan view on a photographic print in Figure 2.9B. Rabben and others (1960, p. 135) made the following points regarding the amount of relief displacement on an aerial photograph:

1. It is directly proportional to the height of the object h. For objects A and C (Figure 2.9A) at equal distances from the principal point, d is greater for A, which is the taller object.
2. It is directly proportional to the radial distance, r, from the principal point to the top of the displaced image of the object. From objects A and B, of equal height, d is greater for A, which is located farther from the principal point.
3. It is inversely proportional to the height H of the camera above the terrain.

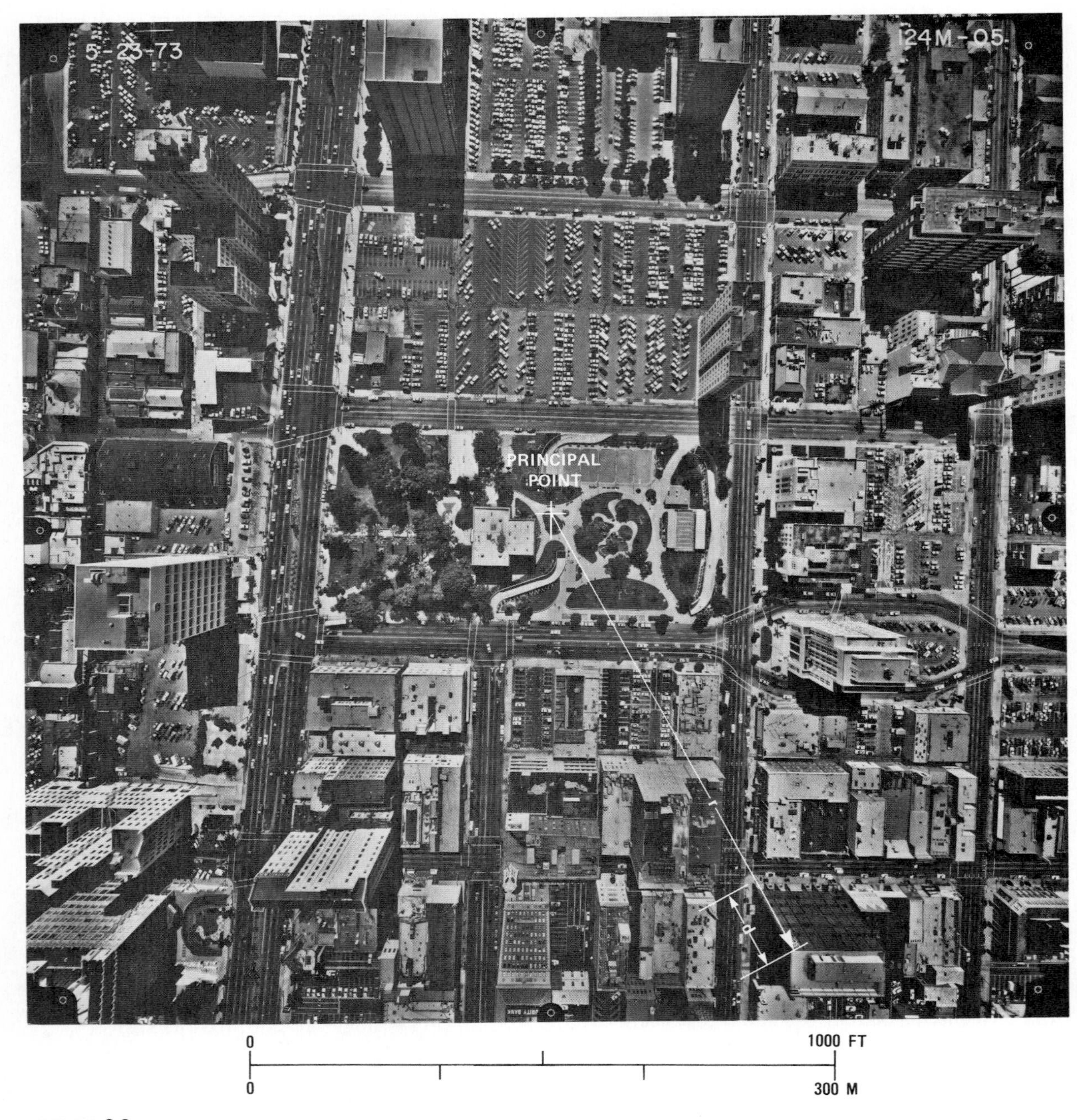

FIGURE 2.8
Vertical aerial photograph of Long Beach, California, showing relief displacement. Photograph acquired May 23, 1973 at a camera height of 212 m and lens focal length of 88 mm. Courtesy J. Van Eden, VTN Consolidated, Inc., Irvine, California.

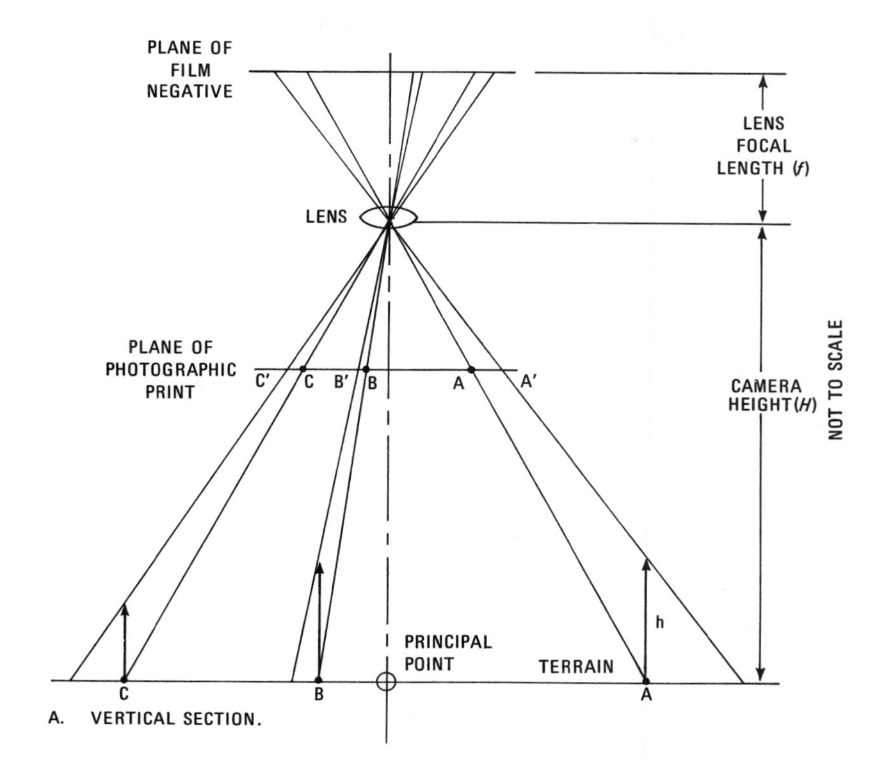

A. VERTICAL SECTION.

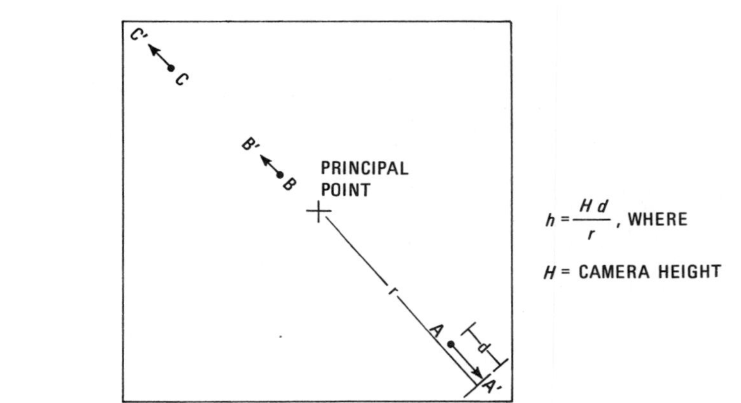

B. PLAN VIEW OF PHOTOGRAPHIC PRINT.

FIGURE 2.9
Geometry of relief displacement on a vertical aerial photograph.

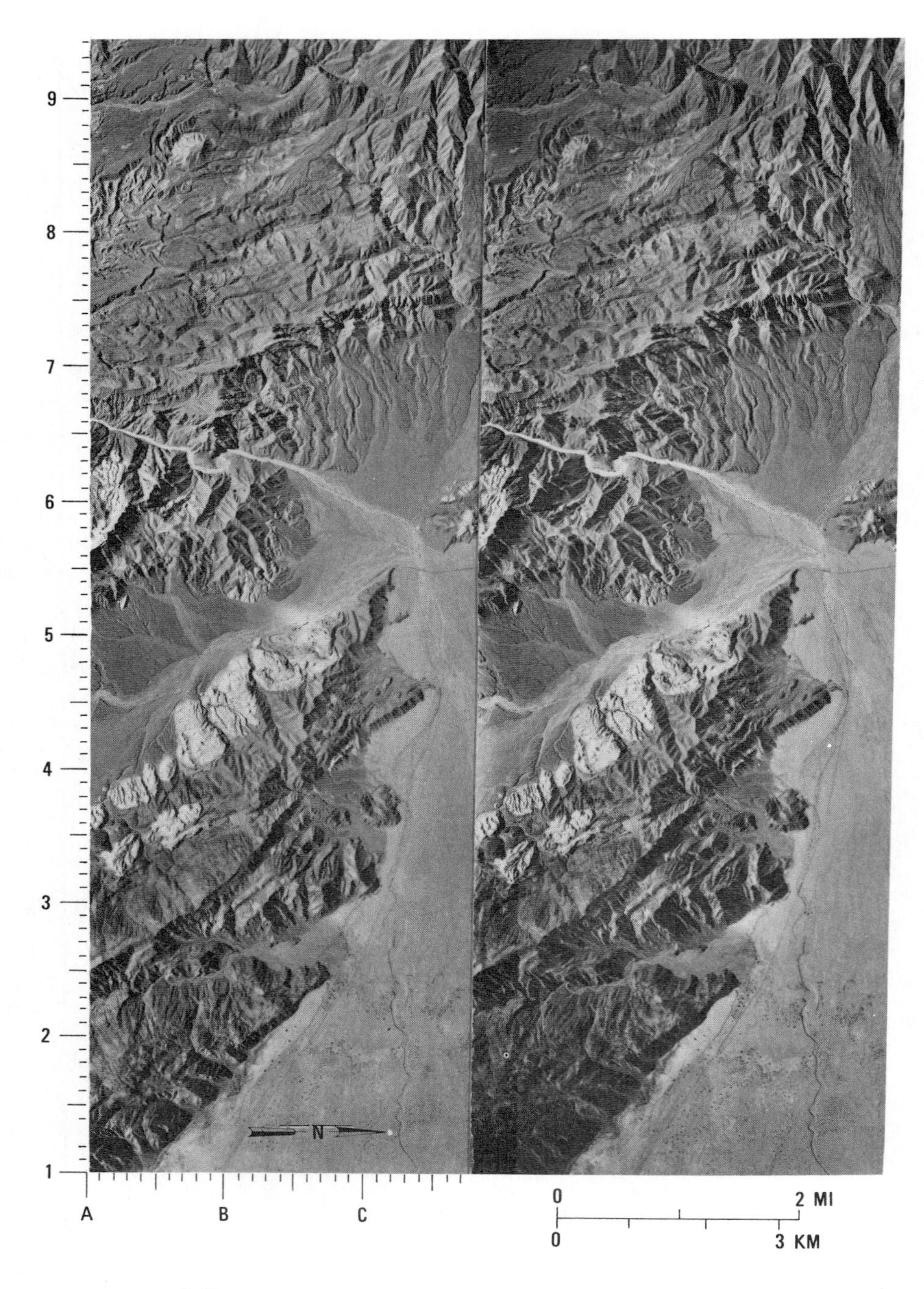

FIGURE 2.10
Stereo pair of aerial photographs, Fish Creek Mountain and Split Mountain, southern California.

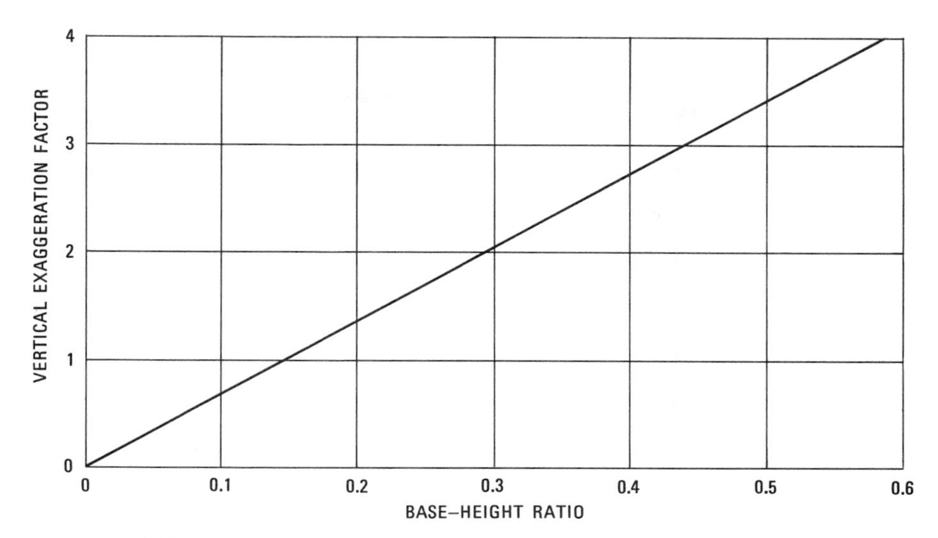

FIGURE 2.11
Relationship between base-height ratio and vertical exaggeration factor for stereo models. From Thurrell (1953, Figure 5).

These relationships are expressed mathematically as

$$d = \frac{hr}{H}$$

which may be transposed to

$$h = \frac{Hd}{r} \tag{2.3}$$

This equation may be used to determine the height of an object from its relief displacement on an aerial photograph. For the building in the lower right corner of Figure 2.8, d and r are measured using the scale of the photograph, and the height is calculated from Equation (2.3) as

$$h = \frac{212 \text{ m} \cdot 40 \text{ m}}{260 \text{ m}}$$

$$h = 32.6 \text{ m}$$

Stereo Photography

Most aerial photography is acquired with 60 percent forward overlap between successive photographs along the flight line. This allows the overlapping photographs to be viewed with a stereoscope to produce a three-dimensional image called a *stereo model.*

Vertical Exaggeration Figure 2.10 consists of the overlapping portions of successive aerial photographs arranged for viewing with a pocket-lens stereoscope to produce a stereo model. The area includes portions of Fish Creek Mountain and Split Rock Mountain in the Peninsular Range province of southern California. The photographs were acquired at a height of 9,000 m with a 152-mm lens, resulting in a photographic scale of 1: 60,000. The first impression on viewing Figure 2.10 with a stereoscope is of extreme topographic relief caused by vertical exaggeration of the stereo model. *Vertical exaggeration* is the ratio of the vertical scale to the horizontal scale of a stereo model and is determined by the geometry of the camera system and of the stereo viewing system. *Air base* is the ground distance between centers of successive overlapping photographs. The ratio of air base to aircraft height, called *base-height ratio*, determines the vertical exaggeration of the stereo model. The relationship of base-height ratio to vertical exaggeration is shown in Figure 2.11. Note that the vertical exaggeration factor can be less than 1.0 when the base-height ratio is small.

The 60 percent overlap of the photographs in Figure 2.10 resulted in an air base of 5,400 m between successive photographs. At the camera height of 9,000 m, the resulting base-height ratio is 0.6. By referring to the graph of Figure 2.11, it can be seen that this ratio results in a vertical

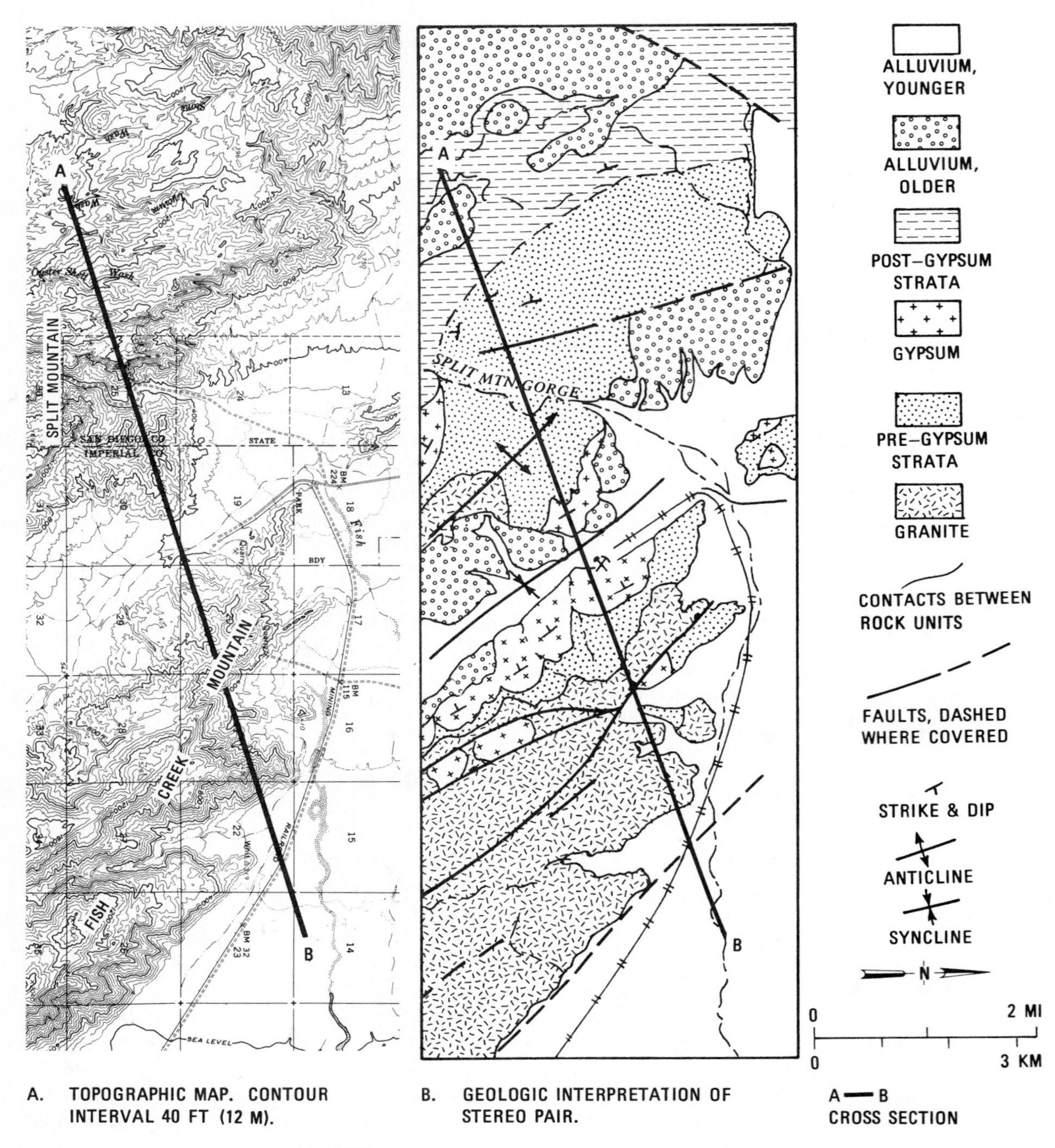

FIGURE 2.12
Maps of Fish Creek Mountain and Split Mountain stereo pair.

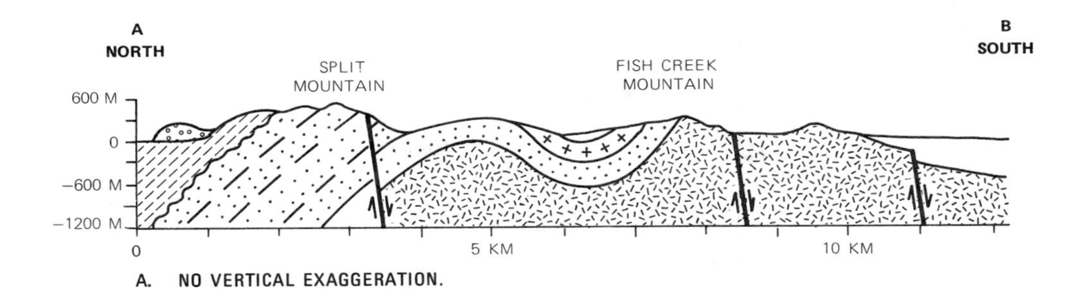

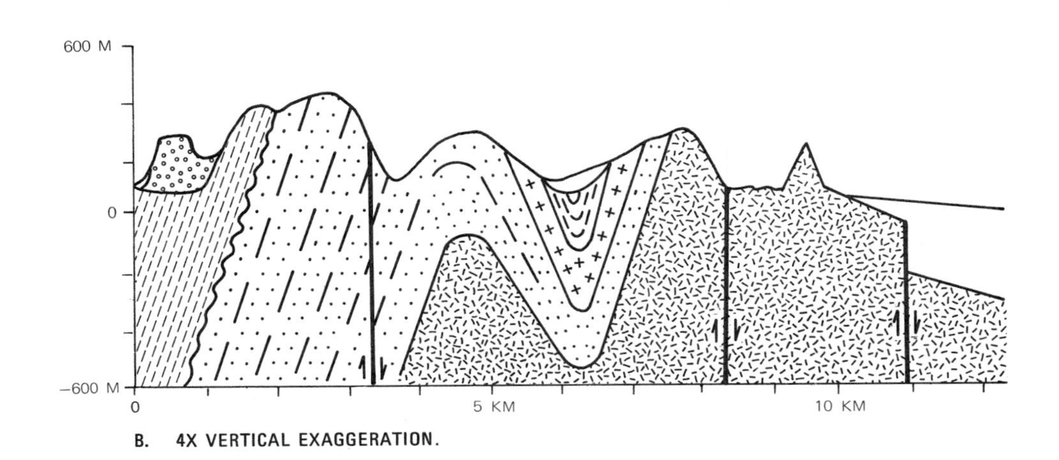

FIGURE 2.13
Geologic cross sections of Fish Creek Mountain and Split Mountain stereo pair, showing effects of vertical exaggeration.

exaggeration factor of four when the photographs of Figure 2.10 are viewed stereoscopically. To illustrate the effect of this vertical exaggeration, a topographic profile with a geologic cross section was constructed along line AB shown on the maps of Figure 2.12. The cross section in Figure 2.13A was constructed with the same vertical and horizontal scales. This is a vertical exaggeration factor of one, which means there is no vertical exaggeration. The cross section in Figure 2.13B was constructed with the vertical scale four times that of the horizontal scale; the resulting vertical exaggeration factor of four is the same as that of the stereo model in Figure 2.10. Comparison of the cross sections gives an evaluation of the effect of vertical exaggeration.

Although vertical relief is exaggerated by a factor of four on the stereo model, the angles of topographic slope and structural dip are not increased by a factor of four. Slope exaggeration is illustrated in Figure 2.14, which diagrams a hill with height BC of 270 m and width AB of 1000 m. The tangent of the true slope angle CAB is BC/AB = 0.27 = 15°. On the stereo model the exaggerated height DB is 1,080 m, and the tangent of the exaggerated slope angle DAB is BD/AB = 1.08 = 47°. The effect of the vertical exaggeration factor of four on structural dips and topographic slopes is illustrated by the cross section in Figure 2.13B.

Interpretation In making a geologic interpretation (Figure 2.12B) of the stereo model the first step is to recognize the geologic units. In this area granite is characterized by its dark tone, intense jointing, and lack of stratification. The distinctive white Fish Creek Gypsum is a useful marker for dividing the sedimentary rocks into three mapping

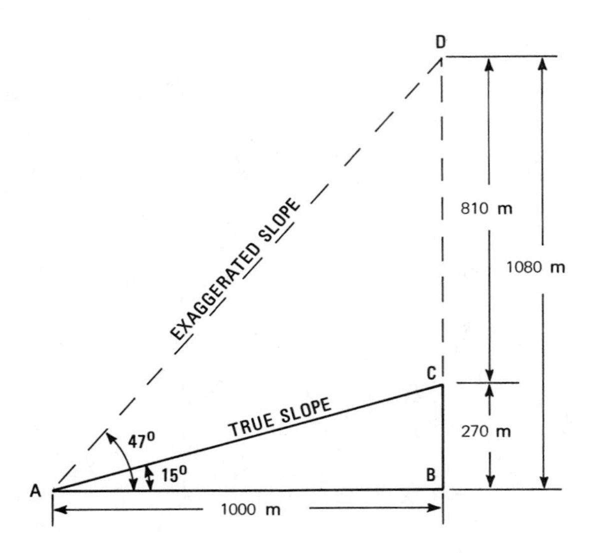

FIGURE 2.14
Slope exaggeration on a stereo model having vertical exaggeration factor of four.

units. The pre-gypsum unit of sandstone and conglomerate occurs between the granite and the gypsum. The post-gypsum strata are conglomerates and shales that overlie the gypsum. These are mapping units for photogeologic purposes. In the field more detailed stratigraphic subdivisions can be recognized. The gypsum bed pinches out at Split Mountain Gorge; northwest from the gorge in Figure 2.12B the pre-gypsum and post-gypsum units are in contact. Older alluvium consists of gravel deposits on the east and west flanks of Split Mountain. The darker tone and erosional dissection of the older alluvium distinguish it from the lighter toned younger alluvium that fills the drainage channels.

Geologic structure is inferred from physiographic features and attitudes of the rocks, both of which are well expressed on the stereo model. Direction of strike and dip can be determined from trends of the upper and lower contacts of the gypsum unit. Dip slopes exposed in Split Mountain also can be used for this purpose. Although direction and relative degree of dip can be estimated, the amount of dip is exaggerated. The northwest-trending valley between Fish Creek Mountain and Split Mountain marks the axis of a syncline shown on the map (Figure 2.12B) and cross sections (Figure 2.13). Note that the gypsum beds on both sides of the valley dip into the valley. Attitudes of the gypsum outcrops in Split Mountain define a northwest-plunging anticline parallel with the syncline. The steep scarp along the northeast front of Fish Creek Mountain that extends from locality 1.0, A.4 to 2.5, B.8 in Figure 2.10, is evidence for a fault that is concealed by younger alluvium. Parallel faults along the crest of Fish Creek Mountain are marked by similar scarps and by offset rock units (Figure 2.12B). The steep eastern scarp of Split Mountain (from locality 6.6, A.0 to 7.3, C.7) is formed by a fault that separates bedrock from older alluvium. This scarp and the fault scarp along Fish Creek Mountain are recognizable on the topographic map but are much more pronounced on the exaggerated stereo model.

Additional geologic details could be added by interpreting larger scale photographs, but more stereo models would be required to cover the area. The additional stereo models would lack the lateral continuity that is a key to recognizing linear geologic features such as fault scarps. A photogeologic interpretation should not be considered complete until the critical areas have been checked in the field. This may be impossible, especially for foreign areas; in that case the legend should note that the map has not been field checked.

Mosaics

In addition to forward overlap, photographs from adjacent flight lines typically have 30 percent sidelap. The photographs may be matched and mounted on a base to form a *photomosaic*, such as the example from the Desert Hot Springs area (Figure 2.15). Mosaics are valuable because they provide broader coverage than the individual photographs. In addition to the geographic and topographic information, geologic interpretations may be made from mosaics (Figure 2.16). The active San Andreas and Mission Creek faults are readily identified by the linear topographic scarps and ridges and by the vegetation anomalies. Vegetation, which has a dark signature on the photomosaic, is concentrated along the northeast side of the San Andreas fault. The fault is a barrier to the subsurface movement of ground water; therefore the water table is nearer the surface on the northeast side and provides moisture that supports salt cedar, mesquite, and creosote bush. Edom Hill in the southeast part of the mosaic is a symmetrical anticline marked by dip slopes formed by the folded sedimentary rocks.

Low-Sun-Angle Photography

Aerial photographs are normally acquired between 10:00 A.M. and 3:00 P.M., because the sun is at a high angle above the horizon, and shadows have a minimum areal extent. This illumination is desirable for topographic mapping, which requires unobscured terrain, but for geologic interpretation, photographs acquired with lower sun angles may be useful. Hackman (1967) photographed a topographic relief model at various angles of illumination and reached the following conclusions:

1. Photographs acquired with the sun 10° or less above the horizon show subtle differences in relief and textural pattern that are otherwise unrecognizable. These photographs are more useful in areas of low relief than in areas of high relief where shadows of large topographic features would obscure much of the area.
2. Tonal differences are less apparent on low sun-angle photographs than on those acquired with high sun angles. It is desirable to acquire both high-angle and low-angle illumination photographs of the same area, but this is rarely done because of the additional expense.
3. If only one set of photographs can be acquired, those with a sun angle of 20° to 30° above the horizon are the most satisfactory.

Acquisition of low-sun-angle photographs is complicated by the relatively limited hours in the morning and evening when the desired illumination occurs. Illumination values are low and change rapidly during these times; therefore, proper camera exposures may be difficult to achieve. The Landsat satellites, described in Chapter 4, acquire images of most of the earth between 9:30 and 10:00 A.M. local sun time, which results in a useful effect of shadows and highlights.

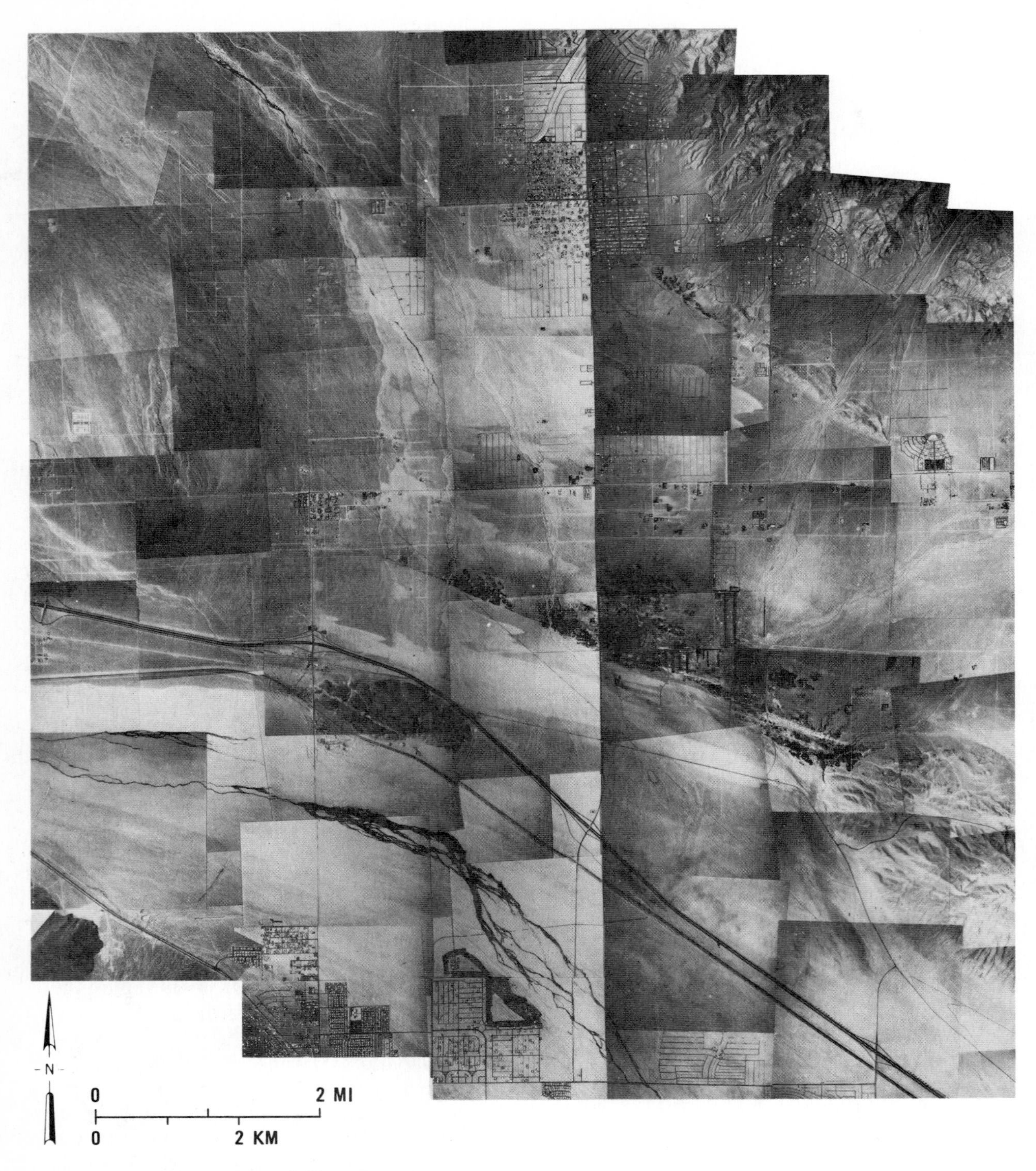

FIGURE 2.15
Photomosaic of Desert Hot Springs area, California, acquired May, 1969 at 2,250 m above terrain. From Sabins (1973A, Figure 6).

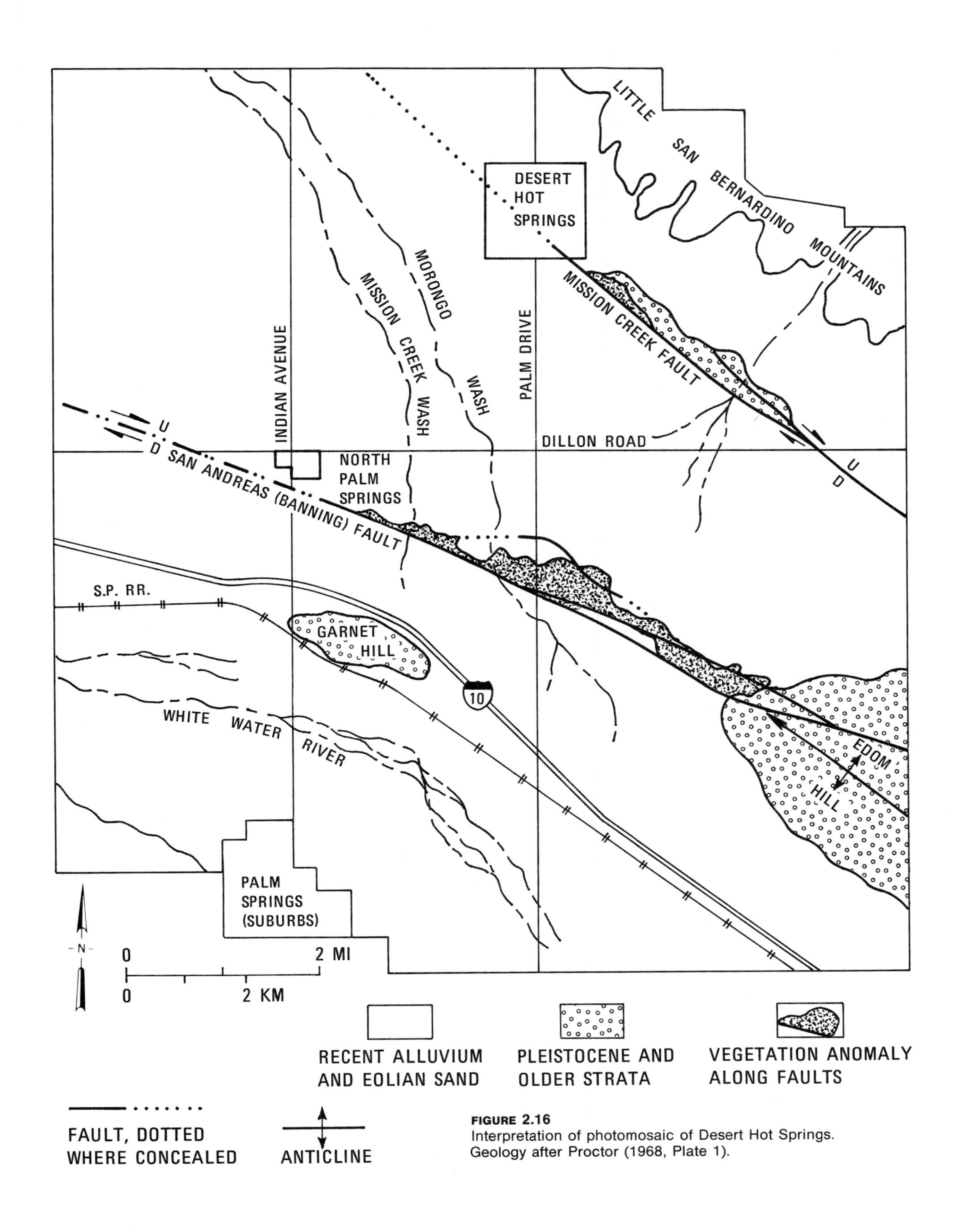

FIGURE 2.16
Interpretation of photomosaic of Desert Hot Springs. Geology after Proctor (1968, Plate 1).

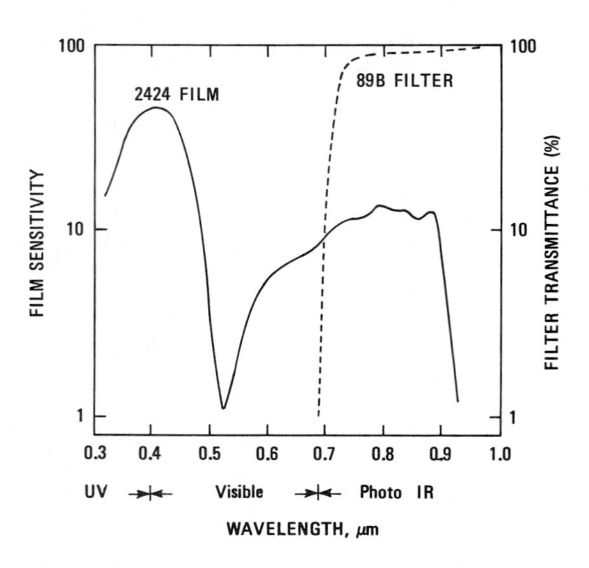

FIGURE 2.17
Spectral sensitivity of Kodak IR Aerographic film 2424 with spectral transmittance of Kodak Wratten filter 89B. From Vizy (1974, Figure 7).

BLACK-AND-WHITE PHOTOGRAPHY

This is the original form of aerial photography and may be accomplished at wavelengths ranging from UV through visible and into the photographic portion of the IR region.

Conventional Black-and-White Photography

Black-and-white aerial photographs are normally acquired with a Kodak Wratten 12 or equivalent filter over the lens to eliminate UV and blue wavelengths that are selectively scattered by the atmosphere (Figure 2.4). These are called *minus-blue* photographs, and examples are shown in Figures 2.7, 2.8, and 2.12. Black-and-white photographs are a widely used and readily available remote-sensing product. Stereo coverage of most of the United States is available at modest prices from the agencies listed in the section called Sources of Aerial Photography in this chapter. These photographs are the basis for compiling topographic maps, geologic surveys, engineering studies, and crop inventories. Color photographs are superior for many applications, but black-and-white photographs remain the standard form of aerial photography.

IR Black-and-White Photography

By using the film and filter combinations shown in Figure 2.17, photographs can be acquired that record energy at wavelengths ranging from 0.7 to 0.9 μm. This spectral band represents reflected solar radiation, not emitted thermal energy, and is called *photographic IR energy*. The following advantages of IR photographs are illustrated in Figure 2.18:

1. Improved haze penetration, because the filter eliminates the severe atmospheric scattering that occurs in the visible and UV regions. The elimination of most scattered light results in a higher contrast ratio and therefore higher spatial resolution on the IR photograph, as described earlier in the section on Effects of Scattering on Aerial Photographs.
2. Maximum reflectance from vegetation occurs in the photographic IR region, as shown by the bright tones in the IR photograph, but this is not always an advantage. Spectral differences between hardwoods and conifers are generally at a maximum in the photographic IR region.
3. IR radiation is absorbed by water, which causes water to have a dark tone on IR photographs. For this reason boundaries between land and water are clearly shown on IR photographs. In the tidal flat area on the east side of Figure 2.18 the tide line and individual tidal channels are clearly distinguished on the IR photograph. Such distinctions are not possible on the conventional photograph because light penetrates the shallow water and does not differentiate the submerged area from the land.

IR color film, described later in this chapter, combines these properties of IR black-and-white film with the advantages of color.

A. CONVENTIONAL PHOTOGRAPH.

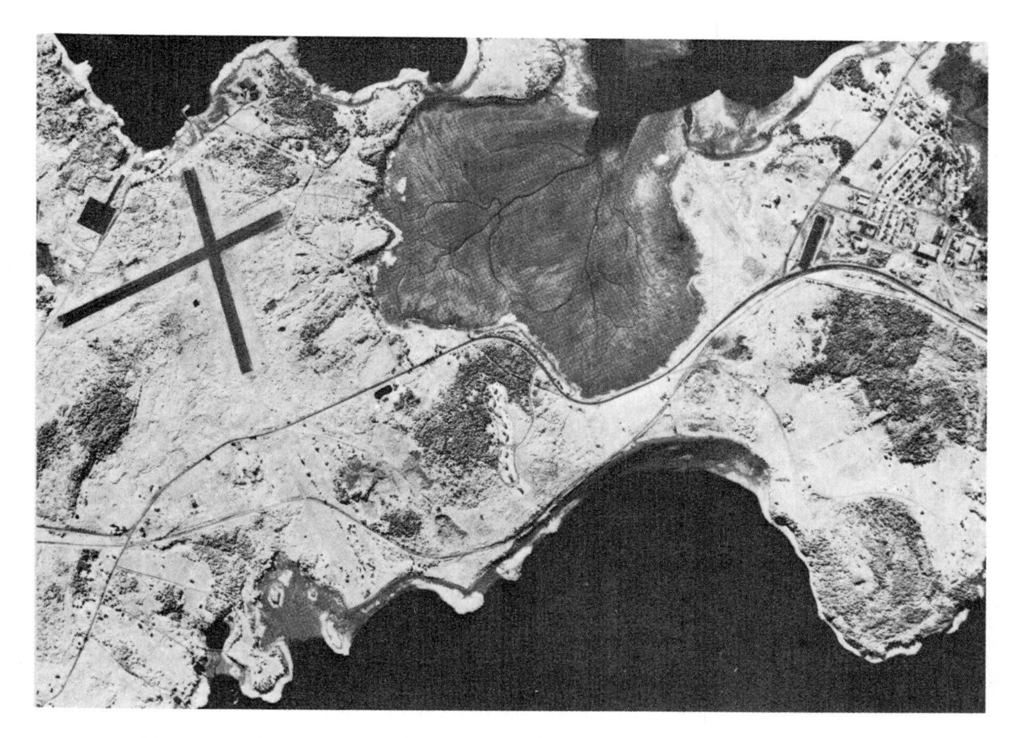

B. IR BLACK–AND–WHITE PHOTOGRAPH.

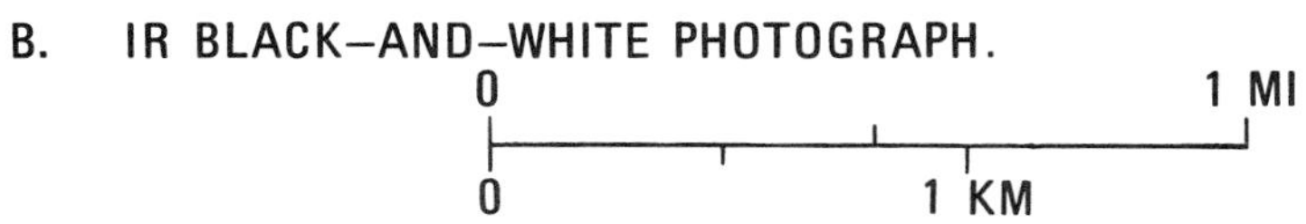

FIGURE 2.18
Comparison of conventional and IR black-and-white aerial photographs of the Massachusetts coast.

UV Photography

Photographs may be acquired in part of the UV spectral region with the film and filter combinations shown in Figure 2.19. Kodak Plus-X Aerographic film 2402 and the Kodak Wratten 18A filter are specific for the photographic UV region. Only a small amount of energy reaches the film, however, because of the narrow spectral transmission range (0.3 to 0.4 μm) of the filter and the high absorption. The typical camera exposure setting of about 1/125 sec with an *f*/4.0 lens aperture is poor for aerial photography because of possible blurring due to image motion. The Kodak Wratten 39 filter transmits both UV and blue energy and provides a typical exposure of about 1/500 sec at *f*/5.6. For detecting oil films on water, the Wratten 39 filter is almost as good as the Wratten 18A and has the advantage of higher energy transmission. The glass of most camera lenses absorbs UV radiation of wavelengths less than about 0.35 μm. Special quartz lenses can transmit shorter wavelengths but provide little advantage because incoming solar radiation of wavelengths shorter than 0.3 μm is absorbed by the ozone layer of the upper atmosphere (Figure 2.19).

The photographs in Figure 2.20 were acquired simultaneously with two cameras, one equipped with a filter that transmits only UV energy, and the other with a minus-blue filter. The effect of selective atmospheric scattering is apparent in the low contrast ratio which results in low resolution on the UV photograph. The bright and dark tones have similar patterns on both photographs. There is a tendency for carbonate rocks to have bright signatures on UV photographs because of fluorescence stimulated by sunlight. Atmospheric scattering effects can be reduced by acquiring UV photographs at lower altitudes. Because of atmospheric scattering, little remote sensing has been done in the UV region with the exception of monitoring oil films on water. Solar radiation causes many oils to fluoresce, with the maximum fluorescence occurring at UV wavelengths. This application is illustrated in Chapter 9 on Environmental Applications, together with a description of optical-mechanical scanners that are capable of acquiring UV images.

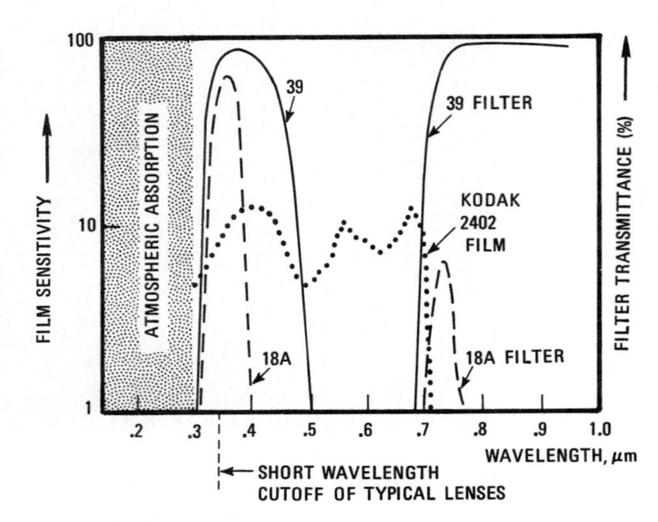

FIGURE 2.19
Film and filter combinations for UV photography. From Vizy (1974, Figure 6).

COLOR SCIENCE

The average human eye can discriminate many more shades of color than it can tones of gray. This greatly increased information content is the major factor favoring the use of color photography. The major drawback is that existing black-and-white photographs are available for most areas of the United States at nominal cost from government agencies, whereas color photographs generally must be acquired specially for most areas. Where new photography is to be acquired, however, color photographs are only moderately more expensive than black and white and are generally well worth the extra cost.

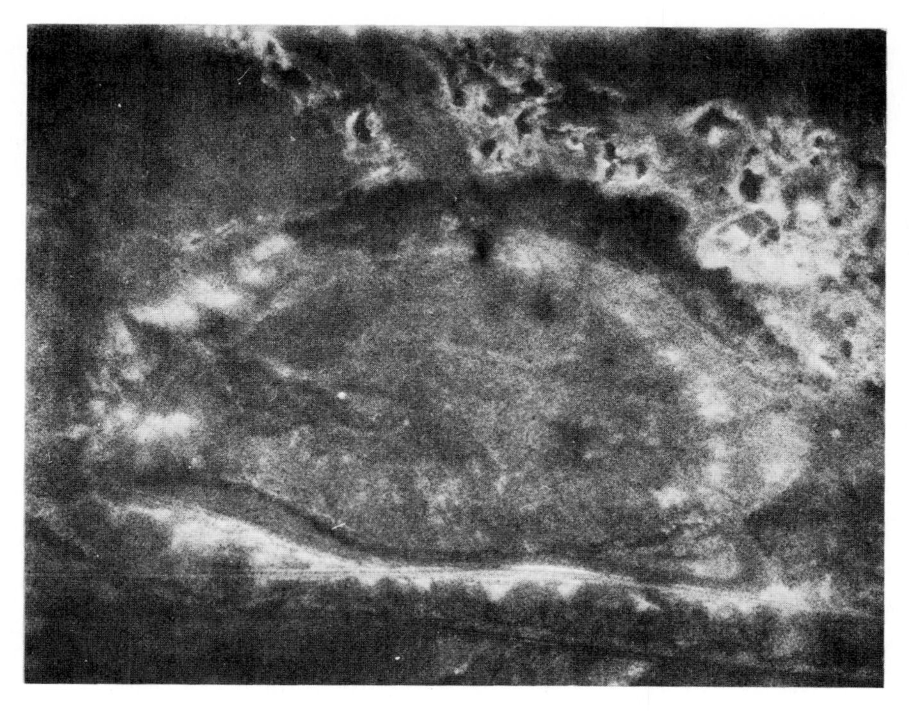

A. UV PHOTOGRAPH USING KODAK WRATTEN 18–A FILTER AND TRI–X FILM.

B. CONVENTIONAL PHOTOGRAPH USING WRATTEN 12 FILTER AND PANATOMIC–X FILM.

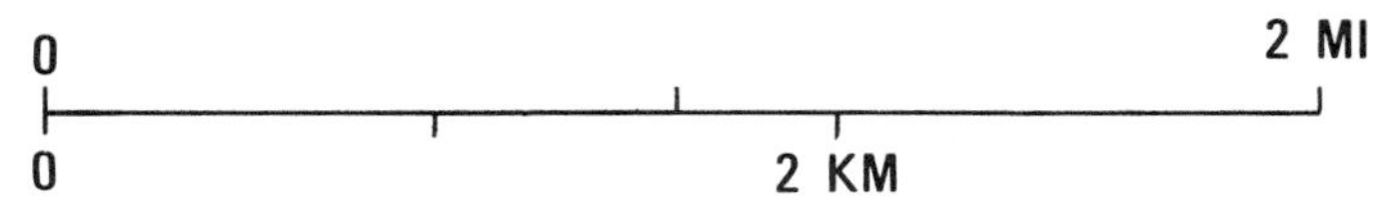

FIGURE 2.20
UV and conventional black-and-white aerial photographs of Thornton anticline, Weston County, Wyoming. Photographs were acquired simultaneously on October 19, 1966 at 3,350 m above terrain.

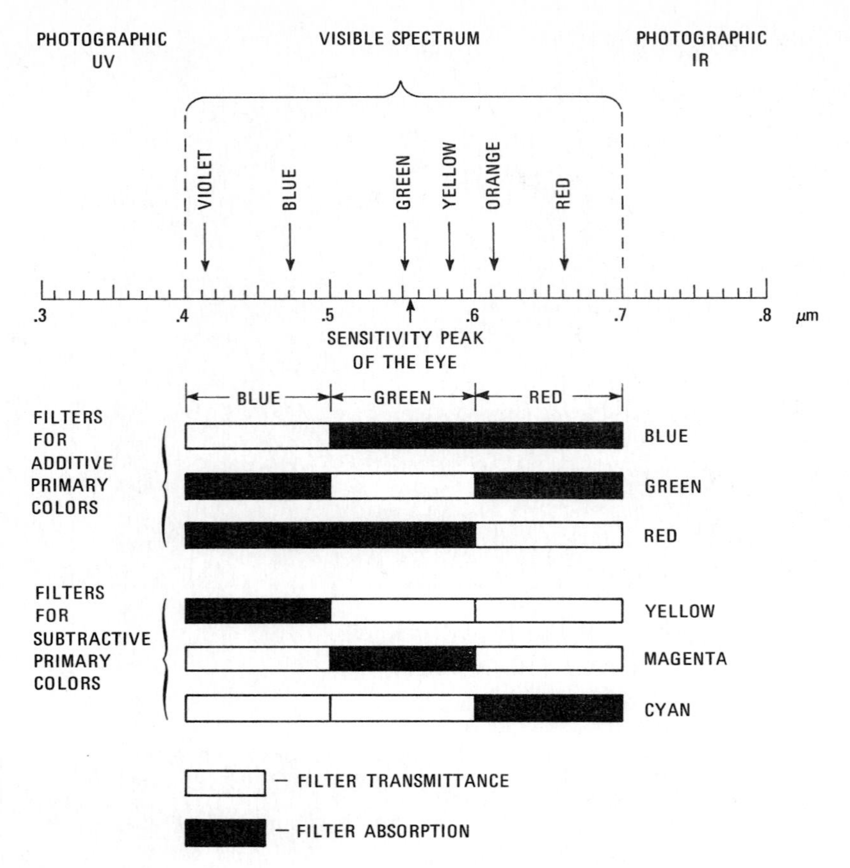

FIGURE 2.21
Visible spectrum with transmission and absorption characteristics of filters of the additive and subtractive primary colors.

Visible Spectrum

White light can be divided into the six gradational colors shown in the upper part of Figure 2.21 with the wavelength indicated for the center of each color. The eye responds as though it were a three-receptor system and many colors can be synthesized by adding different proportions of blue, green, and red. These are called the *additive primary colors* because white light is formed when blue, green, and red lights are added by superposition.

Additive Primary Colors

The wavelength band of each additive primary color is arbitrarily designated to span one-third of the visible spectrum. Figure 2.21 shows the characteristics of theoretical filters, each of which transmits one additive primary color and absorbs the other two. A color image can be produced by acquiring separate black-and-white pictures through each of the three primary filters. Positive films of the three pictures can then be registered on a screen with three projectors using the respective primary colors to produce a color image.

For most purposes it is impractical to use three projectors to produce color photographs, and additive primary colors cannot be mixed by superposition of the films. As shown in Figure 2.21, each additive primary filter absorbs two-thirds of the spectrum, and where any two filters are superposed no light is transmitted. Color television employs additive primary colors, however, by juxtaposing blue, green, and red specks small enough to blend together when observed by the eye.

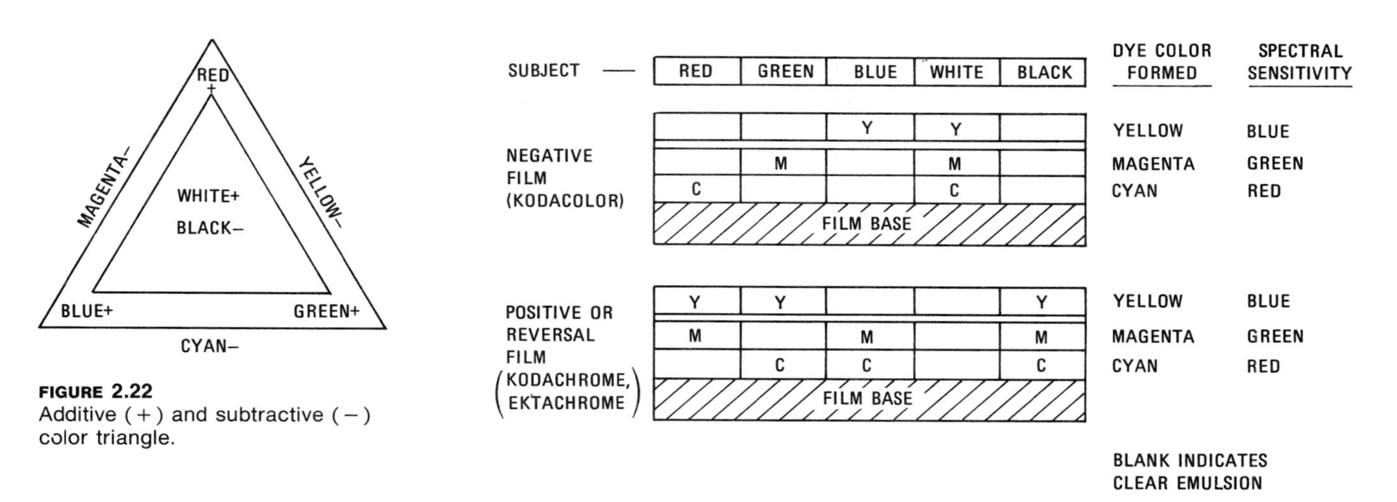

FIGURE 2.22
Additive (+) and subtractive (−) color triangle.

FIGURE 2.23
Formation of images on positive and negative color films.

Subtractive Primary Colors

In order to mix colors by superposition of films the three *subtractive primary colors*—yellow, cyan, and magenta—are used. As shown in the lower part of Figure 2.21, each subtractive primary filter absorbs one-third of the visible spectrum and transmits about two thirds.

The yellow subtractive primary filter absorbs blue light. The magenta primary is a bluish-red filter that absorbs green. Cyan is a greenish-blue filter that absorbs red. The color triangle of Figure 2.22 relates the additive primaries at the corners to the subtractive primaries along the sides. Each subtractive primary absorbs the additive primary at the opposite corner and transmits those at the adjacent corners. For any two superimposed subtractive primary filters illuminated with white light, the additive color at their common corner on the triangle is the color transmitted. For example, white light projected through overlapping yellow and cyan filters appears green. Superposition of all three subtractive filters absorbs all light and results in black. *Complementary colors* are pairs of colors that produce white light when added together, such as magenta and green. On the color triangle (Figure 2.22), complementary colors are located opposite each other. A color-film system using a single projector is possible with subtractive primary colors as described in the following section.

Color Film Technology

The term *color print* refers to color photographs with an opaque base. *Color film* refers to a transparent medium which may be either positive or negative. On conventional negative film the color present is complementary to the color of the subject photographed. For example, a green subject is portrayed by a magenta image on negative film. Negative film is normally employed in making color prints. On positive color film (except IR color film) the subject is represented by an image in its approximate true color. Kodachrome and Ektachrome are positive color films manufactured by Kodak that are familiar to most amateur photographers. Color film consists of a transparent base coated with three emulsion layers (Figure 2.23). The emulsion layers are similar to those on black-and-white film (Figure 2.5), but with the following differences: (1) each layer is sensitive to one of the additive primary colors—blue, green, or red; (2) during developing, each emulsion layer forms a color dye that is complementary to the primary color that exposed the layer. The blue-sensitive emulsion layer forms a yellow positive image; the green-sensitive layer forms a magenta image; the red-sensitive layer forms a cyan image. The silver halide salts of the green-sensitive and red-sensitive layers are also sensitive to blue light. A yellow filter layer beneath the upper emulsion prevents blue light from exposing the green-sensitive and red-sensitive layers. The yellow filter layer is dissolved during film processing. Figure 2.23 illustrates a color subject and the manner in which its image is formed on negative and positive film. On the processed negative or positive film, each layer either will be clear or will form an image of the subject in the subtractive color of that layer.

On *negative color film* the red-sensitive bottom layer produces a complementary cyan image of a red subject. Green and blue subjects produce magenta and yellow images, respectively. The white subject exposes all three layers, resulting in an image that transmits no light; the black subject results in a clear image because none of the layers are exposed. The resulting negative color film is projected or contact printed onto photographic paper coated with sensitive emulsion to produce a color print.

The process of producing color film in positive, or reversal, form has been described by Kodak (1962). The exposure and development of a normal black-and-white film produces a *negative transparency* in which the silver deposit is heaviest in areas corresponding to the brightest parts of the scene and lightest in areas corresponding to the darkest parts of the scene. For direct viewing and projection it is advantageous to process the film in such a manner that it becomes a *positive transparency*. This *reversal* processing is based on the fact that development of a negative image leaves in the emulsion a small amount of unexposed silver halide grains where the negative density is highest, and a large amount of silver halide where the negative density is lowest. The silver halide not used to form the negative image has the gradations of a positive image.

In the Kodak Ektachrome and Kodachrome processes the reversal method is used to produce positive color transparencies. First the film is developed to produce a negative silver image in each of the three emulsion layers. The film is then reexposed with white light to render the remaining silver halide developable. This latent positive image is then chemically *coupled*, or combined, with color dyes to produce a positive color image in each of the three layers. The film is next treated in a bleach which, without affecting the dyes, converts the silver to soluble salts that are removed by subsequent processes. As shown in the diagram of positive film in Figure 2.23, a red subject forms a clear image on the red-sensitive cyan-colored layer. The red subject also forms a magenta and a yellow image, respectively, on the green-sensitive and blue-sensitive layers. When viewed with transmitted white light, the yellow and magenta images absorb blue and green, respectively, and allow a red image to be projected. A white subject forms clear images on all three layers.

An HF-3 haze filter is normally employed in aerial color photography to absorb UV radiation that is strongly scattered by the atmosphere (Figure 2.4). Blue light is not removed by the HF filter because this would destroy the color balance of the film. The bluish appearance of some aerial color photographs is caused by the additional blue light that is selectively scattered into the lens.

With positive color film, the original film transparency may be used for interpretation on a light table. This provides maximum resolution, but the film rolls require special handling and viewing equipment and are not suitable for use in the field. Black-and-white prints, color prints, or transparencies can be made from negative color film. Despite their slightly lower resolution, paper prints are more versatile and are easily used in the field. The photographs described in the preceding paragraphs are called *normal color* photographs to distinguish them from other types of photographs in which the emulsion layers are sensitive to wavelengths other than the three additive primary colors. A print of a normal color aerial photograph is shown in Plate 1A.

IR COLOR PHOTOGRAPHY

The spectral sensitivity of each of the color emulsion layers may be changed to respond to other photographic spectral regions, including photographic IR. Such an IR-sensitive color film was originally designed for camouflage detection, but is now widely used for nonmilitary applications. This film was originally known as "camouflage-detection film" and is occasionally referred to as "false-color film," but the name *IR color film* is most widely used. The manufacturer's official name is Kodak Aerochrome infrared film, type 2443, which is available only as positive film. An IR color aerial photograph is illustrated in Plate 1B.

Film Characteristics

IR color film is best described by comparing it with normal color positive film. Figure 2.24 shows that for normal color film the spectral sensitivities of the three layers correspond to blue, green, and red light. The lower diagram of Figure 2.24 shows that the spectral sensitivity of each layer has been changed to produce IR color film. The cyan-forming layer is primarily sensitive to photographic IR energy of wavelengths from 0.7 to 0.9 μm. Note that this corresponds approximately to the sensitivity of IR black-and-white film with an 89B filter (Figure 2.17). The magenta-forming layer is sensitive to red light and the yellow-forming layer is sensitive to green light. The three layers are also sensitive to blue light, which is eliminated by placing a yellow (minus blue) filter over the lens. The elimination of UV and blue wavelengths by the yellow filter is shown by the diagonal line pattern in Figure 2.24. Removal of these strongly scattered wavelengths improves the contrast and resolution on IR color film. A new IR color film for aerial photography has a yellow

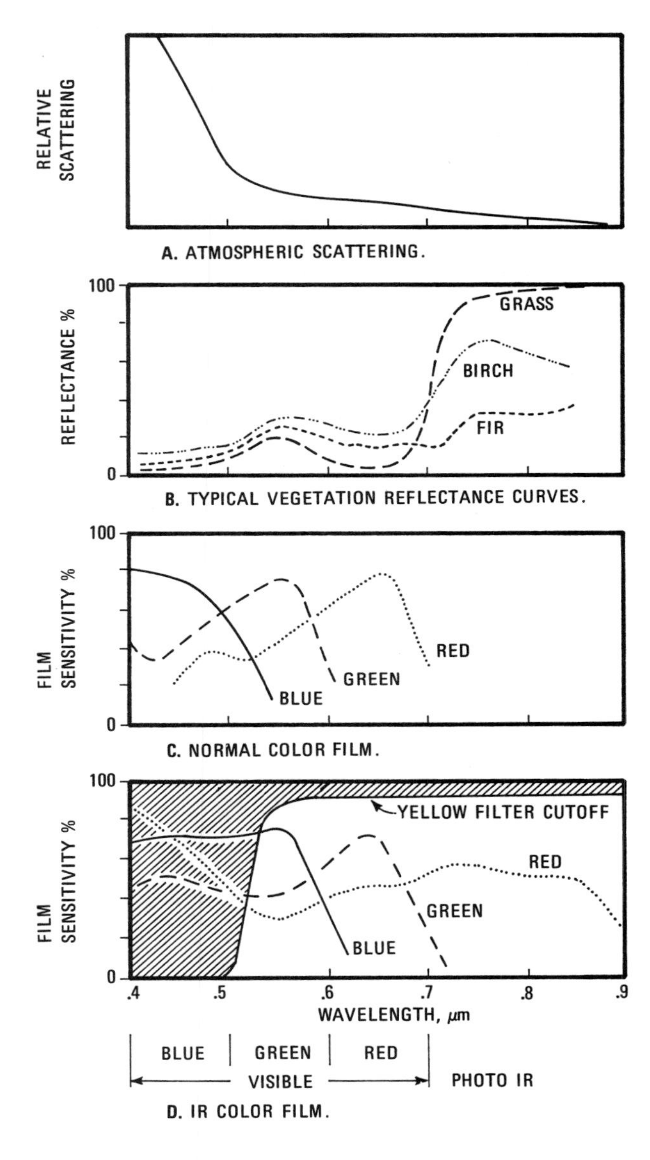

FIGURE 2.24
Spectral sensitivity of normal color film and IR color film, together with atmospheric scattering diagram and vegetation reflectance spectra. From Sabins (1973B, Figure 2).

TABLE 2.4
Terrain signatures on normal color and IR color film

Subject	Signature on normal color film	Signature on IR color film
Healthy vegetation:		
Broadleaf type	Green	Red to magenta
Needle-leaf type	Green	Reddish brown to purple
Stressed vegetation:		
Previsual stage	Green	Darker red
Visual stage	Yellowish green	Cyan
Autumn leaves	Red to yellow	Yellow to white
Clear water	Blue-green	Dark blue to black
Silty water	Light green	Light blue
Damp ground	Slightly darker	Distinct dark tones
Shadows	Blue with details visible	Black with few details visible
Water penetration	Good	Green and red bands: good. IR band: poor
Contacts between land and water	Poor to fair discrimination	Excellent discrimination
Red bed outcrops	Red	Yellow

layer over the upper emulsion layer, which eliminates the need for a yellow filter over the lens.

The term IR suggests heat, and some users mistakenly assume that the red shades on IR color film record variations in temperature. This is *not* the case as a few moments of thought should reveal. If the IR-sensitive layer were sensitive to ambient heat, it would be exposed by the warmth of the camera body itself. As described in Chapter 1, thermal radiation occurs at wavelengths longer than approximately 3 μm, which is beyond the sensitivity range of IR film.

Comparison of IR Color and Normal Color Photographs

Comparison of the simultaneously acquired IR color and normal color photographs of Plate 1 provides a useful way to understand their different characteristics. Table 2.4 compares the color signatures of common subjects on the two types of photographs. The most striking difference is the red color of healthy vegetation in the IR color photograph. This is explained by the spectral reflectance diagram of vegetation shown in Figure 2.24. Blue and red light are absorbed by the leaf structure in the process of photosynthesis. Up to 20 percent of the incident green light is reflected, causing the familiar green color of leaves. The spectral reflectance increases abruptly in the photographic IR region, which is the wavelength that exposes the red-imaging layer in IR color film. Vegetation has a wider range of reflectance values in the photographic IR region than in the green region, which makes it easier to discriminate vegetation types on IR color than on normal color film. In addition to forestry and agronomy this film is useful for geology because vegetation patterns can indicate the nature of the underlying rock and soil units.

The high IR reflectance of leaves is not caused by chlorophyll but is due to multiple internal reflections from the interfaces between hydrated cell walls and intercellular air spaces of spongy mesophyll tissue. When vegetation is stressed by disease, insect infestation, or other factors that cause a lack of water in the leaves, the internal cell structure may begin to collapse and the IR reflectance drops. This causes the red color to be darker than normal on IR color photographs. This loss of IR reflectance is called a *previsual* symptom of plant stress because it commonly occurs days or even weeks before the visible green color begins to change. Advanced plant stress produces a cyan color on the film. The previsual effect may be used for early detection of disease and insect damage in crops and forests. As leaves turn red and brown in the fall, they appear white or yellow on IR color film. The green chlorophyll has decayed so that the red and yellow pigments of the leaf are unmasked and additional ones are formed. The resulting reflectance values in the green, red, and photographic IR bands are nearly equal, resulting in the white to yellow signature on IR color photographs.

Because of its high absorption of photographic IR radiation, clear water has a dark blue or black signature on IR color photographs and contrasts with the red signature of vegetation. This ability to enhance the difference between vegetation and water is especially valuable for mapping drainage patterns in heavily forested terrain. In the northern part of Plate 1B the lighter blue colors of the small ponds are caused by different concentrations of suspended clay. Damp ground may be recognized on IR color photographs by its relatively darker signature, which is caused by absorption of IR energy. Shadows are darker on IR color photographs than on normal color photographs because the yellow filter eliminates scattered blue light.

The IR color photograph in Plate 1 has a better contrast ratio than the color photograph for two reasons: (1) elimination of scattered blue light improves contrast; (2) reflectance differences are commonly greater in the photographic IR region for vegetation, soils, and rocks. The higher contrast ratio results in improved spatial resolution. This is evident on the tree-covered slope in the northern portion of Plate 1 where individual trees are more readily resolved on the IR color photograph.

Thornton Anticline, Wyoming

The photographs in Plate 1 cover the Thornton anticline in the eastern part of the Powder River Basin, Wyoming. These were acquired simultaneously with the UV and minus-blue photographs of Figure 2.20. For overall usefulness in interpreting vegetation, geology, and terrain features the four types of photographs rank as follows:

1. IR color photographs (most useful).
2. Normal color photographs.
3. Minus-blue, black-and-white photographs.
4. UV photographs.

The concentric ridges and valleys indicate that the bedrock has been folded and consists of stratified sedimentary rocks with different degrees of resistance to erosion. Mappable rock units are readily designated on the photographs but the rock types cannot be identified without field checking (preferably), or referring to geologic maps and reports. Erosion has removed the resistant rocks over the crest of the anticline, exposing nonresistant rocks that form the topographic depression in the core of the structure. Two dry test wells for oil (wildcat wells) were drilled on the anticline. A multi-lane interstate highway and a railroad are recognizable in the south part of the photographs. Vegetation is sparse in the autumn season in this semi-arid climate.

The rock units shown on the interpretation map (Figure 2.25) were identified by field checking and

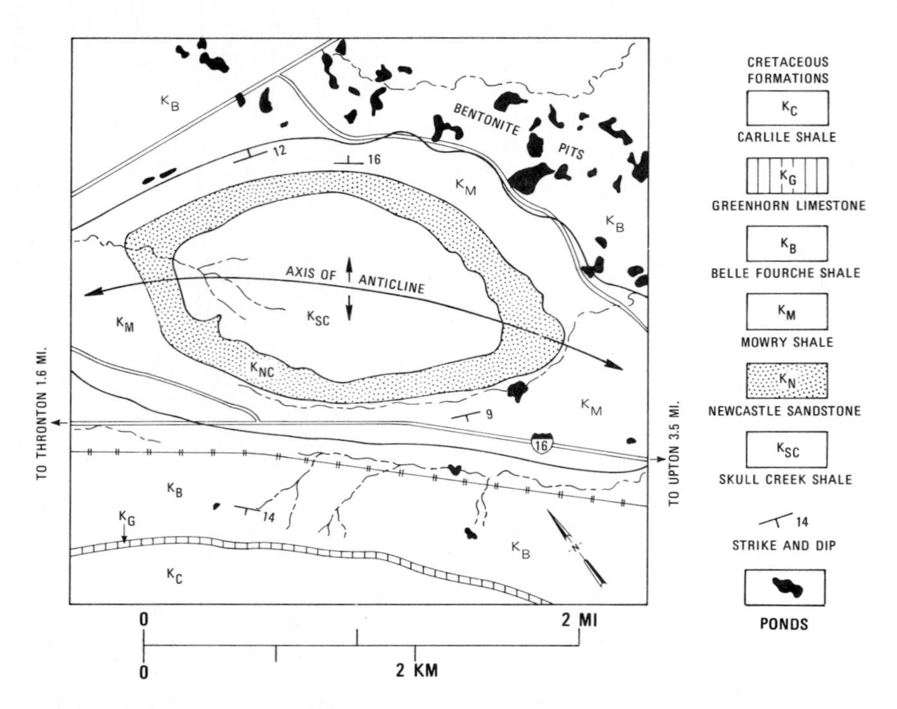

FIGURE 2.25
Geologic interpretation of aerial photographs of Thornton anticline, Weston County, Wyoming. Geology after Mapel and Pillmore (1964, Plate 1).

by referring to published geologic maps. The oldest rock unit is the Skull Creek Shale exposed in the core of the anticline. The grayish-black shale is covered by a thin dark soil that supports little vegetation, as shown by the absence of red hues on the IR color photograph. The Newcastle Sandstone forms a poorly exposed light-colored outcrop belt. The siliceous Mowry Shale forms the ridge that rims Thornton anticline. On the north and northeast flanks of the structure the dip slopes of Mowry Shale support a dense growth of small pine trees with a bright red signature. The Belle Fourche Shale consists of grayish-black nonresistant shale with interbeds of bentonite, a light-colored clay formed from altered volcanic ash that expands when wet. The numerous ponds in the upper part of Plate 1B are pits from which bentonite has been excavated for processing and sale as drilling mud. The resistant Greenhorn Formation consists of calcareous shale with limestone concretions and forms a strike ridge in the south part of the photographs. The dark gray Carlile Shale forms the slopes in the extreme southern part of the photographs.

Experimental Use

In addition to large sizes for aerial cameras, IR color film is available in 35-mm size for use in ordinary cameras. Cost of the 20-exposure casettes and processing is comparable to normal color films. The characteristics and applications of IR color film can be evaluated at minimal expense with this format. For purposes of comparison, it is particularly instructive to acquire normal color photographs simultaneously with the IR color photographs. Unexposed or undeveloped IR color film may deteriorate with time and with excessive heat; if the film is to be kept for more than a few weeks it should be stored in a freezer. The frozen film should be allowed to reach room temperature before the sealed container is opened to prevent moisture from condensing on the cold film. These precautions also apply to the larger sizes of this film.

A yellow filter, such as Kodak Wratten 12, is used to take pictures with IR color film. This film and filter combination has an approximate speed of ASA 100. This speed does *not* apply to cameras

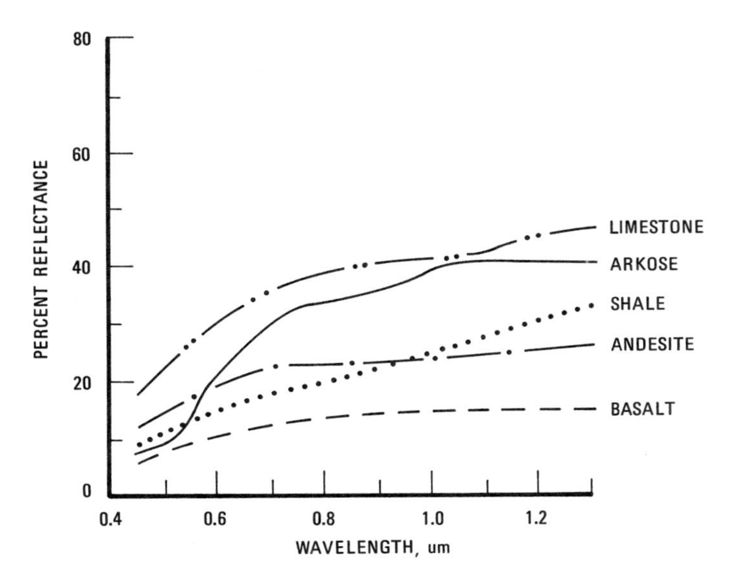

FIGURE 2.26
Reflectance spectra measured in the field of typical volcanic and sedimentary rocks. From Goetz (1976, Figure 1).

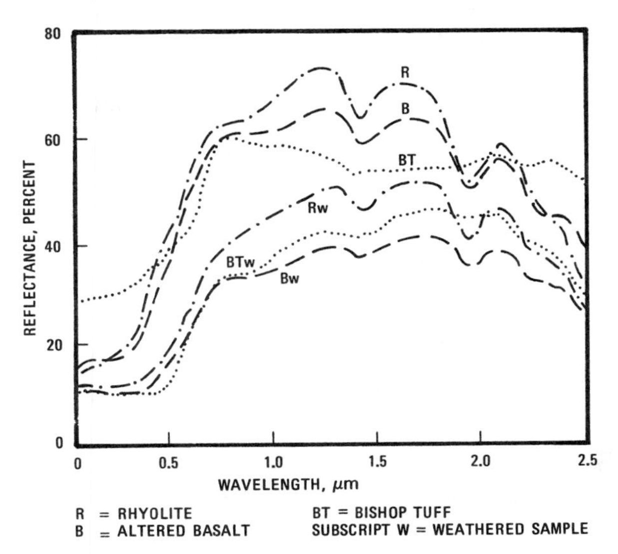

FIGURE 2.27
Reflectance spectra of fresh and weathered rock samples. From Lyon (1970, Figure 4).

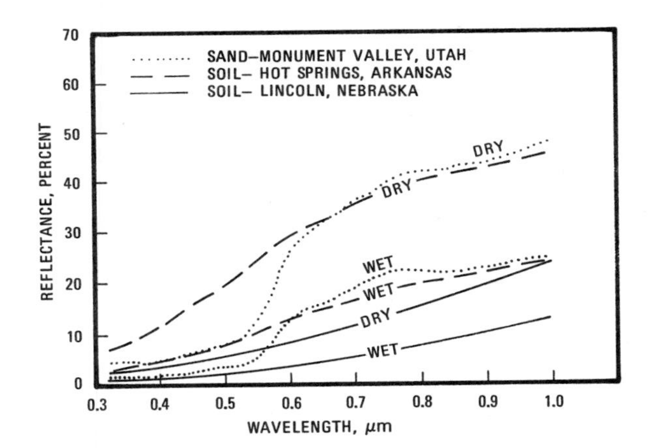

FIGURE 2.28
Reflectance spectra of wet and dry sandy soils. From Condit (1970, p. 955).

with a built-in light meter that measure the light after it has passed through the yellow filter over the lens. In these cameras the ASA setting should be increased to 200 to compensate for the effect of the filter. Some experimentation is necessary to determine the optimum exposure because conventional light meters do not measure the same spectral region to which the film is sensitive. Some cameras have an IR setting on the focusing ring which is intended for IR black-and-white film. This setting should not be used for IR color film because two of the three emulsion layers are sensitive to visible wavelengths.

SPECTRAL REFLECTANCE

Spectral-reflectance curves represent the percent of incident light reflected by materials as a function of wavelength. Figure 2.26 shows spectral-reflectance curves in the visible and reflected IR regions for some typical volcanic and sedimentary rocks. One practical use of such curves is to provide a comparison standard for identifying spectra of unknown materials. Another use is to recognize spectral regions in which various materials can be differentiated. For example, the basalt, andesite, shale, and arkose of Figure 2.26 have very similar reflectance values in the blue and green region (0.4 to 0.6 μm). In the red and photographic IR region (0.6 to 0.9 μm), however, these rocks are readily differentiated. Figure 2.27 compares the spectral reflectance for weathered and unweathered igneous rocks, and illustrates that weathering of these rocks reduces the overall reflectance but does not change their relative spectral properties. Some soils may be distinguished by spectral differences as shown in Figure 2.28, which also illustrates that moisture reduces the reflectance, particularly in the photographic IR region.

Until recently most spectral-reflectance curves

of rocks and minerals were measured in the laboratory using small pulverized samples of unweathered material. These spectra were valuable for understanding the relationship between mineralogy and reflectance. In the real world, however, rocks are weathered to different degrees, have varying moisture contents, and are partially or wholly obscured by vegetation and soils that have distinctive spectral signatures of their own (Figures 2.24 and 2.28). The laboratory spectra are based on a few square centimeters of homogeneous sample, whereas a photograph or image, acquired at altitudes ranging from thousands of meters to hundreds of kilometers, integrates reflectance from a much larger area on the ground. Recently, portable spectrometers have been developed that enable investigators to acquire spectra in the field of natural outcrops up to 200 cm^2 in area. The spectra of Figure 2.26 were acquired in this fashion and represent an important step toward practical field application of reflectance information. An example of this application is shown for the Goldfield, Nevada mining district in Chapter 8 on Resource Exploration.

MULTISPECTRAL PHOTOGRAPHY AND IMAGERY

As an alternative to normal color and IR color film, photographs may be acquired in the *multispectral* mode. A simple system employs four cameras aimed at the same centerpoint on the ground with the shutters linked to trigger simultaneously. Each camera has a black-and-white film and filter combination sensitive to a particular spectral band, typically blue, green, red, and photographic IR. The photographs may be compared by individual inspection. Some satellite photographs have been acquired in this fashion and an example from the Apollo 9 mission is given in Chapter 3 on Manned Satellite Imagery. The term "multiband" is commonly used as a synonym for multispectral, especially for photographs.

Another multispectral system employs a single camera body and shutter for the four lenses and has two film supplies, one for each pair of lenses. The resulting negatives are made into black-and-white positive transparencies that are registered and projected with colored lights in a special viewer to produce normal color or IR color pictures. Various color combinations can be used to enhance specific features. Some systems have employed nine lenses and three types of film.

Gilbertson, Longshaw, and Viljoen (1976) tested multispectral normal color, and IR color aerial photographs for mineral exploration in South Africa. They concluded that multispectral photographs have no significant advantage over normal color and IR color photographs in this area. Their multispectral photographs lack the ground resolution of conventional photographs and are 10 times as expensive to produce. The major advantage of multispectral photography is that the individual images provide more options for displaying the data, which may be useful for research purposes.

Optical-mechanical scanning devices acquire images at various wavelengths, including the photographic region. Color composite images may be prepared from the scanner images in the same manner as from multispectral photographs. Multispectral scanner images acquired from satellites are illustrated in Chapter 3 on Manned Satellite Imagery and in Chapter 4 on Landsat Imagery. Multispectral scanner images acquired from aircraft are illustrated in Chapter 11 on Comparison of Image Types.

SOURCES OF AERIAL PHOTOGRAPHS

The primary source for aerial photographs acquired by the U.S. Geological Survey and NASA is U.S. Geological Survey, EROS Data Center, Sioux Falls, South Dakota 57198. An inquiry to the EROS Data Center should specify the latitude and longitude boundaries of the area and the type of photography that is required. In addition to black-and-white aerial photographs at conventional altitudes (3 to 9 km), EROS has high-altitude (18 km) black-and-white, normal color, and IR color photographs of many portions of the United States. Much of the United States has been photographed by the Agricultural Stabilization and Conservation Service. A set of state index maps and ordering instructions may be obtained from Western Aerial Photography Laboratory, ASCS–USDA, P. O. Box 30010, Salt Lake City, Utah 84115. Black-and-white photographs of national forests are available from regional offices of the U.S. Forestry Service. Many local aerial photography contractors have negatives and can furnish prints of areas they have flown. For color and IR color photography it is generally necessary to hire an aerial photography contractor to fly the area of interest.

COMMENTS

Aerial photography is a versatile and useful form of remote sensing for the following reasons:

1. The spatial resolution and information content of film are very high.
2. Acquisition of photographs is relatively simple and inexpensive.
3. Numerous types of film provide a sensitivity range from the UV spectral region through the visible and into the reflected IR region.

The principal drawbacks of aerial photography are:

1. Daylight and good weather are needed to acquire photographs.
2. In the shorter wavelength regions atmospheric scattering reduces the contrast ratio and resolving power of aerial photographs.
3. Variations in reflectance are recorded in uncalibrated fashion, which precludes quantitative interpretation.

The advantages outweigh the disadvantages and aerial photography should be evaluated as a possible data source for any remote-sensing investigation.

REFERENCES

Condit, H. R., 1970, The spectral reflectance of American soils: Photogrammetric Engineering, v. 36, p. 955–966.

Gilbertson, B., T. C. Longshaw, and R. P. Viljoen, 1976, Multispectral aerial photography as exploration tool—IV–V; An application in the Khomas Trough region, South West Africa; and cost effectiveness analysis and conclusions: Remote Sensing of Environment, v. 5, p. 93–107.

Goetz, A. F. H., 1976, Remote sensing geology—Landsat and beyond: Proceedings of Caltech/JPL Conference on Image Processing Technology, Data Sources, and Software for Commercial and Scientific Applications, Jet Propulsion Lab. SP 43–30, p. 8–1 to 8–8.

Hackman, R. J., 1967, Time, shadows, terrain, and photo-interpretation; U.S. Geological Survey Professional Paper 575-B, p. B155–B160.

James, T. H., 1966, The theory of the photographic process: Third Edition, Macmillan Co., New York.

Jones, R. C., 1968, How images are detected: Scientific American, v. 219, p. 111–117.

Kodak, 1962, Color as seen and photographed: Second Edition, Eastman Kodak Co., Rochester, New York.

———, 1970, Kodak filters for scientific and technical uses: Eastman Kodak Tech. Pub. B-3, Rochester, New York.

Lyon, R. J. P., 1970, Multiband approach to geological mapping from orbiting satellites: Remote Sensing of Environment, v. 1, p. 237–244.

Mapel, W. J. and C. L. Pillmore, 1964, Geology of the Upton Quadrangle, Crook and Weston Counties, Wyoming: U.S. Geological Survey Bulletin 1181-J.

Proctor, R. J., 1968, Geology of the Desert Hot Springs–Upper Coachella Valley area, California: California Division Mines and Geology, Special Report 94.

Rabben, E. L., E. L. Chalmers, E. Manley, and J. Pickup, 1960, Fundamentals of photo interpretation *in* Colwell, R. N., ed., Manual of photographic interpretation: ch. 3, p. 99–168, American Society of Photogrammetry, Falls Church, Va.

Rosenblum, L., 1968, Image quality in aerial photography: Optical Spectra, v. 2, p. 71–73.

Sabins, F. F., 1973A, Aerial camera mount for 70-mm stereo: Photogrammetric Engineering, v. 39, p. 579–582.

———, 1973B, Engineering geology applications of remote sensing *in* Moran, D. E., ed., Geology, seismicity, and environmental impact: Association of Engineering Geologists, Special Publication, p. 141–155, Los Angeles, Calif.

Slater, P. N., 1975, Photographic systems for remote sensing *in* Reeves, R. G., ed., Manual of remote sensing: ch. 6, p. 235–323. American Society of Photogrammetry, Falls Church, Va.

Thurrell, R. F., 1953, Vertical exaggeration in stereoscopic models: Photogrammetric Engineering, v. 19, p. 579–588.

Vizy, K. N., 1974, Detecting and monitoring oil slicks with aerial photos: Photogrammetric Engineering, v. 40, p. 697–708.

ADDITIONAL READING

Cravat, H. R. and R. Glaser, 1971, Color aerial stereograms of selected coastal areas in the United States: U.S. Department of Commerce, National Oceanic and Atmospheric Administration, Washington, D.C.

Lattman, L. H. and R. G. Ray, 1965, Aerial photographs in field geology: Holt, Reinhart, and Winston, New York.

Ray, R. G., 1960, Aerial photographs in geologic mapping and interpretation: U.S. Geological Survey Professional Paper 373.

Smith, J. T. and A. Anson, eds., 1968, Manual of color aerial photography: American Society of Photogrammetry, Falls Church, Va.

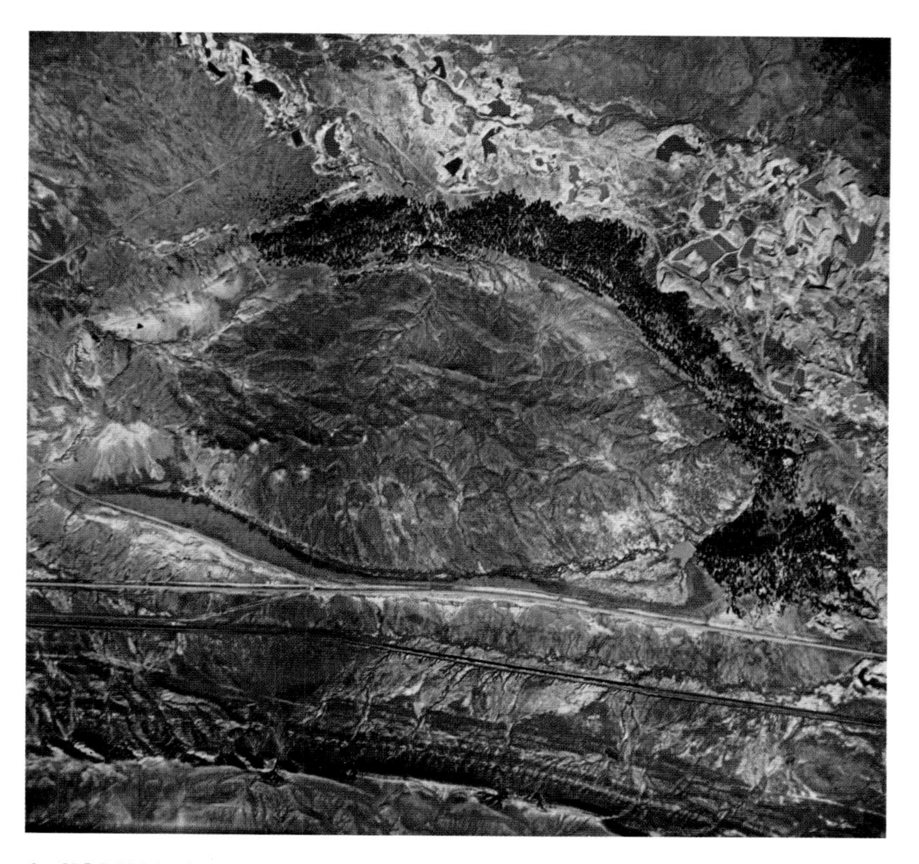

A. NORMAL COLOR.

PLATE 1
Normal color and IR color aerial photographs of Thornton anticline, Wyoming, acquired October 14, 1966 at 3,350 m above terrain.

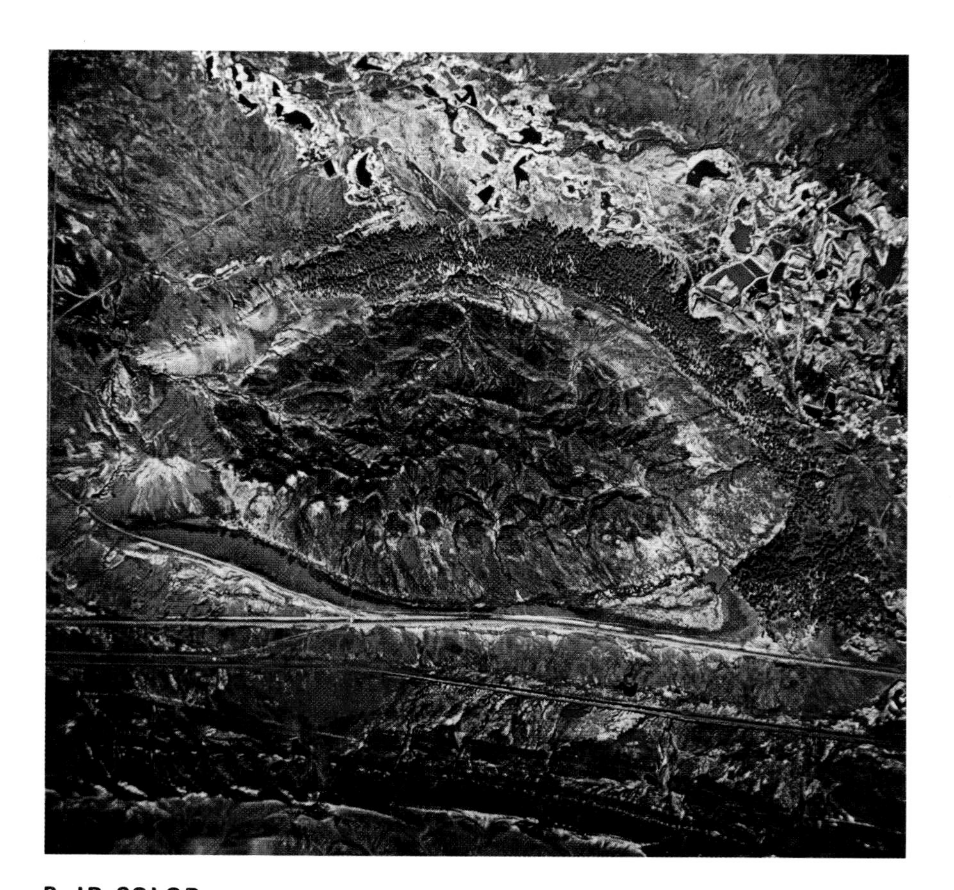

B. IR COLOR.

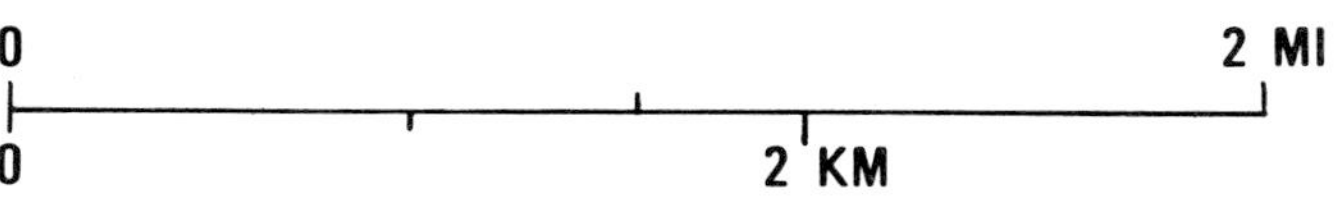

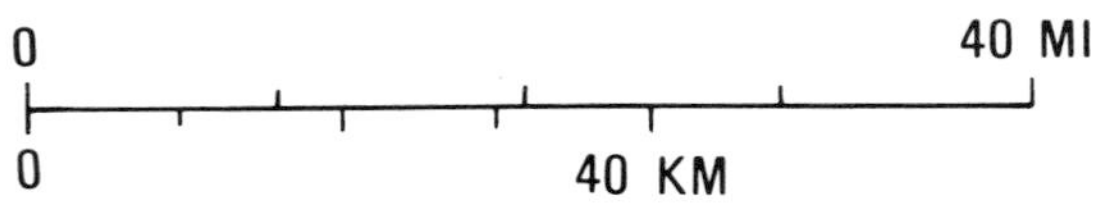

PLATE 2
Color-composite Landsat image of southern California, Landsat 1090-18012, acquired October 21, 1972. Digitally processed by Jet Propulsion Laboratory. Courtesy R. J. Blackwell, Jet Propulsion Laboratory, California Institute of Technology.

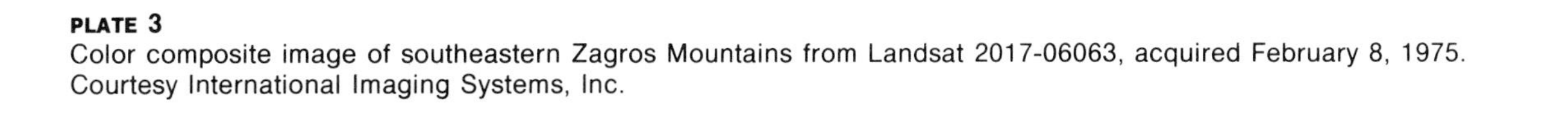

PLATE 3
Color composite image of southeastern Zagros Mountains from Landsat 2017-06063, acquired February 8, 1975. Courtesy International Imaging Systems, Inc.

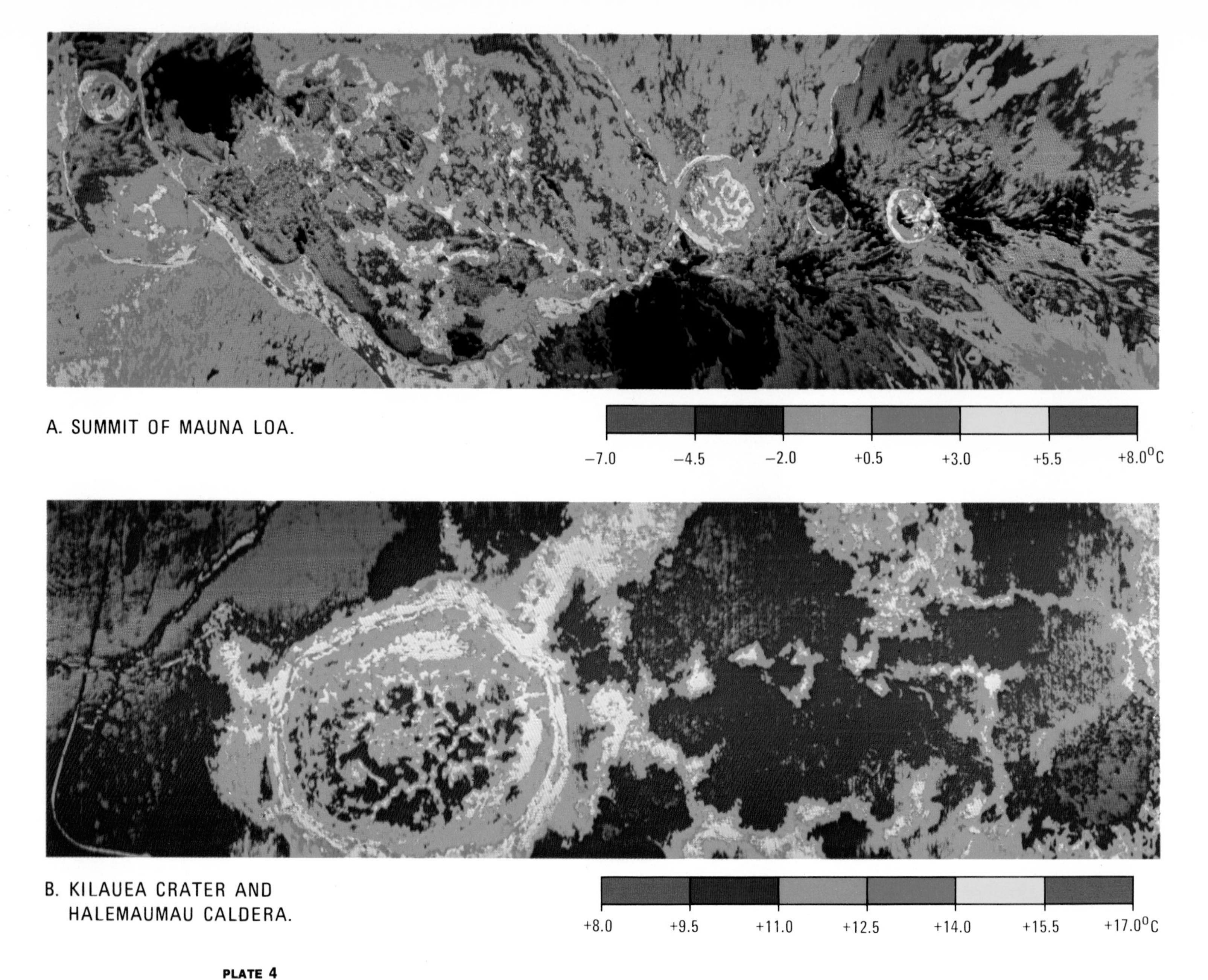

PLATE 4
Quantitative nighttime thermal IR images (8 to 14 μm) of Hawaiian volcanoes, acquired February, 1973. Digitally processed and displayed in Digicolor® by Daedalus Enterprises, Inc. Courtesy National Science Foundation and Daedalus Enterprises, Inc.

A. IR COLOR COMPOSITE IMAGE.

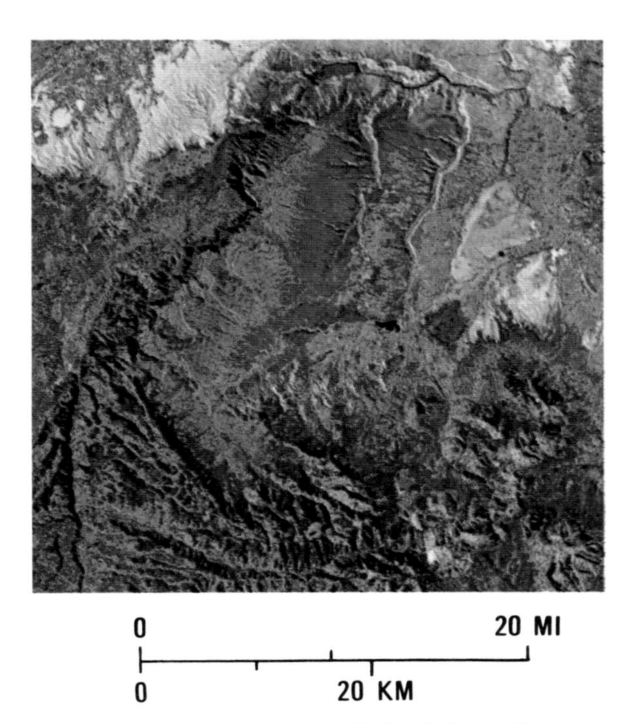

0 20 MI

0 20 KM

B. SIMULATED NORMAL COLOR IMAGE
PREPARED BY U.S. GEOLOGICAL SURVEY, FLAGSTAFF.

PLATE 5
IR color image and simulated normal color image of Gunnison River, Colorado. Portion of Landsat 1407-17190 acquired September 13, 1973.

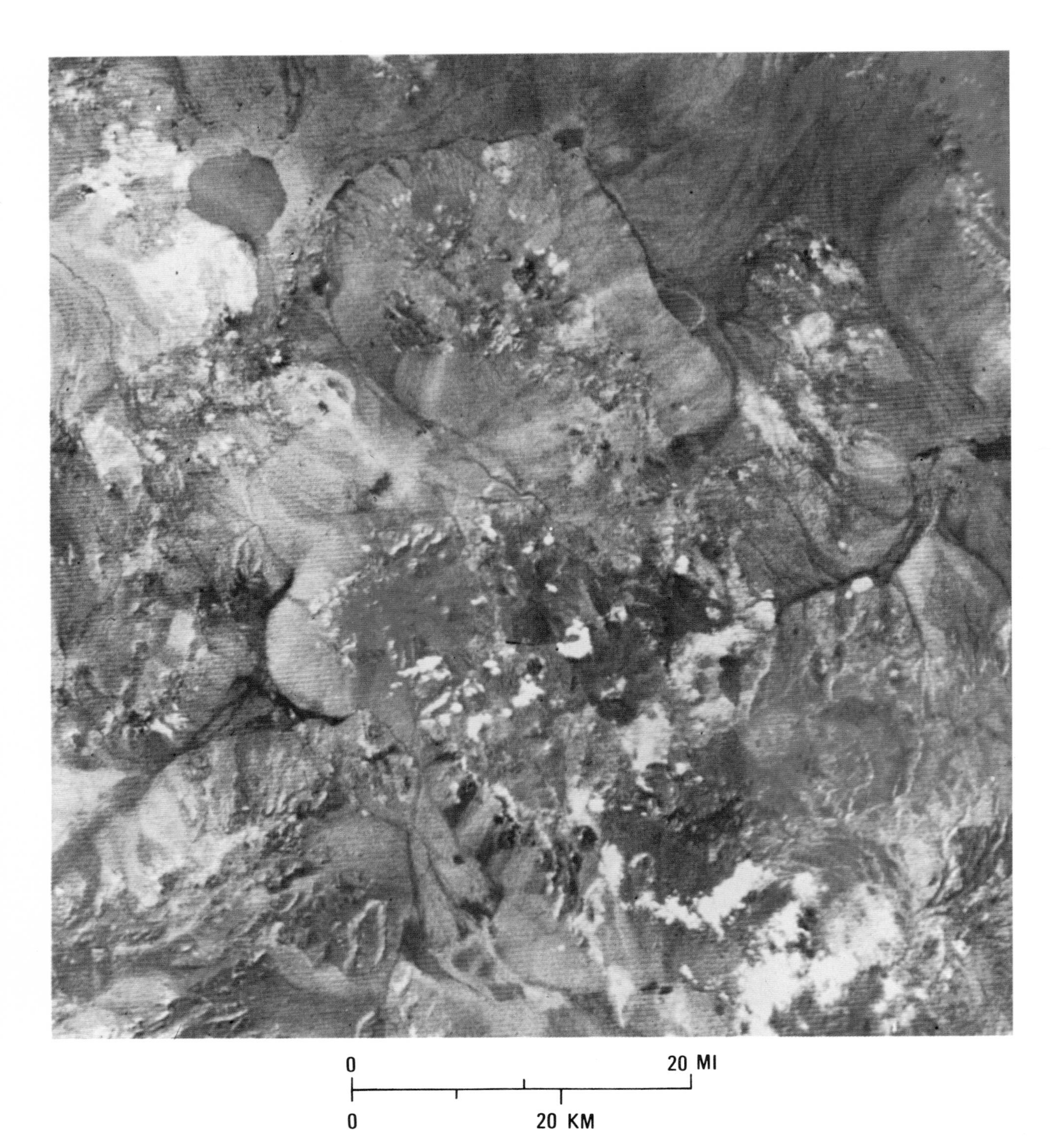

PLATE 6
Color-ratio composite image of Goldfield, Nevada, and vicinity. Digitally processed by Jet Propulsion Laboratory from Landsat 10702-18001, acquired October 3, 1973. The ratio 4/5 image was projected with blue light, 5/6 with yellow, and 6/7 with magenta. From Rowan and others (1974, Figure 17). Courtesy L. C. Rowan, U.S. Geological Survey.

A. SIMULATED NORMAL COLOR IMAGE.

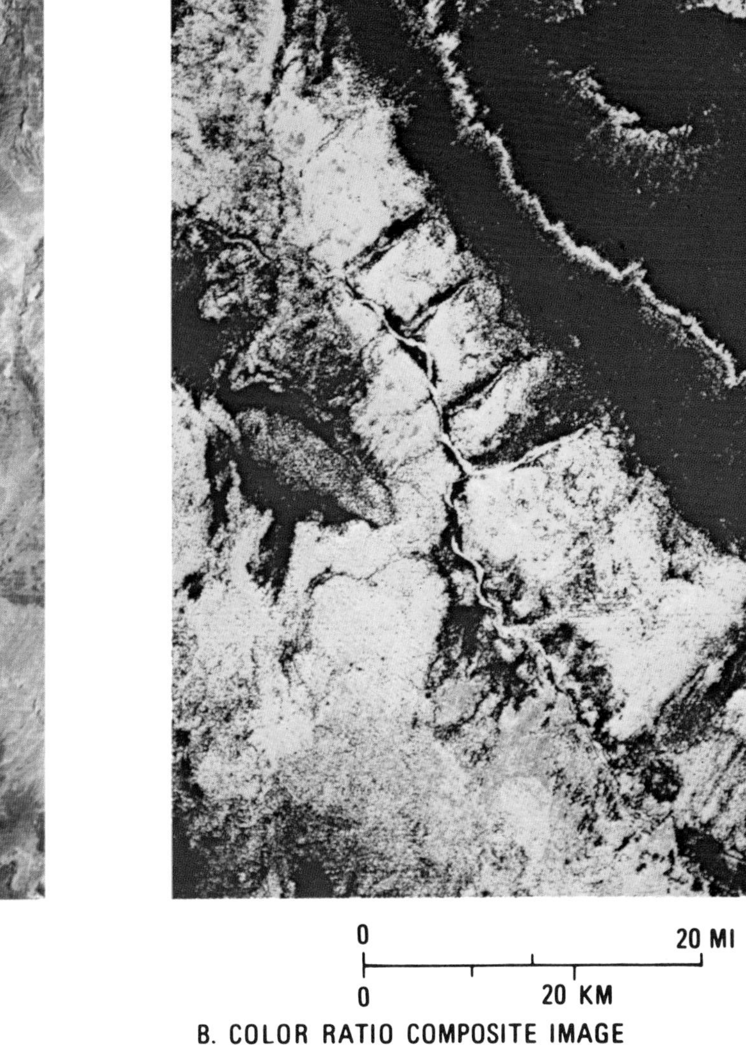

B. COLOR RATIO COMPOSITE IMAGE
BLUE COLOR CORRESPONDS TO SURFACE
ALTERATION ZONES.

PLATE 7
Cameron, Arizona, uranium district. Portion of Landsat 1463-17300, acquired November 5, 1973. Digitally processed by U.S. Geological Survey, Flagstaff, Arizona. From Spirakis and Condit (1975, Figures 4 and 6). Courtesy C. D. Condit, U.S. Geological Survey.

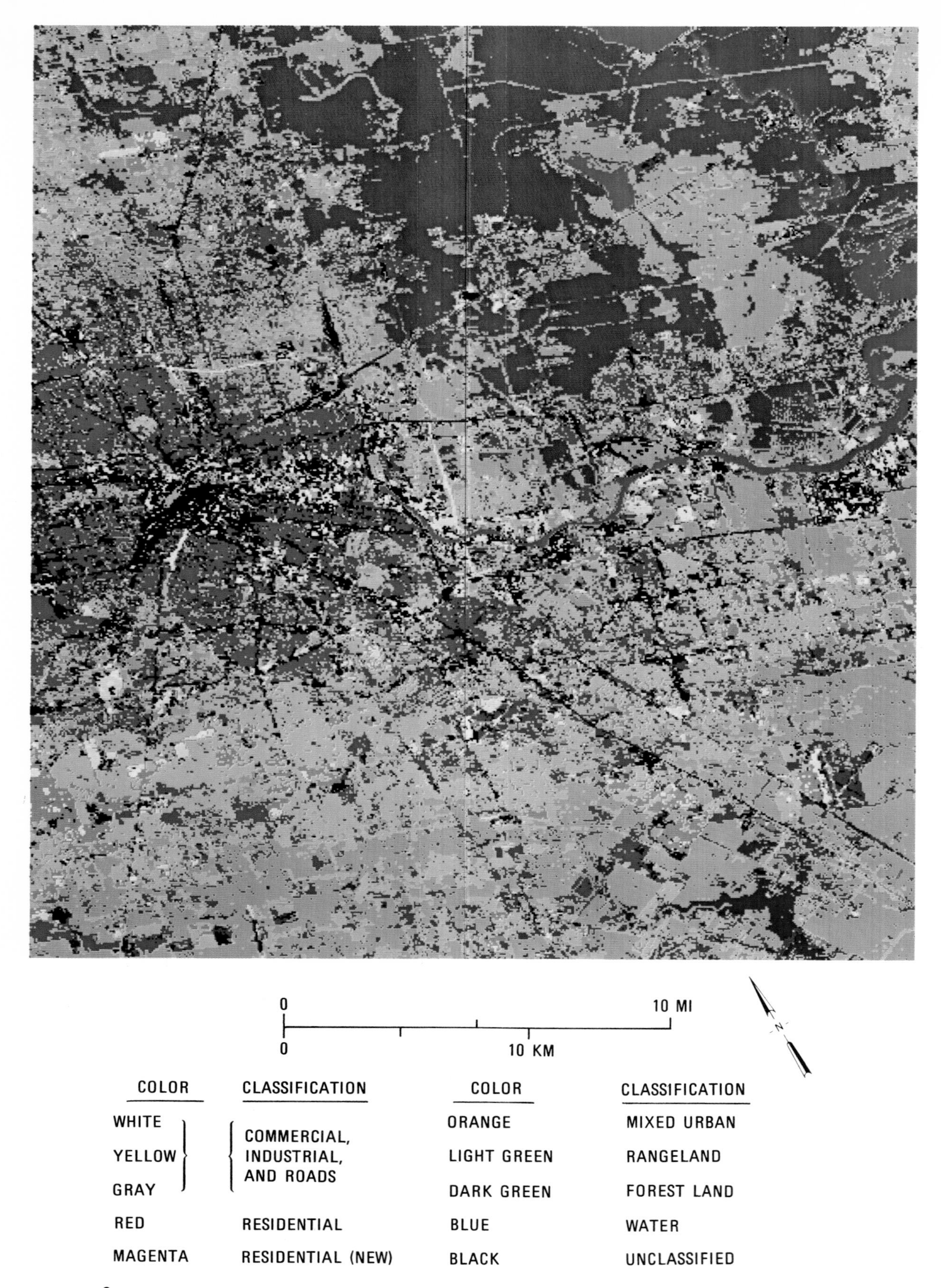

PLATE 8
Multispectral land-use classification of Houston and vicinity from Landsat 1037-16244, acquired August 29, 1972. From Erb (1974A, Figure 7-1). Courtesy O. V. Smistad and E. O. Zeitler, NASA Johnson Space Center.

3
MANNED SATELLITE IMAGERY

The first photographs of the earth from very high altitudes (160 to 320 km) were acquired after World War II by small automatic cameras on sounding rockets launched from White Sands Missile Range, New Mexico. P.M. Merifield (1964) was the first investigator to report on the ultra-high altitude photographs. In Figure 3.1 the northwest-trending mountains and valleys of the Basin and Range physiographic province occupy most of the scene. Part of the Mogollon Rim at the south margin of the Colorado Plateau is shown in the northeast part of the scene. The two prominent parallel north-trending linear features in the northeast (lower right) corner are located immediately west of the Arizona–New Mexico border. Although these are not shown on the Arizona State Geologic Map, they may be previously unmapped faults.

The rocket photographs stimulated interest in orbital photography and a terrain photography experiment was included on the final two Mercury flights in 1962 and 1963. These one-man missions produced usable photographs, chiefly of Tibet. Earlier the unmanned Mercury 4 mission produced several hundred high-angle oblique color photographs in 70-mm format, including many of North Africa. Although the quality was not as good as later photographs, they were used to study the western Sahara, which was the first regional application of orbital photography. Probably the greatest value of these early terrain experiments was the interest they created in orbital photography of the earth's surface.

GEMINI MISSIONS

Because of the encouraging earlier results, terrain photography experiments were included in most of the Gemini two-man missions that orbited the earth in 1965 and 1966.

Photographs

Gemini orbit paths were restricted to the region between latitudes 35°N and 35°S because of tracking, communication, and re-entry requirements. Altitudes normally ranged from about 160 to 320 km. On Gemini 11 the orbit was changed to form two high revolutions; most of the 102 photographs of this mission were acquired from these altitudes of 640 to 1360 km. The Gemini astronauts took the photographs with hand-held, 70-mm cameras through the spacecraft windows. All photographs were acquired on normal color film, except for one magazine of IR color film that produced unsatis-

FIGURE 3.1
IR black-and-white photograph acquired in 1955 from Viking 12 rocket at an altitude of about 224 km. View is slightly north of west with southeast Arizona in foreground. Distance across lower edge of photograph is 218 km. Courtesy P. D. Lowman, NASA Goddard Space Flight Center.

factory results. A collection of color photographs from Gemini 3, 4, and 6 was published by NASA (1967). The primary objective of the Gemini missions was to test men, equipment, and procedures for the Apollo lunar missions; terrain photography was a subordinate objective. Three main problems were encountered in the photography experiments:

1. Fuel limitations prevented the crew from pointing the spacecraft straight down; therefore many photographs were acquired at oblique angles. The resulting distortion presents problems for the interpreter.
2. Cloud cover, especially over jungles, obscured many of the primary target areas.
3. Deposits on the spacecraft windows from rocket exhaust and other sources caused degradation of some photographs, especially on Gemini 7.

Despite these problems, approximately 1,100 photographs proved useful for the study of geology, geography, or oceanography. Figure 3.2 is a black-and-white reproduction of a typical Gemini 4 oblique color photograph with readily identifiable geologic and topographic features. The linear expression of the Hillside fault between the towns

FIGURE 3.2
Gemini 4 photograph no. S-65-34697 looking southwest across west Texas. The Hillside fault between Van Horn (VH) and Sierra Blanca (SB) was proposed as the type locality for the Texas lineament by Albritton and Smith (1956). Courtesy P. D. Lowman, NASA Goddard Space Flight Center.

of Van Horn and Sierra Blanca in west Texas is especially pronounced. Rocks south of the fault near Van Horn are steeply dipping Precambrian metamorphic rocks that form aligned ridges, dragged to the west along the fault. Gently dipping Cretaceous strata are exposed north of the Hillside fault.

Geographic and Geologic Results

Lowman and Tiedeman (1971) summarized geographic and geologic applications of the Gemini terrain photographs and pointed out the following accomplishments:

1. A previously unmapped Quaternary volcanic field was discovered in northern Mexico that includes over 30 volcanoes with associated basalt flows.
2. Photographs of the Agua Blanca fault zone in northern Baja California indicate that it is not a strike-slip fault and is not one of the transform faults of the Gulf of California.
3. West of the area shown in Figure 3.2, the Texas lineament in southwest New Mexico and south-

east Arizona was shown to be a broad zone of folds and faults rather than a major wrench fault. The significance of lineaments is summarized in Chapter 4 on Landsat Imagery.

4. A very large area in North Africa was found to be eroded primarily by deflation and wind erosion, suggesting that wind erosion is more important in the Sahara than in the North American and other deserts.
5. The unexpected views of regional geologic structure stimulated speculative thinking about regional tectonics. Abdel-Gawad (1969) recognized two major shear zones in the crystalline rocks on opposite sides of the Red Sea; the offset of these distinctive features is evidence for northward movement of Arabia 150 km relative to Africa.

APOLLO PROGRAM

Although best known for its lunar accomplishments, the Apollo program produced several hundred terrain photographs including the first multispectral orbital photographs. The earth orbits covered the same 35°N to 35°S strip as the earlier Gemini program.

Conventional Photographs

Apollo 6 was an unmanned mission to test various equipment. An automatic 70-mm camera acquired high-quality, overlapping vertical color photographs on an orbit across the southern part of the United States and central part of Africa. The manned Apollo 7 and 9 missions acquired several hundred useful hand-held 70-mm photographs similar to those from the Gemini missions.

Multispectral Photographs (SO-65 Experiment)

The Apollo 9 earth-orbiting mission included the SO-65 experiment, which acquired the first orbital multispectral photographs. Four 70-mm cameras were mounted in the window of the command module and triggered simultaneously. Three cameras were equipped with film and filter combinations to produce black-and-white photographs in the green, red, and photographic IR spectral regions (Figure 3.3). The fourth camera used IR color film with a yellow filter. Ninety relatively cloud-free sets of photographs cover portions of northwest and southern Mexico, the Caribbean-Atlantic area, and the United States south of latitude 35°N. Careful positioning of the spacecraft provided vertical coverage without the distortion of earlier oblique photographs. The sites had been selected in advance to represent a wide variety of terrain for which supporting information was available. For some of the SO-65 experimental sites, photographs and other data were acquired from aircraft at the time of the Gemini overflight. A number of investigators interpreted the SO-65 photographs to evaluate the usefulness of multispectral photography for a variety of applications.

Lowman (1969) compared the Apollo 9 multispectral photographs of a desert area in New Mexico with those from a forested area in northeastern Alabama. In the desert area there was little difference in relative film response of different rock units on the green, red, and IR black-and-white photographs. The IR color photograph showed fewer hues and less contrast between rock units than normal color orbital photographs acquired earlier. In the heavily forested Alabama area, however, the multispectral photographs (Figure 3.3) were definitely superior to normal color photographs for mapping geologic structure

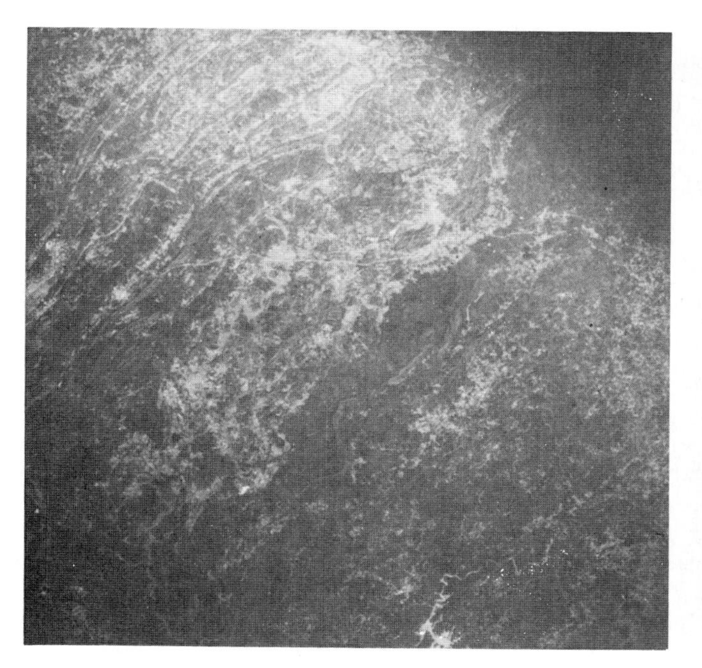

A. GREEN (0.47 TO 0.61 μm).

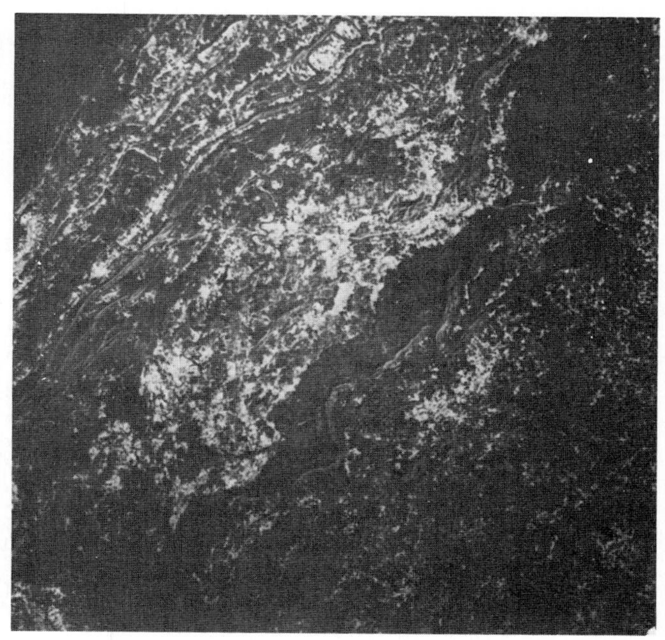

B. RED (0.59 TO 0.72 μm).

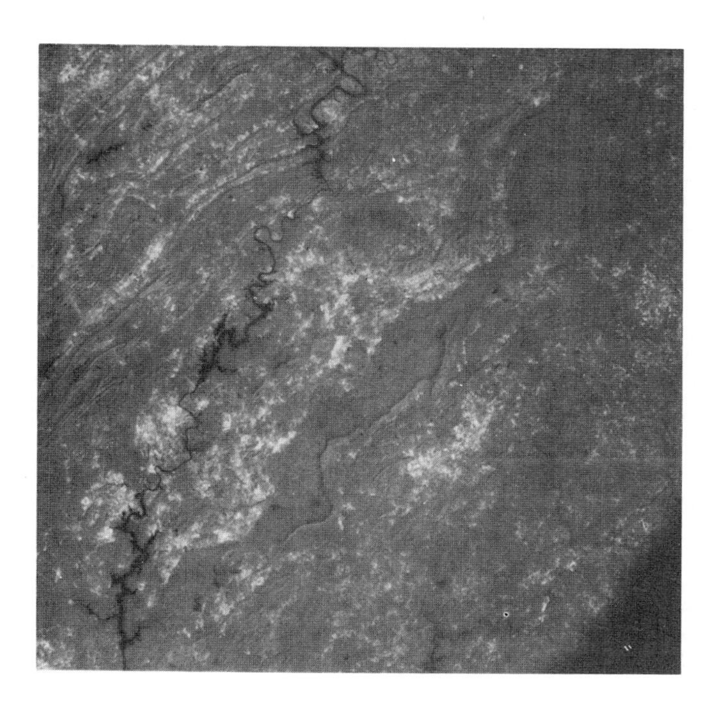

C. PHOTOGRAPHIC IR (0.68 TO 0.89 μm).

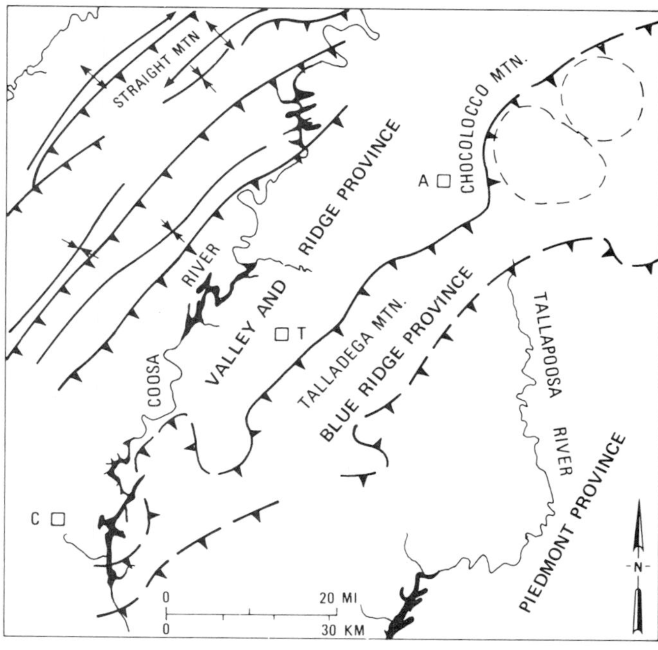

D. INTERPRETATION MAP, A=ANNISTON, C=COLUMBIANA, T=TALLADEGA.

FIGURE 3.3
Apollo 9 multispectral photographs of northeast Alabama. Photographs 9-26-3790-B, C, D acquired March 11, 1969. Courtesy P. D. Lowman, NASA Goddard Space Flight Center.

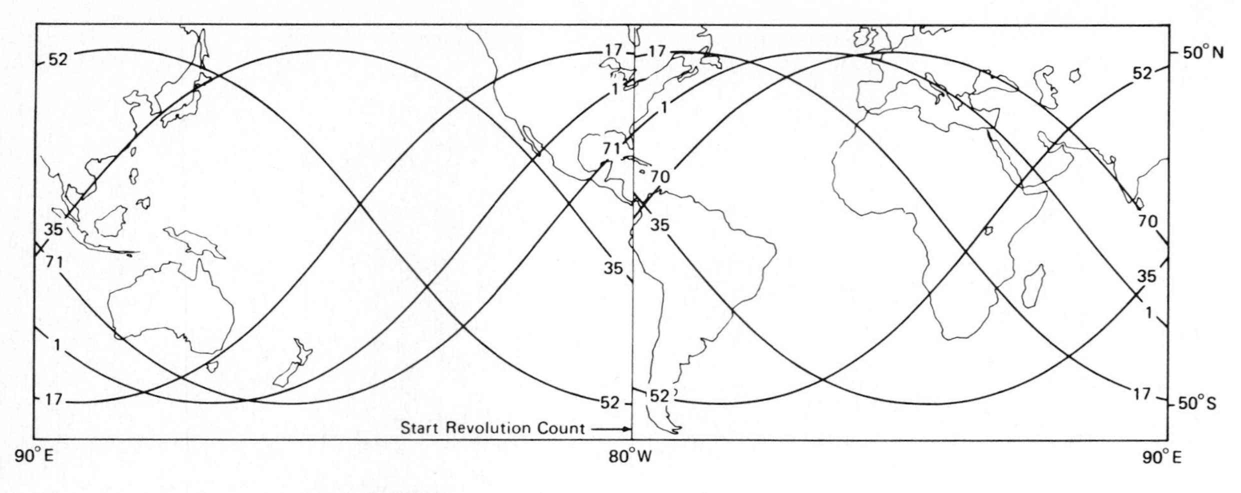

FIGURE 3.4
Typical Skylab orbit paths. From NASA (1974, Figure 4).

and for differentiating among vegetation types, water, rock, and soil. As illustrated on the geologic map of Figure 3.3D the multispectral photographs of northeast Alabama cover parts of the Valley and Ridge, Blue Ridge, and Piedmont physiographic provinces. The folds and thrust faults strike northeast with the displacement toward the northwest. Much regional geologic information can be interpreted from the multispectral images, despite the extensive cover of forest and farmland.

The green band (Figure 3.3A) has the lowest contrast of the three black-and-white photographs, but the major vegetation patterns are visible and some geologic trends are shown in the Valley and Ridge province. Drainage patterns, such as the Coosa River, are very obscure. This is definitely the least useful band. The red band (Figure 3.3B) has the highest tonal contrast with the lighter tones corresponding to cleared or cultivated land, such as the valley of the Coosa River, and the darker tones corresponding to woodlands. The contact between the forested Blue Ridge province and the agricultural land of the Valley and Ridge province is marked by the contrast between dark and light tones. Topography is expressed best on this band, as shown by Talladega Mountain and the strike ridges of folded strata in the Valley and Ridge province.

The IR black-and-white photograph (Figure 3.3C) has the same advantages described for this spectral region in Chapter 2 on Aerial Photography. This band has the highest spatial resolution and a tonal contrast that is intermediate between that of the green and red images. The IR color photograph (not shown) acquired simultaneously by the fourth SO-65 camera combines the advantages of the individual spectral bands. Some investigators composited the green, red, and IR photographic bands with blue, green, and red light, respectively, to produce the equivalent of IR color photographs.

SKYLAB

This was the largest manned space station ever placed into orbit. The Orbital Workshop was the third stage of a Saturn launch vehicle converted to living and working quarters for the astronauts. The astronaut crews used Apollo vehicles to dock with Skylab and return to earth.

Mission Plan

Skylab was launched unmanned and was occupied sequentially by three crews, each consisting of three men. Observations of earth were made from May 25, 1973 through February 8, 1974. As shown in Figure 3.4, the 50° inclination of the orbit path provided coverage of the earth between latitudes 50°N and 50°S at an altitude of 435 km. The ground track was repeated every five days, but image coverage of the earth is incomplete because of clouds and the astronauts' schedules. Astro-

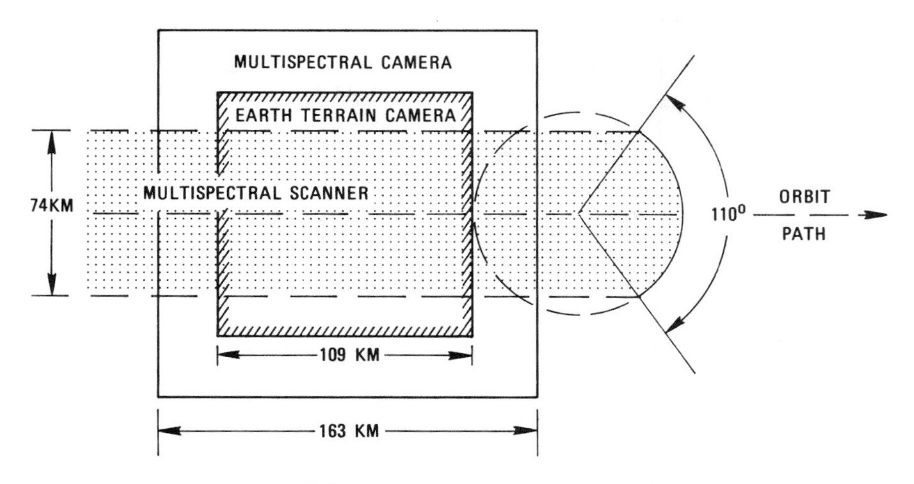

FIGURE 3.5
Ground coverage of Skylab imaging systems.

nomical, biological, and materials experiments were conducted, together with the Earth Resources Experiment Package (EREP), which acquired remote sensor imagery of the earth. The magnetic tape records and exposed camera film were returned to earth by the crews. The following discussion of EREP is summarized from the "Skylab Earth Resources Data Catalog" (NASA, 1974). Spatial resolution of the EREP photographs is the best obtained to this time by NASA satellites. Many photographs were acquired with 60 percent forward overlap and may be viewed stereoscopically.

Multispectral Camera (S-190A Experiment)

This system is similar to the Apollo 9 multispectral photographic experiment. Six cameras were boresighted to the same center point (*nadir*) and triggered simultaneously. The 57-mm square photographs are recorded on 70-mm film. The 21.2° field of view of the 152-mm focal-length lenses provided ground coverage 163 km on a side as depicted in Figure 3.5. The original film has a scale of 1:2,850,000 and can be enlarged more than 10 times with little loss of detail. Significant characteristics of the film and filter combinations are given in Table 3.1. The ground resolutions on this table were measured from second-generation photographs that were somewhat lower in resolution than the original film.

Three of the black-and-white multispectral photographs of the Salton Sea and Imperial Valley, California are shown in Figure 3.6. This is a useful scene for analysis because it includes a large water body, agriculture, desert, and mountains, as identified on the location map (Figure 3.6D). The green band has remarkably good contrast and resolution and provides the maximum information on turbidity distribution within the Salton Sea. As in the Apollo 9 multispectral photographs, the Skylab photograph acquired in the red band (Figure 3.6B) has maximum tonal contrast. The differences in agriculture between the United States and Mexican parts of the Imperial Valley are pronounced on this band. On the IR photograph (Figure 3.6C), the strong reflectance of vigorous vegetation produces the light tones in the

TABLE 3.1
Skylab multispectral camera (S-190A experiment)

Camera	Spectral band, μm	Film type	Ground resolution, m**
1*	0.7 to 0.8	IR black and white	145
2	0.8 to 0.9	IR black and white	145
3	0.5 to 0.9	IR color	145
4	0.4 to 0.7	Normal color	85
5*	0.6 to 0.7	Black and white	60
6*	0.5 to 0.6	Black and white	60

*Illustrated in Figure 3.6.
**From Welch (1974).

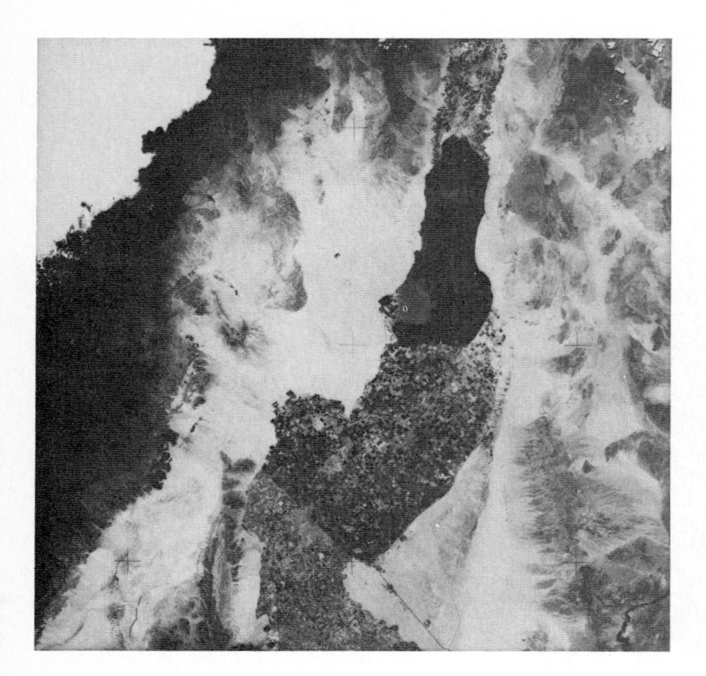

A. GREEN (0.5 TO 0.6 μm), CAMERA 6.

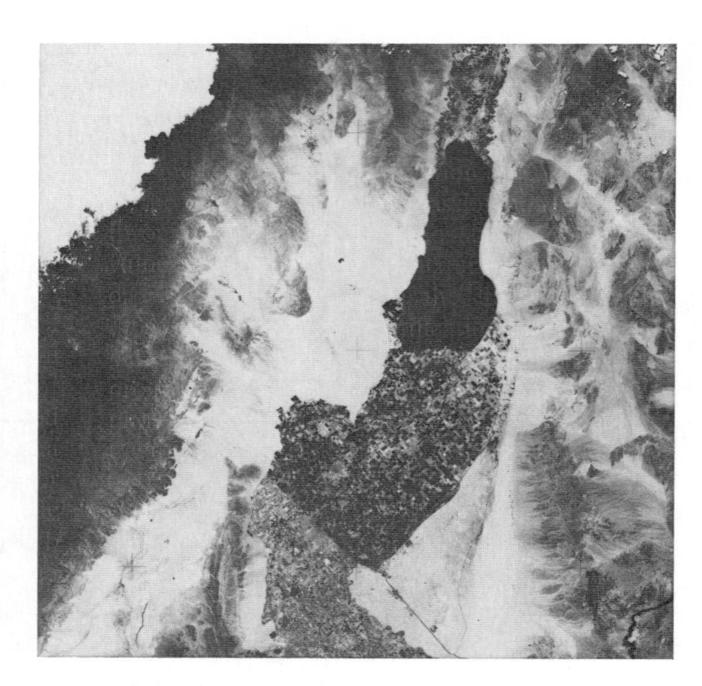

B. RED (0.6 TO 0.7 μm), CAMERA 5.

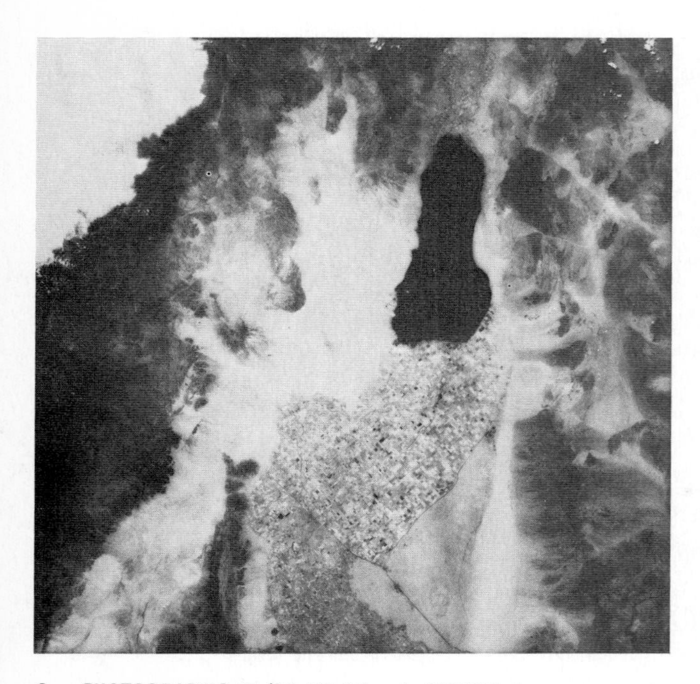

C. PHOTOGRAPHIC IR (0.7 TO 0.8 μm), CAMERA 1.

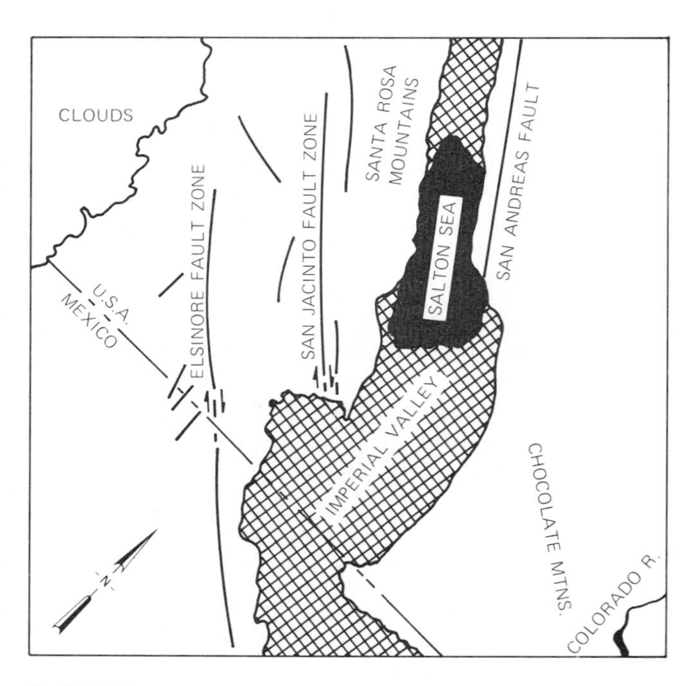

D. LOCATION MAP SHOWING MAJOR FAULTS.

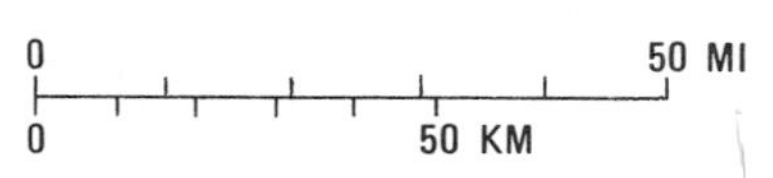

FIGURE 3.6
Skylab multispectral camera (S-190A) photographs of southern California and Mexico. Acquired June 2, 1973.

Imperial Valley that are similar to the tone of the desert on either side of the valley. Vegetation is readily distinguished from desert on the green and red bands. The very dark rectangles in the Imperial Valley are reservoirs and fields that are flooded with water and can be distinguished only on the photographic IR band. This capability to distinguish targets in different spectral bands is a major advantage of multispectral photography and imagery. On all three bands the south-southeast trending Elsinore fault is prominent, especially in Mexico. Several smaller south-trending faults cross the international boundary west of the Elsinore fault zone and are marked by prominent linear tonal anomalies.

The transparencies from cameras 5, 6, and 1 or 2 can be composited using green, blue, and red light, respectively, to produce an IR color photograph. The optimum color balance can be obtained during the compositing process.

Earth Terrain Camera (S-190B Experiment)

This system consisted of a single camera with a 45.7-cm focal-length lens. As illustrated in Figure 3.5, ground coverage is a 109 by 109 km square centered within the larger field of view of the multispectral cameras. Normal color, IR color, and minus-blue, black-and-white photographs were acquired at a scale of 1:950,000 on 11.4-cm square film. Ground resolution of second-generation images is 15 m for the black-and-white photographs and 30 m for the normal color and IR color photographs. These photographs can be enlarged to scales of 1:50,000 (20×) and 1:100,000 (10×), respectively, before blurring occurs (Welch, 1976).

Figure 3.7 is a typical photograph acquired with the earth terrain camera that covers the Kaiparowits Plateau and Lake Powell in south-central Utah. The major features are shown in the location map of Figure 3.8. The photograph clearly

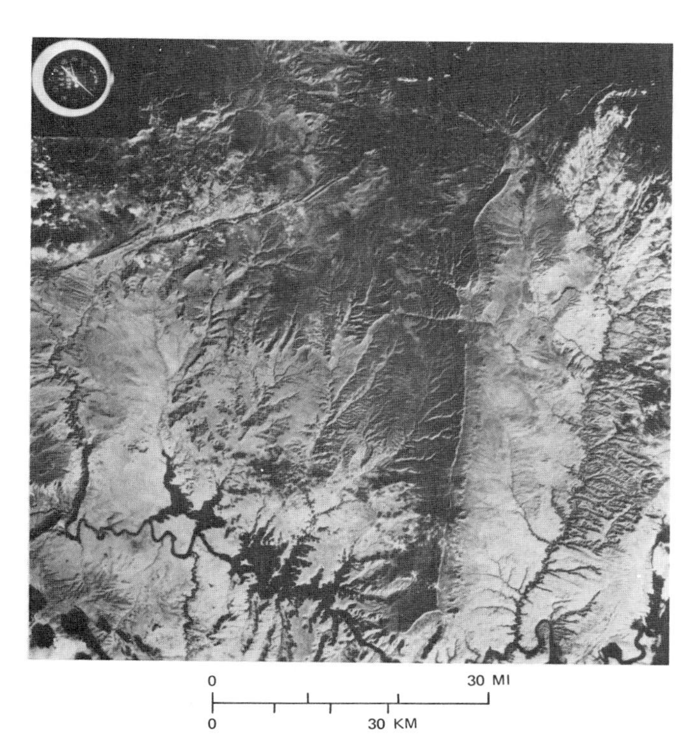

FIGURE 3.7
Skylab 3 earth terrain camera (S-190B) photograph of southern Utah, acquired August 12, 1973 on minus-blue, black-and-white film. The low sun elevation enhances topographic and structural features. Courtesy I. C. Bechtold, Bechtold Satellite Technology Corporation.

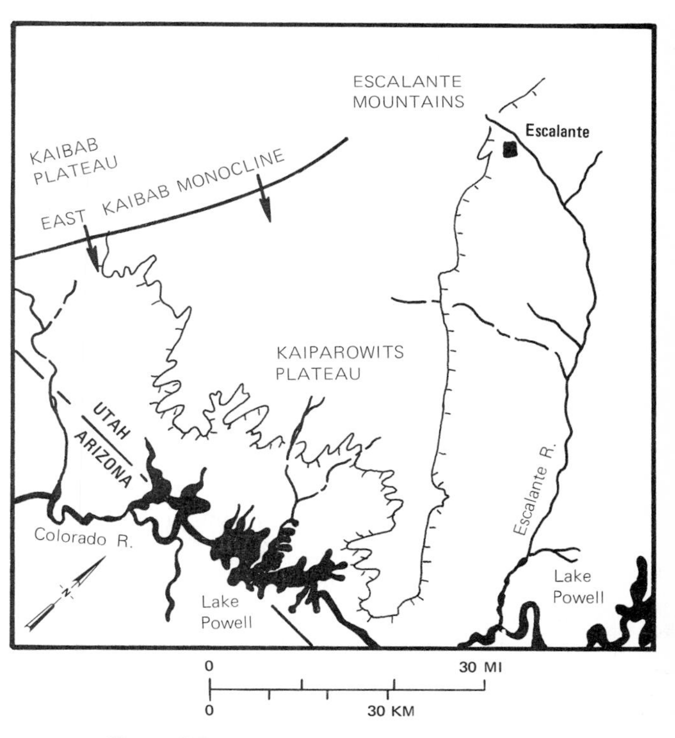

Figure 3.8
Location map for Skylab S-190B photograph of southern Utah.

shows the gentle dips of the Mesozoic strata in this part of the Colorado Plateau. Figure 3.7 also shows the steep east dips of strata along the East Kaibab monocline that separates the Kaibab Plateau on the west from the Kaiparowits Plateau. The high spatial resolution of the earth terrain camera system is shown by the expression of the badland topography along the Escalante River in the east part of the photograph.

Multispectral Scanner (S-192 Experiment)

Images may be acquired with optical-mechanical systems called *scanners* that are described in Chapters 4 and 5. In the Skylab scanner system a faceted mirror assembly rotated 360° about a vertical axis, but only the forward 110° portion of the conical sweep was recorded. The resulting semicircular swath on the terrain is 74-km wide, as shown in Figure 3.5. The 0.182-mrad instantaneous field of view covered a square spot on the ground 79 m on a side, which is the same ground resolution as the Landsat multispectral scanner. Energy reflected and radiated from the terrain was separated into the 13 spectral bands listed in Table 3.2, which were recorded on digital magnetic tapes. Selected portions of the tapes have been played back to produce black-and-white film images of individual spectral bands. The films may be combined into color images by projecting them with suitable colored lights. Many of the tape records are marred by noise that must be removed by digital processing, and only a limited number of image strips have been produced. Images of representative spectral bands for an area in southern Mexico are shown in Figure 3.9. The circular scan pattern is particularly evident on the thermal IR image of band 13. On all bands shown in Figure 3.9 the lakes are dark and the small patches of clouds over the mountains in the northwest part of the image have bright signatures. Strike ridges of folded marine Cretaceous strata are prominent in the southeast. The oval mountainous area south of Lake Tequesquitengo has a dark tone on bands 3, 12, and 13; a light tone on band 10; and an intermediate tone on band 6. These mountains consist of volcanic rocks of middle Tertiary age.

TABLE 3.2
Spectral bands of Skylab multispectral scanner (S-192 Experiment)

Band number	Wavelength, μm	Spectral band
1	0.41 to 0.45	Blue
2	0.44 to 0.52	Blue-green
3*	0.49 to 0.56	Green
4	0.53 to 0.61	Green
5	0.59 to 0.67	Red
6*	0.64 to 0.76	Red-photographic IR
7	0.75 to 0.90	Photographic IR
8	0.90 to 1.08	Photographic IR
9	1.00 to 1.24	Photographic IR
10*	1.10 to 1.35	Photographic IR
11	1.48 to 1.85	Photographic IR
12*	2.00 to 2.43	Photographic IR
13*	10.20 to 12.50	Thermal IR

*Illustrated in Figure 3.8.

Imagery acquired in the blue spectral region (band 1, not shown) is largely unusable because of atmospheric scattering. The scanner system was operated during daylight hours only. Therefore, the thermal IR imagery (band 13) records primarily differential solar heating and shadowing caused by topography. As discussed in Chapter 5 on Thermal IR Imagery, such imagery should be acquired at night to be useful for most earth science applications.

Other Skylab Data

In addition to the vertical metric photography of the multispectral and terrain cameras, a number of hand-held 70-mm photographs were acquired and are listed on NASA index maps. An IR spectrometer, microwave scatterometer, and microwave radiometer were included in the Earth Resource Experiment Package. These systems did not acquire images and are not discussed here. Brief descriptions are given by NASA (1974).

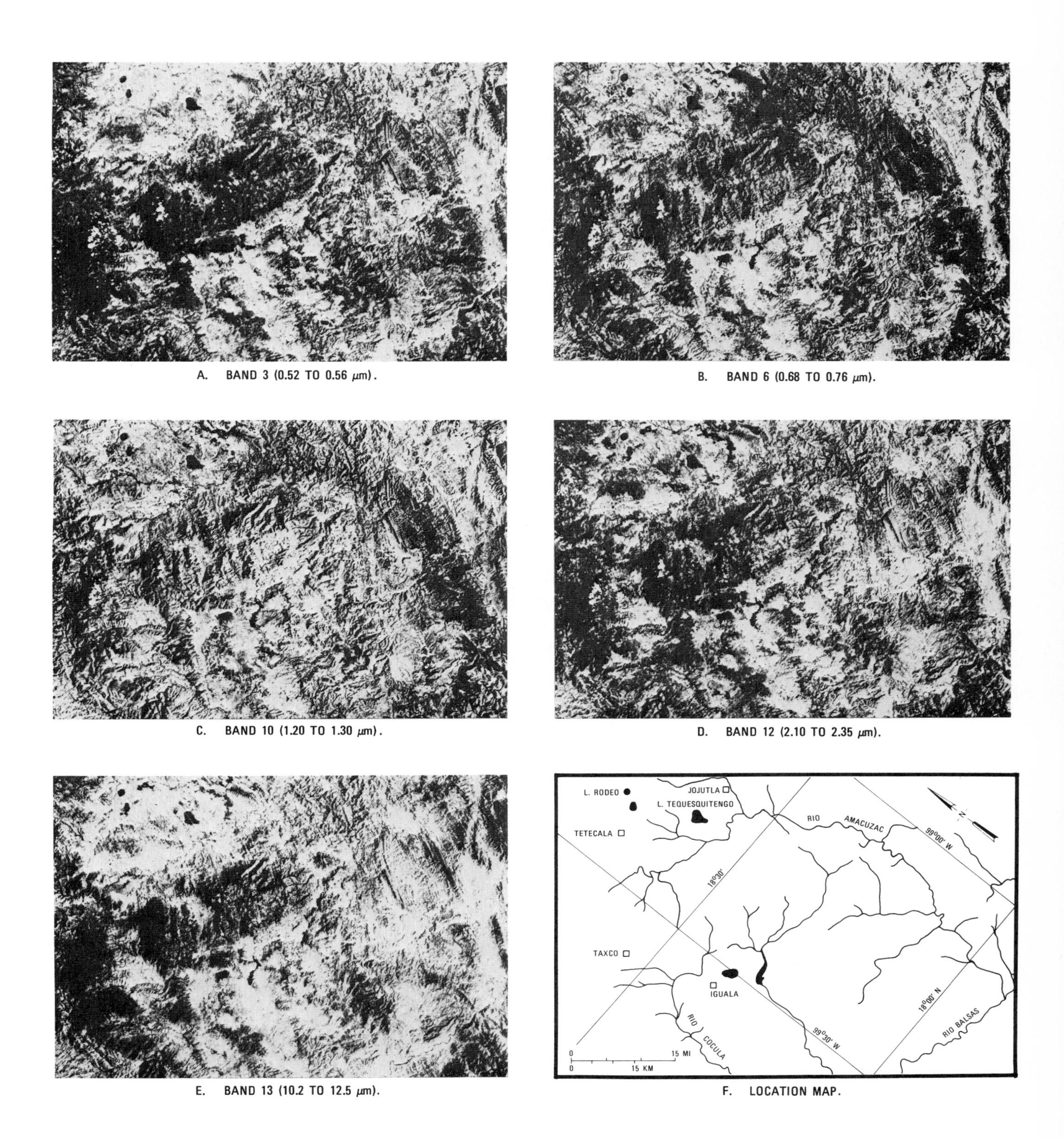

Figure 3.9
Skylab 4 multispectral scanner (S-192) images of Taxco, Mexico, and vicinity. Images acquired December 2, 1973.

SPACE SHUTTLE

Space Shuttle is an orbiting workshop program planned for the 1980s that is designed to accommodate a crew of nine, most of whom will be scientists rather than astronauts. The reusable space shuttles will rotate the crews and resupply the workshop. Earth observations are an important part of the proposed program for the Space Shuttle missions.

SOURCES OF PHOTOGRAPHS

The U.S. Geological Survey EROS Data Center in Sioux Falls, South Dakota 57198 is a repository for photographs and images acquired by the manned satellite missions. An inquiry to the EROS Data Center should specify the type of photograph or image desired and the latitude and longitude boundaries for the area of interest. For Skylab photographs, the "Skylab Earth Resources Data Catalog" (NASA, 1974) provides a complete listing. Index maps showing worldwide Skylab coverage are available. Gemini and Apollo photographs also may be purchased from the Technology Application Center, University of New Mexico, P. O. Box 181, Albuquerque, New Mexico 87106, which provides catalogs of the available photographs.

COMMENTS

The most important contribution to remote sensing of the manned spacecraft missions from Mercury through Apollo was to demonstrate the potential value and advantages of acquiring images from orbital altitudes. Despite the limited coverage, those images provided justification for Skylab and, more significantly, for the unmanned Landsat program, which is described in Chapter 4. The photographs acquired from Skylab demonstrated the utility of high-resolution photographs acquired from space.

With the advent of automated imaging systems, such as Landsat, humans are no longer essential for acquiring orbital imagery. Indeed, the presence of humans on a spacecraft greatly increases the cost of the mission and limits the global coverage because of communications and other constraints. Manned spaceflight will continue, however, with the Space Shuttle program in the 1980s.

REFERENCES

Abdel-Gawad, M., 1969, New evidence of transcurrent movements in Red Sea area and petroleum implications: American Association Petroleum Geologists Bulletin, v. 53, p. 1466–1479.

Albritton, C. H. and J. F. Smith, 1956, The Texas lineament *in* Tomo de relaciones entre la tectonicas y la sedimentacion: Proceedings 20th International Geological Congress, sec. 5, p. 501–518, Mexico, D. F.

Lowman, P. D., 1969, Geologic orbital photography - experience from the Gemini program: Photogrammetria, v. 24, p. 77–106.

——— and H. A. Tiedeman, 1971, Terrain photography from Gemini spacecraft—final geologic report: NASA Goddard Space Flight Center Report X-644-71-15, Greenbelt, Md.

Merifield, P. M., 1964, Photo interpretation of White Sands rocket photography: Lockheed California Company, Report no. LR 17666, Burbank, Calif.

NASA, 1967, Earth photographs from Gemini 3, 4, and 5: NASA SP-129.

———, 1974, Skylab earth resources data catalog: NASA, JSC 09016, U.S. Government Printing Office Stock no. 3300-00586, Washington, D.C.

Welch, R., 1974, Skylab-2 photo evaluation: Photogrammetric Engineering and Remote Sensing, v. 40, p. 1221–1224.

———, 1976, Skylab S-190B ETC photo quality: Photogrammetric Engineering and Remote Sensing, v. 42, p. 1057–1060.

ADDITIONAL READING

Colwell, R. N., 1971, Monitoring earth resources from aircraft and spacecraft: NASA SP-275.

Kent, M. I., E. Stuhlinger, and S. T. Wu, eds., 1977, Scientific investigations on the Skylab satellite: American Institute of Aeronautics and Astronautics, v. 48, N.Y.

Lowman, P. D., 1969, Apollo 9 multispectral photography - geologic analysis: NASA Goddard Space Flight Center Report X-644-69-423, Greenbelt, Md.

Nicks, O. W., ed., 1970, This island earth: NASA SP-250.

4
LANDSAT IMAGERY

This unmanned system was originally called ERTS (Earth Resources Technology Satellite) but the name was changed to *Landsat* two years after the first satellite was launched. Two Landsats have been placed in orbit, both launched from Vandenberg Air Force Base, which is located between Los Angeles and San Francisco. Landsat 1 was launched on July 23, 1972 and acquired data through mid 1976 when operations ceased. The identical Landsat 2 was launched on January 21, 1975 and probably will be operational through 1978. A third Landsat is scheduled for launch in 1978, and additional satellites are being designed for future missions. The major components of Landsat are illustrated in Figure 4.1. The solar arrays generate electric power from sunlight. The antennas receive commands from earth and transmit data to earth receiving stations. Landsat has a nonimaging system called *data collection system* (DCS) that relays information from sensors on earth to the receiving stations. Flood gauges, seismometers, and tiltmeters on active volcanoes are typical sensors that have been installed in areas that are inaccessible for normal data transmission links. Each sensor has a small antenna and power supply for transmitting data to the satellite, which relays it to receiving stations. A *return-beam vidicon* (RBV) system on Landsat consists of three cameras that acquire green, red, and photographic IR images. Instead of using film to record an image, the image is formed on a photosensitive camera tube that is scanned by an electron beam. The resulting video signal is transmitted to earth. Only a few RBV images had been acquired from Landsat 1 when an electrical component failed. Most images have been acquired with the multispectral scanner (MSS) on both Landsat 1 and 2. It is generally agreed that the MSS images are equal or superior in quality to those from the RBV. Landsat has proven to be a highly successful system for imaging the earth. Spatial resolution and spectral ranges will be improved in future satellite systems, but the basic concepts of Landsat will be employed for some time to come.

MULTISPECTRAL SCANNER SYSTEM

All Landsat images in this book are produced from data acquired by the MSS, which is located on the base of the satellite (Figure 4.1). The MSS

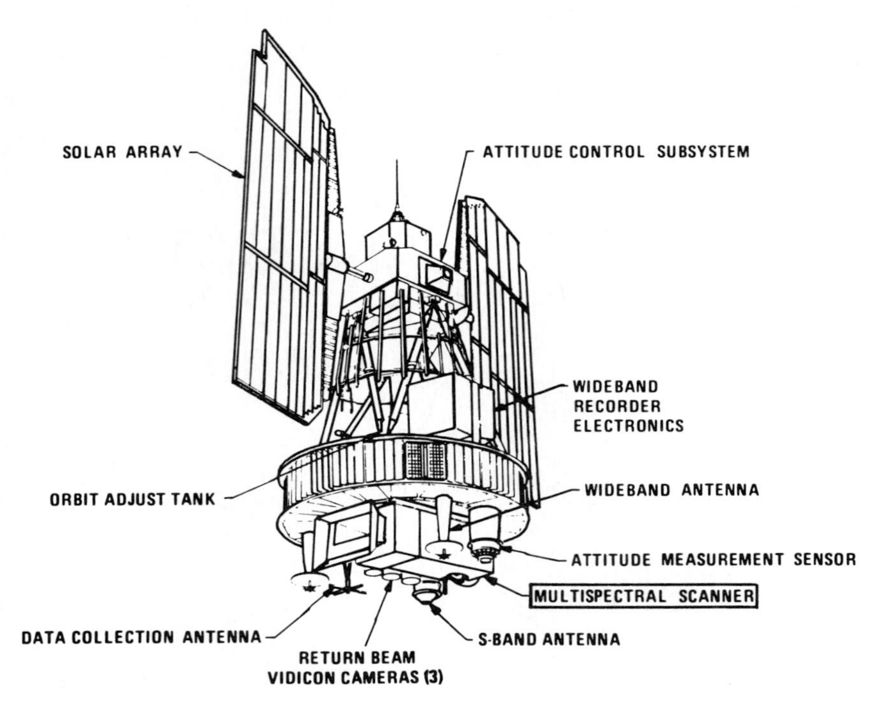

FIGURE 4.1
Diagram of Landsat 1 and 2. The multispectral scanner has produced almost all the imagery from these satellites. From NASA (1976, Figure 2-2).

mirror oscillates through a scan angle of 11.56°. At a nominal 918-km altitude the mirror scans a swath 185-km wide normal to the orbit path (Figure 4.2). Image data are recorded only during the eastbound mirror sweep. Data are acquired continuously along the orbit path and are transmitted to a ground receiving station for recording on magnetic tape. The tapes are processed to produce images, such as those illustrated in Figure 4.3, that cover a 185 by 185 km area on the ground. The processing method provides 10 percent forward overlap between successive images. The instantaneous field of view of each detector is a 0.086 by 0.086 mrad square, which at the 918-km altitude produces a 79-by-79 m ground resolution cell that determines the spatial resolution of the system. A technical description of the entire Landsat system is given by NASA (1976).

Sunlight reflected from the terrain is separated into the four wavelengths, or spectral bands, shown in Figure 4.2. There are six detectors for each spectral band; thus for each sweep of the mirror six scan lines are simultaneously generated for each of the four spectral bands. The energy sensed by the detectors is converted into an electrical signal for recording and transmission as image data. A blue image is not acquired, because it would be severely degraded by the effects of atmospheric scattering. Table 4.1 lists the MSS band number that designates each of the wavelength bands. Band numbers 1, 2, and 3 are used to designate images from the RBV system, which was used only during the first few days of Landsat 1. Table 4.1 also lists the projection colors that are normally used with each MSS band to produce the equivalent of an IR color photograph. Other combinations of colors may be used, but those listed in Table 4.1 represent the standard combination employed by EDC and other processing facilities.

When Landsat is within radio range of one of the earth receiving stations the image data are transmitted directly to the station. When it is out of range a limited number of scenes may be stored on magnetic tape for transmission when the satellite comes within radio range of a receiving station. At the stations, the data for the four MSS bands are recorded on magnetic tape. One of the two tape recorders on Landsat 1 failed early in the mission, and the other recorder eventually failed after meeting its design lifetime of 500 hours. One

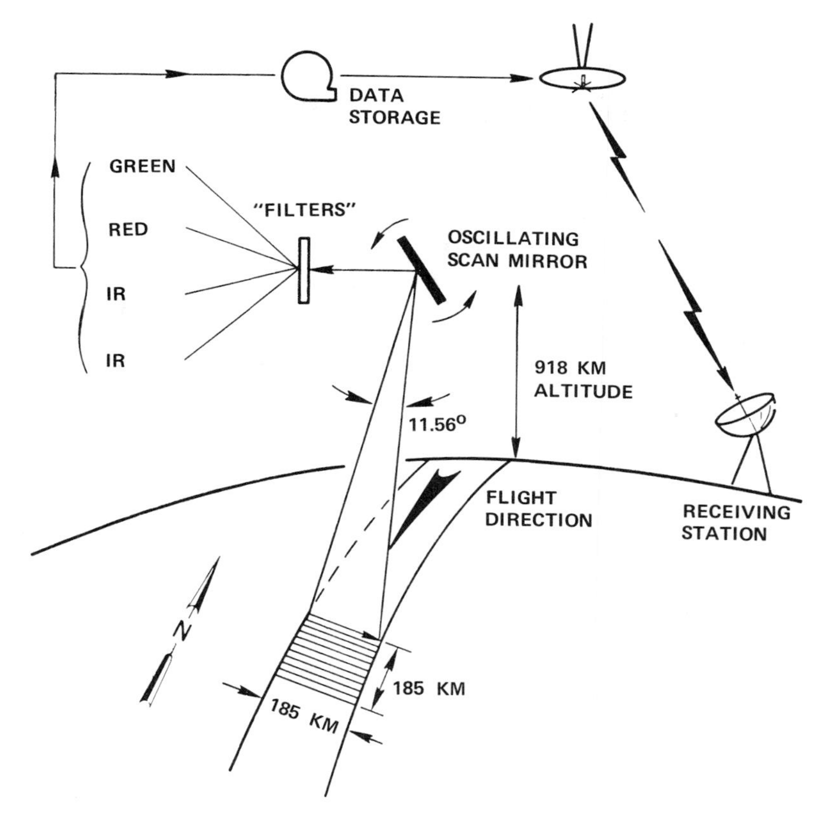

FIGURE 4.2
Landsat multispectral scanner. For each terrain scene, four images are transmitted to a receiving station. From NASA (1976, Figure 2-4).

of the recorders on Landsat 2 has failed and the second recorder is being used sparingly. After the two tape recorders on a Landsat vehicle fail, image acquisition is restricted to localities within range of the receiving stations. Tapes from the United States receiving stations at Goldstone, California and at Fairbanks, Alaska are forwarded to the Goddard Space Flight Center in Maryland, which is also a receiving station. At GSFC the tapes are played back on an electron-beam film recorder to produce a 1:3,369,000 scale black-and-white archival image on 70-mm film for each of the four spectral bands. Duplicates of these images are sent to the U.S. Geological Survey EROS Data Center, or EDC, in Sioux Falls, South Dakota, which sells copies of the images to the public. Beginning with Landsat 3 this system is scheduled to change. The data tapes will be forwarded to EDC where 1:1,000,000 scale original images will be produced. Black-and-white and color composite images will be reproduced from these originals.

TABLE 4.1
Landsat multispectral scanner bands

MSS band	Wavelength, μm	Color	Color normally used for projection
4	0.5 to 0.6	Green	Blue
5	0.6 to 0.7	Red	Green
6	0.7 to 0.8	Photo. IR	Red
7	0.8 to 1.1	Photo. IR	Red

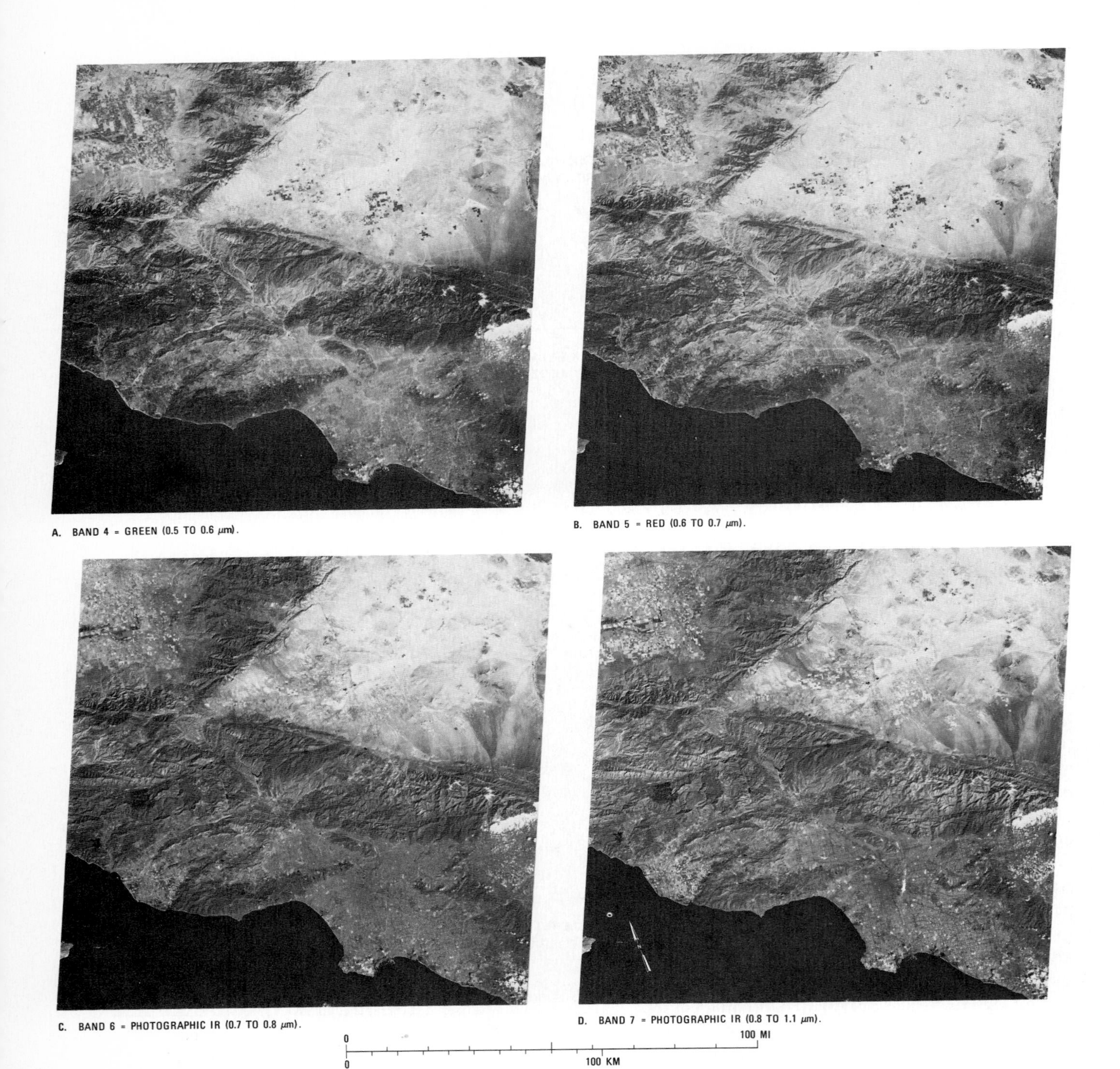

FIGURE 4.3
Individual MSS bands of southern California from Landsat 1090-18012 acquired October 21, 1972. Images were digitally processed by Jet Propulsion Laboratory. Courtesy R. J. Blackwell, Jet Propulsion Laboratory, California Institute of Technology.

IMAGE CHARACTERISTICS

The four spectral bands for an image of southern California are shown in Figure 4.3. During playback of the data tape to produce an image, the scan lines are successively offset to the west to compensate for earth rotation during the 28 sec required to scan the terrain. This accounts for the parallelogram shape of the images. The four bands seen in Figure 4.3 were digitally processed at the Jet Propulsion Laboratory to produce optimum contrast in the land areas. No attempt was made to improve contrast in the water areas, which appear black. The digital processing methods are described in Chapter 7.

Image Annotation

Individual black-and-white Landsat images have an annotation strip with useful information. The following list explains the annotation strip for an image of southern California:

21 Oct 72	Date image was acquired.
C N34—33/W118—24	Geographic centerpoint of the image, in degrees-minutes.
N N34—31/W118—19	Nadir of the spacecraft.
MSS	Multispectral scanner image.
4, 5, 6, or 7	MSS spectral band.
SUN EL39	Sun elevation, in degrees above horizon.
AZ148	Sun azimuth, in degrees clockwise from north.
190	Spacecraft heading, in degrees.
1255	Orbit revolution number.
G, A, or N	Ground recording station: G = Goldstone, California; A = Alaska; N = Network Test and Training Facility at GSFC.
1090-18012	The unique frame identification number composed as follows:
1 or 5	Landsat 1 (5 denotes Landsat 1 for days greater than 999 since launch).
090	Days since launch. This was October 21, 1972. (Beginning in early 1977 this part of the identification number is expanded to a four-digit number, such as 1104, to accommodate 1000 and higher days.)
18	Hour at time of observation, GMT.
01	Minutes.
2	Tens of seconds.

Image Resolution and Quality

Spatial resolution of MSS images is determined by several factors including the 79 by 79 m ground resolution cell, atmospheric conditions, playback and reproduction of the imagery, and contrast ratio of the scene. In general, the spatial resolution ranges from about 200 to 250 m; however smaller features can be detected on many images. Narrow highways and canals can be detected where the contrast is optimum.

Unlike aerial photographs, Landsat images have very little geometric distortion because of the narrow scan angle. The images are typically interpreted at scales ranging from 1:1,000,000 to 1:250,000, but greater enlargements are used for many applications. The images and interpretation maps are readily registered to base maps constructed with a Lambert conic conformal projection. The 1:500,000 scale "Sectional Aeronautical Charts" are especially useful for comparison with Landsat images and interpretations at that scale. The maps are published by the U.S. Department of Commerce, and may be purchased at most airports.

Some Landsat images are marred by *sixth-line banding* in which every sixth scan line is lighter or darker than the others. Some images are marred by *sixth-line dropouts* in which every sixth scan line is completely black due to an equipment problem. Random-line dropouts are another problem. These and other defects are illustrated, and correction methods are shown in Chapter 7. The 70 mm black-and-white positive transparencies supplied by EDC may be very dense (dark) and have a low contrast ratio. Before these transparencies can be used to produce color composite images, the contrast and density must be improved by photographic or digital procedures. The photographic method described in the following section is effective and inexpensive.

Photographic Enhancement of Contrast and Density

The following materials are employed to enhance the low-contrast, high-density Landsat positive transparencies:

Film	Kodak high-speed duplicating film 2575 (ESTAR Base).
Developer	Kodagraph developer, diluted 1:3.

By using the 25-by-30-cm film size, twelve 70-mm Landsat transparencies may be processed together. The transparencies are first sorted into groups with approximately the same density. A light meter may be used to measure the average light transmission of each image or the images may be sorted by visual inspection. The emulsion side of the image transparency is placed against the emulsion side of the film in a contact photographic printer. Newton rings may form at the interface between the film and the glass cover of the printer; this can be prevented by placing a sheet of drafting film, with frosted surface on both sides, between the transparencies and the glass cover of the printer. Using a 20-volt, 100-watt incandescent point source lamp at an intensity of 50 percent, exposure times range from 20 sec for the least dense images to 60 sec for the most dense images. The average exposure is 30 sec. Processing time in the diluted Kodagraph developer is two minutes at 20°C (68°F).

This method has been used to produce acceptable copies from images that were unusable because of low contrast ratio or excessive density. There is some loss of image detail at the high- and low-density ranges, but this is acceptable for most applications. An alternate method for correcting low-contrast images is digital contrast stretching, which is described and illustrated in Chapter 7.

Color Composite Images

Just as IR color aerial photographs are superior to the black-and-white version, color composite Landsat images are superior to the black-and-white individual bands for most applications. For many scenes EDC has prepared reproducible color composite images and can supply color prints and transparencies. The 70-mm or 19-cm positive transparencies may be converted into inexpensive IR color positive transparencies by the diazo process. Diazochrome® color film (manufactured by Scott Graphics Incorporated, Holyoke, Massachusetts 01040) produces a positive color transparency from a Landsat black-and-white positive transparency. The Landsat image is placed on the diazo film, and the two are inserted in a diazo processing unit that exposes the film with a UV light source and develops the image with ammonia gas. The exposure time is determined by experimentation. Yellow diazo film is used with MSS band 4, magenta film with band 5, and cyan with an IR band, usually band 7. The resulting subtractive color images are superimposed in registra-

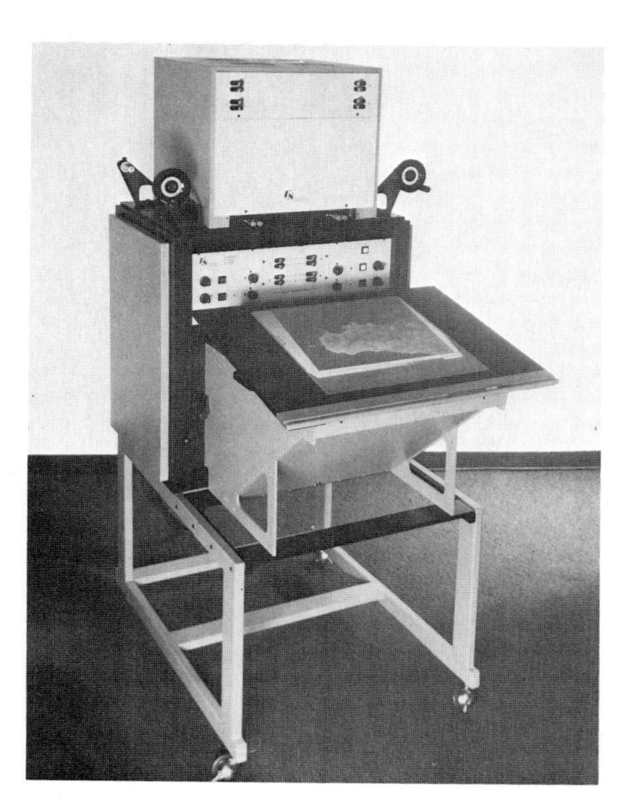

FIGURE 4.4
Color compositor-viewer for Landsat images. Courtesy International Imaging Systems.

tion to produce an IR color transparency that may be projected or viewed on a light table. Inexpensive desk-top diazo processors are available. Some investigators expose the film with sunlight and process it in a box containing a sponge saturated with ammonia.

Color images may also be produced on compositor-viewer systems, such as the model illustrated in Figure 4.4, that projects a 1:500,000 scale color image onto a viewing screen. Photographically enhanced 70-mm black-and-white positive MSS transparencies are placed in a film holder and inserted into the viewer. The images are registered using controls for rotation and for X and Y translation. Each band may be projected with blue, green, red, or white light, but the standard combination is shown in Table 4.1. Regardless of how it is produced, a standard Landsat color image has the spectral color characteristics of an IR color photograph. Bands, 4, 5, and 7 of Figure 4.3 were composited into the IR color image of Plate 2 by the Jet Propulsion Laboratory. Typical color signatures on this and other Landsat images are as follows:

Vegetation	Red
Water	Dark blue
Suspended sediment	White to light blue
Red beds	Yellow
Bare soil	Blue
Eolian sand	White to yellow
Cities	Blue
Clouds and snow	White

Interpretation of Southern California Image

The color image of Plate 2 includes portions of several physiographic provinces that are shown on the interpretation map in Figure 4.5. The Mojave Desert province is represented by the Antelope Valley, a wedge of light-colored terrain in the northeast part of the image that is bounded by the Garlock left-lateral fault on the north and San Andreas right-lateral fault on the south. The highly reflective dry silt and clay surface of Edwards and Rosamond Lakes form the prominent white patches. The fault-block mountains of granitic, volcanic, and metamorphic rocks are deeply eroded and inconspicuous. Gray to buff alluvium and yellow eolian sand cover most of the surface. The dark gray Sheep Canyon alluvial fan in the southeast corner of the Antelope Valley consists of dark minerals (biotite and amphibole) eroded from schist outcrops in the San Gabriel

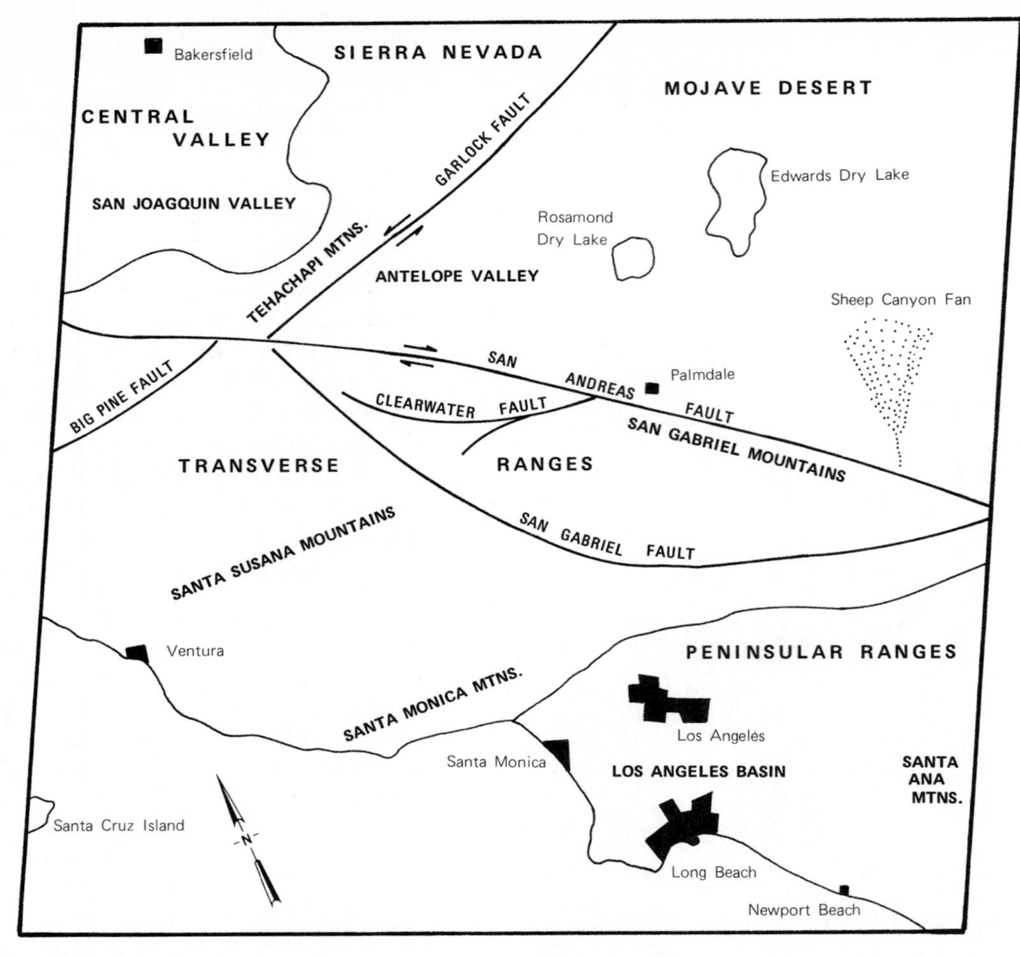

FIGURE 4.5
Major geographic and geologic features of southern California from Landsat 1090-18012. Compare with color composite image in Plate 2.

Mountains. This dark detritus contrasts with the adjacent tan gravels derived from granitic and sedimentary rocks. Some irrigated fields stand out in the southern Antelope Valley. The red circles in the northeast corner of Plate 2 are fields irrigated by sprinklers that rotate about a central pivot.

The San Joaquin Valley in the northwest part of Plate 2 is separated from the Mojave Desert by the Tehachapi Mountains and the south end of the Sierra Nevada. The San Joaquin Valley is the southern part of the Central Valley of California, and is intensively cultivated. The San Gabriel, Santa Susana, and Santa Monica Mountains belong to the northwest-trending Transverse Range province, which has a variety of rock types and complex geologic structure. In addition to the San Andreas fault, the San Gabriel, Clearwater, and Big Pine faults form prominent curved topographic linear depressions. Many smaller faults may be mapped on enlarged versions of this image. The darker shades of red in the mountains are caused by conifers and brush that do not have the high IR reflectance of irrigated crops. A few clouds lap against the east end of the San Gabriel Mountains and the highest peaks are snow covered. The black patch in the Santa Susana Mountains was burned in a brush fire.

The Santa Ana Mountains in the southeast part of the image and the Los Angeles Basin are the north end of the Peninsular Range province. In the basin the blue shades represent commercially developed areas, the largest of which are Los Angeles and Long Beach. The pink to magenta shades are lawns in the suburbs, with parks and golf courses marked by bright red patches.

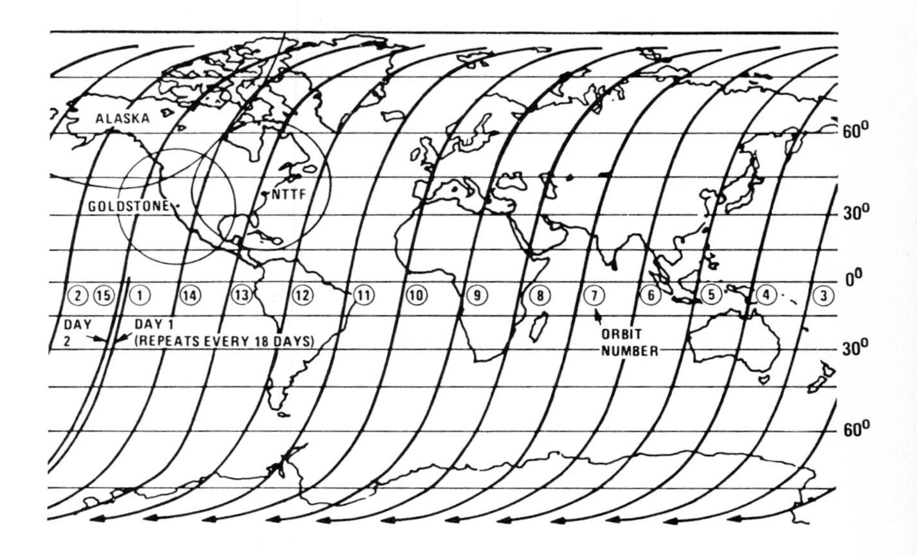

FIGURE 4.6
Typical daytime Landsat orbit paths for a single day. Each day the paths shift westward 160 km at the equator so that every 18 days the paths are repeated. Images are acquired between 9:30 and 10:00 A.M. local sun time, except at high latitudes. Note location and ranges of receiving stations in the United States. From NASA (1976, Figure 2-6).

ORBIT PATHS

Figure 4.6 shows the 14 southbound daytime orbits covered by Landsat during a single day. The northbound orbits cover the dark side of the earth. Rotation of the earth shifts the orbit paths westward each day so that at the end of 18 days, or 252 orbits, the earth has been covered and the cycle begins again. The polar areas above latitude 81° are the only regions not covered by Landsat orbits. As illustrated in Figure 4.7, at latitude 40° there is 34 percent sidelap of the 185-km wide image swaths generated on successive days. The sidelap decreases to 14 percent at the equator and increases to 70 percent at polar latitudes. The sidelapping portions of adjacent images may be viewed stereoscopically, as shown later in the example from the Peten Basin of Mexico and Guatemala.

From January, 1975 through mid 1976 Landsat 1 and 2 had identical orbit patterns that were offset by nine days. By combining the two sets of images, complete coverage could be obtained every nine days. This coverage was acquired only for areas within range of receiving stations because by this time the tape recorders on Landsat 1 were inoperative. The United States stations and their receiving ranges are shown in Figure 4.6. In addition there are stations in Canada, Brazil, and Italy.

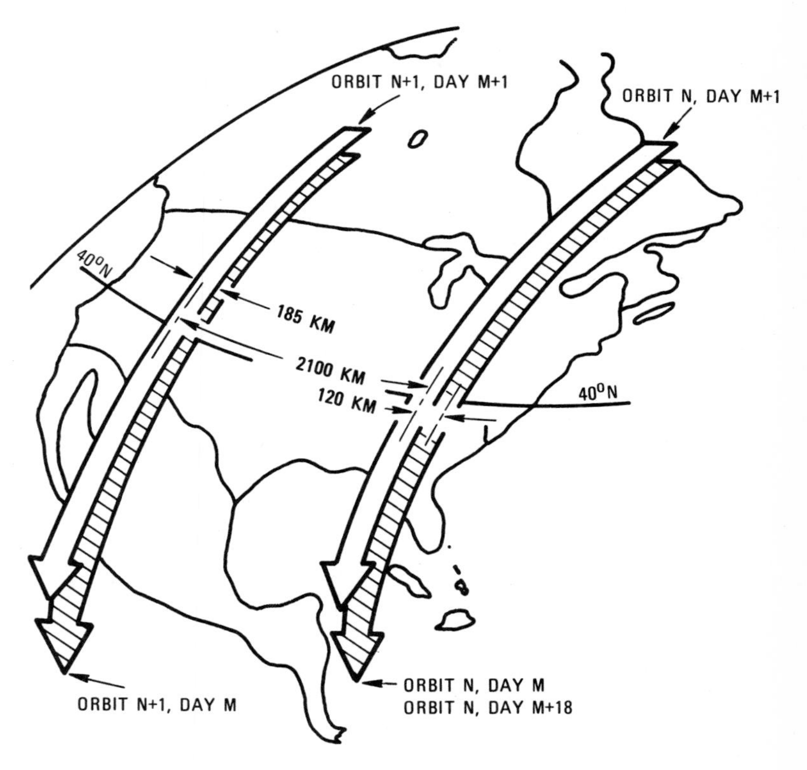

FIGURE 4.7
Landsat orbits over the United States on successive days. Note the 62-km sidelap of successive image swaths at 40°N latitude.

The sun-synchronous orbit pattern was chosen to acquire midmorning imagery at intermediate sun angles. For example, on each southbound pass Landsat crosses the latitude of Los Angeles at 10:00 A.M. local sun time and crosses the equator at 9:42 A.M. Each 18-day repetitive orbit occurs at the same time, facilitating comparison of the images. The intermediate to low sun elevation causes highlights and shadows that enhance subtle linear features, many of which represent faults and fracture zones. The images acquired at very low sun elevations in midwinter are particularly useful for recognizing linear features. Sun azimuth is from the southeast for images in the Northern Hemisphere and from the northeast in the Southern Hemisphere.

INDEXING OF IMAGES

The United States operates Landsat in the *international public domain*, meaning that images of any area in the world may be acquired by Landsat and are available for anyone to purchase. On September 1, 1977 the EROS Data Center had almost 262,000 images (four bands for each) acquired by Landsat 1 and 2. Approximately 5,000 new images are acquired each month. For the user, the first step is to index and select the available images for his areas of interest. Two different approaches to indexing and selection have been developed by EROS Data Center and by Chevron Oil Field Research Company.

EROS Data Center Indexing Services

NASA publishes and the EROS Data Center distributes monthly catalogs listing, by latitude and longitude, the newly acquired Landsat images. In addition to listing images by location and date of acquistion, the image quality is classified. Black-and-white images are graded as follows: 2=poor, 5=fair, 8=good. Color images are graded on a continuous scale from 0 (poorest) to 9 (best). Cloud cover is given in tens of percent. These and other data about the images are maintained on a magnetic core index file at EDC. Users may submit a request to EDC, giving the latitude and longitude of an area of interest; they then receive a computer printout that lists all images covering the area. The printout list is organized so that images with the same nominal centerpoint are grouped together. The list must be manually sorted and plotted onto a map to locate the optimum images. This is satisfactory for small areas, but for regional coverage the volume of computer listings is excessive.

EDC has prepared Landsat index maps of the world. Figure 4.8 is the western portion of the United States map. The maps show nominal image centerpoints, which are the localities at which the image centers should occur. The nominal centerpoints are defined by the intersection of north-to-south orbit paths and east-to-west rows. The centerpoint of the southern California image shown in Plate 2 is designated as path 45, row 36. For each of these centerpoints EDC has selected an optimum image that can be purchased by requesting the desired path and row locality. The ground coverage of an image is shown by the inside dimensions of the 185-by-185 km square in Figure 4.8. The area covered is determined by centering the square on the image point with the east and west sides parallel with the orbit path. On the index maps of foreign areas the minimum degree of cloud cover available is designated by symbols. "*Browse files*," consisting of microfilm viewers and images, are available at U.S. Geological Survey offices and other facilities where microfilm copies of the band 5 image of Landsat scenes may be examined. This enables the user to assess image quality before placing an order.

Computer Index Maps

The EDC index maps are satisfactory for most users of Landsat images. Users of images on a worldwide basis may require an automated system to select images of highest quality, minimum cloud cover, and optimum season. Manual sorting and plotting of the thousands of images from computer listings is a tedious and error-prone task that is ideally suited for a computer. Sabins (1977) described a system for plotting index maps from the magnetic tape index file that was developed at Chevron Oil Field Research Company. Figure 4.9 is the western portion of an index map of the United States plotted by this system. The actual centerpoint of each image, rather than the nominal centerpoint, is shown. This plot is significant because a number of image centers do not coincide

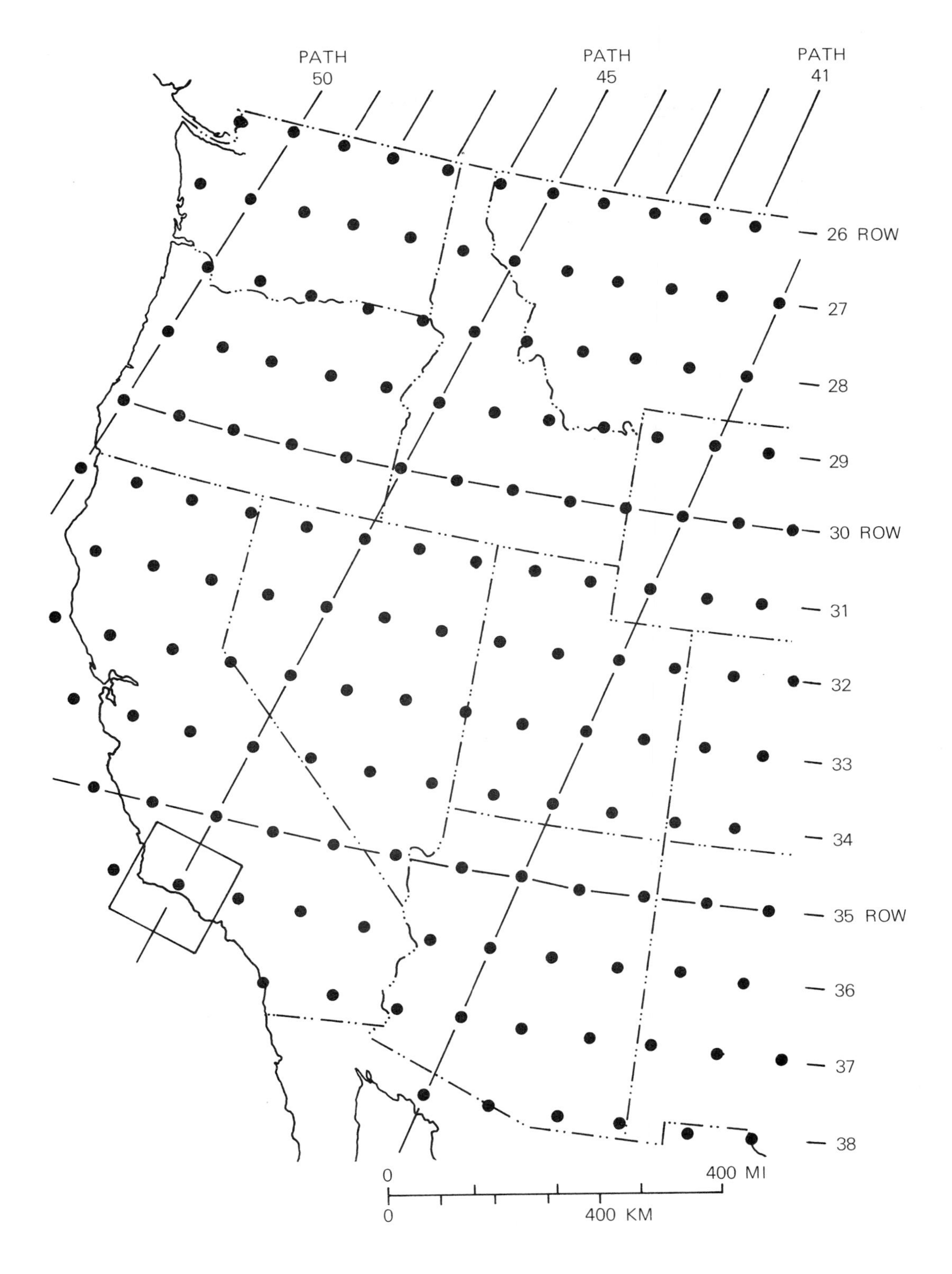

FIGURE 4.8
Landsat index map prepared by EROS Data Center based on nominal image centerpoints. The square shows coverage of a Landsat image. The location of this image is path 45, row 36.

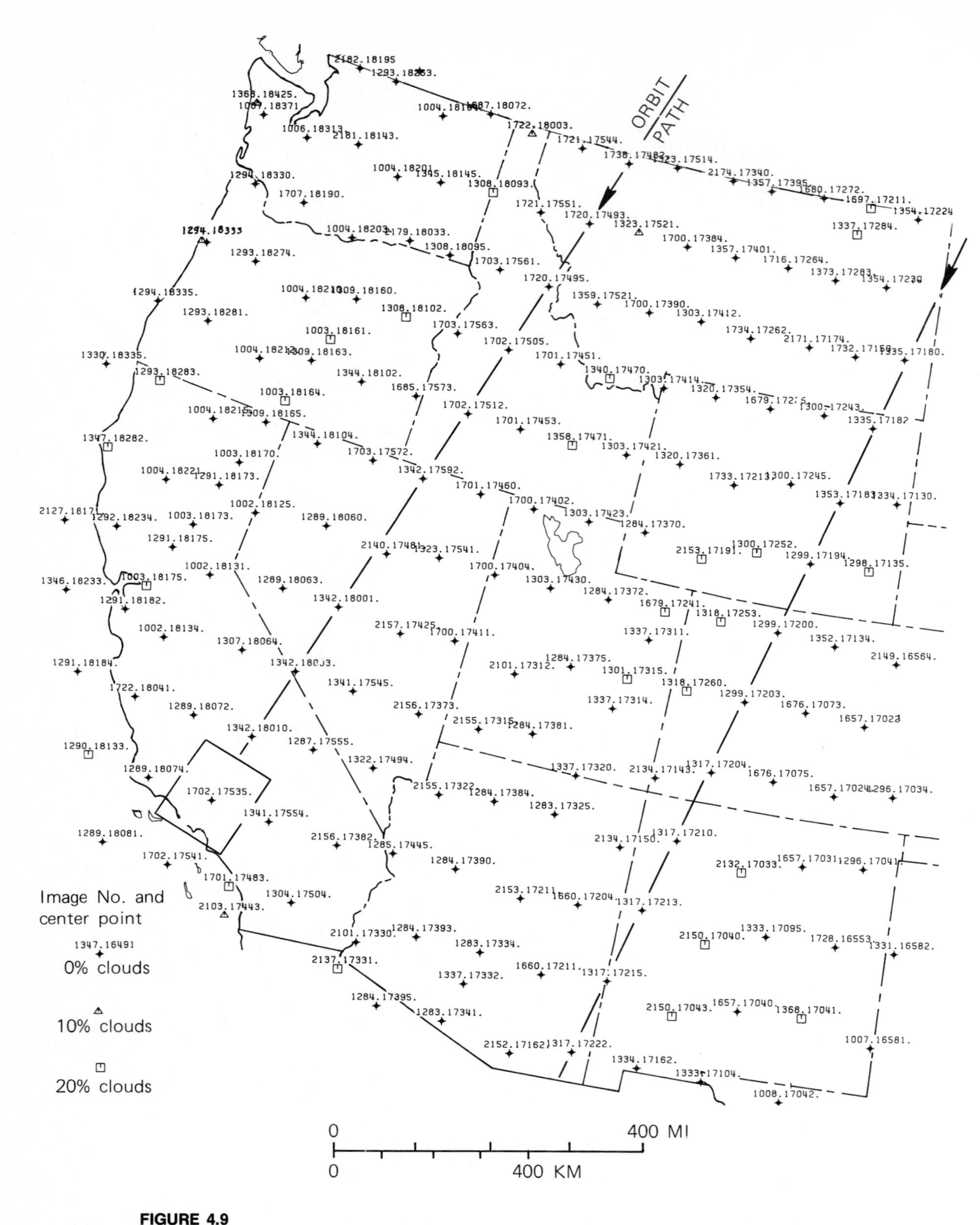

FIGURE 4.9
Typical computer-plotted Landsat index map prepared by Chevron. This map plots optimum Landsat 1 and 2 images acquired during May through July, 1972 to 1975. The square shows coverage of a Landsat image. Courtesy Chevron Overseas Petroleum, Inc.

TABLE 4.2
Formats of Landsat 1 and 2 images available from EROS Data Center

Image size	Scale	Format	Product code	Price*
Black-and-White				
2.2 in. (5.6 cm)	1:3,369,000	Film positive	11	$ 8.00
2.2 in. (5.6 cm)	1:3,369,000	Film negative	01	10.00
7.3 in. (18.5 cm)	1:1,000,000	Film positive	13	10.00
7.3 in. (18.5 cm)	1:1,000,000	Film negative	03	10.00
7.3 in. (18.5 cm)	1:1,000,000	Paper	23	8.00
14.6 in. (37 cm)	1:500,000	Paper	24	12.00
29.2 in. (75 cm)	1:250,000	Paper	26	20.00
Color Composites				
7.3 in. (18.5 cm)	1:1,000,000	Film positive	53	15.00
7.3 in. (18.5 cm)	1:1,000,000	Paper	63	12.00
14.6 in. (37 cm)	1:500,000	Paper	64	25.00
29.2 in. (75 cm)	1:250,000	Paper	66	50.00

*These prices were current in late 1977 but are subject to change. Current prices may be obtained from EROS Data Center.

with nominal centerpoints, as may be seen for Washington and Oregon in Figure 4.9. Images can be selected from the identification numbers plotted on the maps. The computer index maps and programs are not commercially available. The description by Sabins (1977), however, should enable programmers to adapt the system to their computer facility.

ACQUISITION OF IMAGES

Orders for any Landsat images, with prepayment payable to the U.S. Geological Survey, may be sent to EROS Data Center, Sioux Falls, South Dakota 57198. The Landsat image formats and prices listed in Table 4.2 were current in late 1977. Prices are subject to revision and a current list may be obtained from EDC. A fee for generating a color composite is charged if EDC does not have a master color negative on file for a particular scene. Images may also be ordered from U.S. Department of Agriculture, Aerial Photography Field Office, Administrative Services Division, P. O. Box 30010, Salt Lake City, Utah 84115. Color images are also available from General Electric Space Systems, Photographic Engineering Laboratory, 5030 Herzel Place, Beltsville, Maryland 20705 which can supply a catalog and price list. Images of Canada also may be purchased from the following facilities: (1) Center for Remote Sensing, 2464 Sheffield Road, Ottawa, Canada; and (2) Integrated Satellite Information Services Ltd., P. O. Box 1630, Prince Albert, Saskatchewan, Canada. A list and description of outstanding color Landsat images of Canada has been compiled by Slaney (1976). South American images recorded at the Brazilian receiving station may be ordered from: Institute de Pequisas Espaciais (INPE), a/c Direcao, Av. dos Astronautas, 1758, Caixa Postal 515 12.200 Sao Jose dos Campos, S.P., Brazil.

FIGURE 4.10
Landsat mosaic of conterminous United States compiled from band 5 images acquired in the summer season. Mosaic compiled by U.S. Department of Agriculture Soil Conservation Service.

LANDSAT MOSAICS

The broad regional coverage of an individual Landsat image (34,000 km^2) can be extended by combining adjacent images into a mosaic (Figure 4.10). The sidelap of adjacent orbit swaths and the 10 percent forward overlap of consecutive images greatly facilitate mosaic compilation. The uniform scale and minimal distortion of Landsat images also facilitate mosaic compilation, which will be appreciated by anyone who has ever made mosaics of aerial photographs and attempted to match the radially distorted prints. The 18-day repetition cycles have enabled Landsat 1 and 2 to acquire essentially cloud-free coverage of much of the world. Aside from locating the needed images, the major problem is matching the photographic density and contrast ratio of the images to produce a uniform mosaic. This is particularly difficult when mosaics are compiled from images acquired at different seasons of the year.

Mosaic Compilation

The following procedures are useful in compiling mosaics:

1. Select images for the same day along an orbit path, if available.
2. Select images at the same annual season to minimize changes caused by snow cover and vegetation cycles.
3. Bands 5 or 7 are generally superior, as discussed in the section on Landsat Image Analysis.
4. Acquire transparencies or negatives, rather than prints, from EDC. In the darkroom the density and contrast can be controlled to provide uniform prints for the mosaic.
5. First match the images along each orbit path; then fit the adjacent orbit paths together and distribute any mismatch along the entire length.
6. In making a large mosaic, begin with the central orbit path and work outward rather than starting at one edge.

Mosaics of the United States

Figure 4.10 is a greatly reduced copy of the mosaic of the United States that was compiled by the U.S. Soil Conservation Service Photographic Laboratory from approximately 600 band 5 Landsat images. Photographic reproductions of this mosaic and one of Alaska are available at scales from 1:500,000 to 1:10,000,000. For information write to Cartographic Division, Soil Conservation Service, Room G-110, Federal Building, Hyattsville, Maryland 20782. The National Geographic Society (Washington, D.C. 20036) has published a 1:4,560,000 scale color mosaic of the contiguous United States in which the vegetation is portrayed in green. This was accomplished by printing band 7 in green, rather than red. Lithographed copies of the map may be purchased from the National Geographic Society, which will provide price lists. An IR color mosaic of the United States is available from General Electric Space Systems at various scales. A price list is available from the address given earlier in the section on Acquisition of Images. A small-scale version of this IR color mosaic was published by Bishop (1976, p. 144).

Excellent lithographed mosaics of several states are available from the U.S. Geological Survey at reasonable prices. The available mosaics are listed on Table 4.3 and more are in preparation. The geological surveys of many individual states have also prepared state mosaics that are available for purchase.

TABLE 4.3
U.S. Geological Survey Landsat mosaics of states

Locality	Scale	Format
Arizona	1:500,000	Black and white
Florida	1:500,000	IR color
Georgia	1:500,000	Black and white
New Jersey	1:500,000	IR color

SOURCES OF INFORMATION

Excellent collections of Landsat color images and descriptions have been compiled by Williams and Carter (1976) and by Short and others (1977). NASA has convened three symposia devoted exclusively to Landsat. The proceedings are available from the U.S. Government Printing Office, Washington, D.C. 20402 as the following documents:

ERTS-1 Symposium Proceedings, Sept. 29, 1972, NASA X-650-73-10, 165 pp.

Symposium on Significant Results Obtained from ERTS-1, March 5-9, 1973, vol. 1, Technical Presentations, NASA SP-327, 1730 pp.; vol. 2, Summary of Results, NASA X-650-73-127, 198 pp.; vol. 3, Discipline Summary Reports, NASA X-650-73-155, 118 pp.

Third ERTS-1 Symposium, Dec. 10-14, 1973, vol. 1, Technical Presentations, NASA SP-351, 1900 pp.; vol. 2, Summary of Results, NASA SP-356, 179 pp.; vol. 3, Discipline Summary Reports, NASA SP-357, 155 pp.

The journals and remote-sensing symposia volumes listed in Chapter 1 include many Landsat papers. The papers cited at the end of this chapter cover many aspects of Landsat interpretation. More than 300 formal experiments with Landsat 1 data have been funded by the United States government and carried out by the United States and foreign investigators. Reports of these investigators may be obtained from the National Technical Information Service (NTIS), Department of Commerce, 5285 Port Royal Road, Springfield, Virginia 22161. Bidwell and Mitchell (1975) listed reports available through October 15, 1975 and their NTIS numbers.

INTERPRETATION METHODS

Interpretation methods for Landsat images, and other forms of remote-sensing imagery, are determined by the objectives of the interpreter. Geographers, agronomists, and geologists would produce different interpretation maps of the southern California image (Plate 2). The terminology employed by most disciplines is adequate for describing features on Landsat images and there is little need for additional terms. In the case of geology, however, there has been disagreement on terms for describing the linear features that occur on Landsat and other images. The following terms and definitions are used in this book.

Lineaments and Related Features

In the early 1900s the American geologist Hobbs recognized the existence and significance of linear geomorphic features that were the surface expression of zones of weakness or structural displacement in the crust of the earth. Hobbs (1904, 1912) defined lineaments as "the significant lines of landscape which reveal the hidden architecture of the rock basement. . . . They are character lines of the earth's physiognomy." Over the years additional terms have been misused as synonyms of lineament, and the resulting confusion has tended to obscure the geologic significance of lineaments. O'Leary, Friedman, and Pohn (1976) reviewed the origin and usage of the terms linear, lineation, and lineament. Their work and definitions are the basis for the following discussion.

Linear is defined as an adjective that describes the linelike character of an object or an array of objects. Some geologists have misused the adjective linear as a noun substitute and have considered it to be synonymous with lineament. Others have used linear to designate a short lineament, for example one less than 10-km long. It is grammatically incorrect to use the adjective linear as a noun. If we accept the use of linear and linears as nouns, we also have to accept planars, circulars, and so on as real objects (O'Leary, Friedman, and Pohn, 1976). Linear is an indispensable descriptive word (linear valley, linear escarpment, etc.) and it should be used in this sense. *Linear feature* is a good informal term to describe objects in terms of their geometry without any genetic or structural implication.

Lineation is defined as the one-dimensional structural alignment of internal components of a rock that cannot be depicted as an individual feature on a map (O'Leary, Friedman, and Pohn, 1976). Some geologists have used this petrographic term as a synonym for lineament, but this is incorrect and unacceptable.

Lineament is defined as a mappable simple or composite linear feature of a surface, whose parts are aligned in a straight or slightly curving relationship, and which differs distinctly from the patterns of adjacent features and presumably reflects a subsurface phenomenon (O'Leary, Friedman, and Pohn 1976). The surface features making up a lineament may be geomorphic (caused by relief) or tonal (caused by contrast differences). The surface features may be landforms, the linear boundaries between different types of terrain, or breaks within a uniform terrain. Straight stream valleys and aligned segments of valleys are typical geomorphic expressions of lineaments. A tonal lineament may be a straight boundary between areas of contrasting tone or a stripe against a background of contrasting tone. Differences in vegetation, moisture content, and soil or rock composition account for most tonal contrasts. Lineaments may be continuous or discontinuous. In *discontinuous lineaments* the separate features are aligned in a consistent direction and are relatively closely spaced. Lineaments may be simple or composite. *Simple lineaments* are formed by a single type of feature, such as a linear stream valley or aligned topographic escarpments. *Composite lineaments* are defined by more than one type of feature, such as an alignment of linear tonal features, stream segments, and ridges.

Although many lineaments are controlled at least in part by faults, structural displacement (faulting) is not a requirement in the definition of a lineament. On Landsat images of the Precambrian shield in northern Ethiopia, for example, lineaments up to 200-km long are clearly defined by aligned straight segments of valleys and by linear tonal contrasts. On 1:250,000 scale images no offset of the lithologic contacts transected by the lineaments is detectable. These and similar lineaments throughout the world are thought to represent zones of weakness in the crust. Although displacement has not occurred, the rocks may be more highly fractured and susceptible to erosion. Lineaments that are lines or zones of structural offset are called *faults* or *fault zones.* On the Landsat image (Plate 2) and interpretation map (Figure 4.5) of southern California the San Andreas, San Gabriel, and Garlock faults qualify as lineaments, but are designated as faults because they are known to be lines and zones of structural displacement. Linear features newly discovered on images initially may be called lineaments. If field checking establishes the presence of structural offset, they can then be designated as faults. This procedure is illustrated in the example of the Peninsular Ranges, California, later in this chapter.

No minimum length is proposed here as a requirement for lineaments, but significant crustal features are typically measured in tens of kilometers. Lineaments also may be recognized on topographic, gravity, magnetic, and seismic contour maps by aligned highs and lows, steep contour gradients, and aligned offsets of trends. Lineaments are well expressed on Landsat images because of the oblique illumination, suppression of distracting spatial details, and the regional coverage. Linear features were so emphasized in many of the early geologic interpretations of Landsat images that some skeptics claimed the imagery was only useful for that purpose. The recognition of faults, folds, rock types, and mineral deposits shown in this book and by other Landsat investigators (see References) should dispel this misconception.

Landsat Image Analysis

Landsat images may be interpreted in much the same manner as small-scale photographs and images acquired from aircraft and manned satellites. There are some advantages, however, of Landsat images that should be emphasized. Linear features caused by topography may be enhanced or suppressed on Landsat images depending upon their orientation relative to sun azimuth. Linear features trending normal or at a high angle to the sun azimuth are enhanced by shadows and highlights. Those trending parallel with the azimuth are suppressed and difficult to recognize, as are linear features parallel with the MSS scan lines. Scratches and other film defects may be mistaken for natural features, but this can be resolved by determining whether the questionable features appear on more than a single band of imagery. The shadows of aircraft contrails may be mistaken for tonal linear features. These are recognized by checking for the parallel white image of the contrail.

A useful diagnostic procedure is to plot linear features as dotted lines on the interpretation map. The use of field checking and making reference to existing maps will identify some linear features as faults; these are changed to solid lines on the interpretation map. The remaining dotted lines may represent zones of fracturing with no displacement or they may have no structural significance.

The repeated coverage of Landsat enables interpreters to select images from the optimum season for their purpose. Several examples of seasonal effects on images are discussed later in this chapter. Winter images provide minimum sun elevations and maximum enhancement of topographic linear features. Larger structural and topographic features are commonly enhanced on snow-covered images because tonal differences and minor terrain features, such as small lakes, are eliminated or suppressed. Areas with alternating wet and dry seasonal climates should be interpreted on images acquired at the different seasons. In southern California and South Africa, rainy season images are most useful, but this may not be the case everywhere.

Significance of colors on Landsat color composite images was described earlier in this chapter and in the discussion of IR color photography. For special interpretation objectives black-and-white images of individual bands are useful. Bands 4 and 5 are optimum for recognizing water currents containing suspended sediment and for mapping bathymetry in shallow water. Band 5 is an excellent image for general geographic and geologic mapping. Band 7 is optimum for recognizing contacts between land and water, vegetation differences, and differences in soil moisture.

Mosaics are valuable for interpreting regional structural trends that may cross the area being analyzed. It is helpful to project a Landsat image on a screen and study it from a distance. Almost invariably some previously unrecognized features become apparent.

NEVADA LANDSAT INTERPRETATION

Rowan and Wetlaufer (1975) of the U.S. Geological Survey interpreted a 1:1,000,000 scale mosaic of band 5 images of Nevada, shown at reduced scale in Figure 4.11. The following discussion is summarized from their report.

Procedure

After the linear features were interpreted from the mosaic, they were checked and re-evaluated using individual images, an enlarged mosaic, and topographic maps. Linear features caused by cultural features or scan lines were eliminated. The interpretation map (Figure 4.12) shows the correspondence of the major Landsat lineaments to faults shown on the Nevada geologic map. Correlation of a fault and a lineament is confirmed if they are approximately parallel and within 0.5 km of each other. These are shown as solid lines in Figure 4.12. Lineaments that are extensions of mapped faults are indicated with a dashed line and also counted as a correlation. There is an 80 percent correlation between the 367 major lineaments and mapped faults. This high correlation is due in part to the good topographic expression of the Basin and Range boundary faults.

Of the major lineaments, 238 have a single origin, as shown in Table 4.4, which also gives percentage of lineaments. Despite the high correlation of major lineaments with mapped faults, only 29 percent of the faults longer than 10 km on the state geologic map have corresponding Landsat lineaments. Although lineament analysis contributes measurably to regional structural studies, it is not a substitute for field mapping.

TABLE 4.4
Landsat lineaments in Nevada

Origin of lineament	Percent
Boundary between outcrop and alluvium, predominantly that between mountains and basins	56.7
Mountain ridges or series of small ridges	21.8
Mountain canyons	7.6
Stream segments	1.3
Tonal boundaries, alluvial and nonalluvial	12.6

Source: Rowan and Wetlaufer (1975).

Lineament Systems

There are seven major lineament systems, which are shown on the insert map of Figure 4.12. Generally they transect the north to north-northeast-trending topography of the Basin and Range province, and all but one extend for several hundred kilometers. Three of the lineament systems have previously been documented as major crustal features. These are the Walker Lane lineament (A), Midas Trench lineament (B), and Oregon–Nevada lineament (C).

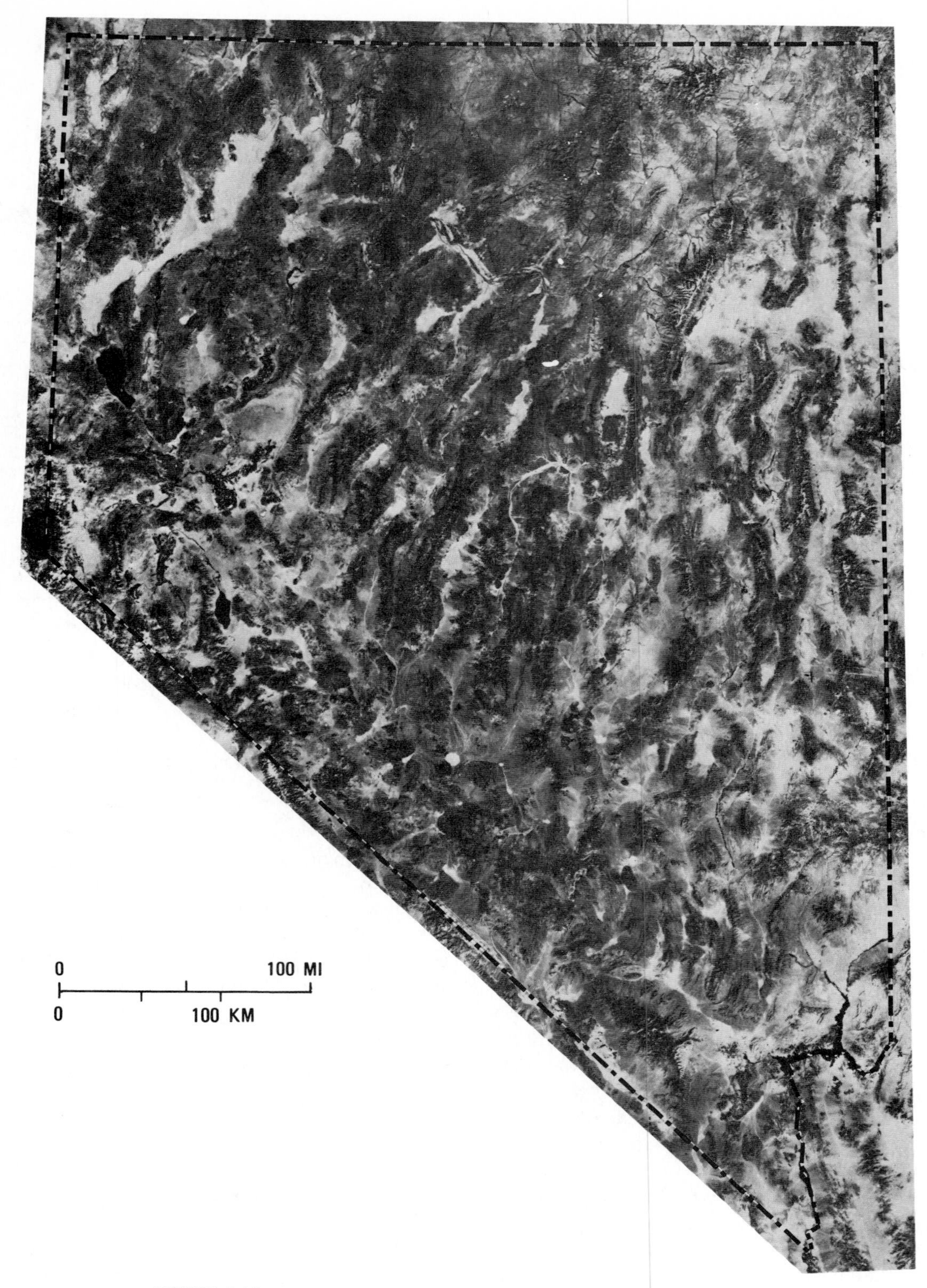

FIGURE 4.11
Landsat mosaic of Nevada compiled from band 5 images. Compare with interpretation map showing lineaments and circular features. From Rowan and Wetlaufer (1975, Figure 4).

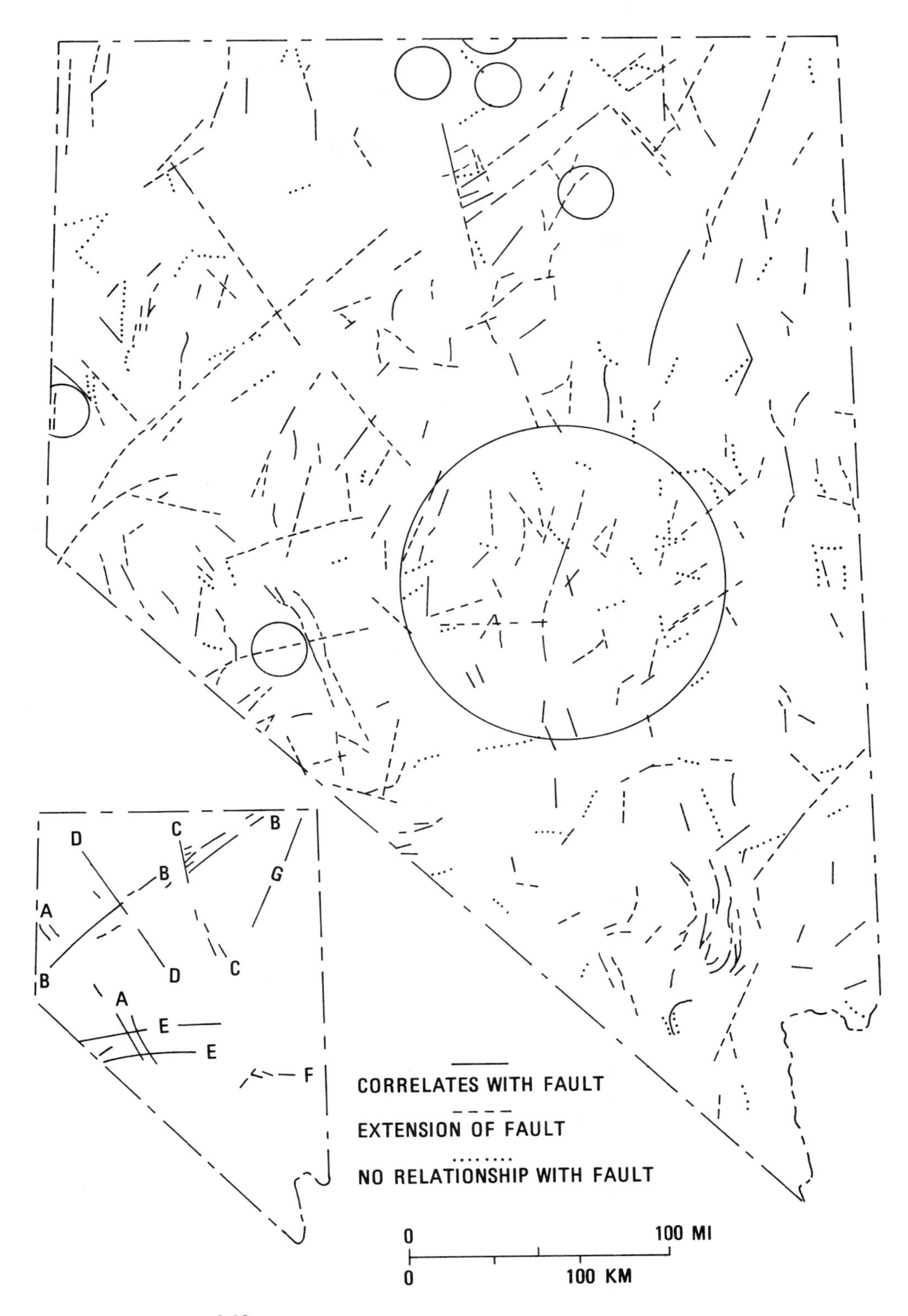

FIGURE 4.12
Major lineaments (>10 km length) and circular features interpreted from Landsat mosaic of Nevada. Insert map shows the following major lineaments: (A) Walker Lane, (B) Midas Trench, (C) Oregon-Nevada, (D) Rye Patch, (E) ENE Lineament, (F) E-W Lineament, and (G) Ruby Mountains Lineament. From Rowan and Wetlaufer (1975).

Walker Lane Lineament This zone of right-lateral transcurrent faulting extends 600 km from Pyramid Lake to Las Vegas, essentially parallel with the San Andreas fault system. The Walker Lane lineament is expressed by a distinct northwest-trending topographic discontinuity, easily visible on the Landsat mosaic of Nevada (Figure 4.11). The regional northeasterly trends of the mountain ranges on either side of the lineament are abruptly changed to northwesterly trends within the Walker Lane zone. In southeastern Nevada the extension of the Walker Lane lineament is the Las Vegas shear zone. Right-lateral displacement is indicated by bending of the mountain ranges and of geologic trends north of the shear zone. Although no single lineament marks the Las Vegas zone, it is clearly defined by these topographic and structural patterns.

Midas Trench Lineament This lineament zone extends northeast from Lake Tahoe for 460 km to the northern state border. It is most conspicuous in north-central Nevada where it is expressed as a linear topographic depression near Midas. To the northwest it is expressed as an escarpment separating the Owhyee Desert from the mountains to the southeast. Near the Nevada-Idaho border the trend is expressed by aligned stream segments in the mountains. The less distinct southwestern extension is expressed as a discontinuous series of linear canyons and ridges in the ranges. Tonal changes in the alluvium of the basins delineate a continuous zone that intersects the Walker Lane lineament just south of Pyramid Lake. Recent lateral movement along faults of the Midas Trench lineament is indicated by offset stream channels.

Oregon-Nevada Lineament This lineament is a 750-km long belt of closely spaced, partly *en echelon* faults extending northwest to north-northwest from central Nevada to central Oregon. The lineament consists of aligned tonal and textural boundaries. The northeast-trending Midas trench is abruptly interrupted where it intersects the Oregon-Nevada lineament. A linear belt of late Miocene lava flows and flow domes mark the Nevada portion of the lineament (Stewart, Walker, and Kleinhampl, 1975).

Rye Patch Lineament This 250-km long system consists of two aligned lineaments that trend northwest, parallel with the Walker Lane lineament. The Rye Patch lineament is somewhat diffuse in appearance, and is indicated predominantly by tonal and textural changes. Ranges appear to be terminated or offset by the lineament, which coincides with regional deep-seated fracture zones proposed by earlier workers.

East-Northeast-Trending Lineament This system terminates or disrupts the north-trending ranges that it intersects. It is expressed physiographically by aligned streams, canyons, and tonal boundaries. The southern of the two parallel lineaments of this system, seen on the insert map of Figure 4.12, consists of individual linear features less than 10 km long; therefore, this trend is not shown on the major lineament map. However, a 45 km stretch of the lineament coincides with a mapped high-angle fault with inferred right-lateral displacement. The trend of the northern segment of this system is parallel with mapped strike-slip faults, one of which formed during a historic earthquake. This lineament zone intersects the Walker Lane lineament with no apparent disruption.

East-West Lineament This is the shortest of the lineament systems, consisting of five east-trending segments with a total length of 70 km. The linear segments form the north margins of the Groom, Pahrangat, and Hiko Ranges and represent the fault-controlled boundaries between bedrock and alluvium. A prominent regional Bouguer gravity anomaly zone corresponds to the lineament zone.

Ruby Mountains Lineament This is the least documented lineament, with a total length of 230 km. The southern part reflects the normal fault along the eastern boundary of the Ruby Mountains, and the northern part aligns with several smaller faults.

Summary of Major Lineament Systems

According to Rowan and Wetlaufer (1975, p. 89) the seven major lineament systems in Nevada have the following characteristics in common:

1. They represent the traces of zones of faulting or fracturing that have substantial vertical and horizontal extent.
2. Where the movement pattern has been determined, strike-slip movement is characteristic although dip-slip movement has also occurred along most of the zones. Except for the Walker Lane lineament, where right-lateral movement has prevailed, contradictory right and left displacements have occurred.
3. The Basin and Range topography is commonly disrupted and terminated where it is intersected by the lineaments.
4. As discussed later, several lineaments coincide with magnetic anomalies, suggesting that the lineaments served as conduits along which magma was intruded.
5. The seven major lineament systems are probably old zones of structural weakness that have been periodically reactivated.

Circular Features

Rowan and Wetlaufer (1975) recognized 50 circular and elliptical features on the Landsat mosaic of Nevada that are presumed to be centers of igneous activity. Comparison with geologic maps suggests that six of these circular features, shown in Figure 4.12, may be previously unrecognized Tertiary volcanic centers. The circular features have not yet been investigated in the field. The major central Nevada circular feature (Figures 4.11 and 4.12) is defined on its eastern and western borders by slightly arcuate basins and ranges. The northern margin is defined by deflection of mountain ranges and drainage networks, but the southern boundary is much less topographically distinct. Within the circular feature there is an especially high density of major lineaments. The lineament trends are anomalously diverse and there is a concentration of lineament intersections within the circular feature. Four of the seven major lineament systems (C, D, E, G) radiate from the approximate center of the circular feature. The occurrence of large volumes of volcanic rocks suggests that the circular feature was a center of igneous activity from 17 to 34 million years ago.

Correlation of Landsat Features and Aeromagnetic Trends

Nevada is an instructive area for comparing Landsat lineaments and circular features with trends on aeromagnetic maps as shown in Figure 4.13. An aeromagnetic map essentially displays variations in the magnetic properties of crustal rocks, expressed in units called gammas. Higher gamma values are typically associated with mafic igneous rocks, such as basalt, that have high concentrations of the mineral magnetite. In addition to showing the presence of bodies of magnetite-rich rocks, aeromagnetic maps also portray the configuration of the crystalline basement rocks. Where the basement rocks are deeply buried beneath sedimentary cover, the magnetic intensity is lower than in areas where the sedimentary cover is thin. Structure and lithologic changes within the crystalline rocks are also shown by trends on aeromagnetic maps. For more information on the preparation and interpretation of aeromagnetic maps see Nettleton (1976).

The eastern third of the Nevada aeromagnetic map (Figure 4.13) is a relatively featureless *quiet zone*, the cause of which is not obvious (Stewart, Moore, and Zietz, 1977, p. 74). The western two-thirds of the map is characterized by narrow linear and arcuate trends of aligned magnetic highs that are separated by broad zones of magnetic lows that include small scattered highs. The Oregon-Nevada lineament correlates with a major linear magnetic high. Stewart, Walker, and Kleinhampl (1975) noted that a linear belt of volcanic rocks of mid-Tertiary age occurs along the magnetic high. They suggested that the magnetic high is caused by the feeder dikes that supplied magma for the volcanic rocks. The feeder dikes may have been intruded along a deep-seated fracture zone that is expressed as the Oregon-Nevada lineament on the Landsat mosaic.

The Walker Lane lineament is marked by a northwest-trending zone of arcuate and slightly sigmoidal aeromagnetic highs. There are two possible causes for the correlation between the Walker Lane lineament and the aeromagnetic pattern (Stewart, Moore, and Zietz, 1977, p. 76): (1) tectonic distortion of pre-existing east–west structural trends by right-lateral displacement along the Walker Lane lineament; (2) emplacement and eruption of Cenozoic igneous rocks along pre-existing curving structures in older rocks. The Midas Trench lineament is marked by an alignment of isolated magnetic highs, but this may be fortuitous because of the abundance of closed magnetic highs in this portion of the map. The two parallel trends of the East-Northeast lineament are both expressed on the aeromagnetic map, but are somewhat obscured by the intersection with the magnetic trends associated with the Walker Lane. The remaining Landsat lineaments are not associated with magnetic trends, which may mean that igneous activity did not occur along all the lineaments. Just as some lineaments are not marked by magnetic trends, some magnetic trends are not associated with Landsat lineaments. A conspicuous north-northwest-trending magnetic high located between the Rye Patch and the Oregon-Nevada lineaments (Figure 4.13) does not correlate with Landsat lineaments. The magnetic trend is largely associated with trends of Mesozoic igneous rocks (Stewart, Moore, and Zietz, 1977, Figure 5); Basin-and-Range faulting in Cenozoic time may have obscured any relationship between this older igneous trend and features seen on Landsat images.

The eastern portion of the central Nevada circular feature occurs within the "quiet zone" of the aeromagnetic map and has no magnetic expression. The western half, however, is marked by a semi-circle of magnetic highs that may represent intrusive bodies around the margin of the circular feature. The western border of the circular feature also marks a boundary between lower magnetic intensities within the feature and higher intensities to the north, west, and south. The thick volcanic cover within the circular feature may mask the magnetic expression of underlying intrusive bodies.

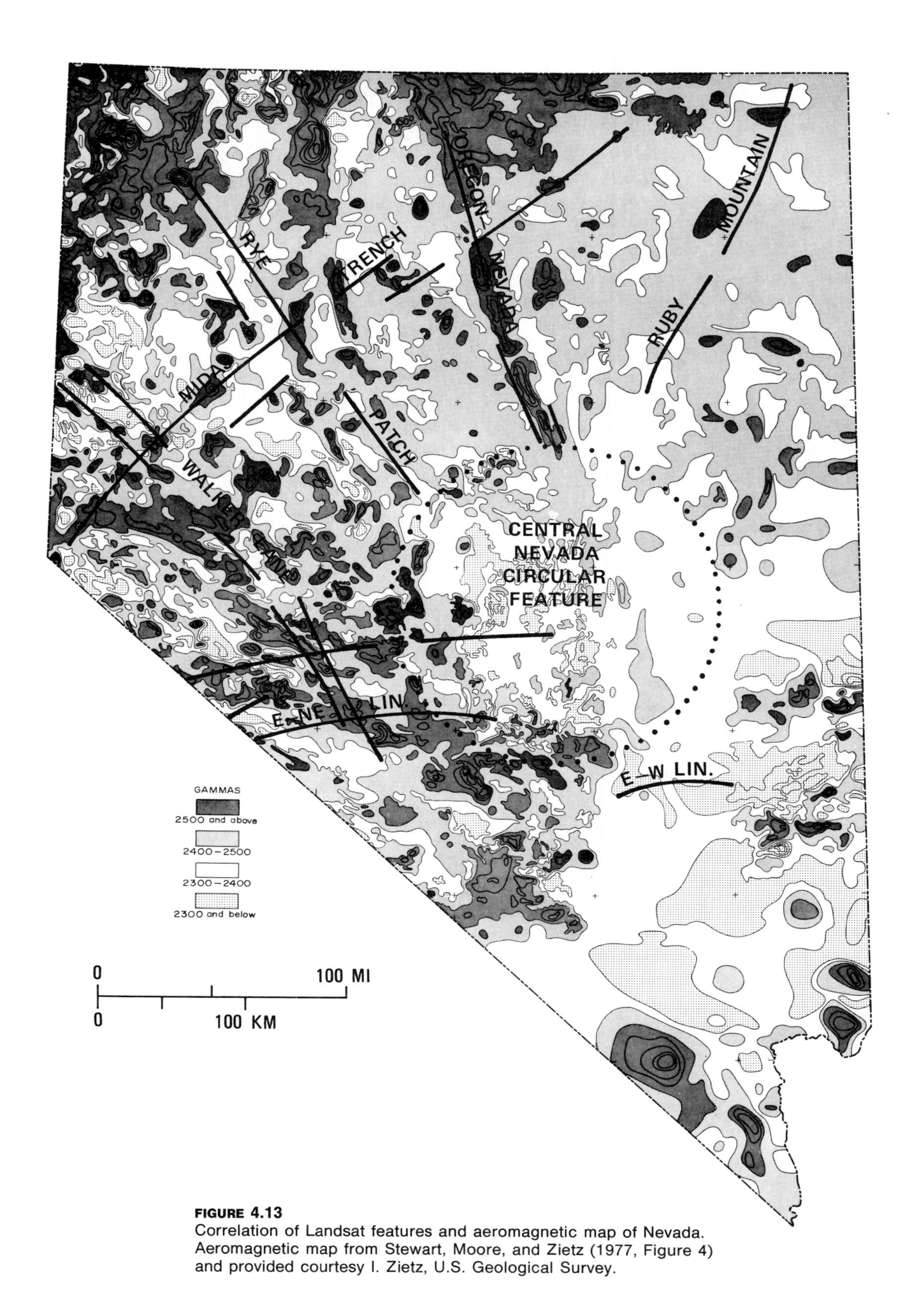

FIGURE 4.13
Correlation of Landsat features and aeromagnetic map of Nevada. Aeromagnetic map from Stewart, Moore, and Zietz (1977, Figure 4) and provided courtesy I. Zietz, U.S. Geological Survey.

FIGURE 4.14
Landsat image 1106-17504 band 5 of Peninsular Ranges, southern California. See Figure 4.15 for identification of major faults and lineaments. From Lamar and Merifield (1975, Figure 4). Courtesy P. M. Merifield, California Earth Science Company.

In Nevada the correlation between Landsat features and aeromagnetic trends largely results from the intrusion of igneous rocks along zones of weakness that are marked by linear and curvilinear features on Landsat images. The magnetite-rich igneous rocks in turn are recorded as high intensities on the aeromagnetic map.

PENINSULAR RANGES, CALIFORNIA

The structure of California south of the Transverse Ranges is dominated by the southwest-trending San Andreas, San Jacinto, and Elsinore strike-slip faults. On Landsat images, however, Sabins (1973) noted prominent lineaments trending north and

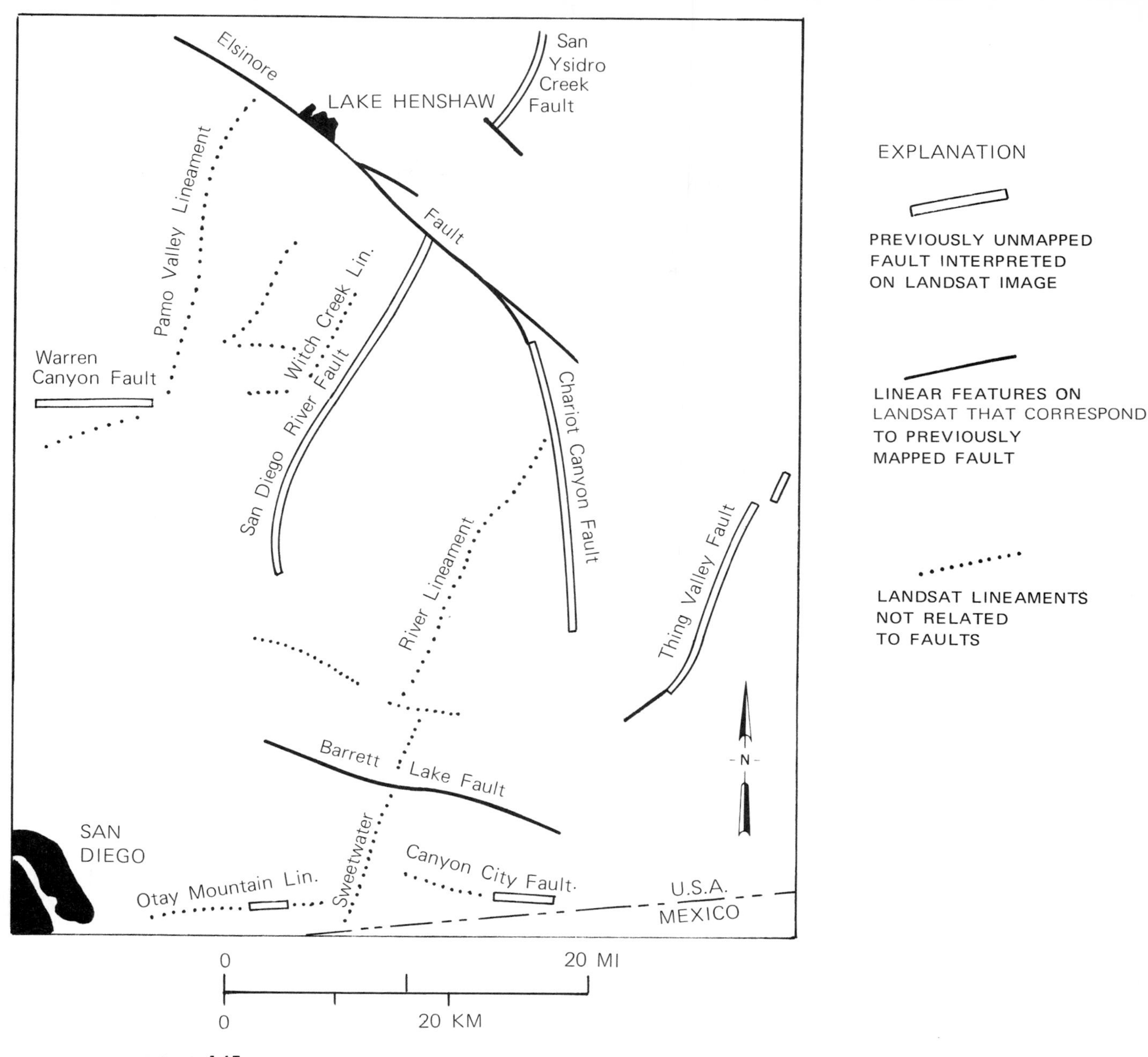

FIGURE 4.15
Field evaluation of lineaments and faults on Landsat 1106-17054 of Peninsular Ranges, California. From Lamar and Merifield (1975, Figure 2).

northeast that do not correspond to mapped faults. Lamar and Merifield (1975) interpreted the Landsat image of Figure 4.14 to produce an interpretation map (Figure 4.15) with three types of lineaments: (1) those marking faults that had not previously been mapped or were of controversial origin, (2) lineaments corresponding to previously mapped faults, and (3) lineaments not caused by faults or of an unknown origin.

A major contribution of Lamar and Merifield was their careful checking and evaluation of the lineaments in the field. This is a difficult task in the rugged Peninsular Ranges with limited access and dense cover of chaparral. Bedrock consists of late Mesozoic plutonic rocks and roof pendants of Paleozoic and Mesozoic metamorphic rocks. Breccia and gouge zones were the main criteria for recognizing faults in the field. Displaced or terminated lithologic contacts were also used, but are scarce in this region.

The field analysis of Lamar and Merifield is summarized in Table 4.5. Four of the north- and northeast-trending lineaments correlate with previously unmapped faults that are recognizable in the field. Three prominent lineaments, however, are not controlled by faults. Jointing, foliation, or aligned stream segments of diverse origin account for these lineaments. West-trending lineaments also have structural and nonstructural origins. The significant fact is that the majority of the Landsat lineaments prove to be faults, either mapped or unmapped. The extensive erosion and lack of recent offsets along the north- and northeast-trending faults indicate that they are not earthquake hazards, in contrast to the active northwest-trending Elsinore, San Jacinto, and related faults.

ADIRONDACK MOUNTAINS, NEW YORK

An instructive contrast to the arid, high-relief terrain of Nevada and California is provided by the Adirondack Mountains of northeastern New York, which were studied by Isachsen, Fakundiny, and Forster (1973 and 1974).

Geologic Setting

This humid area of subdued relief is shown on the Landsat mosaic (Figure 4.16) where the Adirondacks are the dark, highly fractured oval area occupying most of the scene. The Adirondacks are a glaciated dome of Precambrian gneiss, quartzite, and marble with a core of metamorphosed anorthosite. This dome is surrounded by gently dipping sandstones and carbonate rocks of lower Paleozoic age that are lighter in tone and much less fractured than the Adirondack crystalline rocks (Figure 4.16). Physiographically the Adirondack Mountains are bordered on the north and east by the St. Lawrence-Champlain lowlands, on the south by the Mohawk lowlands, and on the west by the Tug Hill Plateau.

Landsat Investigation

The systematic investigation and field checking of the Landsat lineaments by Isachsen, Fakundiny, and Forster (1974) is another excellent example for future investigators. All lineaments visible on Landsat images at the 1:1,000,000 scale were plotted. These were screened to eliminate (1) man-made features such as highways, railroads, powerlines, and canals; and (2) linear features coincident with foliation trends and rock contacts not caused by faulting. Approximately 20 percent of the original lineaments were eliminated and those remaining were compared with the major faults and topographic lineaments shown on the Geologic Map of New York (1:250,000 scale). The majority of the previously mapped faults correspond with Landsat lineaments. Most of the previously mapped faults that are not visible on the mosaic are short and may prove to be visible on larger scale images. In Figure 4.17 the Landsat lineaments that coincide with previously mapped faults are shown with solid lines; those that do not correspond to previously mapped faults are shown with dotted lines.

TABLE 4.5
Field evaluation of lineaments on Landsat image of Peninsular Ranges from Figure 4.14

Lineament	Geologic setting	Nature of linear feature	Remarks
	Lineaments caused by previously unmapped or controversial faults		
San Ysidro Creek fault	Along contact between granitic intrusive and schist.	Fault zone of crushed and sheared gouge up to 7 m wide. Striations indicate predominantly horizontal movement.	The alignment with San Diego River fault to the southwest is probably fortuitous.
San Diego River fault zone	Fault transects terrain of schist and plutonic rocks.	Zone of gouge, sheared and brecciated rock. Up to 630 m of right-lateral separation of lithologic contacts.	Straight course of San Diego River is controlled by the fault zone.
Chariot Canyon fault	Fault follows contact between schist on the west and granitic rocks on the east.	The fault is a broad shear zone of steeply dipping slickensided surfaces separating granite and schist.	Fault is discontinuous and additional field mapping is required.
Thing Valley fault	Predominantly granitic bedrock with inclusions of schist.	Fault is over 20 km long with gouge and breccia zone up to 35 m wide.	Possible right-lateral separation ranges from 100 m to 1 km.
	Lineaments not due to faulting or of unknown origin		
Witch Creek lineament	Bedrock is diorite and schist. Jointing and foliation are not aligned with the lineament.	Aligned remarkably straight canyons parallel with San Diego River fault zone.	Origin is unknown, but did not result from erosion along a fault zone.
Pamo Valley lineament	Jointed and foliated granitic rocks.	Aligned straight canyon segments and saddles. Feature trends parallel with joint and foliation direction. No evidence of faulting, except for short narrow gouge zone.	Apparently due to erosion along foliation and joint direction.
Sweetwater River lineament	Granitic rocks.	Discontinuous or *en echelon* straight stream segments. Second only to San Diego River lineament in prominence on images.	No fault control. The alignment of straight segments of diverse origin is possibly fortuitous.

Source: Summarized from Lamar and Merifield (1975).

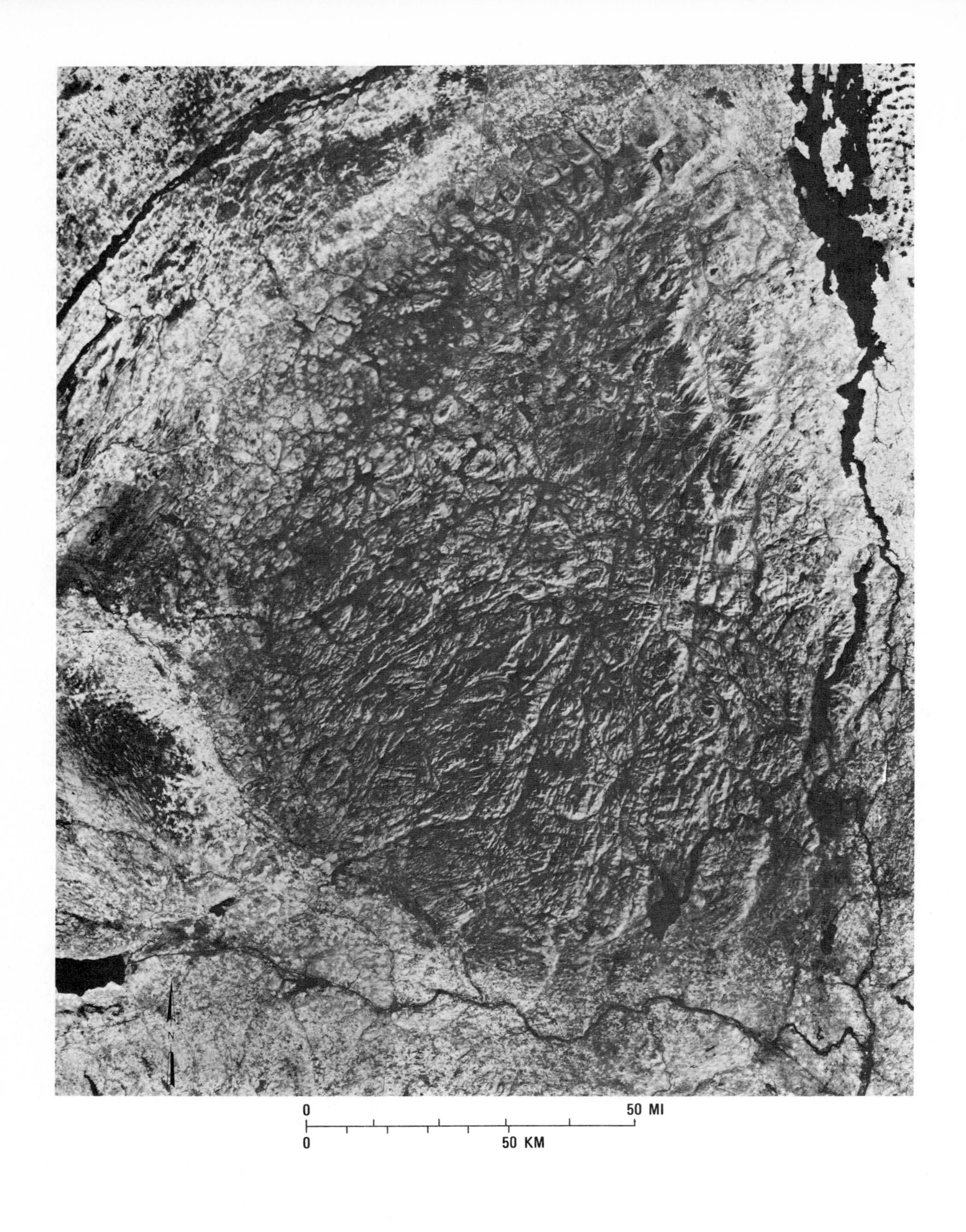

FIGURE 4.16
Landsat mosaic of Adirondack Mountains, New York. Compiled from band 7 images acquired October 10 and 11, 1972. Courtesy Y. W. Isachsen, New York State Geological Survey.

FIGURE 4.17
Interpretation of Landsat mosaic of Adirondack Mountains, New York. From Isachsen, Fakundiny, and Forster (1973, Figure 2). Courtesy Y. W. Isachsen, New York State Geological Survey.

TABLE 4.6
Lineaments on Landsat mosaic of Adirondack Mountains

Lineament	Number	Percent
Straight stream valleys	130	41
Elongate lakes or straight shorelines of lakes	4	1
Edge of topographic high or aligned segments of highs	6	2
Dark vegetation strips	8	3
Natural vegetation borders	6	2
Combinations of two or more of above	129	40
Unexplained	36	11
Total	319	100

Source: Summarized from Isachsen, Fakundiny, and Forster (1974).

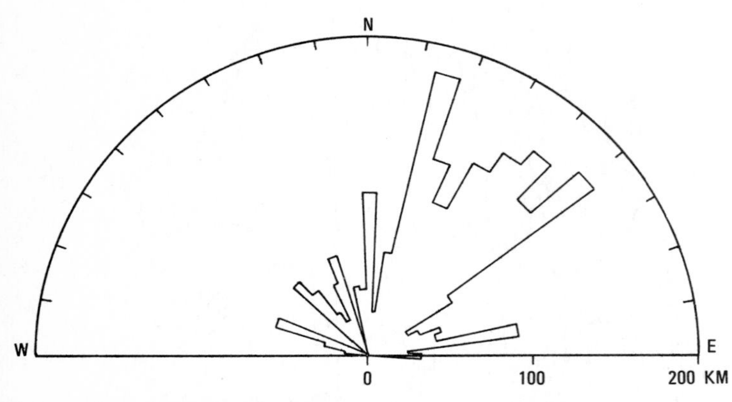

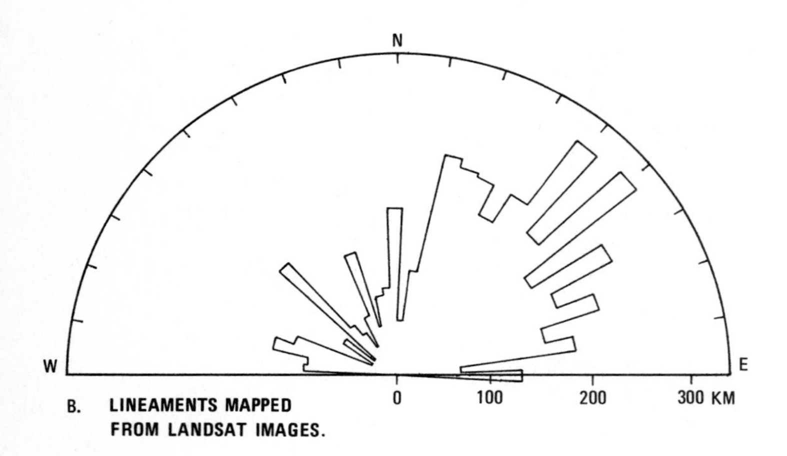

FIGURE 4.18
Radial diagrams for faults, topographic lineaments, and linear features of the Adirondack Mountains. Summed lengths are shown at 5° intervals. From Isachsen, Fakundiny, and Forster (1974, Figures 13B and 15).

Analysis of Lineaments

The major problem encountered in field checking was the difficulty of locating the Landsat lineaments on the ground. Low-level aircraft flights were the most effective and economical method of location, after which key localities were investigated on the ground. On the basis of these investigations the lineaments were classified as shown in Table 4.6.

Straight stream valleys are the predominant category in this relatively humid environment. The radial diagram of Figure 4.18A plots the lengths of all the faults and linear topographic features shown on the Geologic Map of New York for the Adirondack region. Figure 4.18B is a diagram of all the lineaments on the Landsat interpretation map of Figure 4.17 (solid and dotted lines). The close similarity of the two radial diagrams is obvious. The major difference is that the N 15° E maximum for the mapped faults is less prominent for the Landsat lineaments.

In summary, the field expression of the newly mapped Landsat lineaments and their close coincidence with the pattern of previously mapped faults is strong evidence that the lineaments do represent faults and fractures. Because bedrock is so poorly exposed in the Adirondacks, it was suggested by Isachsen, Fakundiny, and Forster (1974) that Landsat images may be the best data available for the regional mapping of fracture zones.

ZAGROS MOUNTAINS, IRAN

Plate 3 is a portion of a color composite Landsat image that covers part of the southeastern Zagros Mountains that form the southwestern part of Iran. These mountains are the site of major oil fields in western Iran, but no oil is produced from the area shown in Plate 3. This image is an especially instructive example for the interpretation of geomorphic, stratigraphic, and structural features.

Geomorphology

Low-lying Qishm Island in the southeast part of Plate 3 is separated from the mainland by Clarence Strait, a narrow arm of the Persian Gulf. The light and medium blue area west of the island is a partially submerged tidal flat with meandering tidal channels. There is little or no coastal plain along the shore of the mainland because mountain building occurred so recently that erosion has not reduced the terrain. The arid climate is indicated by the dry stream channels and scarcity of vegetation. The vegetation (indicated by bright red color) occurs as scattered patches and streaks in the major valleys. Comparison of the image with the geologic interpretation map (Figure 4.19) shows that topography is controlled by geologic structure. The west-trending mountains and valleys are formed by anticlines and synclines, respectively. Long escarpments are formed by eroded edges of resistant beds along the flanks of the folds. The stream channels follow the axes of the synclines and bend around the plunging noses of the anticlines.

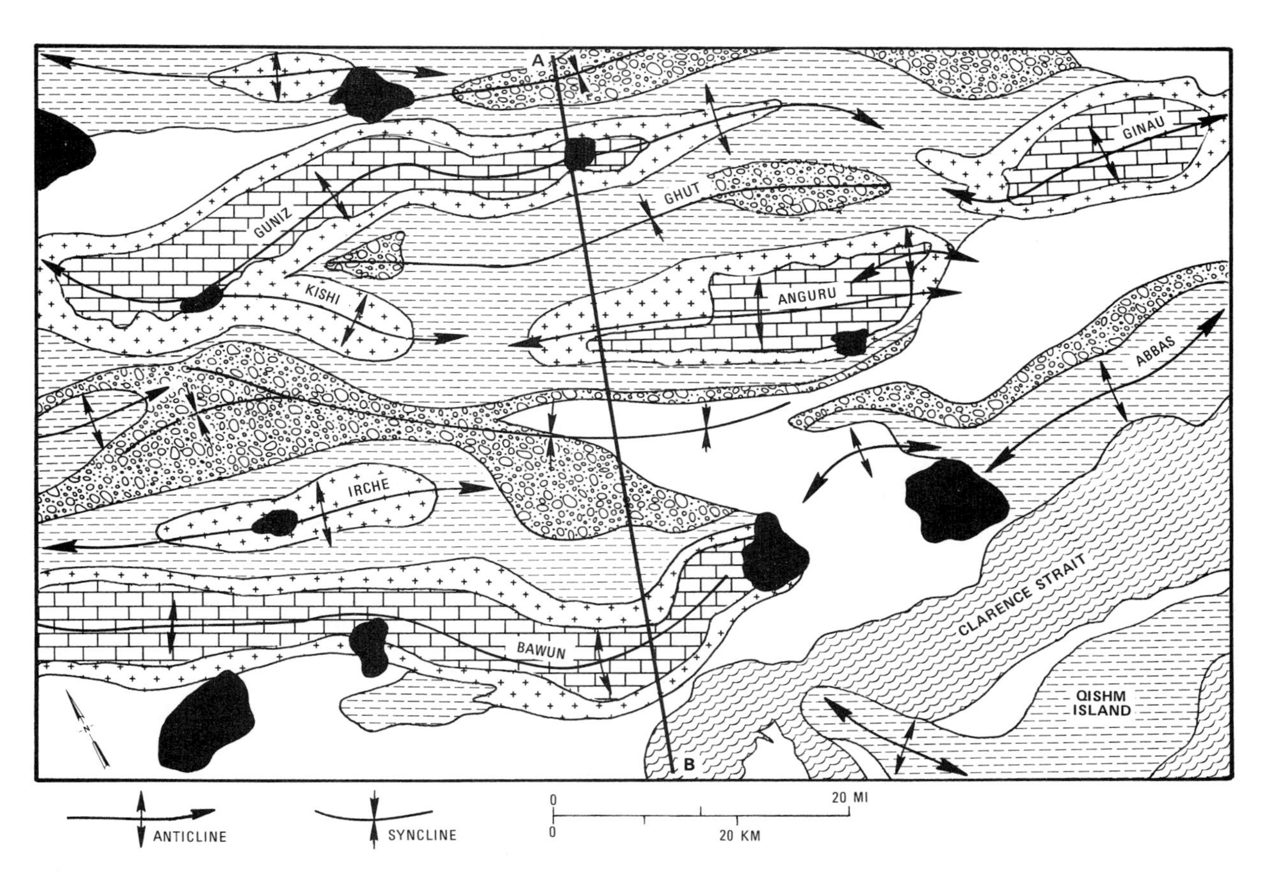

FIGURE 4.19
Geologic interpretation of Landsat image shown in Plate 3 of southeastern Zagros Mountains, Iran. Geologic symbols are explained in Figure 4.20. Geology after British Petroleum Company (1956, Sheet G-40).

TABLE 4.7
Designation of rock units in Zagros Mountains, with and without reference information

With information	Without information
Bakhtiari Conglomerate (Pliocene)	Upper resistant beds
Fars Group (middle and upper Pliocene)	Nonresistant purple beds
Limestone (lower Miocene)	Lower resistant beds
Limestone and marl (lower Tertiary and upper Cretaceous)	Moderately resistant basal beds

Rock Types

The following properties of rocks are expressed on Landsat color composite images:

1. Color—not the true rock color, but the IR color rendition.
2. Internal structure—stratification, metamorphic foliation, jointing, and volcanic flow units.
3. Contacts with adjacent rock units—conformable or unconformable depositional contacts, gradational, or sharp intrusive contacts.
4. Resistance to erosion—controlled in part by climate. Limestones, for example, are resistant in arid climates, but are less resistant in humid climates where chemical weathering occurs.

On many images these characteristics enable the interpreter to recognize the major rock classes: sedimentary, metamorphic, and intrusive and extrusive igneous rocks. Within each major rock class individual rock units may be distinguished and mapped on Landsat images, but the specific rock type (sandstone, limestone, conglomerate, and so forth) cannot be identified without additional information. This results from the fact that diverse rock types may have similar color, internal structure, and resistance to erosion.

On the Zagros Mountains image (Plate 3), the stratification and alternating resistant and nonresistant character of the outcrops indicate that they are sedimentary rocks. Specific types of sedimentary rocks cannot be identified without field checking or reference to geologic reports. The area of Plate 3 is covered by maps of the British Petroleum Company (1956, Sheet G-40) that are used to determine the lithology and age of the rock units. The older rocks exposed in the cores of the Ginau, Anguru, Guniz, and Bawun anticlines (Figure 4.19) are recognizable on the image by their intermediate resistance, indistinct stratification, and light pink to tan color. The published maps show that these are limestones and marl (limy shale) of lower Tertiary and upper Cretaceous age. The eroded anticlines are flaked by a well-stratified resistant unit that forms distinctive triangular facets. At the Guniz anticline this unit can be traced eastward along either flank to the plunge of the structure where the unit extends over the nose of the fold. This resistant unit (a limestone of Miocene age) is largely intact over the crest of the Irche anticline and defines the shape of that fold. The Miocene limestone is overlain by a nonresistant unit with a distinctive light purple color on the Landsat image. According to the published map this is the Fars Group (middle and upper Miocene), which consists of green siltstone with a few limestone beds. The color shift on the IR color composite image causes the green siltstone to appear purple. The Fars Group crops out on the flanks of the folds and in the trough of the Ghut syncline. It also crops out in the eroded Abbas anticline along the coast where the interbedded limestones form distinct strike ridges. The youngest Tertiary formation is a resistant brown unit that crops out in the Ghut syncline and in the syncline between the Anguru and Bawun anticlines. This is the Bakhtiari Conglomerate of Pliocene age, which closely resembles the Miocene limestone on the image. The higher stratigraphic position and slightly darker color of the conglomerate distinguish it from the limestone. The similar appearance of such diverse rock types as conglomerate and limestone indicates the difficulty of identifying rock types from images alone. Table 4.7 shows how the Zagros rock units are designated with and without reference information. The

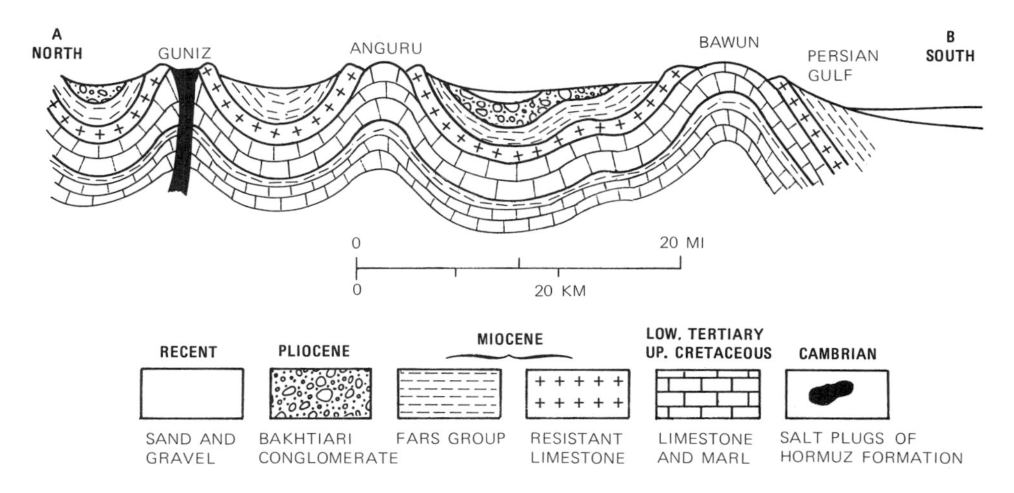

FIGURE 4.20
Geologic cross section of Zagros Mountains along line AB of Figure 4.19. Vertical scale is the same as the horizontal scale.

young surficial deposits consist of light to dark brown gravel fans along the flanks of the anticlinal mountains; light colored sands and possible evaporite deposits occur along the coast together with dark marsh deposits.

The dark circular features with irregular texture are salt plugs of the Hormuz Formation (Cambrian) that have intruded the overlying formations. The salt has carried up detached blocks of Cambrian and other rocks that have been concentrated at the surface as solution removed the salt. Without supporting information these would have been interpreted as probable volcanic necks.

Structural Geology

The geologic structure of this portion of the Zagros Mountains is clearly shown on the Landsat image and the interpretation is shown on the map (Figure 4.19) and cross section (Figure 4.20). The anticlines and synclines are expressed by:

1. Topography. In these young mountains the anticlines are topographic highs and the synclines are lows. In older and more eroded fold belts, such as the Appalachian Mountains, resistant beds preserved in the troughs of synclines cause them to be topographically higher than the anticlines.
2. Attitude of the beds. The dip azimuth can be determined from the image and the angle of dip can be estimated.
3. Outcrop patterns. Older beds crop out in the cores of anticlines and younger rocks in the troughs of synclines. The plunge is shown by the canoe-shaped outcrop patterns around the nose of folds.

A few faults are suggested on the image by aligned canyons cutting diagonally across the western part of the Guniz anticline. No displacement of beds is evident at the scale of the image and any faults are relatively minor. This agrees with the published geologic map. The salt plugs are similar in origin to those of the Gulf of Mexico region and elsewhere. Salt has a lower density than the overlying rocks and this buoyancy enables the salt to migrate upward as narrow cylinders (Figure 4.20) that penetrate the surrounding rocks. The rock fragments that litter the surface exposure of the plugs were picked up during the upward movement of the salt.

STEREO VIEWING OF LANDSAT IMAGES

Stereoscopic viewing of overlapping aerial photographs is possible because of the radial displacement due to topographic relief. The geometry of the Landsat multispectral scanner, and other scanner systems, causes displacement in the scan direction, but not in the flight line direction. The scanner system is directly over the center of each scan line, which means that the image base is zero in the flight direction and stereo viewing is not possible in this direction. There is sidelap between images acquired on adjacent Landsat orbit paths (Figure 4.7), and this provides some capability for stereo viewing.

Stereoscopic Geometry of Landsat Images

The altitude of Landsat 1 and 2 is nearly constant at 912 km. The sidelap of adjacent orbit paths increases from 14 percent at the equator to 85 percent at the 80° latitude. From this information the base-height ratios are calculated and listed in Table 4.8. In Chapter 2 it was pointed out that vertical exaggeration of the stereo model is related to the base-height ratio. The diagram in Figure 2.11 is used to determine vertical exaggeration factor for the different Landsat base-height ratios. The resulting vertical exaggeration factors are listed in Table 4.8 and plotted in Figure 4.21. Note that the maximum exaggeration factor is less than 1.3 in contrast to the typical 4.0 factor for stereo models of aerial photographs. Despite the minimal vertical exaggeration, stereo viewing is a valuable aid for interpreting Landsat images as shown by examples from Mexico and Greenland.

Peten Basin, Mexico, and Guatemala

The procedure for interpreting a Landsat stereo model is the same as that described earlier for aerial photography stereo models. The first step is to map drainage patterns which provide a geographic reference. Major rock units are defined and their boundaries mapped. Structural features such as fold axes, faults, lineaments, strike and dip are recorded. The major differences between interpreting Landsat and aerial photography stereo models are: (1) the typical 1:500,000 scale of Landsat images is an order of magnitude smaller than that of aerial photographs; and (2) the slight vertical exaggeration in Landsat stereo models causes topographic relief and structural attitudes to be approximately correct on the model.

The significance of the stereo pair of Landsat images shown in Figure 4.22 was pointed out by W. V. Trollinger (personal communication) of TGA, Denver. The 10-month time lapse in acquiring the adjacent images was caused by cloud cover and priorities for using the Landsat tape recorders. Note that the image on the right was acquired by Landsat 1 and the image on the left by Landsat 2. To derive maximum benefit from this discussion the reader should view Figure 4.22 with a pocket lens stereoscope, and compare his interpretation with the map and cross section in Figure 4.23. Structural features are shown with remarkable clarity on the stereo model especially considering the vegetation cover of this tropical area. Structure of the area is dominated by east-west trending symmetrical anticlines and synclines. On the stereo model the following features of the folds can be interpreted:

1. Anticlines may be distinguished from synclines by their topographic expression and by the attitude of the dipping strata.
2. Symmetry and relative amplitude of the folds may be determined.
3. Plunge of the folds is mapped by tracing beds that wrap around the plunging noses of the structures.

Some major faults trend parallel with the fold axes and others cut diagonally across the axes. On

TABLE 4.8
Stereoscopic geometry of Landsat images

Latitude, deg	Image sidelap, percent	Image base, km	Base-height ratio	Vertical exaggeration*
0	14.0	159	0.174	1.22
10	15.4	157	0.172	1.20
20	19.1	150	0.165	1.15
30	25.6	137	0.150	1.05
40	34.1	122	0.134	0.94
50	44.8	100	0.110	0.77
60	57.0	80	0.088	0.60
70	70.6	54	0.059	0.40
80	85.0	28	0.031	0.21

*From Figure 2.11.

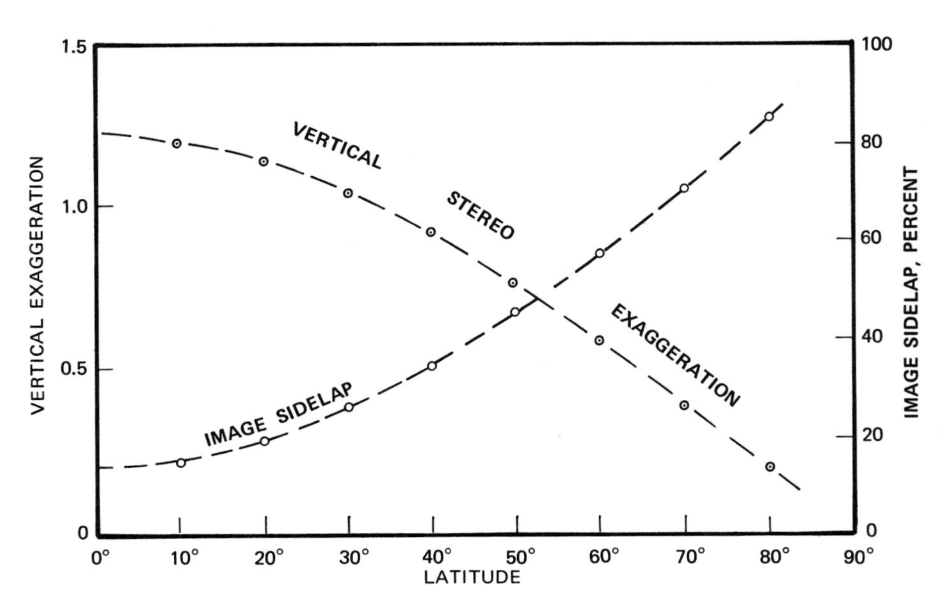

FIGURE 4.21
Image sidelap and vertical exaggeration of Landsat stereo pairs as a function of latitude.

A. LANDSAT 2317–15522, BAND 7, DECEMBER 5, 1975.

B. LANDSAT 1572–15554, BAND 7, FEBRUARY 15, 1974.

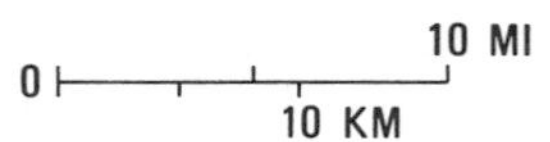

FIGURE 4.22
Stereo pair of Landsat images, Peten Basin, Mexico and Guatemala. At this latitude of 17° 30′N, the vertical exaggeration factor is 1.2.

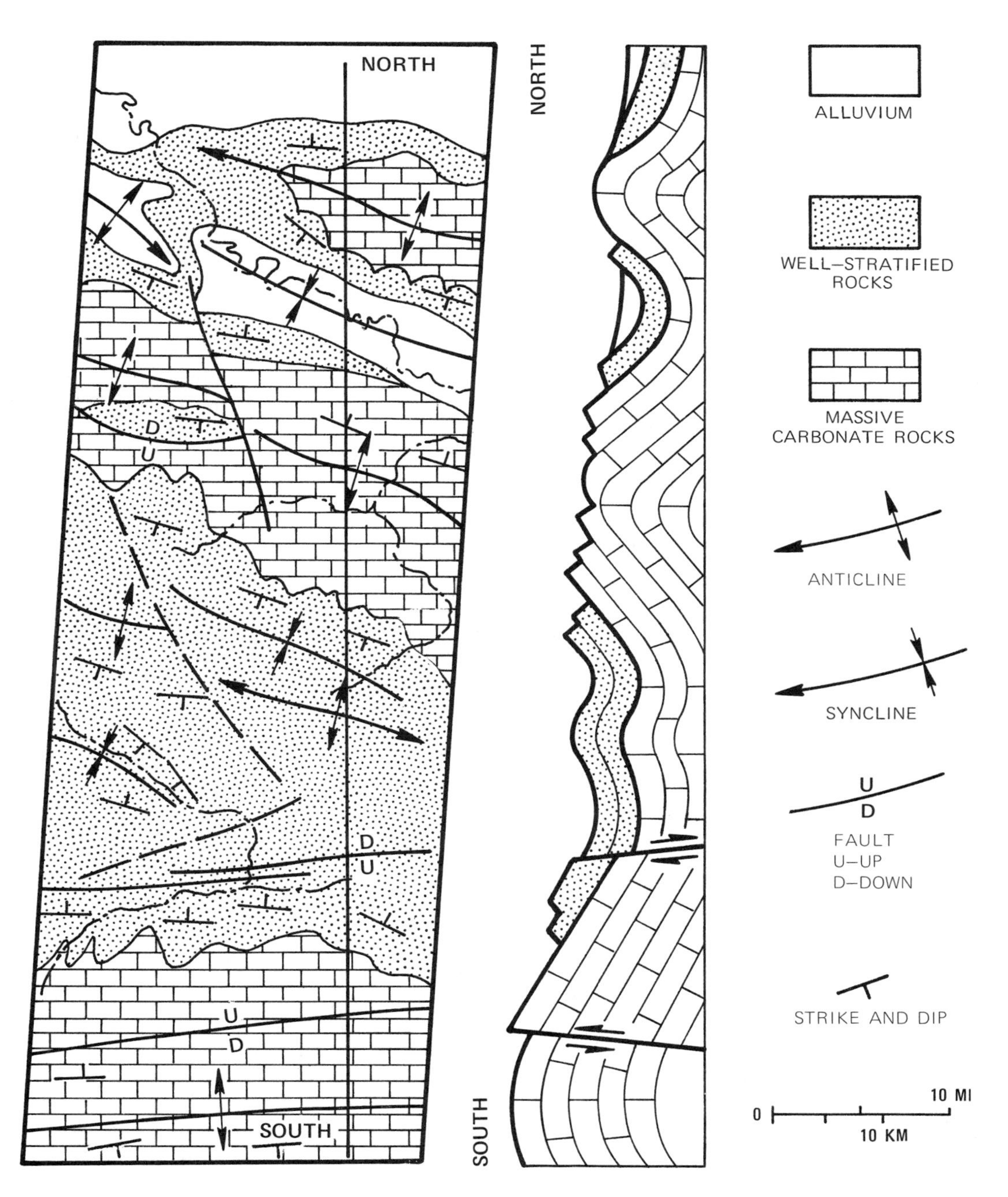

FIGURE 4.23
Generalized geologic interpretation map and cross section of Landsat stereo pair in Peten Basin. Modified from unpublished interpretation by W. V. Trollinger, TGA, Denver.

a single image many of the faults could only be interpreted as possible linear features. On the stereo model, however, the topographic expression and stratigraphic offset can be estimated for many of the faults. The major fault in the south part of the image is particularly well expressed. On a single image this fault appears as a bright lineament against a darker background. Stereo viewing, however, shows that the bright lineament is a steep south-facing scarp illuminated by the sun. The scarp is almost surely a steeply dipping fault with the south side downthrown relative to the north side. Approximately 15 km to the north is a second east-west trending lineament. The eastern portion of this lineament is expressed as a dark linear feature that is interpreted as a shadow on a north-facing scarp. In the northern part of the area, several northwest-trending faults cause lateral offset of stratigraphic units and may have a component of strike-slip displacement.

The sparse stratigraphic information available for this region indicates that the outcrops are marine strata of Cretaceous and Tertiary age. On the Landsat images three major rock units can be distinguished. The oldest unit crops out in the cores of the major anticlines. These rocks are poorly stratified and are characterized by a distinctive fine textured, mottled pattern of irregular dark spots against a bright background. The unit is interpreted as a sequence of massive carbonate rocks with karst topography. Karst topography forms by solution of carbonate rocks in humid climates and is characterized by steep-sided, irregular depressions. The shadows formed in these depressions cause the dark mottled pattern against the bright background of high vegetation reflectance on the Landsat band 7 images of Figure 4.22. Overlying the carbonate unit is a sequence of well stratified rocks that crops out on the flanks of the anticlines and in the troughs of the synclines. The stratification is expressed by variations in tone and resistance to weathering of the beds. Most of the strike and dip information on the map (Figure 4.23) is interpreted from outcrops of this well-stratified unit. Some local areas of karst topography may be formed on carbonate beds interstratified with the unit. The third rock unit consists of featureless alluvial deposits in the stream valleys of the northern part of the area. According to W. V. Trollinger (personal communication) geologic interpretation of Landsat stereo models has been a major factor in the successful oil exploration campaign in the Peten Basin.

Western Greenland

At the 71°N latitude of Figure 4.24 there is 72 percent sidelap, but the decreased base-height ratio provides a vertical exaggeration factor of only 0.40 (Figure 4.21). The snow cover of the Greenland images eliminates tonal differences and finer topographic details of the crystalline bedrock. A major north-trending linear valley across the west end of Nugssuaq Peninsula apparently marks a fault that is enhanced on the stereo model. The depression trending west across Ubekendt Island is a probable fault.

Future Requirements for Stereo Imagery

The limited stereo coverage with minimal vertical exaggeration provided by Landsat 1 and 2 has proven useful for interpretation. Worldwide stereo coverage with Landsat imagery having optimum vertical exaggeration should be very useful. Unfortunately, areas between latitudes 55°N and 55°S are not completely covered by stereo Landsat images because the sidelap of images from adjacent orbits is less than 50 percent (Figure 2.11). Future satellite systems are under consideration that will provide complete stereo coverage with optimum vertical exaggeration.

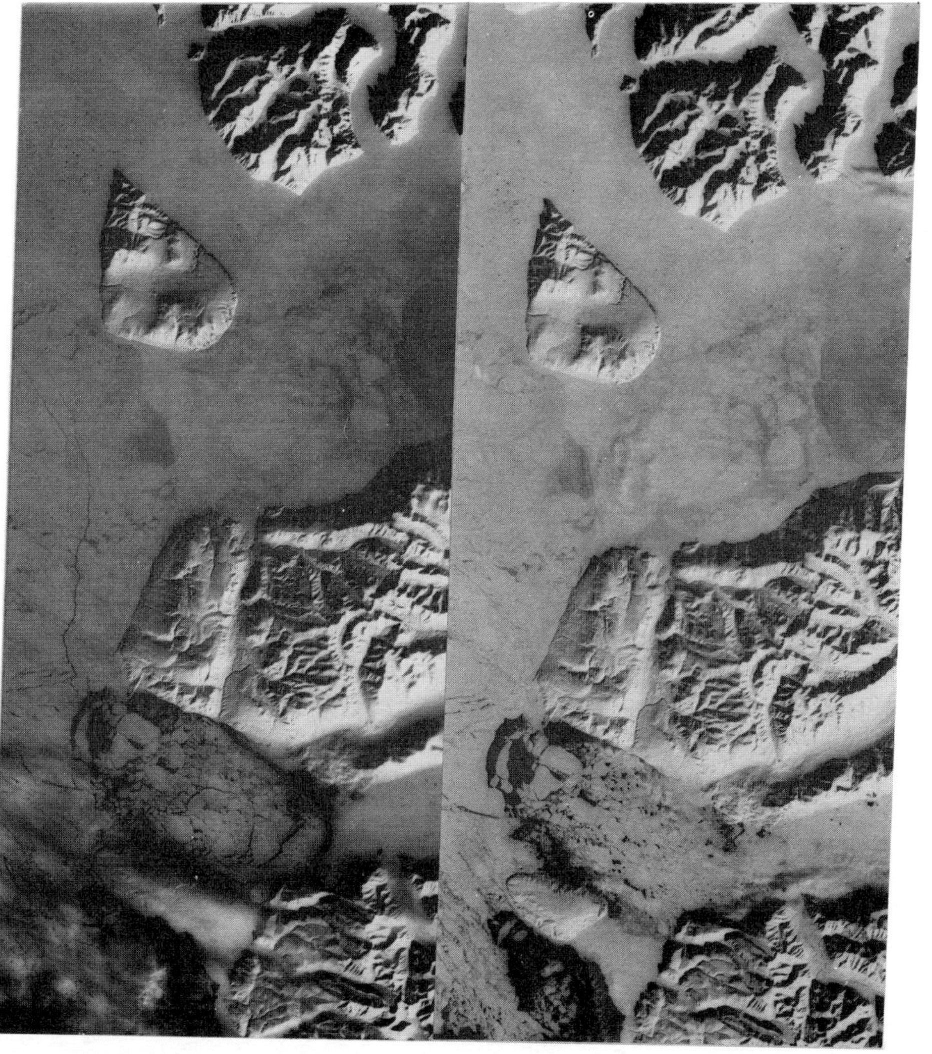

A. LANDSAT 1241–14591, BAND 7, MARCH 21, 1973.

B. LANDSAT 1240–14591, BAND 7, MARCH 20, 1973.

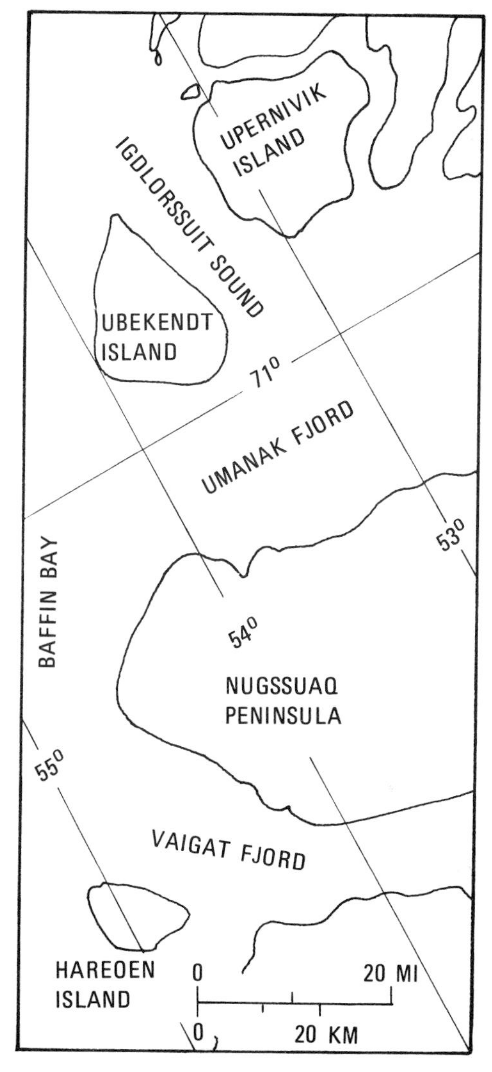

C. LOCATION MAP.

FIGURE 4.24
Stereo pair of Landsat images, west coast of Greenland. At this 71°N latitude, the vertical exaggeration factor is 0.4.

SEASONAL INFLUENCE ON IMAGERY

Many interpreters believe that a single set of good aerial photographs is adequate for mapping an area. The repetitive coverage of Landsat has demonstrated that images acquired at different seasons may enhance different geologic features. In the past occasional aerial photographs acquired in winter have indicated that snow cover may enhance some structural features.

Stratigraphic Mapping in South Africa

In the western Transvaal Province of the Republic of South Africa, Grootenboer (1973) compared an image acquired at the end of the dry winter season with one acquired in the middle of the summer rainfall season (Figure 4.25). The area is a grass-covered plateau where the more resistant rock units form conspicuous ridges.

Geologic Setting The geologic map of Figure 4.26 was interpreted from a 1:500,000 scale IR color composite of the rainy season image using the viewer shown in Figure 4.5. The rock types distinguished on the image were identified by referring to the maps of Grootenboer (1973). The northern two-thirds of the image covers the southwest flank of the Transvaal Basin where the strata dip gently north. A northeast-trending anticline separates the Transvaal Basin from the more complex Potchefstroom Basin in the southeast corner of the image (Figure 4.25). The oldest rocks in the image area are of Archean and Proterozoic age exposed along the axis of the anticline. The Archean rocks are predominantly granite. Proterozoic rocks are the Witwatersrand Quartzite with minor shale and the Ventersdorp Andesite and interbedded sediments. The overlying Transvaal System consists of a thin basal clastic unit (Black Reef Series) overlain by massive dolomitic limestone (Dolomite Series), which is overlain by alternating quartzites and shales with some volcanic layers (Pretoria Series). Intrusive diabase sills are abundant near the top of the Pretoria Series.

The Bushveld igneous complex that occupies the center of the Transvaal Basin occurs in the northern part of the area. The Layered Sequence at the base consists of mafic igneous rocks and is overlain by the Bushveld Granite. At the north edge of the image, the Bushveld Complex is intruded by the Pilansberg Complex, a circular ring structure of alkaline intrusive rocks. Small scattered patches of flat-lying, deeply weathered Karroo sediments (tillite, sandstone, and shale) occur throughout the area. There are numerous dolerite dikes and sills of post-Karroo age.

Most of the area is covered by residual soil resulting in poor outcrop conditions. The only significant outcrops are the quartzites of the Witwatersrand and Transvaal Systems, the Pilansberg Complex, and scattered inselbergs of the Bushveld Complex.

Comparison of Images The dramatic differences between the dry and rainy season images are even more striking on color composites than on the band 5 black-and-white illustrations of Figure 4.25. On all four MSS bands the rainy season image is superior for geologic mapping. When the dry season image (Figure 4.25A) was acquired the area was covered with dry, brown grass, the indigenous vegetation was leafless, and the corn fields were fallow. Black patches mark areas of recent burning. Slight tonal variations enable recognition of the major stratigraphic units to a degree comparable to that on published 1:1,000,000 scale geologic maps (Grootenboer, 1973). The image of Figure 4.25B was acquired at the height of the summer rainy season when the area was covered by green grassland with the vegetation in full leaf. The strong tonal variations are directly related to surface lithology, particularly in the area underlain by the Transvaal System and the Bushveld Complex. Locally the lithologic detail on the image is comparable to that on 1:250,000 scale published geologic maps of the area.

Of particular interest are the seven zones of tonal variations within the outcrop of the Transvaal Dolomite. Field checking by Grootenboer, Eriksson, and Truswell (1973) established that the four darker zones correspond to dark chert-free dolomite. The three lighter zones correspond to lighter colored dolomite with abundant chert. During the previous 90 years of continuous geologic investigation in the area, no such stratigraphic subdivisions had been recognized in the Transvaal Dolomite.

Grootenboer (1973) attributes the superiority of the rainy season image to two factors: (1) Atmospheric haze caused by windblown dust produces severe light scattering on the dry season image. During the wet summer season rainfall removes the dust, producing a clear atmosphere and good image contrast. (2) The higher soil moisture content enhances tonal and color differences. Also the surface dust layer is washed away.

Mapping of Structure and Lineaments in Arctic Canada

Seasonal influence on structural mapping is demonstrated by images of Arctic Canada near Bathurst Inlet. On the early summer image (Figure 4.27A) there is no snow, most of the lakes are thawed, and vegetation is vigorous as shown by the bright tones on band 7. Very little geology can be interpreted. Geologic structure of the area is much more clearly delineated on the snow-covered early spring image with lower sun elevation (Figure 4.27B). A major fold, outlined by strike ridges of sedimentary rocks, is surrounded by highly fractured, unstratified crystalline basement rocks. The fold is a syncline or structural basin with younger strata preserved in the center. Published geologic maps and the report of Gregory and Moore (1976, Figure 5) show that the stratified rocks in the syncline are argillites, sandstones, and quartzites of Proterozoic age that overlie older crystalline rocks of Archean age.

The major lineament trending northwest across

FIGURE 4.25
Seasonal effect on Landsat images of western Transvaal Province, Republic of South Africa.

A. END OF DRY SEASON, SEPTEMBER 11, 1972.
LANDSAT 1050–07355, BAND 5.

B. MIDDLE OF RAINY SEASON, DECEMBER 28, 1972.
LANDSAT 1158–07363, BAND 5.

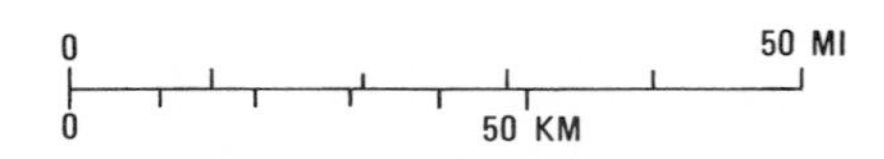

the image is shown on published geologic maps as a major fault. North of the fault is a small syncline of Proterozoic strata. Also north of the fault at the east margin of Figure 4.27B is a prominent ridge that corresponds to a gabbro dike on the published maps. At the west boundary of the image a distinct circular drainage anomaly 8 km in diameter marks a possible ring dike or an igneous plug. These geologic features are mappable only on the early spring image with snow cover, which eliminates the distracting detail of the lakes and vegetation differences. The lower sun elevation also enhances geologic features on the early spring image.

Images acquired at three seasons (Figure 4.28) illustrate the importance of sun elevation and azimuth for mapping lineaments. This area, which is directly west of the area shown in Figure 4.27, is characterized by low relief and numerous lakes, many of which are still frozen on the early summer image (Figure 4.28A). A few lineaments are indicated by straight drainage segments and aligned lakes. The lineament indicated at L in Figure 4.28A is defined by the contrast between the dark tone of the water and light tone of the vegetation. This feature cannot be detected on the winter images when the drainage and lakes are obscured by snow cover. Because feature L trends parallel with the sun azimuth, it is not enhanced by highlights and shadows. Other lineaments trending at

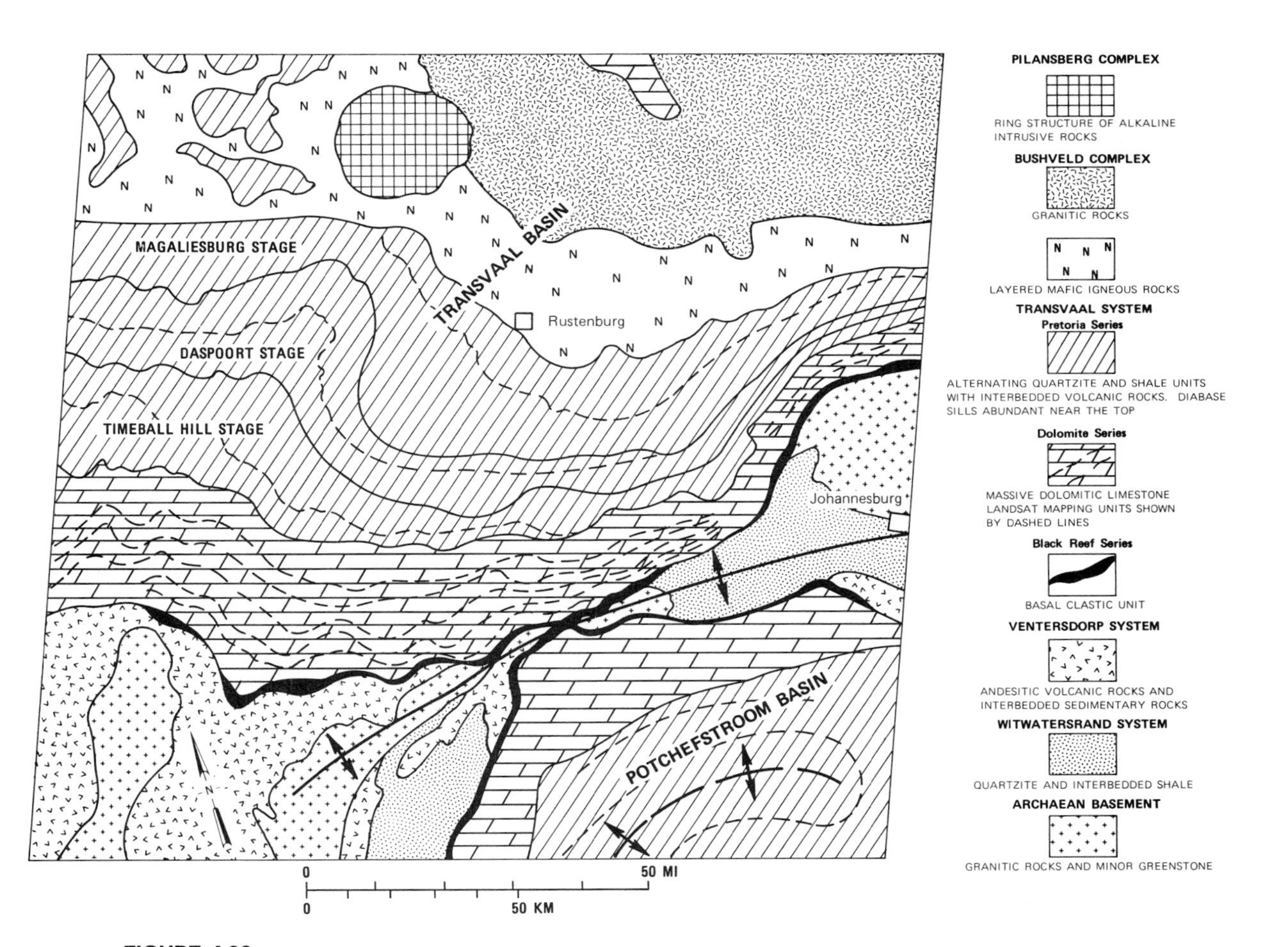

FIGURE 4.26
Generalized geologic map of south flank of Transvaal Basin, interpreted from Landsat image acquired in rainy season. From Grootenboer (1973, Figure 1).

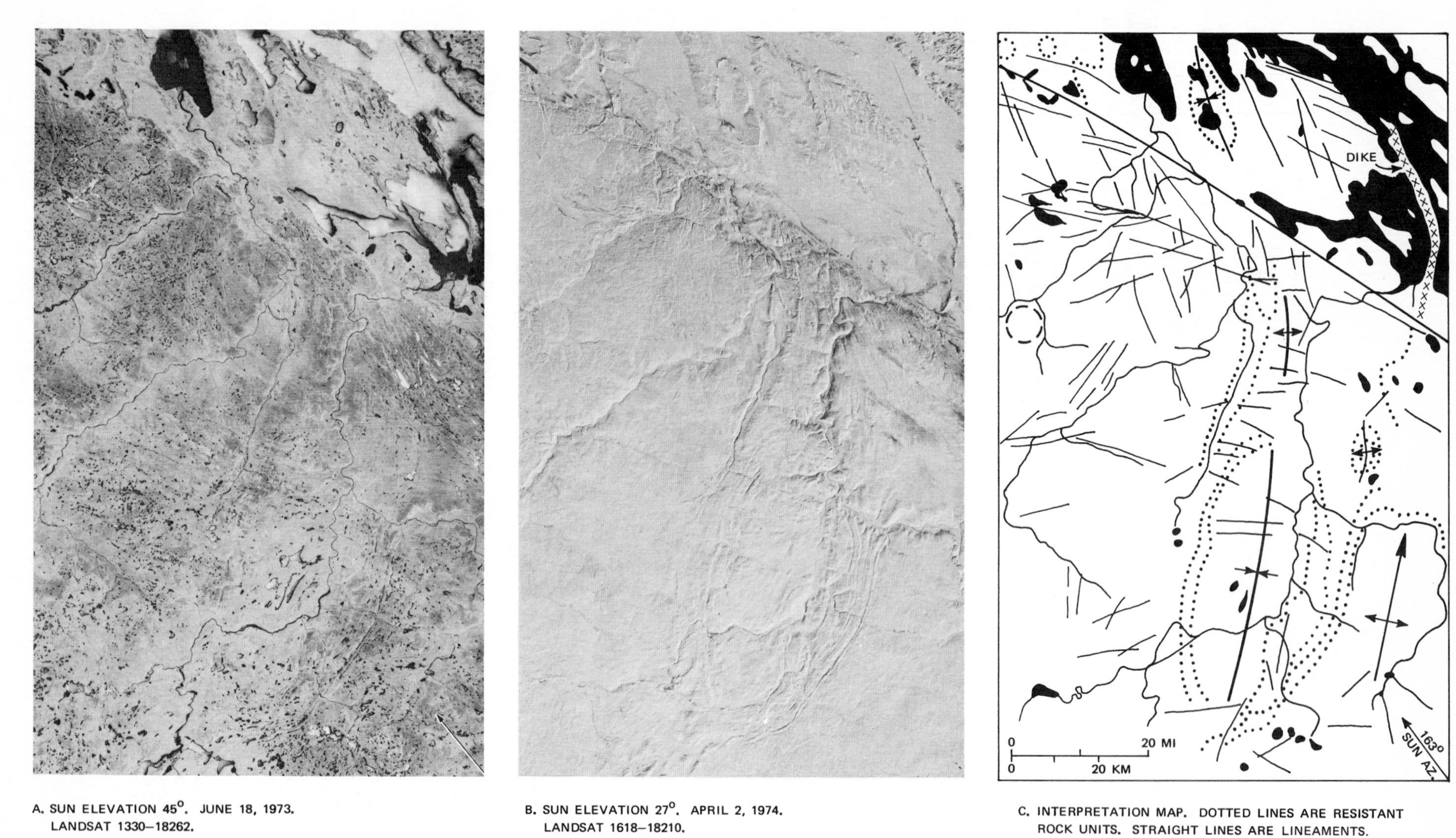

A. SUN ELEVATION 45°. JUNE 18, 1973.
LANDSAT 1330–18262.

B. SUN ELEVATION 27°. APRIL 2, 1974.
LANDSAT 1618–18210.

C. INTERPRETATION MAP. DOTTED LINES ARE RESISTANT ROCK UNITS. STRAIGHT LINES ARE LINEAMENTS.

FIGURE 4.27
Seasonal effects on Landsat band 7 images of Bathurst Inlet, Arctic Canada. Map reprinted with permission from A. F. Gregory and H. D. Moore, "Recent Advances in Remote Sensing from Space," copyright 1976, Figure 5, Pergamon Press.

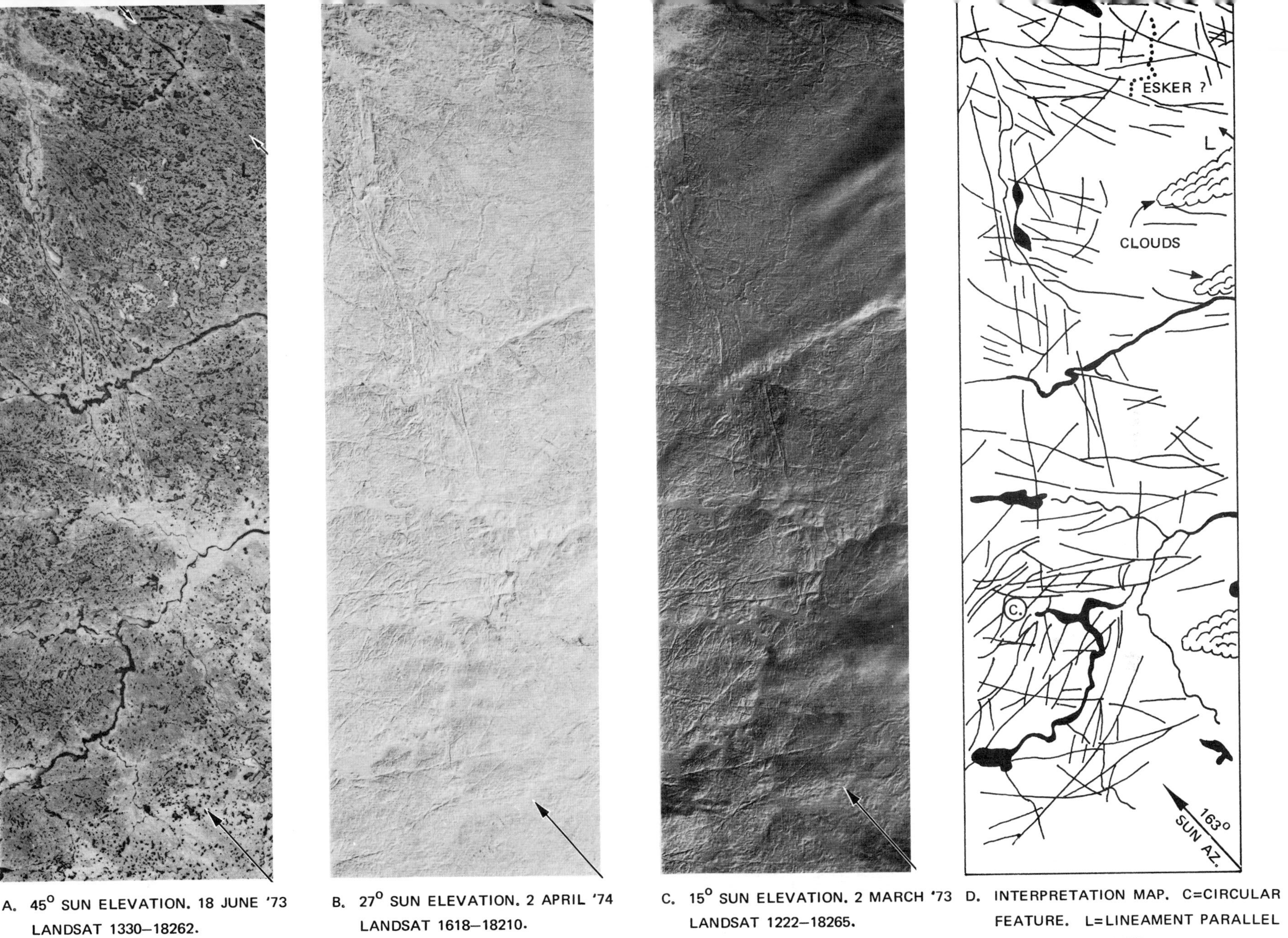

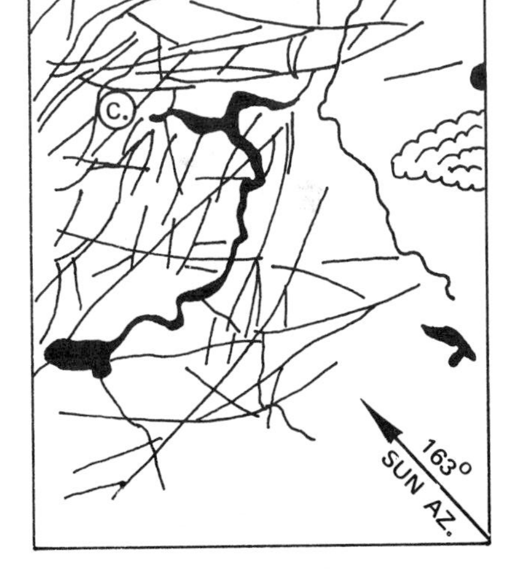

A. 45° SUN ELEVATION. 18 JUNE '73
LANDSAT 1330–18262.

B. 27° SUN ELEVATION. 2 APRIL '74
LANDSAT 1618–18210.

C. 15° SUN ELEVATION. 2 MARCH '73
LANDSAT 1222–18265.

D. INTERPRETATION MAP. C=CIRCULAR FEATURE. L=LINEAMENT PARALLEL WITH SUN AZIMUTH.

FIGURE 4.28
Effects of sun elevation and snow cover on the expression of lineaments. Landsat band 7 images near Bathurst Inlet, Arctic Canada.

an angle to the sun azimuth are much more pronounced on the images acquired at lower sun elevations (Figures 4.28B and C). The 15° sun elevation of the March 2 image (Figure 4.28C) causes maximum enhancement of suitably oriented features. Several Landsat investigators have reported that lineaments oriented parallel with the sun azimuth are difficult to detect. Features oriented parallel with the nearly east-west scan lines on Landsat images also tend to be obscured. An esker appears in the northeast parts of Figures 4.28B and C. This is a sinuous ridge of sand and gravel deposited by a stream flowing along the base of a continental glacier.

PLATE TECTONIC INTERPRETATION OF AFAR TRIANGLE, ETHIOPIA

The regional coverage of Landsat is ideal for studying plate tectonic features exposed on the continents. The imagery also provides information for analyzing regions where published information is lacking or unavailable. The Afar triangle of Ethiopia is the site of the triple junction where the spreading axes of the Red Sea, Gulf of Aden, and Ethiopian rift converge (Figure 4.29, index map). The Arabia, Nubia, and Somalia plates are separating along these spreading axes. Much of the Afar triangle is below sea level, but it is separated from the Red Sea and Gulf of Aden by the Danakil and Aisha horsts of crystalline continental crust. The region is inhospitable and of difficult access, but excellent Landsat images are available.

On the image of Figure 4.30 Lake Abbe marks the present position of the triple junction which has migrated through time. The Gulf of Tadjura is the western end of the Gulf of Aden. The light-toned areas are recent deposits of clay, silt, and evaporite in the depressions. Basalt (dark tone) and andesite (intermediate tone) of Tertiary age form the outcrops. Mount Marsabit is a recent volcano with associated lava flows. Structure of the area is dominated by normal faults that form the boundaries of numerous horsts and grabens. The faults are accentuated by shadows on the image that indicate the topographic relief and the sense of throw. In the southwest part of the image the north-northeast-trending faults belong to the Ethiopian rift system. The northwest-trending faults are parallel with the Red Sea spreading axis. The prominent west-trending faults in the southeast are parallel with the Gulf of Aden spreading axis. Except for an apparent strike-slip fault northwest of Lake Asal, there is no suggestion of transform faulting. The sinuous Gawa graben in the north part of the image is a departure from the regional pattern.

Landsat images have been used effectively to understand patterns of plate tectonics in many other regions. A good example is the analysis by Molnar and Taponnier (1977) of the collision resulting from the northward drift of the Indian subcontinent toward Eurasia. Following the initial collision of the two crustal plates between 40 and 60 million years ago, India has continued to move northward another 2,000 km with respect to Eurasia. A major problem is to account for the vast land area that was displaced by the convergence of the plates. The solution of the problem is complicated by two factors: (1) the geology of Eurasia is complex and appears to present a chaotic jumble of landforms; and (2) the geology is poorly known. Landsat images aided Molnar and Taponnier in resolving these complications. When they viewed Eurasia as a whole on Landsat images the geology seemed to fall into a simple coherent pattern attributable to the collision. Structural information that was previously unknown, to most geologists at least, was interpreted from the images. Major strike-slip faults were recognized and for some the sense of displacement was evident. The Landsat interpretation was combined with earthquake data to determine the sense of motion associated with the principal faults. This analysis indicates to Molnar and Taponnier that east-west trending faults with left-lateral displacement are predominant. Thus China has been displaced laterally eastward out of the way of India. This interpretation is reasonable, because material displaced westward would encounter the resistance of the Eurasian land mass. Eastward motion, however, is easily accommodated by thrusting of China over the oceanic plates along the margins of the Pacific.

ADVANTAGES OF LANDSAT IMAGES

The advantages of Landsat images are summarized as follows:

1. Cloud-free images are available for most of the world with no political or security restrictions.
2. The low to intermediate sun angle enhances many subtle geologic features.
3. Repetitive coverage provides images at different seasons and illumination conditions.
4. The cost is low.
5. IR color composite images are available for many of the scenes. With suitable equipment color composites may be made for any image.
6. Broad synoptic coverage of each scene under uniform illumination aids in the recognition of major features. Mosaics extend this coverage.
7. Image distortion is negligible.
8. Images are available in digital format suitable for computer processing.
9. Limited stereo coverage is available.

FUTURE LANDSAT MISSIONS

In early 1978 NASA plans to launch Landsat C, which will be designated Landsat 3 when it becomes operational. The sensors selected for Landsat C are a modified multispectral scanner system (MSS), a two-camera return beam vidicon (RBV), and a data collection system (DCS). The MSS will be the same as the systems on Landsat 1 and 2 with the addition of a fifth band operating in the thermal IR spectral region from 10.4 to 12.6 μm. The thermal band, designated as band 8, has two detectors with a ground resolution cell of 238 m, which is three times larger than that of the detectors used in bands 4 through 7. Imagery from all five MSS bands will be acquired during daylight hours. In addition, band 8 images will also be acquired at night and during periods of very low sun elevation.

The RBV of Landsat C consists of two identical cameras operating in the 0.5 to 0.75 μm region with a ground resolution cell of 40 m. The image produced by the RBV camera lens and shutter system is stored on a photosensitive surface which is electronically scanned to produce a video signal for transmission to the receiving stations. The two cameras are aligned to view adjacent 98 km square ground scenes with 14 km of sidelap. This produces a scene-pair that measures 98 km in the orbit track direction and 182 km in the across-track direction. Two successive scene-pairs will cover the area of each MSS scene.

In support of Landsat C an improved image processing system is also planned at the EROS Data Center. Magnetic tapes of image data recorded at the Goldstone and Alaska receiving stations will go to Goddard Space Flight Center for preprocessing, together with data recorded at the Goddard receiving station. The data tapes will then be forwarded to the EROS Data Center where they will be used to produce master films

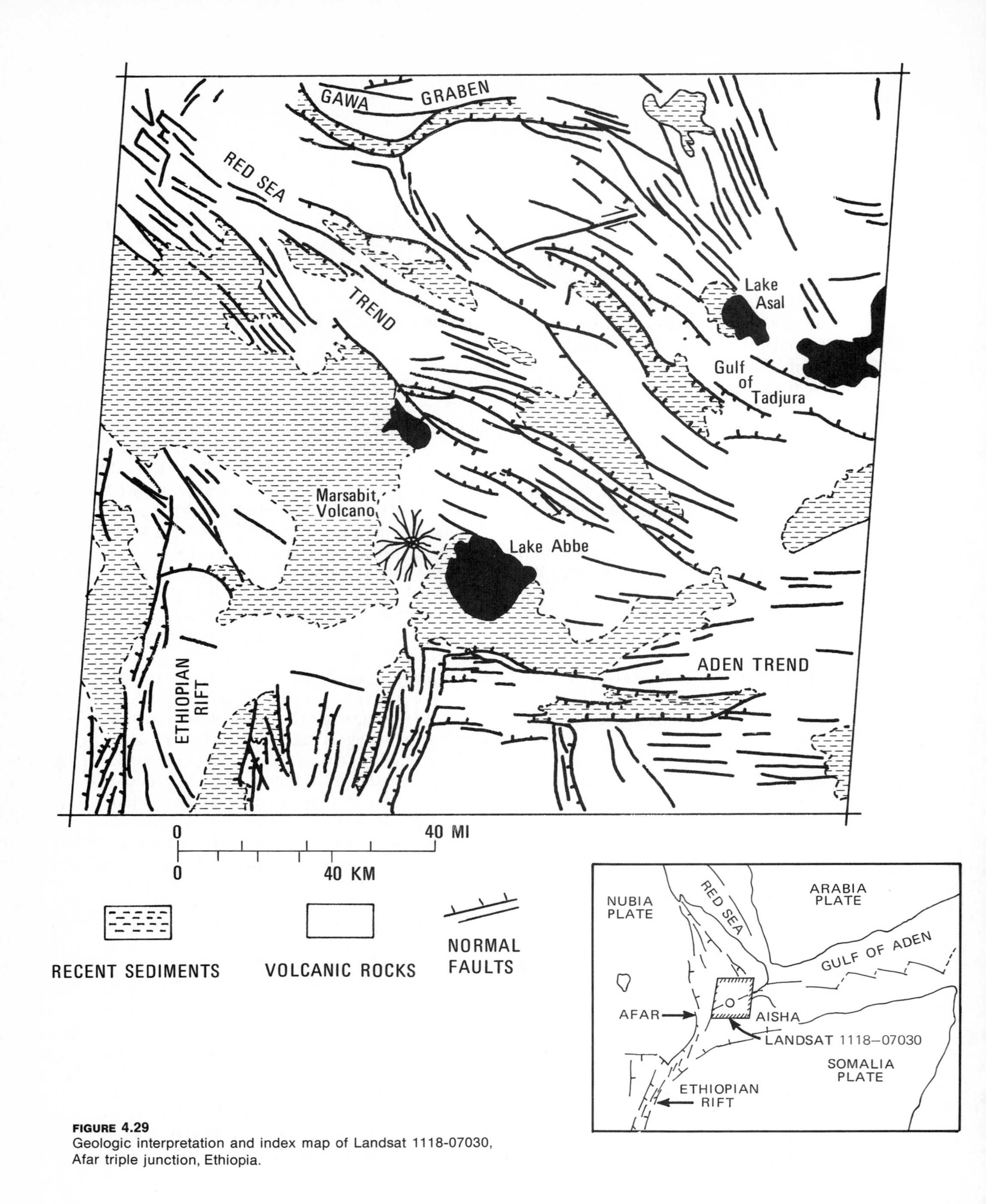

FIGURE 4.29
Geologic interpretation and index map of Landsat 1118-07030, Afar triple junction, Ethiopia.

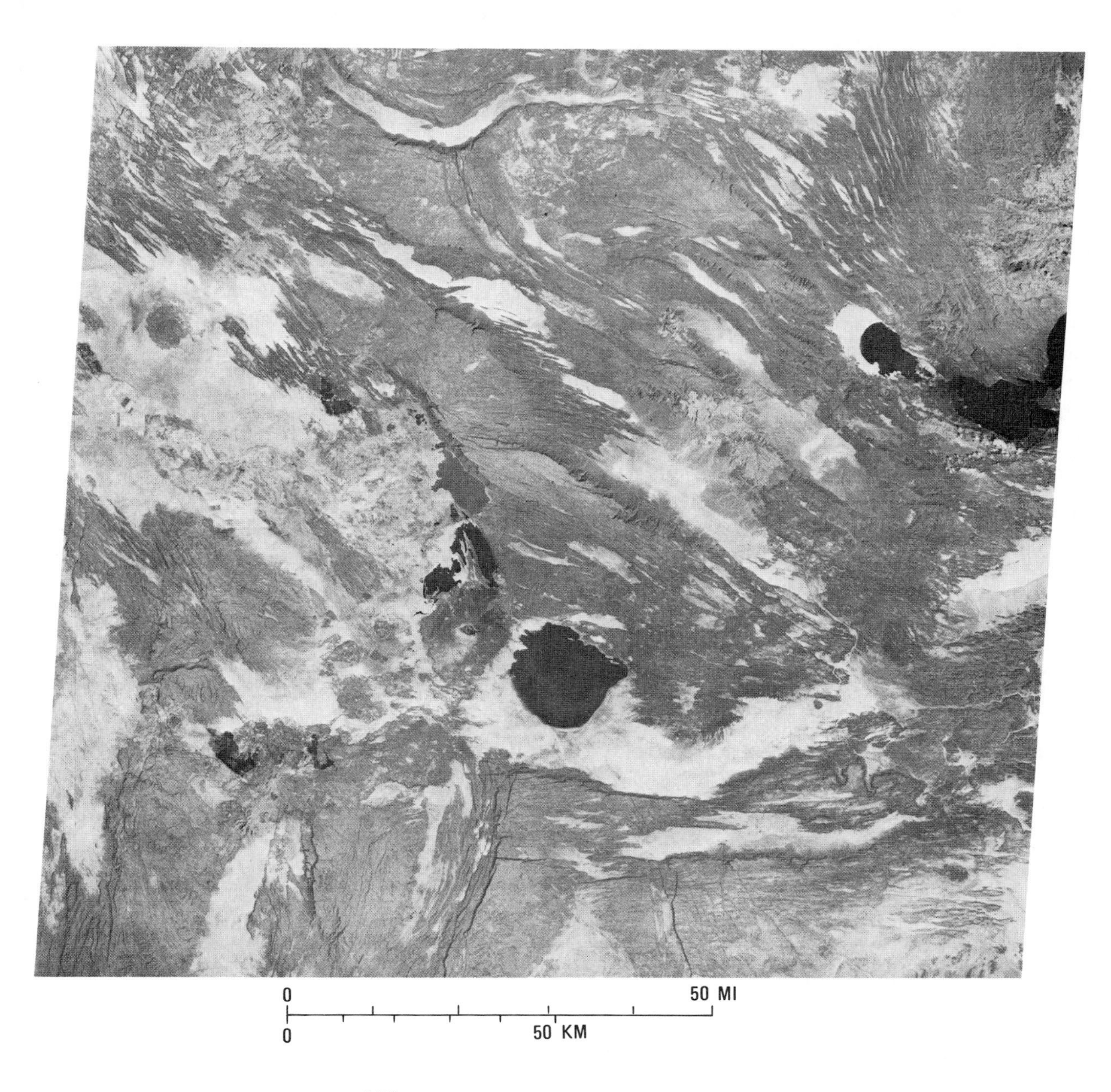

FIGURE 4.30
Landsat 1118-07030, band 6, of Afar triple junction, Ethiopia. Image acquired November 18, 1972.

and CCTs. It is anticipated that the new system will speed up delivery of new images and that image quality will be improved.

Under the title of Landsat Follow-on, or Landsat D, NASA is planning missions to take place in the 1980s. Multispectral imaging systems with additional spectral bands and higher spatial resolution are being considered. These improvements would be especially useful for mineral exploration, as discussed in Chapter 8. Many users have strongly recommended that images be acquired with overlap for stereo viewing. The examples shown here indicate that complete stereo coverage with adequate vertical exaggeration would be valuable. This and other image requirements are discussed more fully in the final chapter of this book.

COMMENTS

The Landsat program is a major advance in the field of remote sensing. The worldwide availability of high-quality, low-cost imagery on a repetitive basis has proven valuable for many applications. In addition to the applications discussed in this chapter, Landsat imagery is valuable for resource exploration, environmental monitoring, land-use analysis, and evaluation of natural hazards. These and other applications are illustrated in Chapters 8, 9, and 10. Landsat is an experimental program, and much has been learned about possible improvements for the design of future generations of satellite imaging systems.

Another major contribution of Landsat is the impetus it has given to digital image processing, described in Chapter 7. The availability of low-cost multispectral image data in digital form has encouraged the application and development of computer methods for image processing that are increasing the usefulness of the data for interpreters in many disciplines.

REFERENCES

Bidwell, T. C. and C. A. Mitchell, 1975, Author index to published ERTS-1 reports: U.S. Department of Commerce, National Technical Information Service, Document PB 248294. Springfield, Va.

Bishop, B. C., 1976, Landsat looks at hometown earth: National Geographic, v. 150, p. 140–147.

British Petroleum Company, 1956, Geological maps and sections of southwest Persia: Proceedings of 20th International Geological Congress, Mexico, D. F.

Gregory, A. F. and H. D. Moore, 1976, Recent advances in geologic applications of remote sensing from space: Proceedings of 24th International Astronautical Congress, p. 153–170. Pergamon Press, New York.

Grootenboer, J., 1973, The influence of seasonal factors on the recognition of surface lithologies from ERTS imagery of the western Transvaal: Third ERTS Symposium, NASA SP-351, v. 1, p. 643–655.

———, K. Ericksson, and J. Truswell, 1973, Stratigraphic subdivision of the Transvaal Dolomite from ERTS imagery: Third ERTS Symposium, NASA SP-351, v. 1, p. 657–664.

Hobbs, W. H., 1904, Lineaments of the Atlantic border region: Geological Society of America Bulletin, v. 15, p. 483–506.

———, 1912, Earth features and their meaning: an introduction to geology for the student and general reader: Macmillan Co., New York.

Isachsen, Y. W., R. H. Fakundiny, and S. W. Forster, 1973, Evaluation of ERTS-1 imagery for geological sensing over the diverse geological terranes of New York State: Symposium on Significant Results Obtained from ERTS-1, NASA SP-327, v. 1, p. 223–230.

———, 1974, Assessment of ERTS-1 imagery as a tool for regional geological analysis in New York state: National Technical Information Service Document E74-10363, Springfield, Va.

Lamar, D. L. and P. M. Merifield, 1975, Application of Skylab and ERTS imagery to fault tectonics and earthquake hazards of Peninsular Ranges, southwestern California: California Earth Science Corporation, Technical Report 75-2, Santa Monica, Calif.

Molnar, P. and P. Tapponnier, 1977, The collision between India and Eurasia: Scientific American, v. 236, p. 30–41.

NASA, 1976, Landsat data users handbook: Goddard Space Flight Center, Document no. 76SDS-4258, Greenbelt, Md.

Nettleton, L. L., 1976, Gravity and magnetics in oil prospecting: McGraw-Hill Inc., New York.

O'Leary, D. W., J. D. Freidman, and H. A. Pohn, 1976, Lineaments, linear, lineation—some proposed new standards for old terms: Geological Society of America Bulletin, v. 87, p. 1463–1469.

Rowan, L. C. and P. H. Wetlaufer, 1975, Iron-absorption band analysis for the discrimination of iron-rich zones: U. S. Geological Survey, Type III Final Report, Contract S-70243-AG.

Sabins, F. F., 1973, Geologic interpretation of radar and space imagery of California (abstract): American Association Petroleum Geologists Bulletin, v. 57, p. 802.

———, 1977, Worldwide retrieval and indexing of Landsat images *in* Woll, T. W. and W. A. Fischer, eds., Proceedings of the first annual W. T. Pecora memorial symposium, October, 1975: U.S. Geological Survey Professional Paper 1015, p. 307–317.

Short, N. M., P. D. Lowman, S. C. Freden, and W. A. Finch, 1976, Mission to earth, Landsat views the world: NASA SP-360, U.S. Government Printing Office, Stock No. 033-000-00659-4, Washington, D.C.

Slaney, V. R., 1976, Landsat imagery—a Canadian listing: Geological Survey of Canada Open File Report 386.

Stewart, J. H., G. W. Walker, and F. J. Kleinhampl, 1975, Oregon-Nevada lineament: Geology, v. 3, p. 256–268.

Stewart, J. H., W. J. Moore, and I. Zietz, 1976, East-west patterns of Cenozoic igneous rocks, aeromagnetic anomalies, and mineral deposits, Nevada and Utah: Geological Society of America Bulletin, v. 88, p. 67–77.

Williams, R. S. and W. D. Carter, ed., 1976, ERTS-1, a new window on our planet; U.S. Geological Survey Professional Paper 929.

ADDITIONAL READING

Otterman, J., P. D. Lowman, and V. V. Salmonson, 1976, Surveying earth resources by remote sensing from satellites: Geophysical Surveys, v. 2, p. 431–467.

5
THERMAL INFRARED IMAGERY

All matter radiates energy at thermal IR wavelengths both day and night. The ability to detect and record this thermal radiation in image form at night removes the cover of darkness and has obvious reconnaissance applications. Accordingly, the early development of thermal IR imaging technology was funded by government agencies, beginning in the 1950s, and was classified for security purposes. Military interpreters recognized that terrain factors greatly influenced the background against which strategic targets were displayed. The potential for terrain mapping by thermal IR systems created interest in the civilian community and in the mid-1960s some manufacturers received approval to acquire imagery with the classified systems for civilian clients. In 1968 the government declassified systems that did not exceed certain standards for spatial resolution and temperature sensitivity. Today excellent scanner systems are available for unrestricted use. The term thermography has been suggested as a replacement for thermal IR imagery but is *not* adopted in this book. Thermography has long been associated with medical applications of thermal IR imagery; any change in the accepted use of this term serves no purpose and would be confusing.

THERMAL PROCESSES AND PROPERTIES

In order to interpret images acquired in the thermal IR spectral region, the basic physical processes that control the interaction between thermal energy and matter must be understood. The thermal properties of matter that determine the rate and intensity of the interaction also must be understood.

Heat, Temperature, and Radiant Flux

Heat is the kinetic energy of the random motion of particles of matter. The random motion causes particles to collide resulting in changes of energy state and the emission of electromagnetic radiation from the surface of materials. The internal, or kinetic, heat energy of matter is thus converted into radiant energy. The amount of heat is measured in calories. A *calorie* is the amount of heat required to raise the temperature of 1 gm of water 1°C. *Temperature* is a measure of the concentration of heat. On the Celsius scale 0° and 100°C are the temperatures of melting ice and boiling water, respectively. On the Kelvin, or absolute, temperature scale 0°K is absolute zero, at which point all molecular motion ceases. The Kelvin and Celsius scales correlate as follows: 0°C = 273°K; 100°C = 373°K. The electromagnetic energy radiated from a source is called *radiant flux, F,* and is measured in watts · cm^{-2}.

The concentration of kinetic heat of a body of material may be called the *kinetic temperature,* T_{kin} and is measured with a thermometer placed in direct contact with the material. The concentration of the radiant flux of the material may be called the *radiant temperature,* T_{rad}. Radiant temperature may be measured remotely by devices that detect electromagnetic radiation in the thermal IR wavelength region. The radiant temperature of materials is less than the kinetic temperature because of a thermal property called emissivity, which is defined later in the section on Thermal Properties of Materials.

Heat Transfer

Heat energy is transferred from one place to another by three mechanisms.

1. *Conduction* is the transfer of heat through a material by molecular interaction. The transfer of heat through a pan to cook food is one example.
2. *Convection* is the transfer of heat through the physical movement of heated matter. The circulation of heated water and air are examples of convection.
3. *Radiation* transfers heat in the form of electromagnetic waves. Heat from the sun reaches the earth by radiation. In contrast to conduction and convection, which can only transfer heat through matter, radiation can transfer heat through a vacuum.

Materials at the surface of the earth receive thermal energy primarily by radiation from the sun and to a lesser extent by conduction from the interior of the earth. There are daily and annual cyclic variations in the duration and intensity of solar heat. Energy from the interior of the earth is called *geothermal energy* and is transferred primarily by conduction. The flow rate of geothermal energy is relatively constant at any locality, although there are regional variations. Locally the energy is transferred by convection at hot springs and volcanoes. The regional heat flow patterns may be altered by geologic features such as salt domes and faults.

IR Region of the Electromagnetic Spectrum

The IR region is generally defined as that portion of the electromagnetic spectrum ranging in wavelength from 0.7 to 300 μm. The terms short, middle, and long have been used to subdivide the IR region but are *not* used here because of confusion about the boundaries. In Chapter 1 the reflected IR band and the thermal IR band were defined. The *reflected IR region* ranges from wavelengths of 0.7 to 3 μm and includes the photographic IR band (0.7 to 0.9 μm) that may be detected by IR-sensitive film. On both IR color photographs and Landsat color composite images the red signature records IR energy that is reflected predominantly by vegetation and has no thermal significance. IR radiation at wavelengths from 3 to 14 μm is called the *thermal IR region.* Thermal IR radiation is absorbed by the glass lenses of conventional cameras and cannot be detected by photographic films. Special detectors and optical-mechanical scanners are used to detect and record images in the thermal IR spectral region. Unfortunately the term IR energy connotes heat to many people; therefore it is important to recognize the difference between *photographic IR* energy and *thermal IR* energy.

Atmospheric Transmission All wavelengths of thermal IR radiation are not uniformly transmitted through the atmosphere. Carbon dioxide, ozone, and water vapor absorb energy at certain wavelengths called *absorption bands.* As depicted in Figure 1.3, IR radiation at wavelengths from 3 to 5 μm and from 8 to 14 μm is readily transmitted through the atmosphere and these regions are referred to as *atmospheric windows.* A number of detection devices have been designed that are sensitive to radiation of these wavelengths. The narrow absorption band from 9 to 10 μm, shown as a dashed curve in Figure 1.3, is caused by the ozone layer at the top of the earth's atmosphere. The effects of this absorption band must be considered in satellite-borne thermal IR systems. Aircraft systems, which operate below the ozone layer, are not affected. IR radiation at wavelengths longer than 14 μm is not employed in remote sensing because it is absorbed by the earth's atmosphere.

Radiant Power Peaks For an object at a constant temperature the *radiant power peak* refers to the wavelength at which the maximum amount of energy is radiated, which is expressed as λ_{max}. As shown earlier in Figure 1.2, the 8 to 14 μm band spans the 9.7 μm radiant power peak of the earth (at 300°K) and is commonly employed for remote-sensing purposes. The 3 to 5 μm region is useful for hotter targets such as forest fires and active volcanoes because at higher temperatures the radiant power peak shifts to shorter wavelengths. This shift to shorter wavelengths with increasing temperature is described by Wien's displacement law, which states

$$\lambda_{max} = \frac{2{,}897\ \mu\text{m}\ ^\circ\text{K}}{T_{rad}}$$

where T_{rad} is radiant temperature in °K and 2,897 μm °K is a physical constant. By substituting the value of T_{rad} for an object, the wavelength in micrometers of the maximum energy peak radiated by the object may be determined. For example, the average radiant temperature of the earth is approximately 300°K (27°C or 80°F). Substituting into Equation (5.1) results in

$$\lambda_{max} = \frac{2{,}897\ \mu\text{m}\ ^\circ\text{K}}{T_{rad}}$$

$$\lambda_{max} = \frac{2{,}897\ \mu\text{m}\ ^\circ\text{K}}{300^\circ\text{K}}$$

$$\lambda_{max} = 9.7\ \mu\text{m}$$

which is the value given in Figure 1.2 for the radiant power peak of the earth.

Thermal Properties of Materials

Radiant energy striking the surface of a material is partly reflected, partly absorbed, and partly transmitted through the material. Therefore

$$\text{Ref} + \text{Abs} + \text{Trn} = 1 \qquad (5.2)$$

where

Ref = reflectivity

Abs = absorptivity

Trn = transmissivity

Reflectivity, absorptivity, and transmissivity depend upon the wavelength of the incident radiant energy and also upon the temperature of the surface. As was discussed in Chapter 2, reflectivity may be expressed as albedo, A, which is the ratio of reflected energy to incident energy. For materials in which transmissivity is negligible, Equation (5.2) reduces to

$$\text{Ref} + \text{Abs} = 1 \tag{5.3}$$

The absorbed energy causes an increase in the kinetic temperature of the material.

Black Body Concept, Emissivity, and Radiant Temperature The concept of a black body is fundamental to understanding heat radiation. By definition a *black body* is a material that absorbs all the radiant energy that strikes it, which means

$$\text{Abs} = 1 \tag{5.4}$$

A black body also radiates the maximum amount of energy, which is dependent on the kinetic temperature. According to the Stefan–Boltzmann law the radiant flux of a black body, F_b, at a kinetic temperature, T_{kin}, is

$$F_b = \sigma T_{\text{kin}}^4 \tag{5.5}$$

where σ is the Stefan–Boltzmann constant, $5.67 \cdot 10^{-12}\ \text{W} \cdot \text{cm}^{-2} \cdot {}^\circ\text{K}^{-4}$ where W is the abbreviation for watt. For a black body with a kinetic temperature, T_{kin}, of 10°C (283°K) the radiant flux may be calculated from Equation (5.5) as

$$F_b = \sigma T_{\text{kin}}^4$$

$$F_b = (5.67 \cdot 10^{-12}\ \text{W} \cdot \text{cm}^{-2} \cdot {}^\circ\text{K}^{-4})\ (283^\circ\text{K})^4$$

$$F_b = (5.67 \cdot 10^{-12}\ \text{W} \cdot \text{cm}^{-2} \cdot {}^\circ\text{K}^{-4})\ (6.41 \cdot 10^9\ {}^\circ\text{K}^4)$$

$$F_b = 3.6 \cdot 10^{-2}\ \text{W} \cdot \text{cm}^{-2}$$

TABLE 5.1
Emissivity of representative samples of various materials determined in the 8 to 12 μm wavelength region

Material	Emissivity, ε
Granite	0.815
Dunite	0.856
Obsidian	0.862
Feldspar	0.870
Granite, rough	0.898
Silica sandstone, polished	0.909
Quartz sand, large grains	0.914
Dolomite, polished	0.929
Basalt, rough	0.934
Dolomite, rough	0.958
Asphalt paving	0.959
Concrete walkway	0.966
Water, with a thin film of petroleum	0.972
Water, pure	0.993

Source: Buettner, K. J. K., and C. D. Kern, Journal of Geophysical Research, v. 70, p. 1333, 1965, copyrighted by American Geophysical Union.

A black body is a physical abstraction, for no material has an absorptivity of 1, Equation (5.4), and no material radiates the full amount of energy in Equation (5.5). For real materials a property called *emissivity*, ε, has been defined as

$$\varepsilon = \frac{F_r}{F_b} \tag{5.6}$$

where F_r is radiant flux from a real material. For a black body $\varepsilon = 1$, but for all real materials $\varepsilon < 1$. Emissivity is wavelength dependent, which means that the emissivity of a material is different when it is measured at different wavelengths of radiant energy.

Table 5.1 lists the emissivities of various materials in the 8 to 12μm wavelength region, which is widely used in remote sensing. The emissivities of materials in Table 5.1 fall within the relatively narrow range of 0.81 to 0.96. Note the high emissivity of water and that a thin film of petroleum lowers the emissivity. In the 8 to 12 μm region, polished metal surfaces have very low emissivities of about 0.06, but a coat of flat black paint increases the emissivity to about 0.97. By combining Equations (5.5) and (5.6) the radiant flux of a real material may be expressed as

$$F_r = \varepsilon \sigma T_{\text{kin}}^4 \qquad (5.7)$$

where ε is the emissivity for that material. Emissivity is a measure of the ability of a material to both radiate and absorb energy. Materials with a high emissivity absorb and radiate large proportions of the incident and kinetic energy, respectively. Materials with lower emissivities absorb and radiate lower amounts of energy. The effect of different emissivities on radiant flux is illustrated in Figure 5.1, where the aluminum block has a uniform kinetic temperature of 10°C (283°K). The portion of the aluminum block that is painted dull black has an emissivity of 0.97. The radiant flux for this material may be calculated from Equation (5.7) as

$$F_r = \varepsilon \sigma T_{\text{kin}}^4$$

$$F_r = 0.97(5.67 \cdot 10^{-12}\ \text{W} \cdot \text{cm}^{-2} \cdot {}^\circ\text{K}^{-4})(283^\circ\text{K})^4$$

$$F_r = 3.5 \cdot 10^{-2}\ \text{W} \cdot \text{cm}^{-2}$$

For the shiny portion of the aluminum block with an emissivity of 0.06, the radiant flux may be calculated from Equation (5.7) as

$$F_r = \varepsilon \sigma\ T_{\text{kin}}^4$$

$$F_r = 0.06(5.67 \cdot 10^{-2}\ \text{W} \cdot \text{cm}^{-2} \cdot {}^\circ\text{K}^{-4})(283^\circ\text{K})^4$$

$$F_r = 2.2 \cdot 10^{-3}\ \text{W} \cdot \text{cm}^{-2}$$

Although the aluminum block has a uniform kinetic temperature of 283°K, the radiant flux from the surface with high emissivity is more than 10 times higher than from the surface with low emissivity.

Most thermal IR remote-sensing systems record the radiant temperature T_{rad} of terrain, rather than radiant flux. In order to determine T_{rad}, consider a black body and a real material that have different kinetic temperatures, but that have the same radiant flux, so that $F_b = F_r$. For a black body $T_{\text{rad}} = T_{\text{kin}}$; therefore Equation (5.5) may be written as

$$F_b = \sigma T_{\text{rad}}^4$$

This equation and Equation (5.7) may then be combined so that

$$F_r = \varepsilon \sigma T_{\text{kin}}^4$$

$$F_b = \sigma T_{\text{rad}}^4$$

$$F_r = F_b$$

Therefore

$$\sigma T_{\text{rad}}^4 = \varepsilon \sigma T_{\text{kin}}^4$$

and

$$T_{\text{rad}} = \varepsilon^{1/4} T_{\text{kin}} \qquad (5.8)$$

Radiant temperatures may be measured with remote-sensing devices called radiometers, which are described later in the section on IR Detection and Imaging Technology. For the aluminum block in Figure 5.1 with a kinetic temperature of 283°K, equation (5.8) is used to calculate the radiant temperature of the portion that has an emissivity of 0.97,

$$T_{\text{rad}} = 0.97^{1/4} \cdot 283^\circ\text{K}$$

$$T_{\text{rad}} = 281^\circ\text{K}$$

For the portion of the block with an emissivity of 0.06,

$$T_{\text{rad}} = 0.06^{1/4} \cdot 283^\circ\text{K}$$

$$T_{\text{rad}} = 140^\circ\text{K}$$

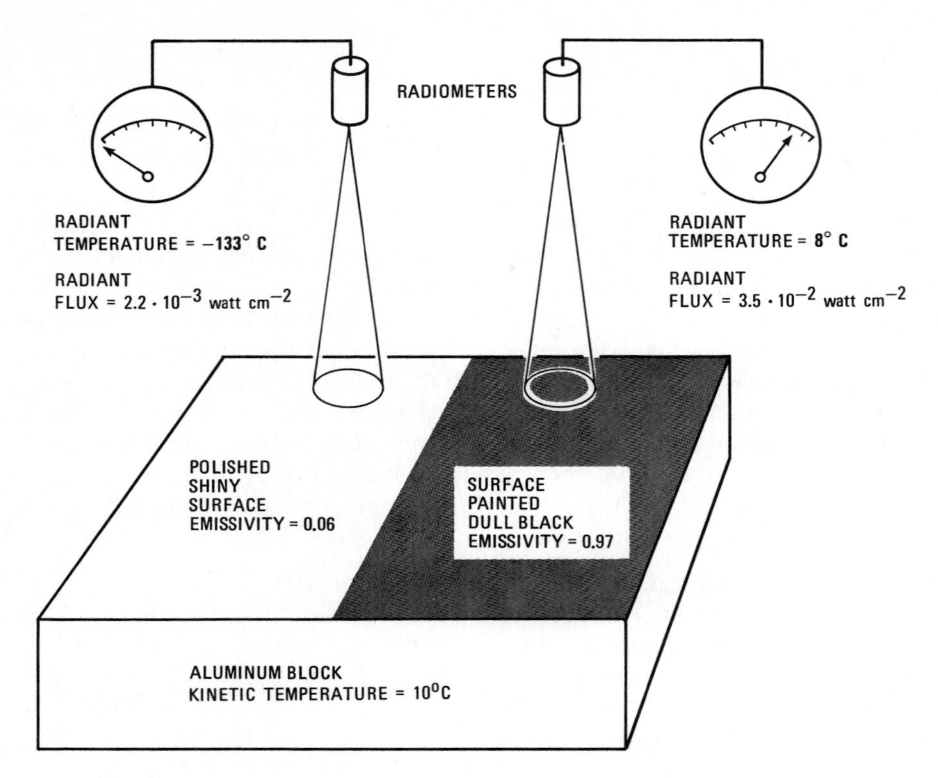

FIGURE 5.1
Effect of emissivity differences on radiant temperature. Kinetic temperature of aluminum block is uniformly 10°C. Different emissivities cause different radiant temperatures that are measured with radiometers.

which is 141°K lower than the radiant temperature for the portion of the aluminum block with higher emissivity. An alternate way to understand the low radiant temperature of the shiny surface involves Equation (5.3). Because the emissivity is low, the absorptivity is also low and the reflectivity is therefore high. Outdoors, with no clouds, the very low temperature of outer space is reflected by the shiny surface and measured by the radiometers. For this reason metallic objects, such as airplanes and metal-roofed buildings, have cold radiant temperatures.

Thermal Conductivity *Thermal conductivity, K,* is the measure of the rate at which heat will pass through a material and is expressed as $cal \cdot cm^{-1} \cdot sec^{-1} \cdot {}^{\circ}C^{-1}$. The number K for a material is the quantity of heat in calories that will pass through a 1 cm cube of the material when two opposite faces are maintained at 1°C difference in temperature. Thermal conductivities for geologic materials are given in Table 5.2. For any rock type the thermal conductivity may vary ± 20 percent from the value given. Thermal conductivity of porous materials may vary up to 200 percent depending on the nature of the substance that fills the pores. Rocks are relatively poor conductors of heat. The average thermal conductivity of the materials in Table 5.2 is $0.006\ cal \cdot cm^{-1} \cdot sec^{-1} \cdot {}^{\circ}C^{-1}$, which is two orders of magnitude lower than the thermal conductivity of metals such as aluminum, copper, and silver.

Thermal Capacity *Thermal capacity, c,* is the ability of a material to store heat. It is the number of calories required to raise the temperature of 1 g of a material by 1°C and is expressed in $cal \cdot g^{-1} \cdot {}^{\circ}C^{-1}$. In Table 5.2 note that water has the highest thermal capacity of any substance. The difference between thermal capacity and temperature is shown diagrammatically in Figure 5.2. Spheres of the same volume made from rhyolite, limestone,

TABLE 5.2
Thermal properties of geologic materials and water at 20°C

Geologic materials	K Thermal conductivity, cal · cm^{-1} · sec^{-1} · °C^{-1}	ρ Density, gm · cm^{-3}	c Thermal capacity, cal · gm^{-1} · °C^{-1}	k Thermal diffusivity, cm^2 · sec^{-1}	P Thermal inertia, cal · cm^{-2} · sec$^{-1/2}$ · °C^{-1}
1. Basalt	0.0050	2.8	0.20	0.009	0.053
2. Clay soil (moist)	0.0030	1.7	0.35	0.005	0.042
3. Dolomite	0.0120	2.6	0.18	0.026	0.075
4. Gabbro	0.0060	3.0	0.17	0.012	0.055
5. Granite	0.0075 0.0065	2.6	0.16	0.016	0.052
6. Gravel	0.0030	2.0	0.18	0.008	0.033
7. Limestone	0.0048	2.5	0.17	0.011	0.045
8. Marble	0.0055	2.7	0.21	0.010	0.056
9. Obsidian	0.0030	2.4	0.17	0.007	0.035
10. Peridotite	0.0110	3.2	0.20	0.017	0.084
11. Pumice, loose	0.0006	1.0	0.16	0.004	0.009
12. Quartzite	0.0120	2.7	0.17	0.026	0.074
13. Rhyolite	0.0055	2.5	0.16	0.014	0.047
14. Sandy gravel	0.0060	2.1	0.20	0.014	0.050
15. Sandy soil	0.0014	1.8	0.24	0.003	0.024
16. Sandstone, quartz	0.0120 0.0062	2.5	0.19	0.013	0.054
17. Serpentine	0.0063 0.0072	2.4	0.23	0.013	0.063
18. Shale	0.0042 0.0030	2.3	0.17	0.008	0.034
19. Slate	0.0050	2.8	0.17	0.011	0.049
20. Syenite	0.0077 0.0044	2.2	0.23	0.009	0.047
21. Tuff, welded	0.0028	1.8	0.20	0.008	0.032
22. Water	0.0013	1.0	1.01	0.001	0.037

Source: From Janza (1975, Table 4.1).

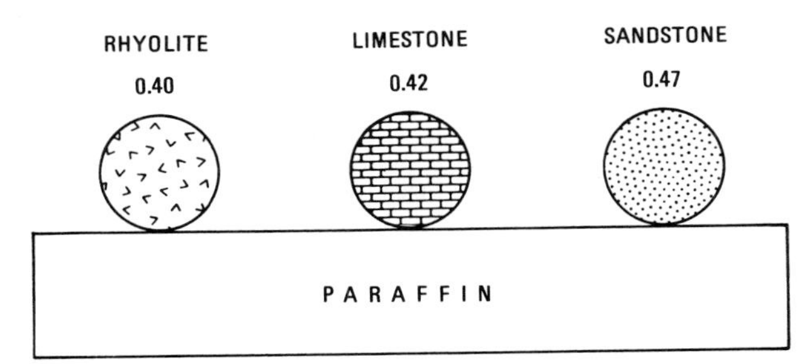

A. SPHERES OF ROCK HEATED TO 100°C AND PLACED ON SHEET OF PARAFFIN. THE VALUE FOR EACH ROCK IS THE PRODUCT OF ITS THERMAL CAPACITY, c, AND DENSITY, ρ, IN CAL · CM^{-3} · °C^{-1}.

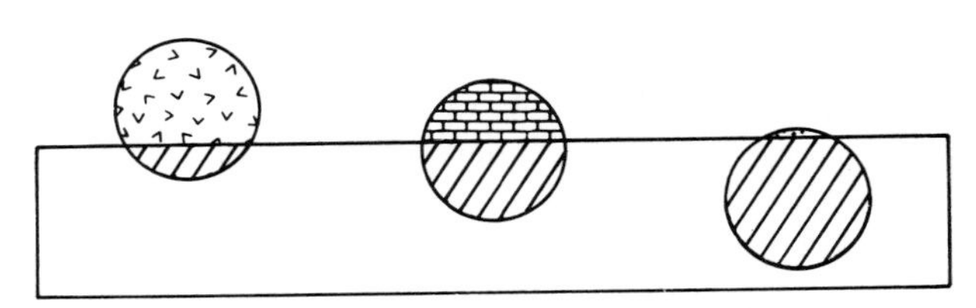

B. AFTER THE ROCKS AND PARAFFIN HAVE REACHED THE SAME TEMPERATURE.

FIGURE 5.2
The effect of differences in thermal capacity of various rock types.

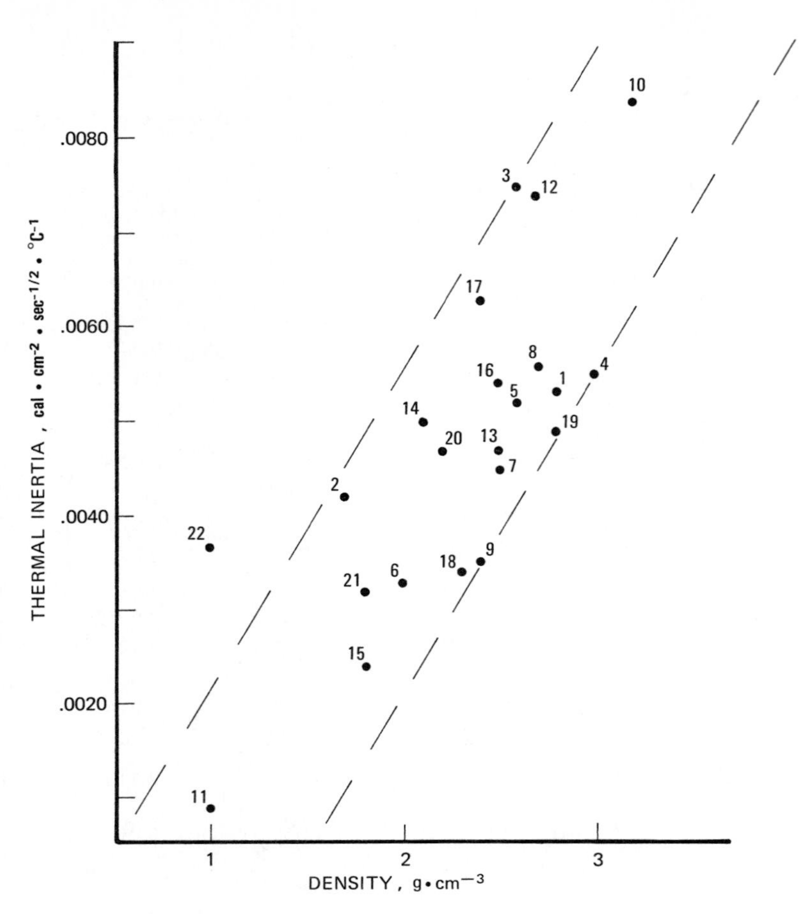

FIGURE 5.3
Relationship of thermal inertia to density for rocks and water. Numbers refer to the materials listed in Table 5.2.

and sandstone are placed in boiling water to reach a uniform temperature of 100°C. It is assumed that these rocks have zero porosity and no water is absorbed. The thermal capacity and density, from Table 5.2, are multiplied to determine the number of cal · cm^{-3} · °C^{-1} that are stored by each rock type. Rocks were selected that have a uniform density of 2.5 gm · cm^{-3}; therefore the different values are determined solely by differences in thermal capacity. As illustrated in Figure 5.2A the greatest amount of calories is stored by the sandstone and the least amount by the rhyolite. At the same instant the heated spheres are placed on a sheet of paraffin. Melting ceases when the spheres and paraffin have reached a uniform temperature. As seen in Figure 5.2B, the amount of melting is related to the thermal capacity of the rocks, and not to their temperature.

Thermal Inertia *Thermal inertia, P*, is a measure of the thermal response of a material to temperature changes and is given in cal · cm^{-2} · sec$^{-1/2}$ · °C^{-1}. Thermal inertia may be calculated from the relationship

$$P = (K\rho c)^{1/2} \qquad (5.9)$$

where K is thermal conductivity, ρ is density, and c is thermal capacity. Of the three properties that determine thermal inertia, *density* is the most important. For most materials thermal inertia increases linearly with increasing density, as shown in Figure 5.3, where the values for the materials listed in Table 5.2 are plotted.

Figure 5.4 illustrates the effect of differences in thermal inertia on surface temperature. During a diurnal solar cycle materials with lower thermal

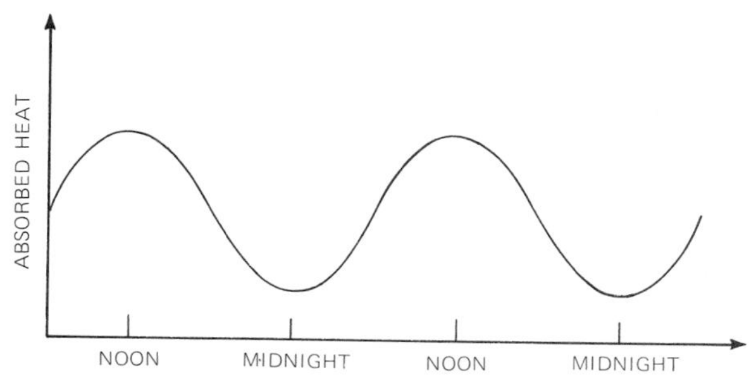

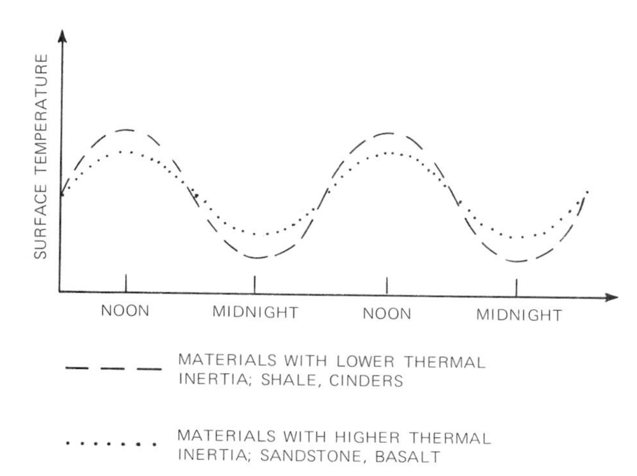

FIGURE 5.4
Effect of differences in thermal inertia on surface temperature during diurnal solar cycles. From K. Watson (unpublished), U.S. Geological Survey.

inertia, such as shale and volcanic cinders, reach a relatively high surface temperature in the daytime. At night these materials cool to a relatively low temperature. In contrast the temperatures of materials with higher thermal inertia, such as sandstone and basalt, are cooler in the daytime and warmer at night. In other words, materials with higher thermal inertia have more uniform surface temperatures day and night than materials with lower thermal inertia.

Thermal Diffusivity *Thermal diffusivity*, k, may be calculated from the relationship

$$k = \frac{K}{c\rho} \tag{5.10}$$

Thermal diffusivity (given in $cm^2 \cdot sec^{-1}$) governs the rate at which temperature changes within a substance. This expresses the ability of a substance to transfer heat from the surface to the interior during a period of solar heating. At night thermal diffusivity expresses the ability of a substance to transfer stored heat to the surface.

Influence of Water and Vegetation

The thermal inertia of water is similar to that of soils and rocks (Table 5.2), but during the day water bodies have a cooler surface temperature than soils and rocks. At night the surface temperatures are reversed with water becoming warmer than soils and rocks. The reason is that convection currents maintain a relatively uniform temperature at the surface of a water body. Convection does not operate to transfer heat in soils and rocks; therefore, heat from solar flux is concentrated near the surface of these solids during the day causing a higher surface temperature. At night the heat is radiated to the atmosphere and is not replenished by convection currents in these solid materials, causing surface temperatures to be lower than in adjacent water bodies (K. Watson, personal communication). Some thermal IR images may not be annotated for the time of day at which they were acquired. The thermal signatures of any water bodies are a reliable index to the time of image acquisition. If water bodies have warm signatures relative to the adjacent terrain, the image was acquired at night. Relatively cool water bodies indicate daytime imagery.

Damp ground is cooler than dry ground, both day and night, because of the cooling effect as absorbed water is evaporated. The evaporation effect produces cool signatures on thermal IR images. Damp areas adjacent to water bodies and areas that are damp from rain have cool signatures on thermal IR images. Areas where ground water reaches the surface and evaporates also have cool signatures. Examples of this effect along the San Andreas fault are illustrated later in the section

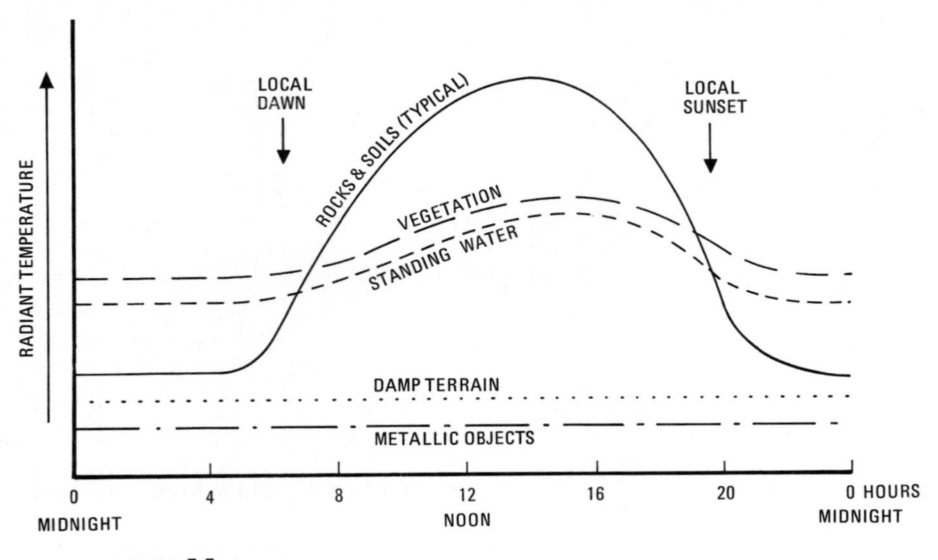

FIGURE 5.5
Diurnal radiant temperature curves (diagrammatic) for typical materials.

describing thermal IR images of the Indio Hills, California.

Green deciduous vegetation has a warm signature on nighttime images. During the day transpiration of water vapor lowers the leaf temperature causing vegetation to have a cool signature relative to the surrounding soil. The relatively high nighttime and low daytime radiant temperature of conifers, however, does not appear to be related to their water content. The composite emissivity of the needle clusters making up a whole tree approaches that of a black body. Dry vegetation, such as crop stubble in agricultural areas, appears warm on nighttime imagery in contrast to bare soil, which is cool. The dry vegetation insulates the ground to retain heat and causes the warm nighttime signature.

Diurnal Temperature Variations

Typical diurnal variations in radiant temperature are shown diagrammatically in Figure 5.5. Note that the most rapid temperature changes occur near dawn and sunset. At the times when two curves intersect, called *thermal cross overs*, there is no radiant temperature difference between the materials. Quantitative data on diurnal changes in radiant temperature for various materials are given later in the section describing the Indio Hills.

IR DETECTION AND IMAGING TECHNOLOGY

Radiant temperature of terrain may be measured at discrete ground points or continuously along flight lines by IR radiometers. The pattern of radiant temperature variations may be recorded as an image by optical-mechanical scanners.

IR Radiometers

An IR *radiometer* (Figure 5.6) is a nonimaging device for quantitatively measuring radiant temperature. An IR-sensitive detector and a filter are used to measure radiation at a specified wavelength, commonly 8 to 14 μm, that corresponds to an atmospheric window. In a typical radiometer a thermistor-controlled, electrically heated cavity provides an internal calibration source against which radiation from the target is referenced by a chopper (Figure 5.7). A *chopper* is a rotating disk with alternating blades and open spaces that resemble the blades of an electric fan. The blades are plated with gold or another polished metal to reflect IR radiation. When a chopper blade is in the field of view, radiant energy from the calibration source is reflected from the blade onto the detector. When the blade rotates out of the field of view, radiation from the target, of unknown temperature, is focused on the detector. The dif-

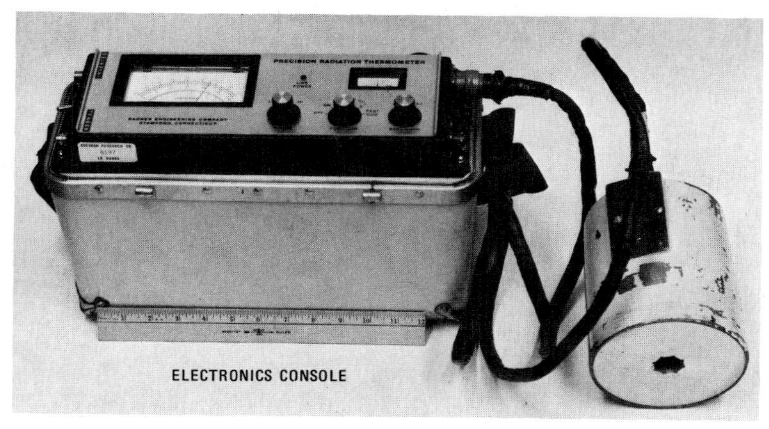

FIGURE 5.6
Battery powered IR radiometer.

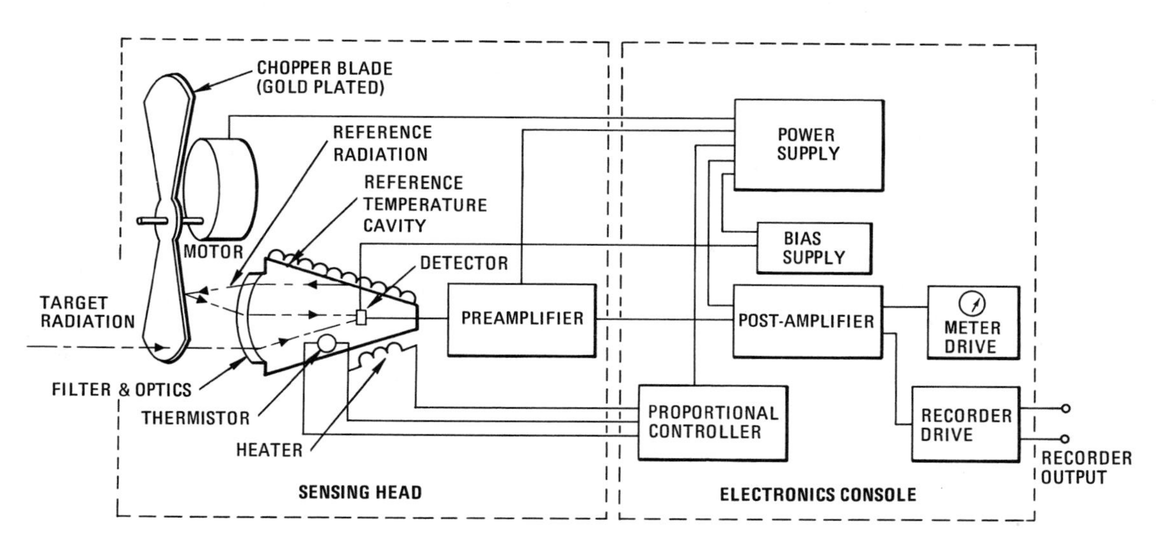

FIGURE 5.7
Schematic diagram of IR radiometer. As the chopper blade rotates, the detector alternately receives radiation from the target and from the temperature reference. From Moxham (1971, Figure 9). Reproduced by courtesy of Barnes Engineering Company.

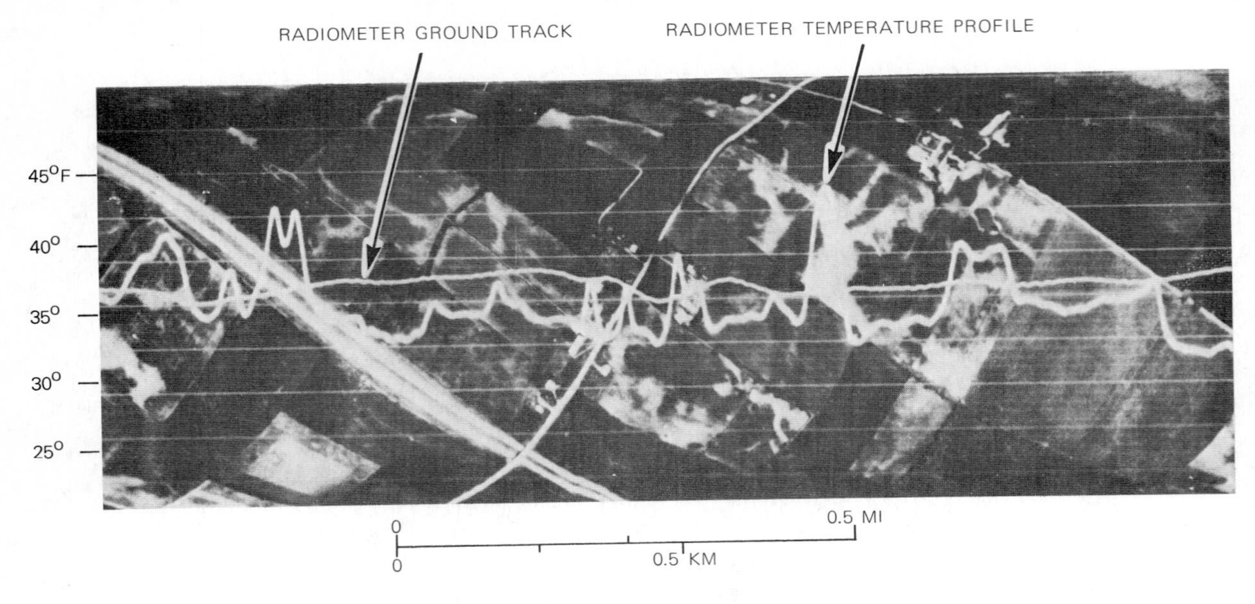

FIGURE 5.8
Radiant temperature profile measured with an airborne radiometer and superimposed on a daytime thermal IR image that was acquired simultaneously. On the image light tones represent warm areas, dark tones represent cool areas. Courtesy Daedalus Enterprises, Inc.

ference between the known and unknown signals is converted into target temperature for display on the dial and the optional chart recorder. Temperature sensitivity is typically 0.2°C. The instantaneous field of view of a typical radiometer is determined by the optical system and typically is either 2° or 20°.

Radiometers are commonly used to record radiant temperatures of the terrain along an aircraft flight path. The radiometer may be boresighted to measure radiant temperatures along the center of a strip of IR imagery that is acquired concurrently. A typical radiant temperature profile is shown in Figure 5.8 superimposed on an IR image. The trace of the ground track on the image shows the exact path of the radiometer and the targets that are measured. Irregularities of the ground track are caused by aircraft roll. Note that the highest radiant temperature is associated with the freeway on the left side of the image and that the cooler median divider strip is resolved on the profile. Radiant temperatures may be measured in the field by portable radiometers equipped with battery packs, such as the model shown in Figure 5.6.

Airborne IR Scanner Systems

Thermal IR images are produced by airborne scanner systems that consist of three basic components: an optical-mechanical scanning subsystem, a thermal IR detector, and an image recording subsystem. As shown in Figure 5.9, the scanning subsystem consists of an electric motor mounted in the aircraft with the rotating shaft oriented parallel with the aircraft fuselage and the flight direction. A front-surface mirror with a 45° facet mounted on the end of the shaft sweeps the terrain at a right angle to the flight path. The angular field of view is typically 90° or 120°. IR energy radiated from the terrain is focused from the scan mirror onto the IR detector that converts the radiant energy into an electrical signal that varies in proportion to the intensity of the IR radiation. Detectors are cooled to 73°K to reduce the electronic noise caused by the molecular vibrations of the detector crystal. In scanners such as the one in Figure 5.9 the detector is enclosed by a vacuum bottle, or dewar, filled with liquid nitrogen. Other detectors employ a closed-cycle mechanical cooler to maintain a low detector

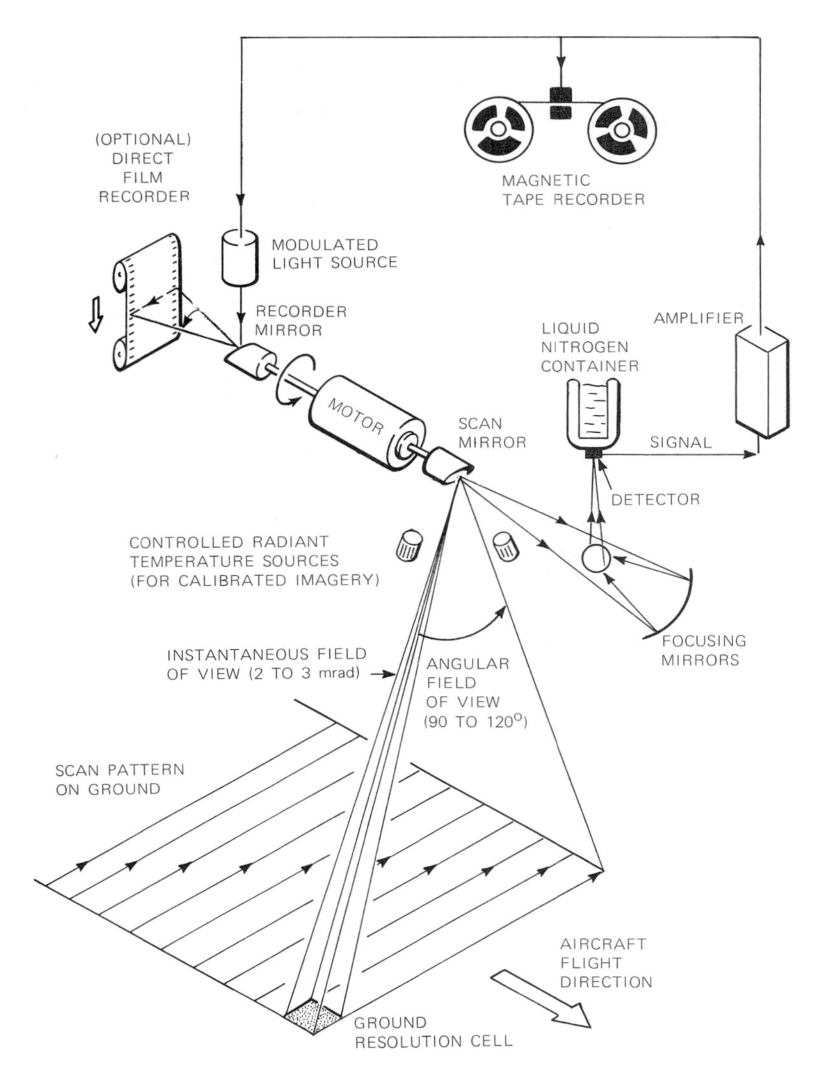

FIGURE 5.9
Diagram of thermal IR scanner system. After Sabins (1969, Figure 2).

temperature. In the simplest scanners of the mid-1960s the amplified signal modulates the intensity of a small light source. The image of this modulated light source is swept across a strip of recording film by a second mirror rotating synchronously with the scanning mirror. The recording film advances at a rate proportional to the aircraft ground speed divided by the altitude, so that each scan line on the ground is represented by a scan line on the film. In the late 1960s, direct film recording was largely replaced by magnetic tape recording. The tapes are played back onto film in the laboratory to produce imagery. The controlled radiant temperature reference sources shown in Figure 5.9 are similar to those used in radiometers. The reference sources are built into the scanner housing so that the scanner mirror views one at the beginning and at the end of each scan. The known temperatures from the reference sources are used to produce calibrated images as discussed later in the section on Temperature Calibration of Images. The components of a typical scanner system are illustrated in Figure 5.10. The gyroscopically controlled roll-stabilization device provides data for correcting for the effect of aircraft roll. The control unit enables the operator to monitor and modify the operation of the scanner.

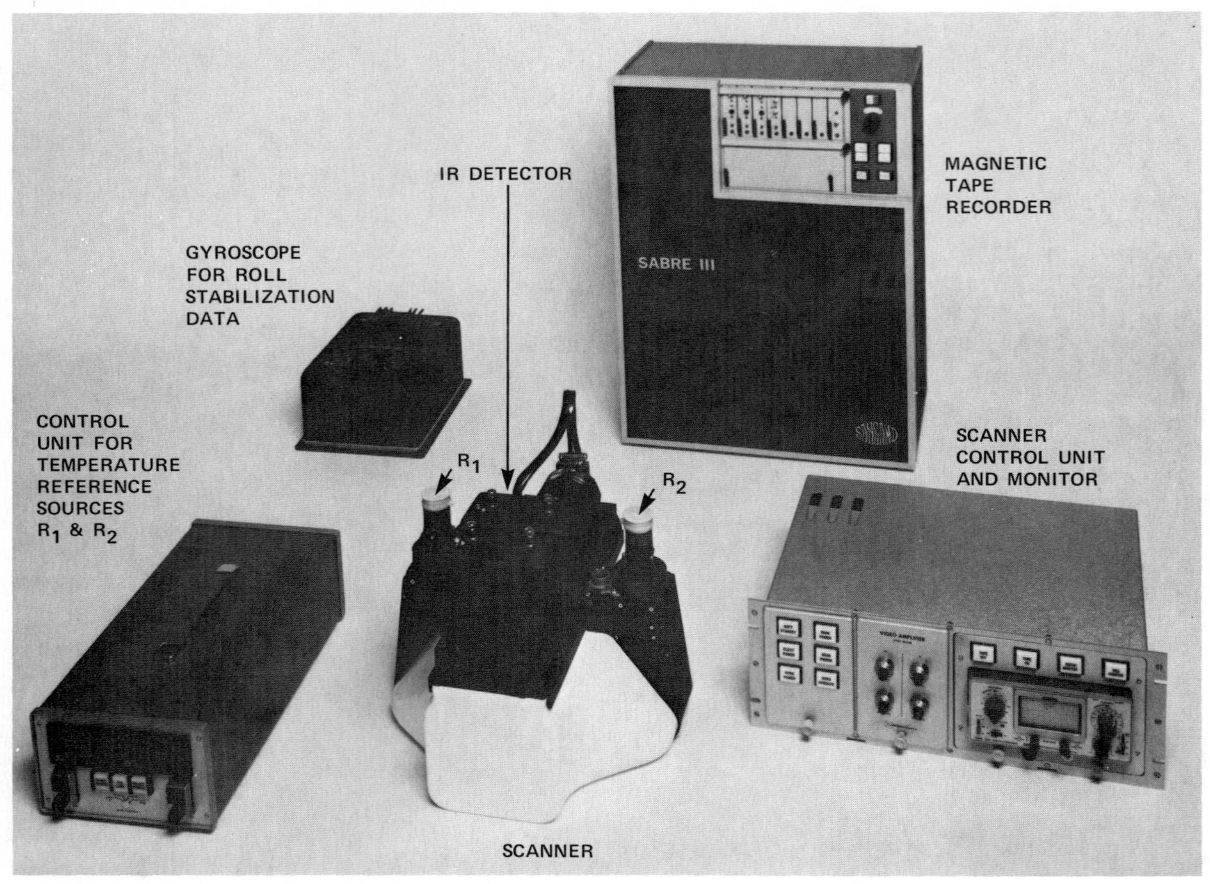

FIGURE 5.10
Typical thermal IR scanner system. Courtesy Daedalus Enterprises, Inc.

Some newer scanner systems provide a processed paper copy of the image continuously during the flight in addition to the magnetic tape record. Although inferior in quality to the film image produced from magnetic tape, the paper print is a valuable in-flight monitor of system performance and a check on aircraft navigation.

In some scanners the faceted mirror rotates about a vertical axis to generate a conical sweep. A segment of the circular ground path in advance of the aircraft is recorded in a fashion similar to that of the Skylab multispectral scanner (Chapter 3). The conical sweep pattern provides a constant path length, which results in a ground resolution cell of constant size. This contrasts with the cylindrical scanner (Figure 5.9) in which the path length at either edge of the scan line is longer than the vertical path directly beneath the aircraft. Therefore at the margins of the image the ground resolution cell is larger than it is directly beneath the aircraft. The specifications of commercially available thermal IR scanners are given by Lowe (1975, Table 8.2).

For most investigations the 8 to 14 μm wavelength band is generally preferred to the other atmospheric window at 3 to 5 μm. For the average surface temperature range of the earth the 8 to 14 μm band includes the maximum intensity of the radiant energy flux. Therefore the maximum intensity of radiation is available for detection in this wavelength band. Mercury-cadmium-telluride detectors are commonly used to detect radiation in the 8 to 14 μm region. At night the detector is generally unfiltered because its response range closely matches the 8 to 14 μm atmospheric window and a filter attenuates some of the IR radiation. The instantaneous field of view of typical detectors ranges from 1 to 3 mrad. At a flight altitude of 1,000 m the resulting ground resolution cell (Figure 5.9) ranges from 1 to 3 m in width. Typical IR detectors are sensitive to temperature differences on the order of 0.1°C. The more commonly used detectors are listed and described by Baker and Scott (1975, Table 7.1).

Stationary Scanners

Airborne scanners are designed for use in aircraft or spacecraft where the forward movement of the vehicle provides coverage along the flight path and the rotating scan mirror provides coverage at right angles to that direction. *Stationary scanners* are designed to acquire images from a fixed position. In these scanners the faceted scan mirror rotates about a vertical axis to provide coverage in the horizontal direction. Coverage in the vertical direction is provided by a plane mirror that tilts about a horizontal axis. Another system employs a television-type camera to electronically scan the scene. On both types of stationary scanner the radiation transmitted through an IR filter is focused onto a detector that converts the radiation into an electrical signal. The resulting image is displayed in real time on a small television-type screen that may be photographed to produce a permanent record. Stationary scanners are mounted on a tripod stand for most applications, although some systems are sufficiently light and compact that the operator can hold the sensor unit in his hand.

One use of stationary scanners is to record the pattern of heat radiated from the human body. The term *thermography* is generally associated with this medical application of thermal IR images, which are called thermograms. Tumors and impaired blood circulation are physiological disorders that have been detected on the images. Stationary scanners are used to monitor industrial facilities for "hot spots" that may indicate potential problems. Anomalous hot spots on the exterior of industrial furnaces may indicate areas where the fire brick lining has eroded and failure is imminent. In electrical transmission facilities, faulty transformers and insulators have been detected by their high radiant temperatures.

CHARACTERISTICS OF IR IMAGES

On conventional thermal IR images (Figure 5.8) the lightest tones represent the warmest radiant temperatures and the darkest tones represent the coolest temperatures. The apparent similarity of IR images to black-and-white aerial photographs results from the fact that both are recorded as gray-scale variations on film. In photography, however, film is both the detecting and recording medium for reflected energy in the 0.4 to 0.9 μm wavelength region. Thermal IR energy is detected by a semiconductor device and film serves as a medium to display the imagery.

Geometric Distortion and Correction

Scanner systems, such as the one shown in Figure 5.9, produce a characteristic geometric distortion on the images. Because of the greater distance from the scanner to the ground, the ground resolution cell of the detector is larger at either end of the scan line than in the center directly beneath the aircraft (Figure 5.11A). The scanner mirror rotates at a constant angular rate, but the imagery is recorded at a constant linear rate so that each ground resolution cell is recorded equally, causing compression toward the edges of the image. This compression results in the geometric distortion shown diagrammatically in Figure 5.11B and with a real image in Figure 5.12A. The S-shaped curvature of straight roads trending diagonally across the flight path is a typical expression of scanner distortion. Images recorded on magnetic tape may be played back onto film with an electronic cor-

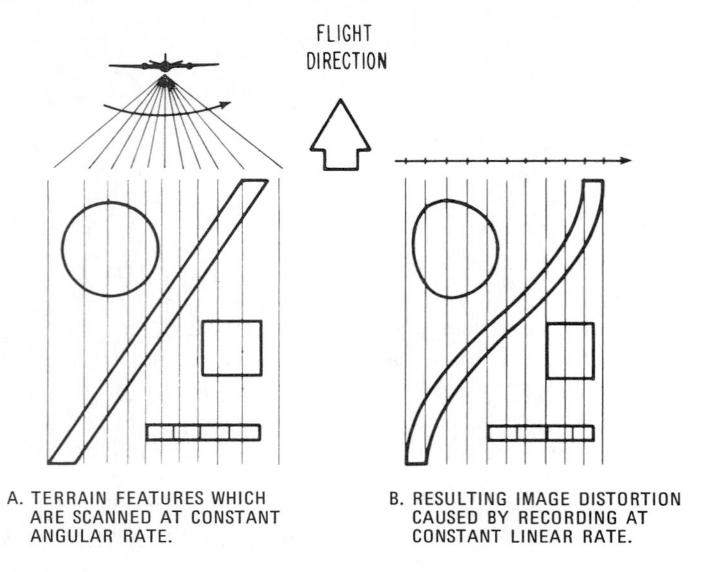

FIGURE 5.11
Distortion characteristics of scanner images. From Sabins (1969, Figure 3).

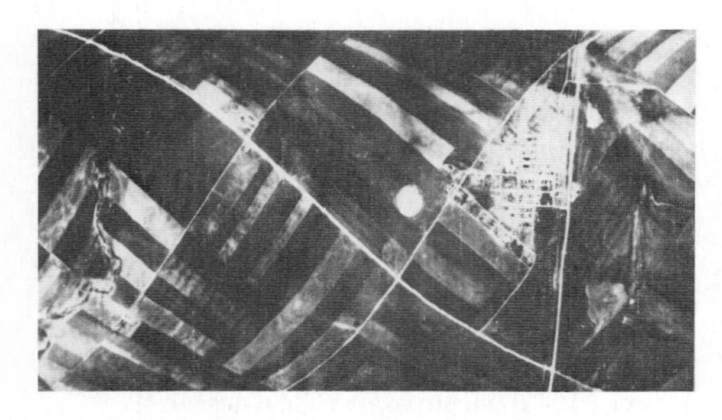

A. IMAGE WITH NORMAL SCANNER DISTORTION.

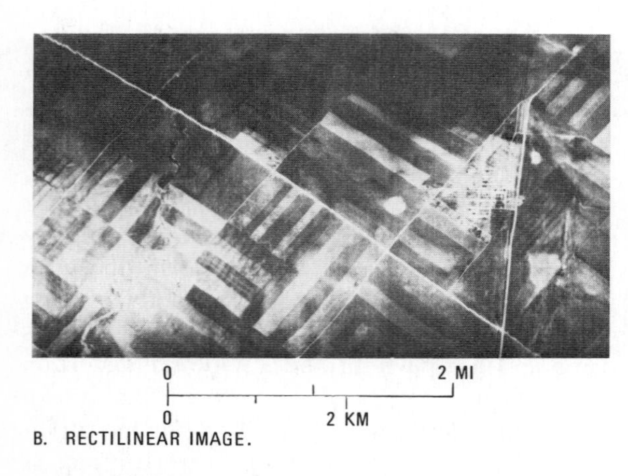

B. RECTILINEAR IMAGE.

FIGURE 5.12
Distorted and rectilinear versions of a thermal IR scanner image. Flight direction is from left to right. From Sabins (1973B, p. 839).

rection to produce rectilinear images free from distortion as shown in Figure 5.12B. There are several advantages to rectilinear images, particularly for compiling mosaics. The full width of a rectilinear image strip may be used for interpretation, which allows the flight lines to be more widely spaced.

Image Irregularities

In addition to the normal scanner distortion, image irregularities may result from various causes that are described here as a guide for interpreters. The overall high yield of good imagery is encouraging, especially considering that scanners are installed in vibrating, often unstable aircraft with numerous potentially interfering electrical fields, and are operated at the mercy of the weather by fallible humans.

Aircraft Motion Distortions Rotation of an aircraft about the longitudinal axis is called *roll.* Most scanners incorporate a gyroscopic system to compensate for aircraft roll, but when this system is not operating, image distortion may occur. In Figure 5.13A the roll-compensation failure is evident in the distortion of the straight road. In other types of terrain it may be more difficult to recognize this form of irregularity. On images of

A. UNCOMPENSATED AIRCRAFT ROLL.

B. CLOUDS.

C. SURFACE WIND SMEAR.

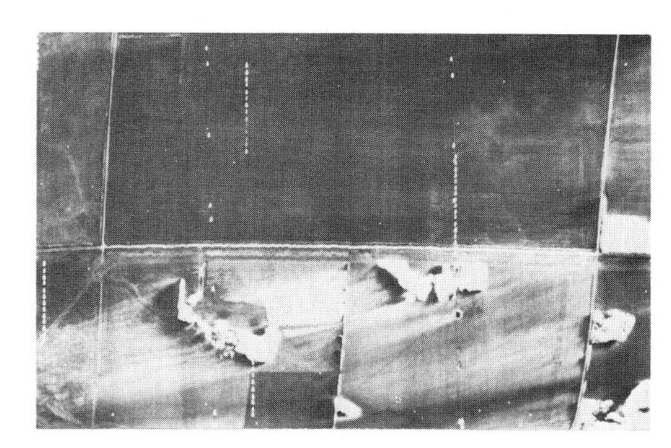

D. SURFACE WIND STREAKS.

E. RADIO TRANSMISSION INTERFERENCE.

F. UNIDENTIFIED ELECTRONIC NOISE.

G. SHIFT IN BASE LEVEL.

H. FILM DEVELOPER STREAK.

FIGURE 5.13
Typical irregularities on thermal IR images. From Sabins (1973B, Figure 2).

dipping sedimentary rock outcrops, roll-compensation failure has produced patterns that resemble plunging folds. Some scanner systems incorporate a warning device that alerts the operator to malfunctions of the roll-compensation system.

Aircraft bank and turn is another cause of image distortion because roll-compensation systems can only correct for about 10° of roll. When an aircraft maneuver exceeds this limit, the image is distorted in a curving pattern. Such maneuvers may occur while the pilot is lining up the beginning of the flight line or breaking off at the end. This is avoided by correct flight planning that programs adequate maneuvering room beyond the ends of the flight line. The flight path is generally assumed to be a straight line on the ground; however, it may have bends and curves that distort the image geometry. This problem is analyzed in the following manner. Draw a line down the center of the image strip and select a number of image features along this line that can be recognized on a base map of the area. Mark the features on the base map and connect them. The resulting line is the actual flight path of the aircraft.

Crosswinds cause the axis of the aircraft fuselage to be oriented at an angle to the flight direction. This attitude is called *yaw*, or crab, and results in distortion throughout the image strip. Some scanners are installed in a camera ringmount that is rotated to correct for yaw.

Effects of Weather on Images Clouds typically have the patchy warm and cool pattern illustrated in Figure 5.13B where the dark signatures are relatively cool and the bright signatures are relatively warm. Scattered rain showers produce a pattern of streaks parallel with the scan lines on the image. A heavy overcast layer reduces thermal contrasts between terrain objects because of reradiation of energy between the terrain and cloud layer. Modern scanner systems are sufficiently sensitive to compensate for this reduction in thermal contrast.

Surface winds produce characteristic patterns of smears and streaks on images. Wind smears (Figure 5.13C) are parallel curved lines of alternating warmer and cooler signature that may extend over wide expanses of the image. Wind streaks occur downwind from obstructions and typically appear as the warm (bright) plumes shown in Figure 5.13D. On this example the obstructions are clumps of trees with warm signatures. The obvious solution to wind effects is to acquire images only on calm nights, but in many areas surface winds persist for much of the year and their effects must be endured.

Electronic Noise Transmissions from aircraft radios may cause strong interference patterns on images. In Figure 5.13E, the interference forms bands of electronic noise that obscure the underlying image pattern. Radio transmissions may also produce a wavy *moiré* interference pattern. Electronic shielding of the scanner equipment may prevent this interference, but the simplest solution is to observe radio silence during image runs and communicate during turns and offset legs. The aircraft radar transponder may also be a source of noise.

Cyclic repetition of discrete signal patterns (Figure 5.13F) is an annoying, but not serious, form of electronic interference. In this example the noise occurs as positive (bright) dots, but it may also have a negative signature or occur as dashes. It has been suggested, but not established, that outside sources such as air traffic radars may be responsible. Electronic shielding of the scanner installation may reduce the effect.

Processing Effects While recording images along a flight line, on either tape or film, there may be a progressive change in the overall radiant temperature level. The operator may have to shift the recording base level to stay within the optimum range, resulting in an abrupt change in image density, as shown in Figure 5.13G. For images recorded directly on film, base-level shifts may be partially compensated by printing the denser and thinner negatives with different exposures. With magnetic tape records, the compensation may be

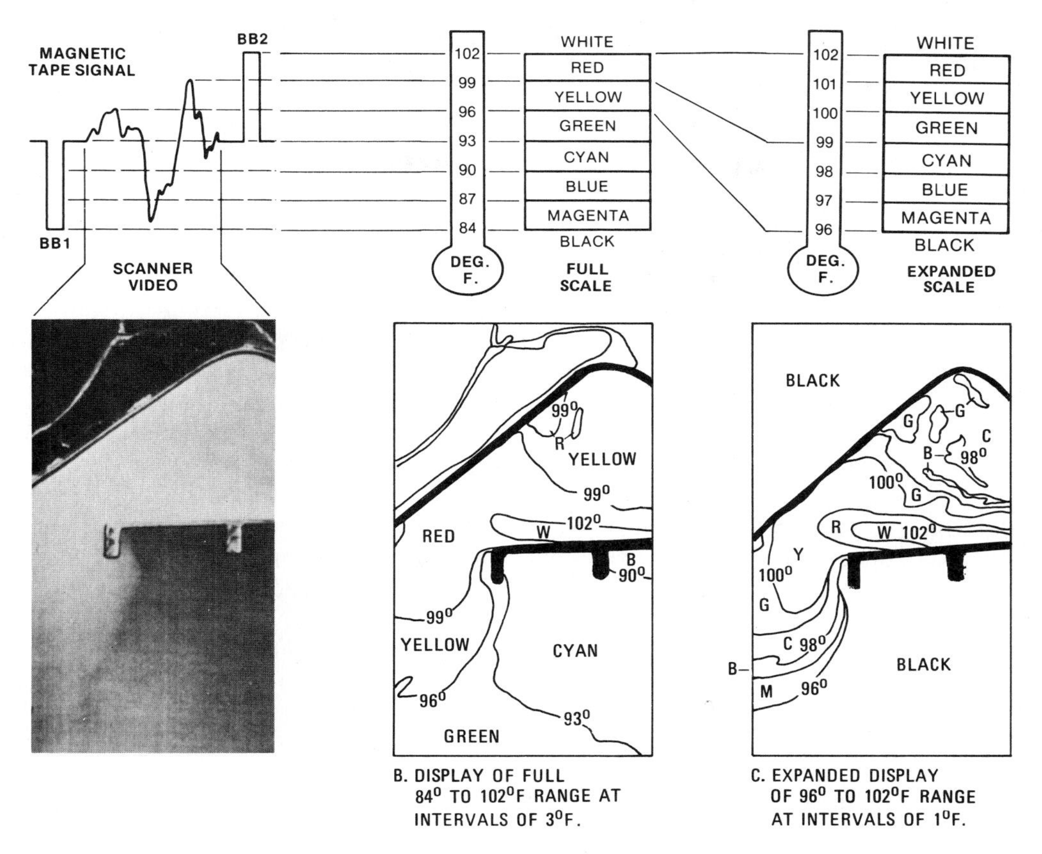

B. DISPLAY OF FULL 84° TO 102°F RANGE AT INTERVALS OF 3°F.

C. EXPANDED DISPLAY OF 96° TO 102°F RANGE AT INTERVALS OF 1°F.

FIGURE 5.14
Temperature-calibrated thermal IR images of power plant discharge. Courtesy Daedalus Enterprises, Inc.

accomplished during playback by monitoring the signal level and adjusting for base-level shifts to obtain a uniform film density. Temperature-calibrated quantitative images, described in the following section, avoid most of these problems.

The photographic development of the image film, whether directly recorded or played back from tape, is another potential source of image irregularities. The developer streak in Figure 5.13H resulted from uneven alignment of a pressure roller in an automatic film processor, which was corrected before further use. Sporadic streaks of this nature may be mistaken for thermal anomalies by an inexperienced interpreter.

Temperature Calibration of Images

Most of the older IR images were qualitative rather than quantitative, because temperature calibration was not provided for scanners, as it is in radiometers. Gray tones of uncalibrated images show relative radiant temperatures rather than quantitative temperatures. For many purposes qualitative images are entirely satisfactory. In the early 1970s, manufacturers began to equip scanners with internal temperature calibration sources that are now standard on most systems. The scanner diagram of Figure 5.9 shows electrically heated temperature calibration sources mounted in the scanner housing on either side of the angular field of view. The scanner records the radiant temperature of the first calibration source, then sweeps the terrain, and finally records the radiant temperature of the second source. The resulting signal is recorded on magnetic tape and has the appearance shown in the upper part of Figure 5.14A. The electrical heater of source 1 was set at a temperature of 84°F, and source 2 was set at 102°F. These reference temperatures provide a scale for determining the temperature at any point along the magnetic tape record of the scan line. During playback of the magnetic tape the temperature range may be *sliced* into intervals and displayed quantitatively. One playback system employs six colors, plus black and white, to display the temperature values. In Figure 5.14B each color is assigned a temperature interval of 3°F. The contact between each color is a temperature contour, or *isotherm*, as illustrated in Figure 5.14B. Finer temperature detail may be displayed by assigning each color to a 1°F interval (Figure 5.14C). Comparison of the calibrated displays with the conventional gray-scale image (Figure 5.14A) illustrates the advantage of quantitative images. Color displays of calibrated images prepared in this manner for Hawaiian volcanoes are shown in Plate 3. Isotherm maps are depicted in Chapter 9 on Environmental Applications.

CONDUCTING IR SURVEYS

Thermal IR imagery is lacking for most areas; therefore special arrangements must be made to acquire imagery for an area of interest. There are two ways to acquire imagery: (1) hire an aerial survey contractor who has the necessary equipment to acquire imagery; (2) the investigator may lease or purchase an IR scanner system and fly his own survey. The acquisition cost of IR imagery is approximately three times that of aerial photography. The cost per square mile decreases as the area of the survey increases.

The following sections, based on a number of years of personal experience, are intended to aid in planning, flying, and evaluating an IR survey. These are intended as general guidelines for surveys and can be modified to meet specific circumstances. For example, the imagery requirements for a regional geothermal survey differ from those for detailed mapping of rock types.

Thermal IR surveys are of two general types, single flight lines and mosaic coverage. A single flight-line over a small test site is simple to plan and conduct. The flight line may be repeated at different times of day to evaluate diurnal thermal variations. Repeated images of the Caliente Range, California are illustrated in Figure 5.15. Mosaic coverage with parallel sidelapping image strips provides regional coverage, and is employed for reconnaissance surveys. As described in the section on Nighttime Navigation, positioning of lines for a mosaic is a much more demanding task than for a single flight line. The guidelines for conducting IR surveys apply to both types of coverage.

Almost any aircraft that can accommodate a pilot, navigator, equipment operator, and several hundred kilograms of equipment is suitable for thermal IR surveys. An open camera port in the bottom of the aircraft is essential. Any glass cover must be removed from the port because glass absorbs thermal IR radiation. There are materials that are transparent to IR radiation, but these are rarely available. On aircraft that lack a camera port, scanners have been mounted externally, but the operator is unable to service the scanner during the mission.

Time of Day

Nighttime imagery is necessary for most geologic applications because the thermal effects of differential solar heating and shadowing are greatly reduced. On daytime images topography is typically the dominant expression because of these differential solar effects. These differences are illustrated on the predawn and postsunrise images of the Caliente and Temblor Ranges, California. On the postsunrise image (Figure 5.15A), the ridges and canyons in the mountain ranges are clearly defined by differential solar heating and shadowing, as is the anticlinal hill in the Carrizo Plain. On the predawn image (Figure 5.15B), topographic effects are largely eliminated and geologic features are emphasized as illustrated on the map of Figure 5.15C. The narrow warm signatures in the Caliente Range are basalt out-

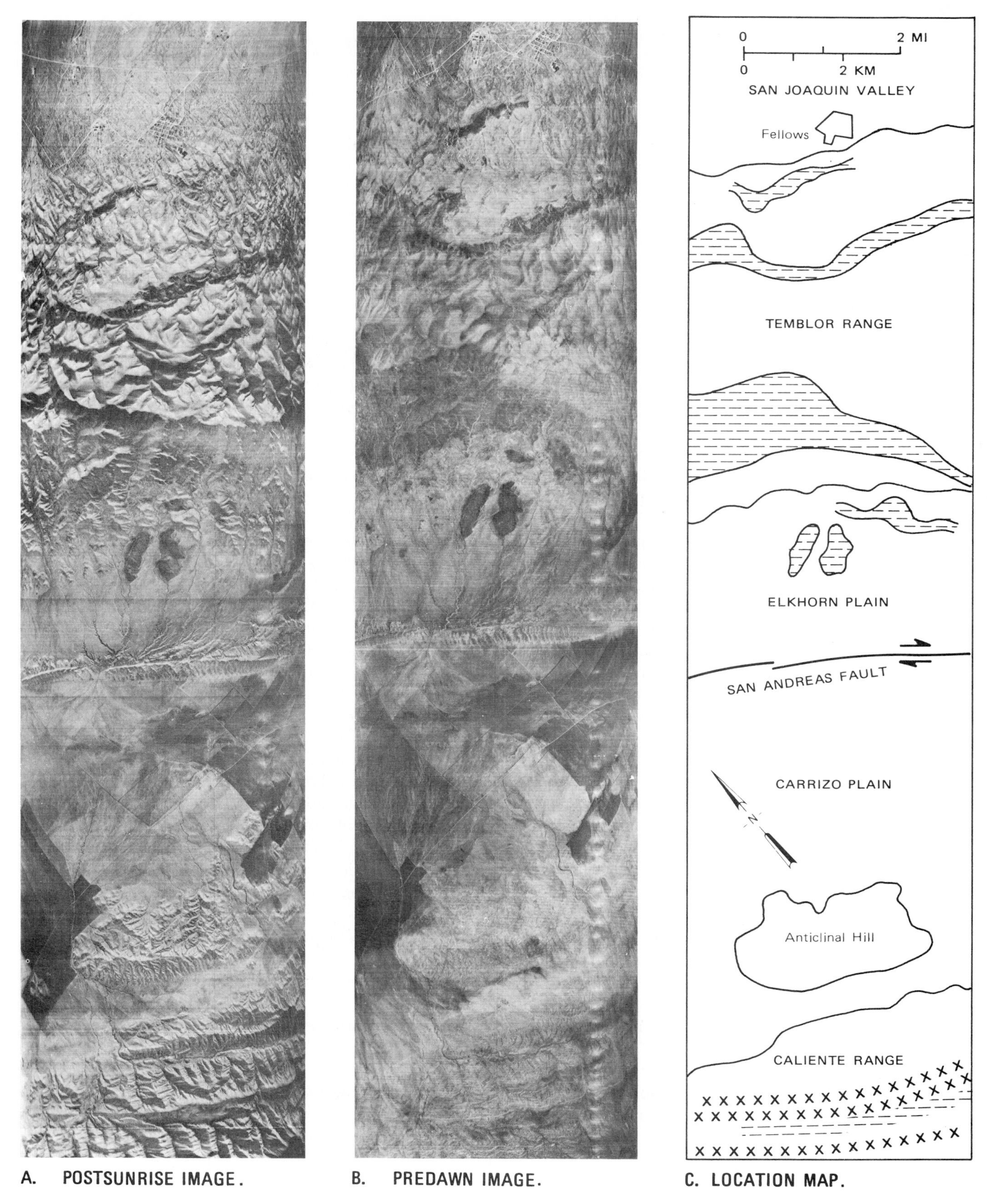

FIGURE 5.15
Comparison of day and night thermal IR images (8 to 13 μm) of Caliente and Temblor Ranges, California. The crosses on the map are basalt outcrops and the dashes are shale outcrops. From Wolfe (1971, Figures 3 and 4). Courtesy E. W. Wolfe, U.S. Geological Survey.

crops. In the Temblor Range the bands with cool signatures are outcrops of shale and siltstone. The broad belts of warm signature are outcrops of sandstone and conglomerate.

In the Arbuckle Mountains of Oklahoma, Rowan and others (1970) concluded that predawn imagery is most useful for identifying rock types and mapping fracture zones. On high-altitude thermal IR imagery of the Colorado Front Range flown two hours after sunrise, Offield (1975, p. 496) noted that the differential heating and shadowing effects aided in recognizing structural features that had subtle topographic expressions. As seen in the curves of diurnal temperatures (Figure 5.5), radiant temperatures are relatively constant in the predawn hours. If a number of lines of imagery are to be acquired, more uniform results can be obtained from predawn flights. The mapping of thermal inertia requires both daytime and nighttime images as described later in this chapter during the discussion of thermal models.

Orientation and Altitude of Flight Lines

For geologic projects the regional structural strike or *tectonic grain* of the area may be known in advance. If the flight lines are oriented normal to the regional strike, the scan-line pattern will be parallel with the strike and may mask linear features. It is preferable to orient flight lines parallel with the regional strike, although scan lines are not a problem on the high-quality images acquired with modern scanner systems. Most imagery is now recorded on magnetic tape and the recording time per tape determines the maximum length of flight lines. The tape recorder shown in Figure 5.10 can record continuously for one hour. Flight altitude controls image scale, lateral ground coverage, and resolution. Flight altitude is meaningful only when expressed as height above average terrain and this notation is used in the accompanying image examples and charts. In some surveys *drape flying* is used to maintain a constant aircraft elevation above variations in terrain height. Generally, however, the survey aircraft maintains a constant altitude that is selected to provide optimum height above the average terrain elevation.

Figure 5.16 illustrates the scale of rectilinear imagery as a function of flight elevation above terrain. For nonrectilinearized imagery a similar chart can be constructed by calculating the average scale for the central two-thirds of the angular field of view of the scanner. The highly compressed outer margins of nonrectilinearized images are not used in compiling mosaics. Lateral ground coverage, normal to the flight line is related to flight elevation on the chart in Figure 5.17.

In order to illustrate images at different scales, the same flight line was repeated at four different altitudes with a scanner having a 120° angular field of view. The images are shown in Figure 5.18, which also illustrates the different ground coverage at different elevations. For many applications imagery acquired at 1,800 m above terrain is generally optimum. This is not advocated as a standard elevation for all imagery, but 1,800 m is a good trade off among the many factors that must be considered. At 1,800 m above the terrain, a scanner with a 120° angular field of view covers a ground swath 6.4 km wide, or 3.2 km on either side of the flight path. By spacing flight lines 4.8 km apart, 1.6 km of sidelap is obtained between adjacent strips of imagery. This allows a margin of error for any navigation problems and covers the project area without flying an excessive number of lines.

Spatial Resolution and Information Content of Images

Spatial resolution of IR imagery is determined by the flight altitude and by the instantaneous field of view of the detector, which typically ranges

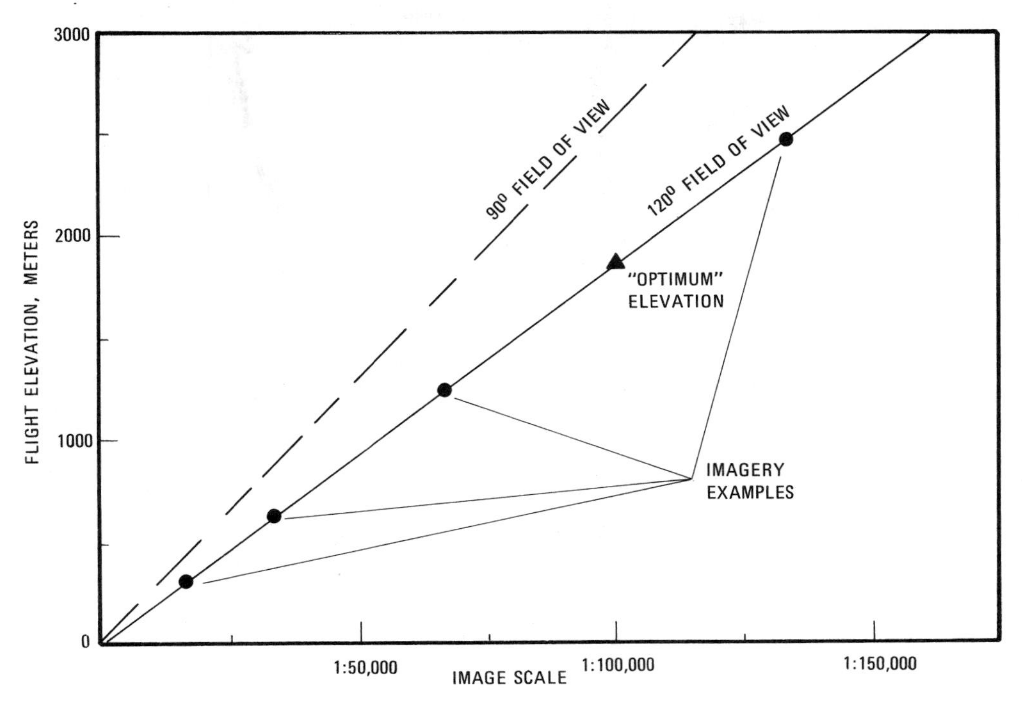

FIGURE 5.16
Image scale related to flight elevation above terrain and to angular field of view of scanner. This is for rectilinear images printed on 70 mm film. See Figure 5.18 for examples of images. From Sabins (1973A, Figure 2).

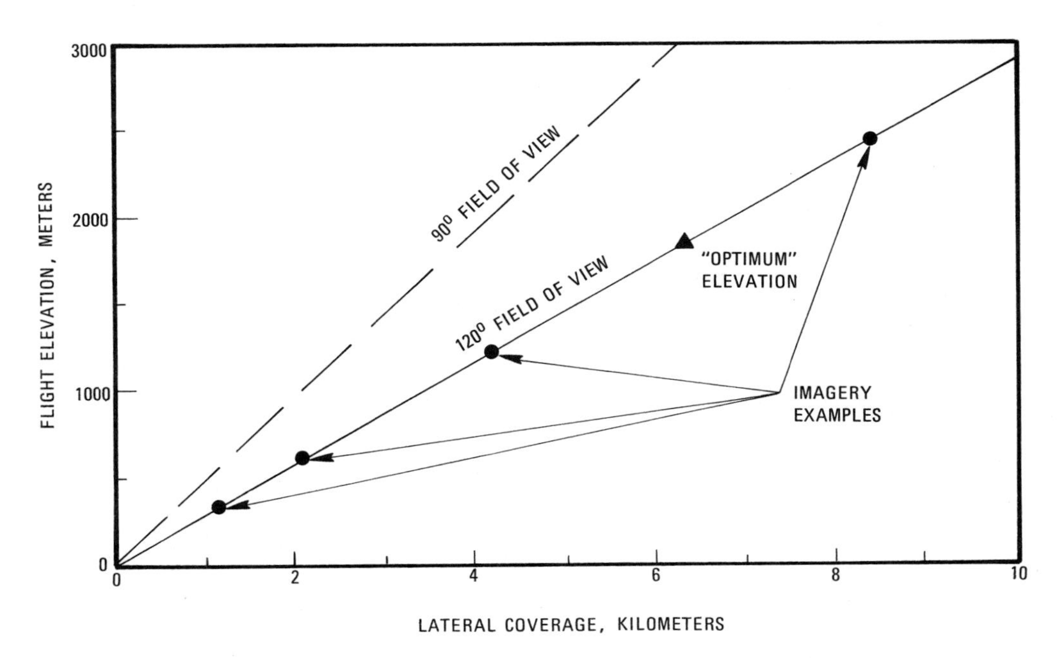

FIGURE 5.17
Lateral coverage of images related to flight elevation above terrain and to angular field of view of scanner. See Figure 5.18 for examples of images. From Sabins (1973A, Figure 4).

A. 2400 M ABOVE TERRAIN.

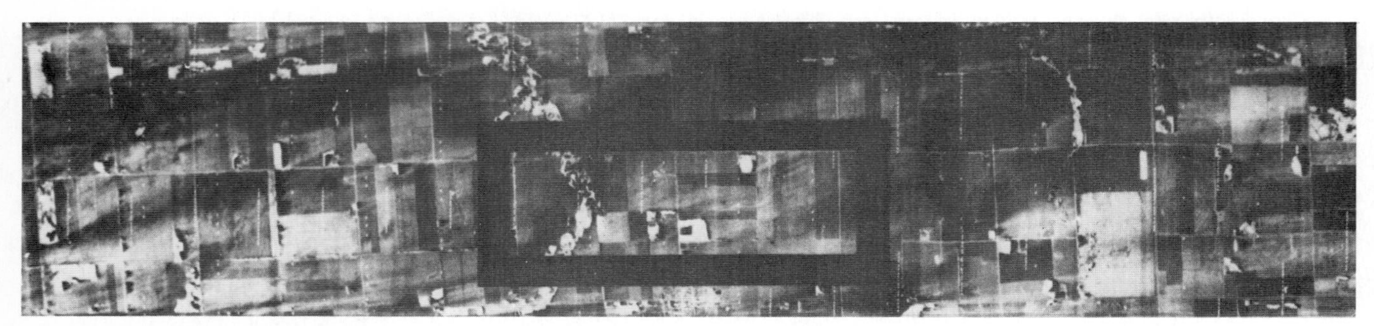

B. 1200 M ABOVE TERRAIN.

C. 600 M ABOVE TERRAIN.

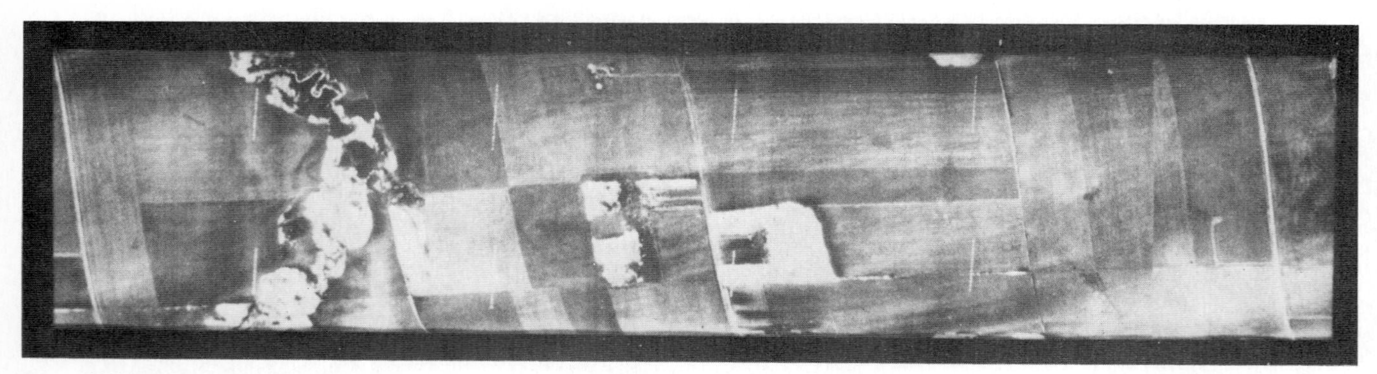

D. 300 M ABOVE TERRAIN.

FIGURE 5.18
Flight line repeated at four different altitudes to illustrate relationship to image scale. The same area is outlined on all images and is 0.8 km wide by 3.2 km long. Nighttime 8 to 14 μm images. From Sabins (1973A, Figure 3).

TABLE 5.3
Navigation methods for nighttime surveys

Method	Comments
Airway navigation aids	Designed for point-to-point navigation; of limited use for surveys.
Dead reckoning	Unsatisfactory because of winds aloft and lack of check points.
Recognition of ground features	May be useful during full moon and good visibility with adequate recognizable ground check points.
Portable light beacons	Requires ground parties with spotlights; satisfactory for single lines, but impractical for mosaics.
Inertial guidance systems and Doppler systems	Accurate and self-contained; expensive.
Very low frequency radio	Accurate and self-contained; moderately priced.

Source: From Sabins (1963A).

from 2 to 3 mrad. At a distance of 1,000 m, 1 mrad subtends an arc of 1 m. Along the flight line, for example, a 3 mrad detector at 2,000 m above terrain is able to resolve targets separated by more than 6 m, providing there is adequate thermal contrast between targets and the background. Resolution becomes lower (poorer) toward the margins of the imagery because the greater viewing distance increases the size of the ground resolution cell. Spatial resolution, however, is not the sole measure of image quality. Detection of thermal patterns is determined primarily by differences in radiant temperature, which are expressed as tonal contrast on the image.

One way to evaluate imagery acquired at different altitudes is to reproduce the imagery at a uniform scale. The imagery within the rectangles in Figure 5.18 was photographically enlarged or reduced to the same scale and illustrated in Figure 5.19. This example shows the progressive loss of detail as the ground resolution becomes poorer at higher altitudes. Good tonal contrast persists to the highest altitude, although the boundaries between adjacent cultivated fields are not as sharp at 2,400 m altitude as at lower altitudes. Imagery acquired at 1,800 m above terrain has ample information content for most applications.

Navigation at Night

The problem of navigating a pattern of regularly spaced parallel flight lines at night can be appreciated only after one has actually attempted it. The two phases of the navigation problem are: (1) fly from the air strip to a starting point in the survey area and (2) fly the mosaic pattern. A typical nighttime survey at 1,800 m above terrain requires a series of parallel lines 64 to 80 km long and spaced 3 to 5 km apart. A number of navigation methods are summarized in Table 5.3.

The very low frequency (VLF) radio navigation system described by Sabins (1973A) has been successfully employed in a number of IR surveys. Figure 5.20A is part of a larger mosaic of IR imagery acquired at night with VLF navigation that provided regularly spaced, straight flight lines. The high quality of this mosaic results not

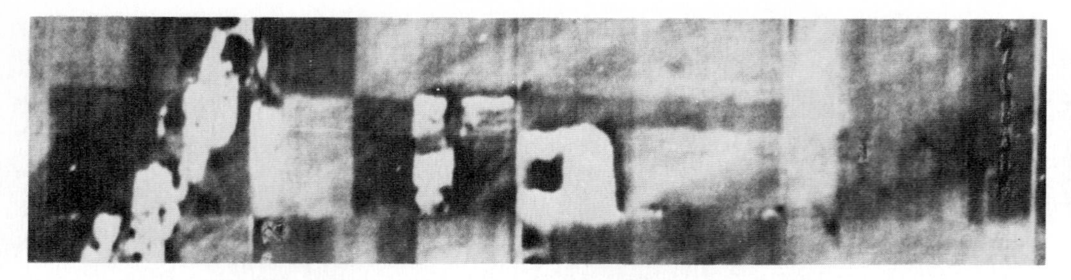

A. 2400 M ABOVE TERRAIN. 3X ENLARGEMENT OF ORIGINAL.

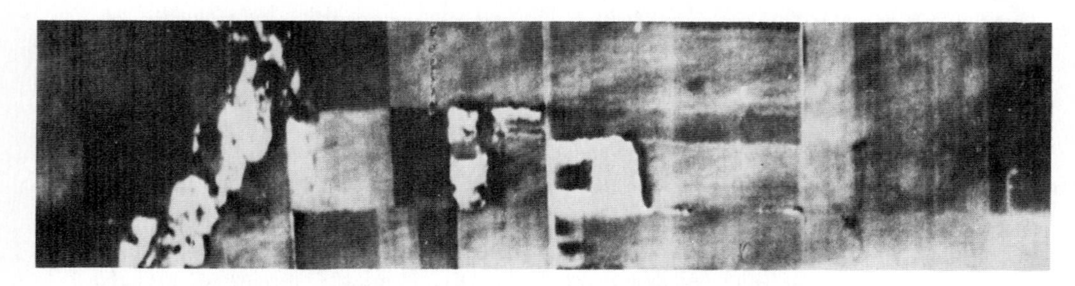

B. 1200 M ABOVE TERRAIN. 2X ENLARGEMENT OF ORIGINAL.

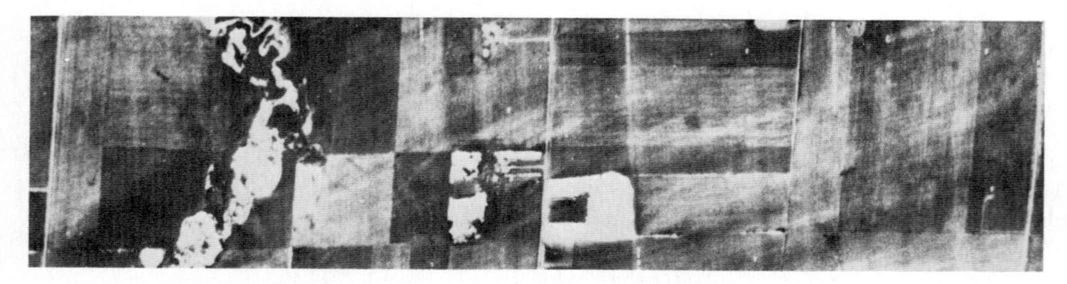

C. 600 M ABOVE TERRAIN. CONTACT PRINT OF ORIGINAL.

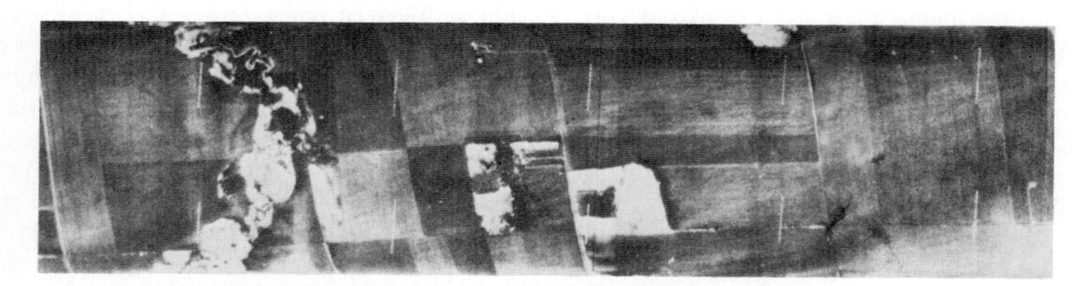

D. 300 M ABOVE TERRAIN. ½X REDUCTION OF ORIGINAL.

FIGURE 5.19
Images acquired at different elevations and photographically enlarged or reduced to a uniform scale. Each image is 0.8 km wide by 3.2 km long and is taken from the areas outlined in Figure 5.18. From Sabins (1973A, Figure 5).

A. MOSAIC OF NIGHTTIME THERMAL IR (8 TO 14 μm) IMAGE STRIPS.

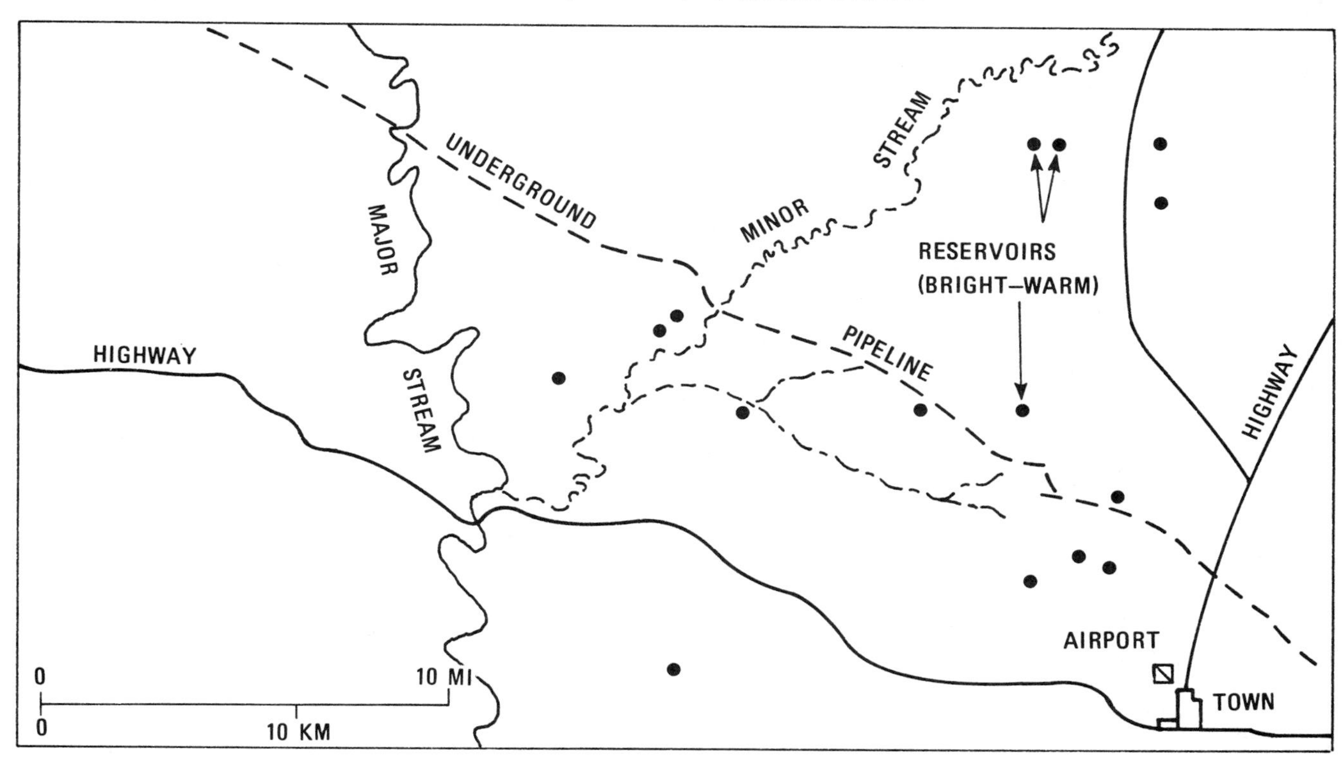

B. BASE MAP OF MOSAIC AREA.

FIGURE 5.20
Mosaic display of thermal IR images acquired at night using VLF navigation system.

only from the good navigation, but also from the magnetic tape recording of the imagery. By carefully adjusting film speed during image playback a constant image scale was obtained, despite wide variations in aircraft ground speed. The map (Figure 5.20B) illustrates the continuity of terrain features across the image strips.

Ground Measurements

Weather and surface conditions play a large role in determining terrain expression on IR images. It may be useful to collect ground information on weather conditions, soil moisture, and vegetation at the time of the IR survey. The term ground truth was coined for these measurements in the early days of remote sensing, but the term has been abandoned by most investigators. There is no reason that measurements made on the ground are more truthful than those made remotely. Ground measurements are most practical and useful for surveys of relatively small areas that can be covered with a single flight line. If repeated flights are made, ground data on changing weather conditions and solar flux may be useful in comparing and interpreting the images.

For larger areas, such as the one shown in Figure 5.20, ground measurements can be made at only a limited number of localities during the 3 to 4 hours needed to acquire the images. Ground measurements are most valuable if they are made at localities that have anomalous image signatures. The measurements may help explain the anomalies or may eliminate possible causes. In practice, however, it is difficult to anticipate where anomalies will occur; therefore contemporaneous ground measurements are omitted in most regional surveys. The availability of calibrated thermal IR images also reduces or eliminates the need for ground measurements.

Ground measurements to help explain image signatures may be made some time after the image flight, as is discussed later in this chapter in reference to the Indio Hills example. Ground information is useful for understanding other forms of remote-sensing imagery in addition to thermal IR images. The type of ground measurements used depends upon the wavelength region of the airborne sensor.

THERMAL MODELS

There are two basic approaches to understanding the significance of temperature signatures on thermal IR images. The first approach is to make an *empirical* correlation between image signatures and the corresponding ground features. Warm and cool areas on the images can be matched with localities on the ground to determine the material responsible for the signature. This empirical method is rapid and direct, but it does not consider the underlying physical causes for the thermal expression of different materials.

In the second approach a simplified mathematical abstraction, or *thermal model*, is calculated to relate the surface temperature to the physical properties of materials. The model is based on the physical properties of the materials and the thermal processes that act on the materials. The advantage of the model approach is that it analyzes temperature variations in terms of physical properties and processes and enables the investigator to understand their interactions.

Derivation of Thermal Models

K. Watson and coworkers of the U.S. Geological Survey defined a mathematical model to predict ground temperature based on physical properties of materials (rocks) and the diurnal solar heating cycle. It is assumed that each material is a semi-infinite solid with homogeneous thermal properties. The following derivation of the model is summarized from Watson (1971).

The radiant flux, F, from the sun has a cyclic daily variation shown by the curve of Figure 5.4A. Solar flux into the ground can be written as

$$F = F_0 \cos \omega t \qquad (5.11)$$

where F_0 is the flux at time $t = 0$, which is the time of maximum solar heating at noon. The diurnal angular frequency ω is defined as

$$\omega = \frac{2\pi}{p} \qquad (5.12)$$

where the period, p, is one day. The kinetic temperature, T_{kin}, at the surface for this periodic flux can be derived as

$$T_{kin} = \frac{F_0}{(P\omega)^{1/2}} \cos\left(\omega t - \frac{\pi}{4}\right) \qquad (5.13)$$

where P is thermal inertia, which was defined earlier.

This statement assumes that all materials are heated in the same manner and ignores the fact that dark materials with low albedos absorb more solar energy than light materials with high albedos. In order to correct this assumption the effect of solar heating, or insolation, must be considered. If atmospheric effects are ignored insolation, I, may be expressed as

$$I = I_0 \cos Z \qquad (5.14)$$

where I_0 is the solar constant, which is the rate at which solar radiation is received outside the earth's atmosphere on a surface normal to the incident radiation and at the earth's mean distance from the sun. The sun's *zenith angle,* Z, is the angle between the sun's rays and the zenith, which is the vertical line from the point of observation. The change of the sun's zenith angle with time for different latitudes and seasons is described by the expression

$$\cos Z = \cos l \cos \delta(\cos \omega t + \tan l \tan \delta) \qquad (5.15)$$

where l is the local latitude and δ is the sun's declination. The model for flux absorbed by the ground is expressed as

$$I = I_0(1 - A) \cos Z \qquad (5.16)$$

where A is the albedo of the ground. Therefore (1–A) denotes the fraction of the insolation that is absorbed. The ground loses heat primarily by radiation, which was given by the Stefan–Boltzmann law (5.5).

The original simple flux boundary condition

of Equation (5.11) is now replaced by the relationships

$$F = I_0(1 - A)\cos Z - \sigma T_{\text{kin}}^4 \qquad \text{daytime} \quad (5.17)$$

$$F = -\sigma T_{\text{kin}}^4 \qquad \text{nighttime} \quad (5.18)$$

where the Stefan–Boltzmann constant, σ, is

$$5.67 \cdot 10^{-12}\,\text{W} \cdot \text{cm}^{-2} \cdot {}^\circ\text{K}^{-4}.$$

The resulting revision of Equation 5.13 is used with a computer to calculate the diurnal variations of surface temperature. Typical results of the model calculations are illustrated in Figure 5.21. The effects of the two rock properties used in the model, thermal inertia and albedo, are shown in the upper two diagrams of this figure. For materials of the same albedo those with a high thermal inertia go through a smaller temperature change than those with low thermal inertia (Figure 5.21A). During the day thermal contrast is greatest one hour after noon; at night the maximum contrast occurs just before dawn. For materials with the same thermal inertia but different albedos (Figure 5.21B) the maximum thermal contrast occurs near noon and the minimum contrast at dawn. Note that dark materials (low albedo) have higher overall temperatures and a greater temperature range than light materials (high albedo). Therefore the predawn hours are the optimum time to record temperature contrasts due to differences in thermal properties of the materials because the effects of thermal inertia are at a maximum and the effects of albedo are at a minimum.

The kinetic surface temperatures calculated from the thermal model may be converted into radiant temperatures by using Equation (5.8) and measured emissivity values for the rocks in Figure 5.21C. Thermal models have been used in two ways: (1) To predict temperatures for materials, based on known values of albedo and thermal inertia for the materials. This approach was used with imagery of the Mill Creek area, Oklahoma. (2) To determine thermal inertia values of materials based on observed maximum and minimum radiant temperatures during a diurnal cycle. Imagery of Pisgah Crater, California was analyzed in this fashion, as described in Chapter 11.

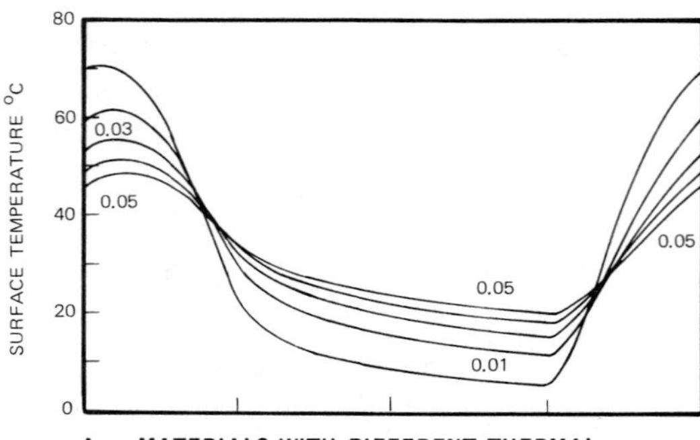

A. MATERIALS WITH DIFFERENT THERMAL INERTIAS.

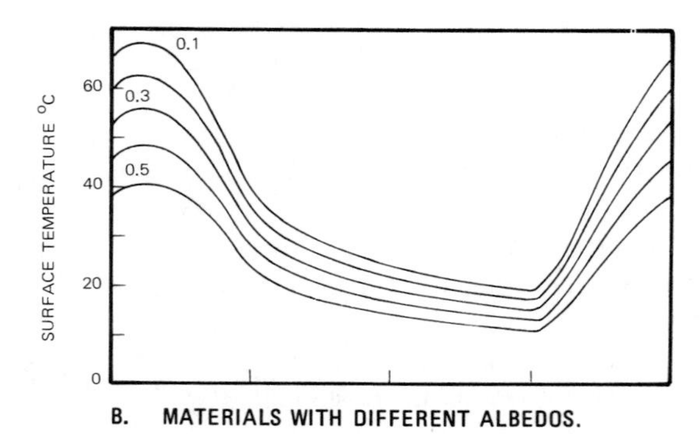

B. MATERIALS WITH DIFFERENT ALBEDOS.

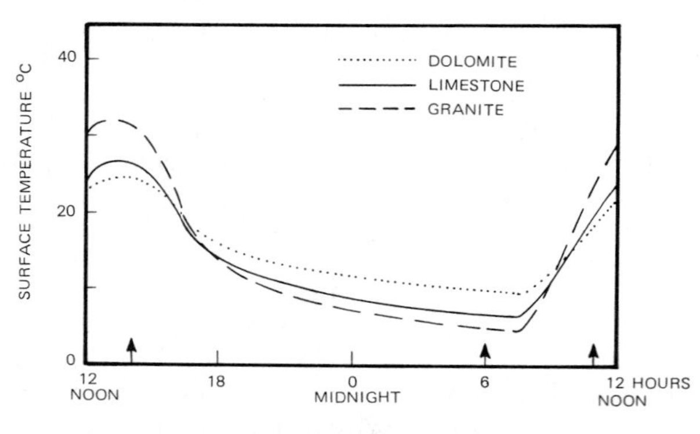

C. LIMESTONE, DOLOMITE, AND GRANITE OUTCROPS FROM MILL CREEK, OKLAHOMA. ARROWS SHOW TIMES AT WHICH IMAGES OF FIGURE 5.22 WERE ACQUIRED.

FIGURE 5.21
Mathematical models showing diurnal variation in surface temperatures calculated for different thermal inertias, albedos, and rock types. From Watson (1971).

Thermal Model and Images of Mill Creek Area, Oklahoma

In December, 1968 the U.S. Geological Survey acquired thermal IR images of the Mill Creek area, Oklahoma. The images shown in Figure 5.22 were acquired at 6 A.M., 11 A.M., and 2 P.M. The rock outcrops are limestone, dolomite, and granite that are well exposed with some lichen cover. Watson (1971) used his thermal model to calculate the diurnal variations in surface temperature for these rock types. The rock properties listed in Table 5.4 were used together with the site latitude of 34.5°N and sun declination of −23.3° to calculate the model curves shown in Figure 5.21C. The arrows indicate the times at which the images of Figure 5.22 were acquired.

TABLE 5.4
Rock properties used in thermal model at Mill Creek, Oklahoma

Rocks	*P* Thermal inertia, $cal \cdot cm^{-2} \cdot sec^{-1/2} \cdot {}^{\circ}C^{-1}$	*A* Albedo (no units)
Granite	0.023	0.19
Limestone	0.036	0.22
Dolomite	0.058	0.15

Source: From Watson (1971, Table 1).

The accuracy of the temperatures predicted by the model may be evaluated by comparison with the temperature signatures on the corresponding images. On the 6 A.M. image (Figure 5.22A) dolomite has the warmest signature, granite the coolest, and limestone has an intermediate temperature. These values agree with the 6 A.M. temperatures predicted for these rocks on the curves of Figure 5.21C. On the 11 A.M. and 2 P.M. images, however, the rock types have similar temperature signatures. Any temperature differences caused by rock type are obscured by the larger temperature differences caused by topography. This finding reinforces the earlier arguments for acquiring imagery in predawn hours in order to discriminate types of surface materials.

The theoretical model was later used to generate a thermal-inertia map from the emittance and reflectance data acquired by the Nimbus 3 and 4 satellites by Pohn, Offield, and Watson (1974). Their technique is based on the concept that thermal inertia values can be derived from measurements of the amplitude of the diurnal temperature variation in conjunction with reflectance data. The low-resolution (8 km) satellite data precluded the need to correct for topographic slope effects. The day to night temperature difference used to estimate the amplitude of the diurnal temperature variation was shown to be reasonably insensitive to elevation changes. The technique was subsequently automated (Watson, 1975) and a theoretical basis for correction of topographic effects was developed. The technique was subsequently applied to produce a thermal inertia image of the Raft River area, Idaho (Sawatzky, Kline, and Watson, 1975).

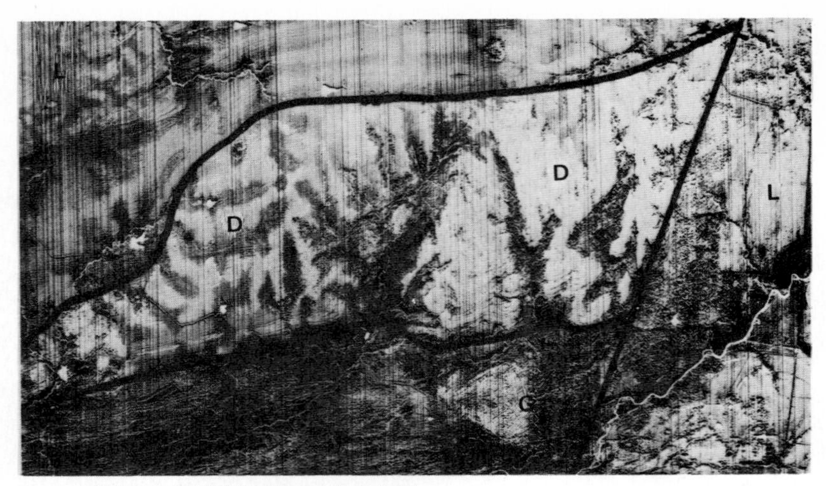

A. IMAGE ACQUIRED AT 6 A.M.

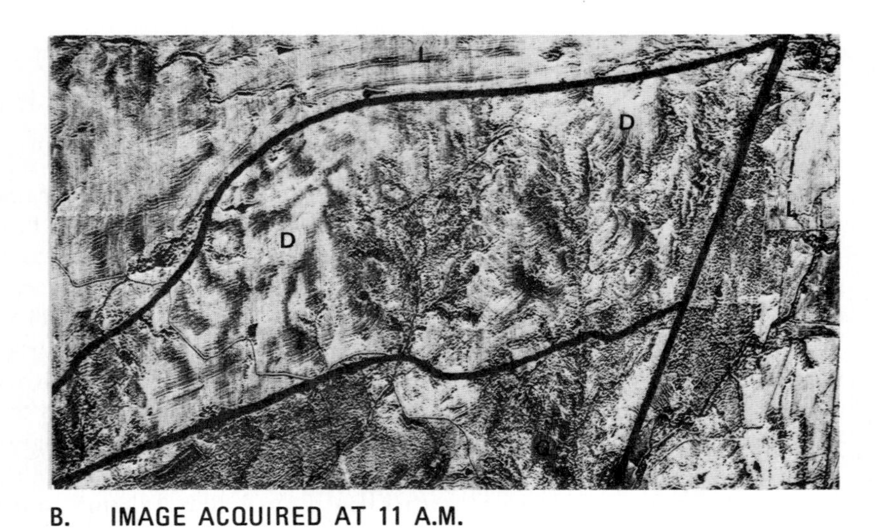

B. IMAGE ACQUIRED AT 11 A.M.

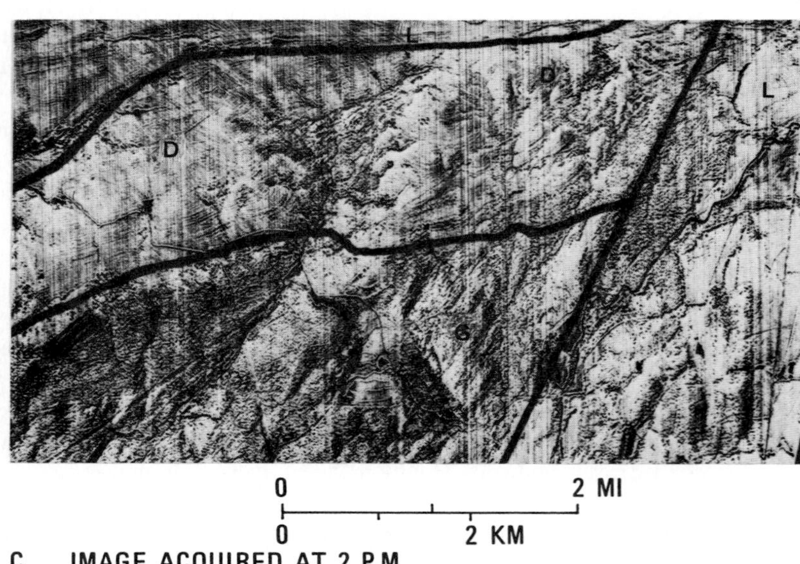

C. IMAGE ACQUIRED AT 2 P.M.

FIGURE 5.22
Thermal IR images (8 to 14 μm) of Mill Creek area, Oklahoma. L = limestone, D = dolomite, G = granite. From Watson (1971). Courtesy K. Watson and T. W. Offield, U.S. Geological Survey.

TABLE 5.5
Comparison of thermal inertia values for basalt and cinders from Mount Lassen and Pisgah Crater

Locality	Thermal inertia, $cal \cdot cm^{-2} \cdot sec^{-1/2} \cdot {}^{\circ}C^{-1}$	
	Basalt	Cinders
Mount Lassen, calculated from K, ρ, and c	0.038 to 0.049	0.011 to 0.012
Pisgah Crater, estimated from thermal inertia map	0.05 ± 0.005	0.02 ± 0.005

Thermal Inertia Mapping of Pisgah Crater, California

Kahle, Gillespie, and Goetz (1976) of the Jet Propulsion Laboratory developed a model to calculate thermal inertia when data are given for the diurnal temperature range ΔT, albedo, and topography. Values for ΔT are determined by comparing radiant temperature values from calibrated IR images acquired during the day and night. Albedo is measured from images of the visible band (0.4 to 0.7 μm). The model also considers sky and ground radiation, heat conducted into the ground, and the computed thermal effect of measured wind velocity and air temperature.

This model was used to calculate thermal inertia of surface materials at Pisgah Crater, a young cinder cone surrounded by basaltic lava flows 64 km east of Barstow in the California desert. Figure 5.23 shows the albedo image and the day and night thermal IR images acquired March 30, 1975 by a NASA remote-sensing aircraft. The value of ΔT was determined by digitally registering the day and night IR images and subtracting the radiant temperature value for each 3-by-3 m ground resolution cell. The albedo, the ΔT, and meteorologic data were processed using the thermal model to calculate thermal inertia, which is illustrated by the map of Figure 5.24B. On this map the black pattern indicates surface material with low thermal inertia values of about 0.02 cal $\cdot$ $cm^{-2} \cdot sec^{-1/2} \cdot {}^{\circ}C^{-1}$. Comparison with the geologic interpretation map (Figure 5.24A) shows that these areas of low thermal inertia correlate with alluvium, eolian sand, and the cinders at Pisgah Crater. The light and intermediate tones of gray correspond to higher thermal inertia values of about 0.05 cal $\cdot$ $cm^{-2} \cdot sec^{-1/2} \cdot {}^{\circ}C^{-1}$ and correlate with the basalt flows. Laboratory measurements of thermal inertia are not available for the rocks at Pisgah Crater. At Mount Lassen, however, similar basalt and cinders occur and thermal inertia values have been calculated from data given by Quade and others (1970). In Table 5.5 the calculated values for Mount Lassen cinders and basalt agree very well with those determined from the thermal model for similar rocks at Pisgah Crater.

The Jet Propulsion Laboratory thermal model is designed to remove the effect of topography on solar heating, but digital topographic data were not available when the thermal inertia map was made. For that reason the shadowed area on the north flank of Pisgah Crater (Figure 5.24B) has an anomalously high thermal inertia value. Topographic effects also occur in the area of older alluvium west of the road (Figure 5.24A). Comparison of the thermal inertia map (Figure 5.24B) with the albedo image shows that the north-facing shaded slopes of the gullies have an apparent higher thermal inertia than the south-facing, sunlit slopes. Areas that are shadowed have a lower daytime temperature; hence the ΔT between their day and night temperatures is lower and gives the effect of a higher thermal inertia for the material. This emphasizes the importance of correcting for topographic effects in thermal models. Thermal inertia information alone does not identify geologic units but can be useful in separating materials of similar albedo and different thermal properties, such as basalt and cinders.

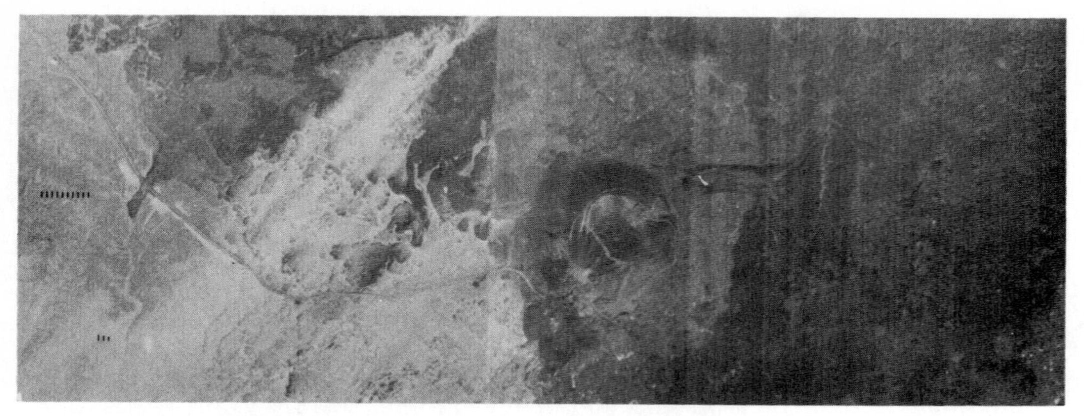

A. DAYTIME REFLECTANCE, OR ALBEDO, IMAGE (0.4 TO 0.7 μm) ACQUIRED AT 2:00 P.M.

B. DAYTIME THERMAL IR IMAGE (8 TO 14 μm) ACQUIRED AT 2:00 P.M.

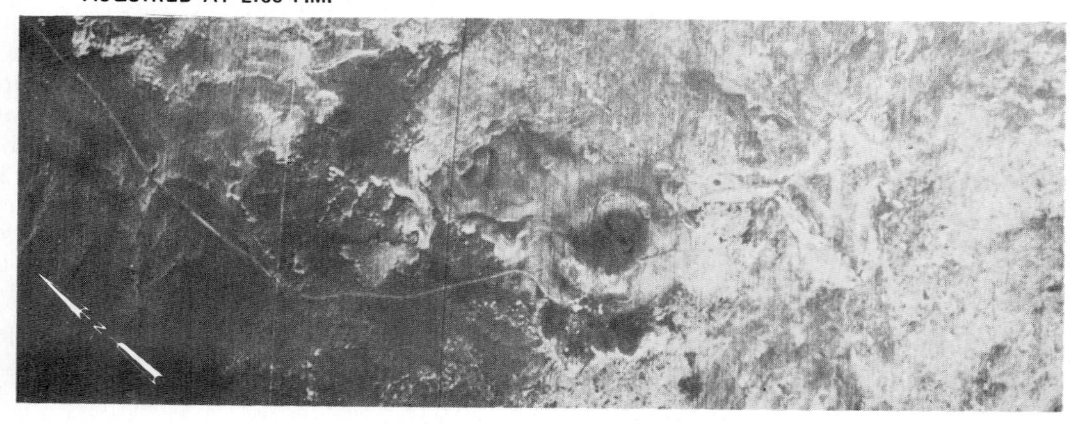

C. NIGHTTIME THERMAL IR IMAGE (8 TO 14 μm) ACQUIRED AT 5:00 A.M.

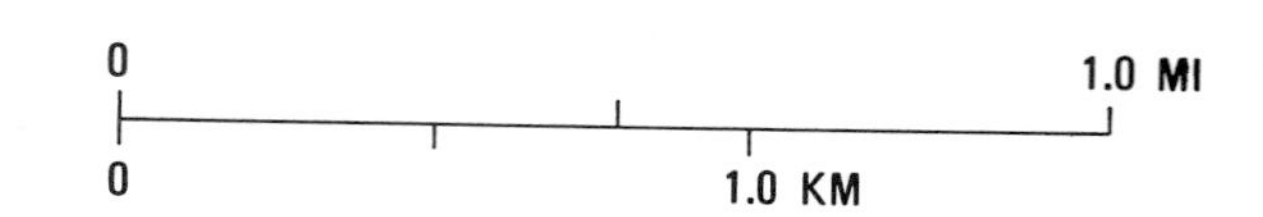

FIGURE 5.23
Thermal IR and visible images of Pisgah Crater, California acquired March 30, 1975 by NASA and processed by Jet Propulsion Laboratory. Courtesy A. B. Kahle, Jet Propulsion Laboratory, California Institute of Technology.

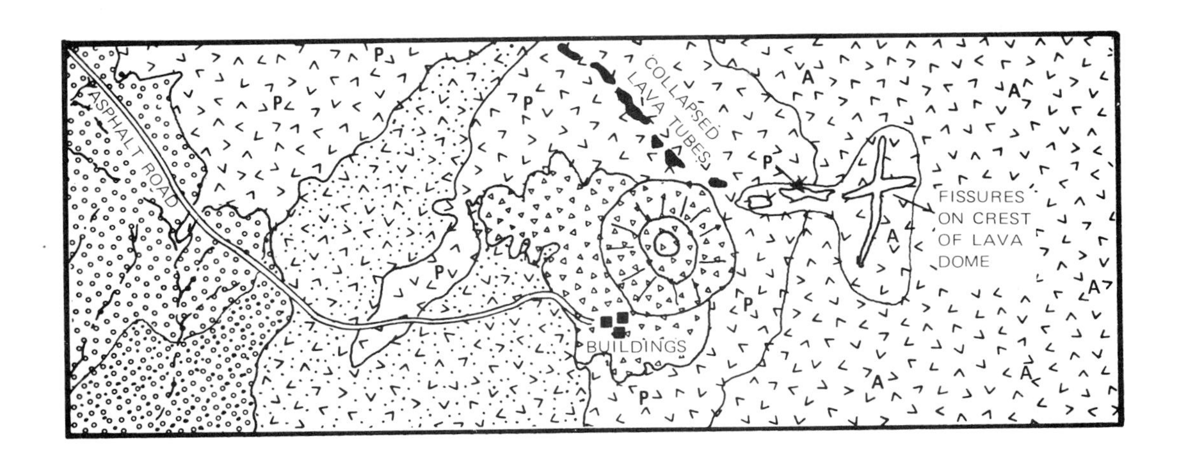

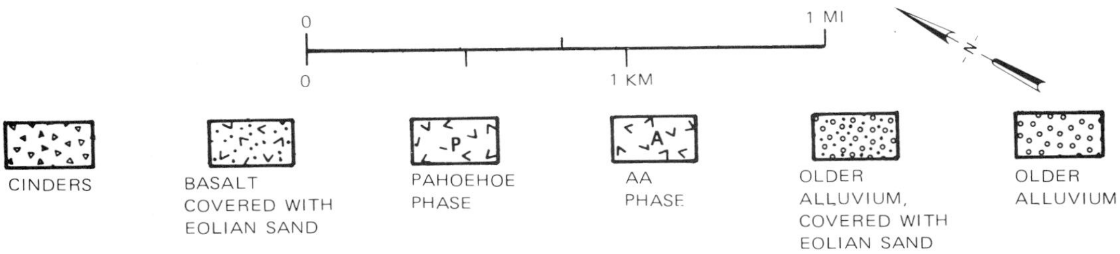

A. GEOLOGIC INTERPRETATION OF THERMAL IR IMAGES.

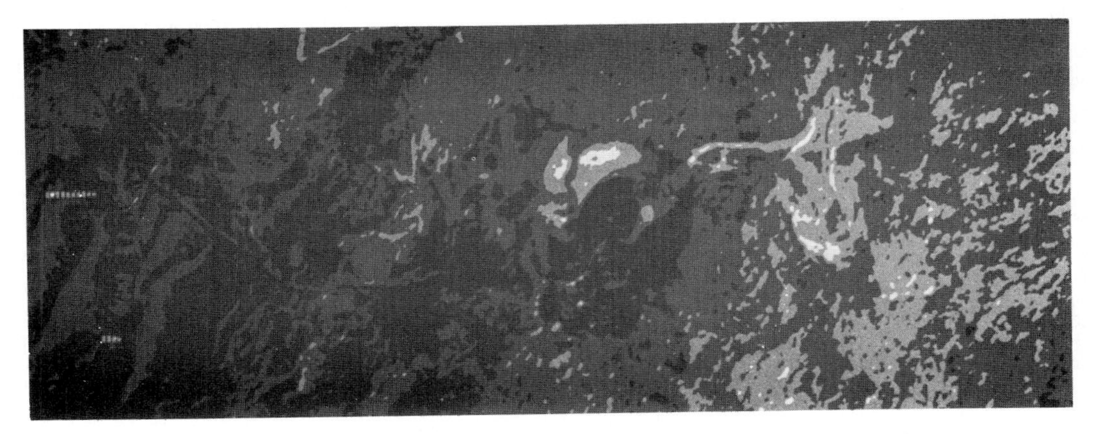

B. THERMAL INERTIA MAP. BLACK PATTERN REPRESENTS LOW THERMAL INERTIA OF ABOUT 0.02 CAL CM^{-2} $SEC^{-1/2}$ $^{o}C^{-1}$. GRAY PATTERNS ARE ABOUT 0.05 CAL CM^{-2} $SEC^{-1/2}$ $^{o}C^{-1}$. COURTESY JET PROPULSION LABORATORY.

FIGURE 5.24
Geologic interpretation map and thermal inertia map of Pisgah Crater. Thermal inertia map courtesy A. B. Kahle, Jet Propulsion Laboratory, California Institute of Technology.

INDIO HILLS, CALIFORNIA

The Indio Hills, in the eastern Coachella Valley, are a low range of deformed clastic sedimentary rocks of late Tertiary age trending southeastward parallel with the San Andreas fault zone. A geologic map (Figure 5.25), aerial photograph, and nighttime IR image (Figure 5.26) cover the south end of the hills. This arid terrain with little vegetation and well-exposed bedrock is ideal for acquiring and analyzing IR images.

Thermal IR Signatures of Rock Types

The alluvium surrounding the hills has a relatively cool and featureless signature on the nighttime image. Two types of bedrock may be readily distinguished. One type has a relatively warm and uniform signature and consists of poorly stratified, moderately to poorly consolidated conglomerate of the Ocotillo, Canebrake, and Mecca Formations (Figure 5.25). The outlying hill in the lower center of the image (locality 1.3, A.8 to D.5 of Figure 5.26B) is a good example. The other bedrock type is the Palm Spring Formation, consisting of well-stratified alternating beds of resistant conglomeratic sandstone and nonresistant siltstone up to 12-m thick that weather to the ridge and slope topography illustrated in Figure 5.27. This well-stratified bedrock has a distinctive pattern of alternating warm and cool bands on the nighttime IR image. Careful correlation of the image with outcrops in the field established that the warm signatures are sandstones and the cool signatures are siltstones. This observation was later verified by radiometer studies.

Radiometer Investigations

In January, 1969 a portable radiometer was used to measure radiant temperatures of eight sandstone and eight siltstone outcrops at locality 3.5, B.7 of Figure 5.26B, part of which is illustrated in Figure 5.27. Temperatures were measured at four different times, but only the midnight and daytime data are shown in Figure 5.28. At midnight the sandstones are warmer than the siltstones, but when the outcrops are fully illuminated by 8:35 A.M., the siltstones are much warmer (Figure 5.28). Additional measurements indicated that the maximum temperature changes occurred at dawn within an hour's time. The field localities were selected for uniformity of exposure, but the association of sandstones with ridges and siltstones with slopes suggested that topography might control the radiant temperature. This point was investigated by placing large slabs of sandstone and siltstone on the ground and making day and night radiometer measurements. These temperatures matched those measured at the outcrops and correlated with the nighttime image signatures. This experiment proved that the physical properties of these rocks, rather than their topographic expression, determine the radiant temperature at night.

Thermal properties for siltstone are not given in Table 5.2, but shale (no. 18) is a similar rock type. Thermal inertia is lower for shale (0.034) than for sandstone (0.054), which is consistent with their relative densities. The daily variation in radiant temperature for sandstones of the Palm Spring Formation follows the curve for materials with higher thermal inertia shown in Figure 5.4B. The siltstones follow the curve for lower thermal inertia because their radiant temperature reaches higher and lower extremes than materials with higher thermal inertia.

The sparse patches of vegetation have conspicuous warm signatures on the nighttime image. At locality 2.4, G.2 in Figure 5.26B the irregular bright rings are large patches of mesquite that are alive and green at the margin but dead in the center. In contrast to the warm nighttime expression, vegetation typically appears cool in daytime images. In order to evaluate this phenomenon quantitatively, radiometer measurements were made at locality 3.1, F.3 in Figure 5.26B where three salt cedar trees have distinctly warmer signatures than the surrounding bare soil. Beginning at noon, radiometric temperature measurements were made for the salt cedars plus three smaller creosote bushes, and the values were averaged. For each observation period, radiometric temperature measurements of six soil exposures were also averaged. The results are plotted in Figure 5.29 together with air temperature readings. This diagram shows that at night vegetation is consistently warmer than the soil with a maximum temperature difference of 4°C. The temperature relationships are reversed during the day when the soil is much warmer than vegetation. Note the thermal crossovers that occur in the evening and morning. These diurnal temperature relationships of soil and vegetation have since been confirmed on day and night IR images at many localities.

Folds and Faults

Geologic structures are well expressed on the image. The plunging anticlines and synclines in the Palm Spring Formation are indicated by the pattern of the warm and cool signatures. The San Andreas fault borders the west side of the Indio Hills; to the south it passes along the east side of the outlier of Ocotillo Conglomerate; farther south the fault trace is concealed by alluvium and has no topographic expression. On the aerial photograph the concealed trace of the fault is marked on the northeast side by a vegetation anomaly, which is a concentration of vegetation that abruptly terminates at the fault trace (locality 1.6, D.2 to 1.3, G.3 in Figure 5.26A). On the IR image the fault trace is expressed by an alignment of very cool (dark) anomalies along the northeast side (locality 1.6, E.1 to 1.6, G.3 in Figure 5.26B). The cool anomalies are not related to the vegetation distribution because vegetation appears warm on the nighttime image. The cool anomalies are probably related to the barrier effect of the San Andreas fault on ground water movement. In the spring of 1961, which was a few months before the IR survey, Cummings (1964, p. 34) reported a 15 m difference in elevation of the water table across the fault in this vicinity. The shallower water table and higher moisture content on the east side of the fault could have caused evaporative cooling. Similar anomalies appear on nighttime images of the San Andreas fault in the Carrizo Plain 320 km to the northwest, which are described by Wallace and Moxham (1967). On IR images of the northern Indio Hills (Sabins, 1967, Figure 2) the trace of the Mission Creek fault is indicated by cool anomalies at Thousand Palms Oasis where the fault is a barrier to ground water movement.

Some anomalies on thermal IR images are difficult to explain. An example is the trough of a syncline at locality 3.3, A.0 to 3.7, A.9 in Figure 5.26B that is marked by a distinct tonal anomaly that is cool on the south side and warm on the north. The straight interface coincides with the axis of the syncline. Careful field checks show that there is no faulting or unusual fracturing nor is there any apparent stratigraphic or topographic cause of this anomaly. One possible explanation is that the synclinal structure influences moisture content in the rocks, but this is difficult to verify.

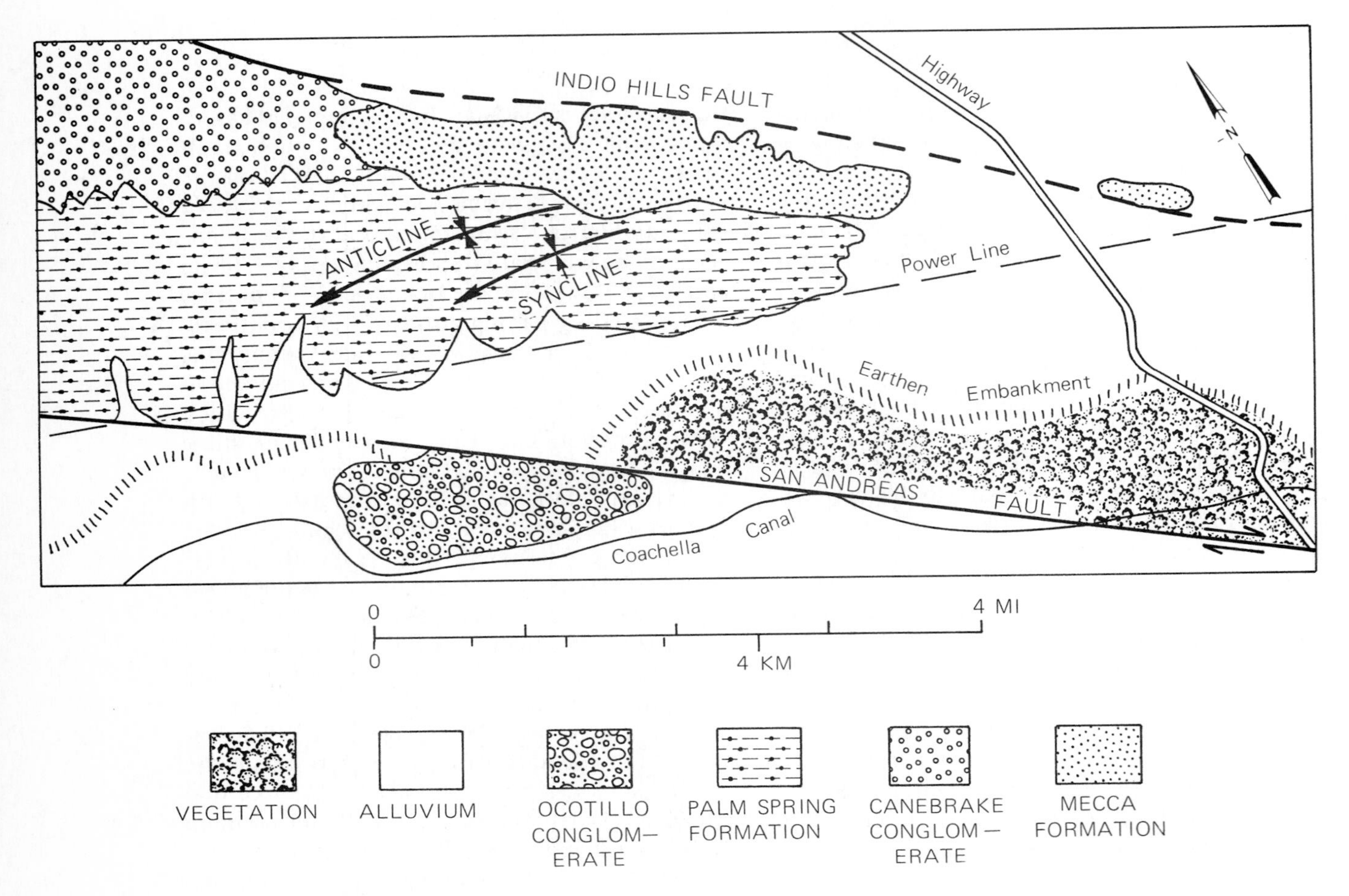

FIGURE 5.25
Geologic map of southern part of Indio Hills, Riverside County, California.

A. AERIAL PHOTOGRAPH ACQUIRED MAY 5, 1953.

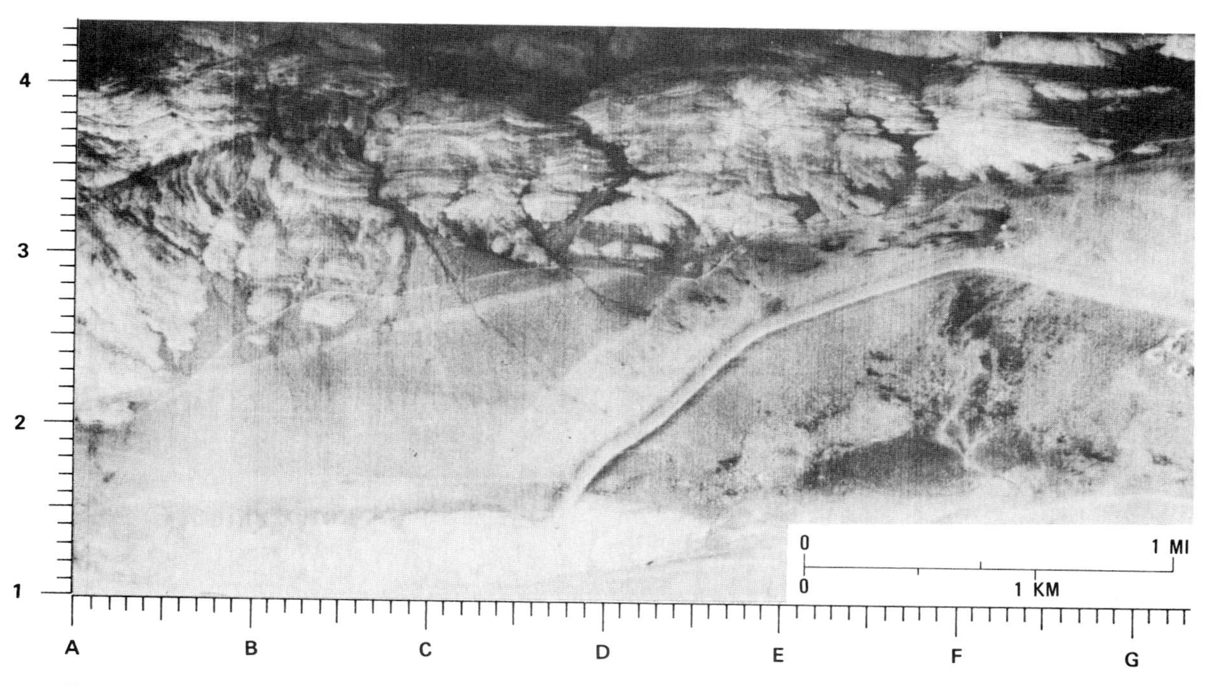

B. NIGHTTIME THERMAL IR IMAGE (8 TO 14 μm) ACQUIRED OCTOBER, 1963.

FIGURE 5.26
Southern part of Indio Hills, California. From Sabins (1967, Figures 3 and 4).

FIGURE 5.27
Ledge-forming sandstones (Ss) and slope-forming siltstones (St) of Palm Spring Formation at locality 3.4, B.7 of Figure 5.26B. Radiant temperatures of these outcrops are shown in Figure 5.28. Courtesy Eugene Borax, Pennzoil.

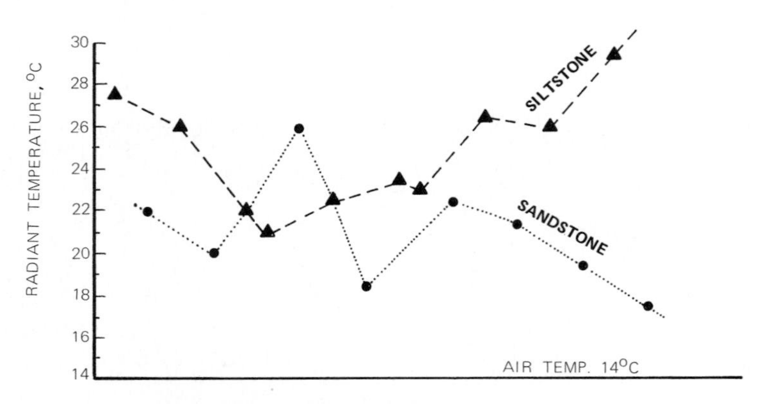

A. DAYTIME TEMPERATURES (8:45 A.M.). THE SILTSTONES ARE WARMER THAN THE SANDSTONES, WITH ONE EXCEPTION.

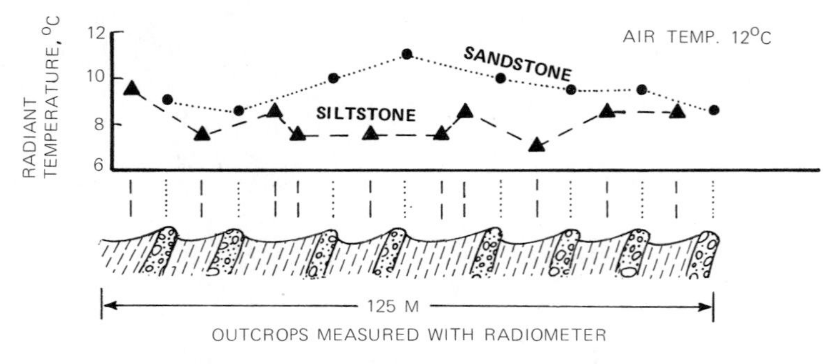

B. MIDNIGHT TEMPERATURES. THE SANDSTONES ARE WARMER THAN THE SILTSTONES, WITH TWO EXCEPTIONS.

FIGURE 5.28
Day and night radiant temperatures of sandstones and siltstones of Palm Spring Formation at locality 3.4, B.7 of Figure 5.26B.

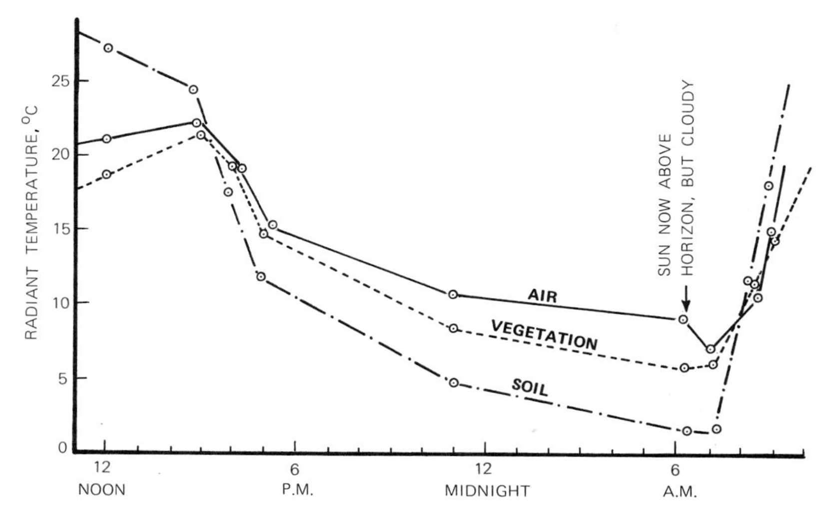

FIGURE 5.29
Diurnal radiant temperatures of vegetation and soil at Indio Hills. Locality 3.1, F.3 of Figure 5.26B.

IMLER ROAD ANTICLINE AND SUPERSTITION HILLS FAULT, CALIFORNIA

A nighttime IR image and an aerial photograph of this area on the west margin of the Imperial Valley are depicted in Figure 5.30 together with an interpretation map (Figure 5.31). There is a striking increase in geologic information on the IR image relative to the aerial photograph.

Terrain Expression

The irrigated fields in the south portion of the area are part of the Imperial Valley agricultural area. The Fillaree irrigation canal at the south edge of area has the typical warm nighttime signature of standing water. The very warm field area north of the canal (locality 1.8, C.0 in Figure 5.30A) was probably flooded with standing water to leach salt from the soil at the time the image was acquired. The very cool fields were probably damp from recent irrigation, resulting in evaporative cooling.

Most of the area is very low-relief desert terrain with sparse clumps of vegetation that localize patches of eolian sand on their east margin, downwind from the prevailing westerly winds. Larger sand dunes stabilized by mesquite trees appear warm on the image and dark on the aerial photograph. These are indicated on the interpretation map (Figure 5.31). The very warm Y-shaped feature at locality 3.2, B.1 in Figure 5.30A is a thick accumulation of eolian sand lodged against an earthen embankment. Imler Road in the northern part of the area is actually straight, but appears curved because of distortion caused by the IR scanner. The road was surfaced with hard packed sand when the image was acquired, but today it is paved with asphalt.

Bedrock in the area is the Borrego Formation (Pleistocene age), which consists of brownish gray siltstone with thin interbeds of well-cemented, brown sandstone. Flaggy pieces and concretions of sandstone litter the surface where it crops out. Light-colored, nodular, thin layers within the siltstone help define bedding trends within this monotonous sequence. On the nighttime image the siltstone is relatively cool and the sandstone is warm, corresponding to thermal signatures of similar rock types at the Indio Hills.

In addition to the dunes and patches of sand, a thin layer of eolian sand covers much of the area. The sand has a warmer signature on the image than the bedrock exposures. During Quaternary time, pluvial Lake Coahuilla covered this area. Except for travertine coatings on large boulders, any Lake Coahuilla deposits have been reworked by the wind.

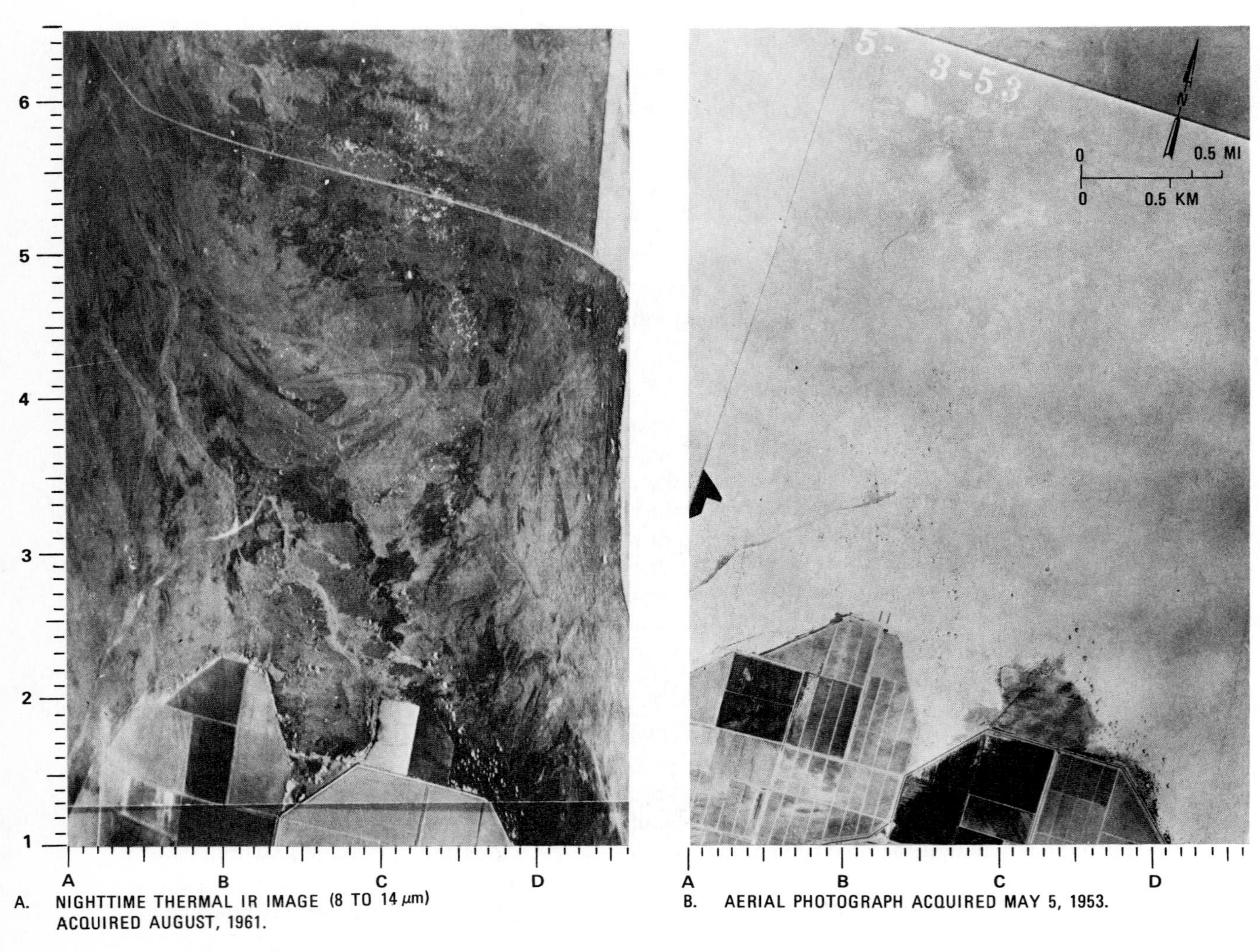

A. NIGHTTIME THERMAL IR IMAGE (8 TO 14 μm) ACQUIRED AUGUST, 1961.

B. AERIAL PHOTOGRAPH ACQUIRED MAY 5, 1953.

FIGURE 5.30
Imler Road anticline and Superstition Hills fault, Imperial County, California. From Sabins (1969, Plate 1).

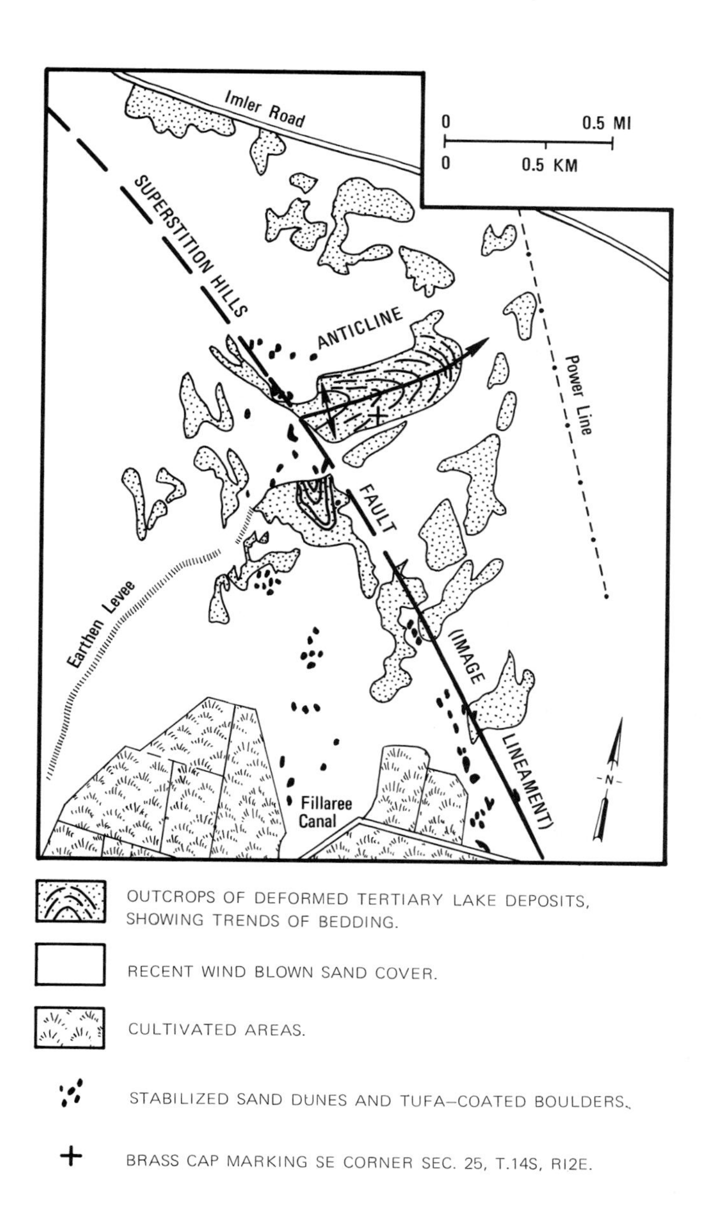

FIGURE 5.31
Interpretation of IR image of Imler Road anticline and Superstition Hills fault. From Sabins (1969, Figure 5).

Structural Geology

The following structural features are well expressed on the IR image but are absent or obscure on the aerial photograph: (1) the Imler Road anticline in the center of the area, which was not known prior to the image interpretation by Sabins (1969) and (2) the southeast extension of the Superstition Hills fault.

Anticline The east-plunging anticline in the center of the IR image (locality 4.2, B.7 in Figure 5.30A) is 1.2 km long and 0.4 km wide. Had this structure not been observed first on the image, one could have walked across the anticline in the field without recognizing it, for there are no conspicuous lithologic or topographic patterns. After the anticline has been located by referring to the image, the limbs and plunge can be defined by carefully walking along the outcrop of individual beds. Structural attitudes are obscure in the siltstone, but dips of up to 45° were measured in outcrops of the sandstones and the dip reversal across the fold axis was located. A broad, low ridge with up to 9 m of relief coincides with the general area of the anticline.

In Figure 5.31 the anticline is mapped as solid bedrock, but there are numerous thin patches of eolian sand that cause the local gray tones on the image. The core of the anticline consists of contorted siltstone with a very cool signature. The alternating warm and cool pattern outlining the anticline correlates with outcrops of sandstone (warm) and siltstone (cool) respectively. The west end of the anticline is abruptly truncated at the southeastward projection of the Superstition Hills fault. The inferred trace of the fault is obscured by sand, but siltstone outcrops in the immediate vicinity of the fault are strongly deformed, suggesting drag folding. South of the anticline there is a small arcuate pattern of alternating dark and light bands on the IR image (locality 3.6, B.5 in Figure 5.30A). Field inspection showed this to be an exposure of gently dipping siltstone and sandstone bedrock with an outcrop pattern generally resembling that on the image. The pattern is more pronounced on the image than on the ground.

Superstition Hills Fault This right-lateral, strike-slip fault was named for exposures in the Superstition Hills, 14 km northwest of the Imler Road area. The fault alignment shown in Figure 5.31 differs from that shown on the El Centro sheet of the "California Geologic Atlas." In addition to truncating the anticline, the fault is marked in the southeast part of the image by a southeast-trending linear feature that is cooler on the east and warmer on the west (locality 2.2, C.5 to 1.2, D.0 in Figure 5.30A). The trend of the linear tonal anomaly is parallel with, and about 0.2 km to the east of, the row of prominent sand dunes. On April 9, 1968 the Borrego Mountain earthquake caused surface breaks along the trace of the Superstition Hills fault that were mapped by A. A. Grantz and M. Wyss (in Allen and others, 1972, Plate 2). The portion of the map that covers the area of Figure 5.30A shows a series of breaks with less than 2.5 cm of right-lateral displacement. The trend of the breaks closely coincides with the linear anomaly on the image. The image, which was acquired four years prior to the earthquake, located an important fault that is obscure both on aerial photographs and in the field.

Comparison of IR Image and Aerial Photograph

The striking difference in tonal contrast and geologic information content between the IR image and the aerial photograph (Figure 5.30) is not caused by the eight-year difference in the dates when the images were acquired. This desert area is a relatively stable environment in which natural changes occur very slowly. During annual field trips over the past 10 years no significant changes have been noted in the area. The contrast and resolution of the aerial photograph are good and accurately record the low contrast of this area in the visible spectral region. Normal color and IR color oblique aerial photographs of the anticline are not significantly better than the black-and-white aerial photographs. For the rocks and surface materials in this area, the differences in thermal properties are greater than the differences in visible reflectance. This accounts for the greater contrast on the thermal IR image.

CALIENTE RANGE AND CARRIZO PLAIN, CALIFORNIA

The northwest-trending Caliente and Temblor Ranges are in the southern part of the Coast Range province and are separated by the Carrizo and Elkhorn Plains (Figure 5.32). This area is suited for IR investigations because of the semi-arid climate and moderate vegetation cover. In June, 1965 the U.S. Geological Survey acquired predawn and postsunrise thermal IR images along a northeast to southwest flight line across the trend of the ranges. These images provide an excellent opportunity to compare day and night thermal signatures of rock units. The geology and image expression of the Temblor Range are similar to that of the Indio Hills. Therefore this discussion concentrates on the Caliente Range, where basalt and mudstone are exposed.

Geologic Setting

The images of Figure 5.33 cover the northeast flank of the Caliente Range which consists of steeply northeast-dipping marine clastic strata and interbedded basalts, all of late Tertiary age (Figure 5.32). There are some minor faults and folds. Topography is shown on the postsunrise image and consists of parallel strike ridges and valleys formed by tilted, alternating resistant and nonresistant geologic units. The rectangular drainage pattern is formed by subsequent streams along the valleys and consequent streams cutting across the ridges. The stream channels were dry when the images were acquired. Brush and small trees are concentrated along the streams. Soil and debris mantle the range and extensive exposures of bedrock are rare. The gently sloping, alluvium-covered Carrizo Plain is used for farming and ranching. This setting contrasts with that of the Indio Hills, which are arid with bare outcrops that are devoid of soil and vegetation.

Image Interpretation

The images (Figure 5.33) were recorded directly on film and are geometrically distorted, which accounts for the curvature of the roads. Topography is clearly expressed on the postsunrise image and geologic units cannot be differentiated. Brush and trees along the dry streams are cool (dark tone) on the postsunrise image and warm (light tone) on the predawn image, consistent with the radiometer measurements of vegetation in the Indio Hills (Figure 5.29). Significant features are numbered on the images and described in the following discussion, which applies to the predawn image unless otherwise noted.

1. Subparallel warm linear trends are outcrops of steeply dipping sandstone beds and basalt flows.
2. Bright line is the bare outcrop of the uppermost of three basalt flows.

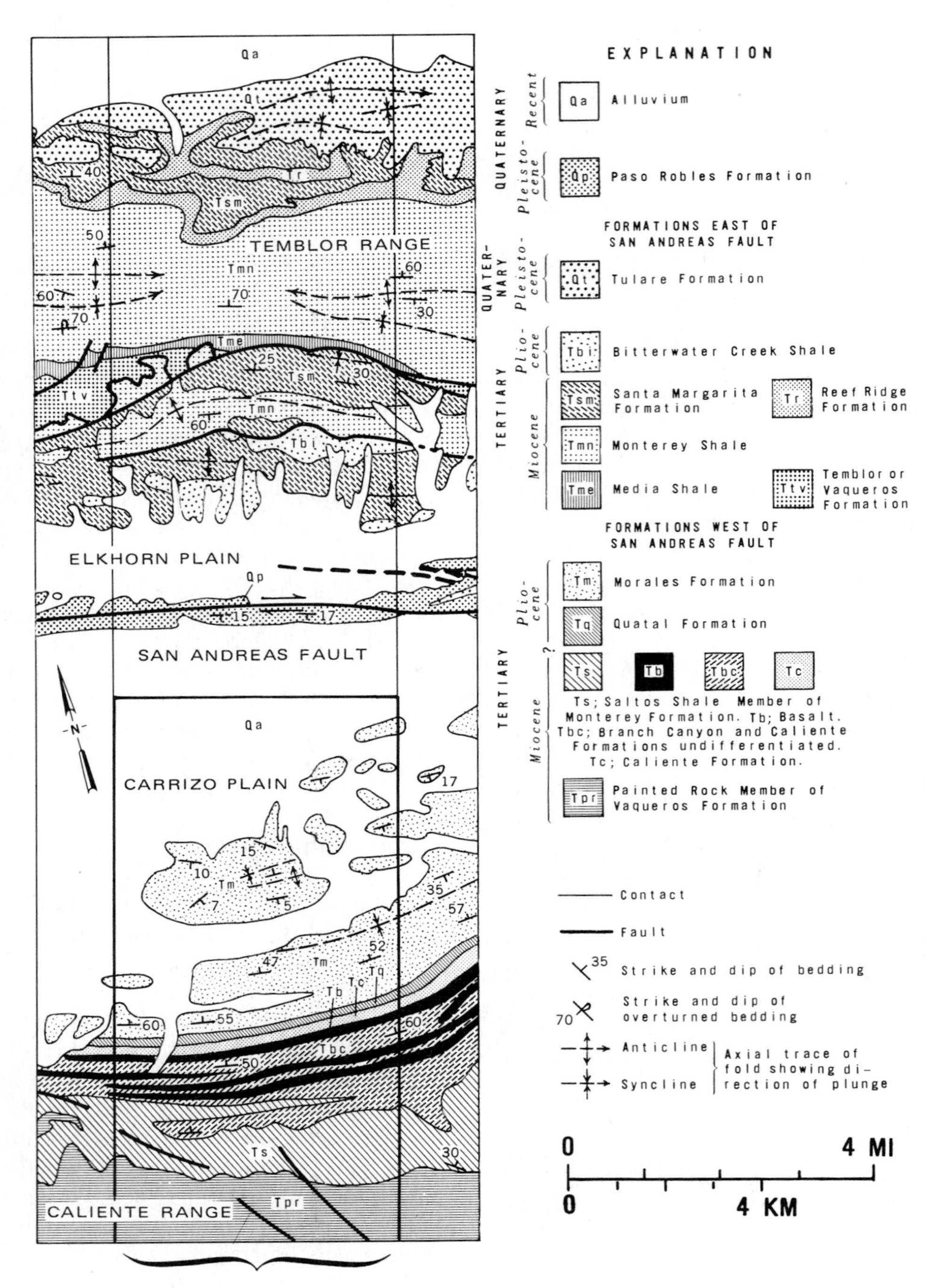

FIGURE 5.33

FIGURE 5.32
Geologic map of Caliente and Temblor Ranges, California. From Wolfe (1971, Figure 2). Courtesy E. W. Wolfe, U.S. Geological Survey.

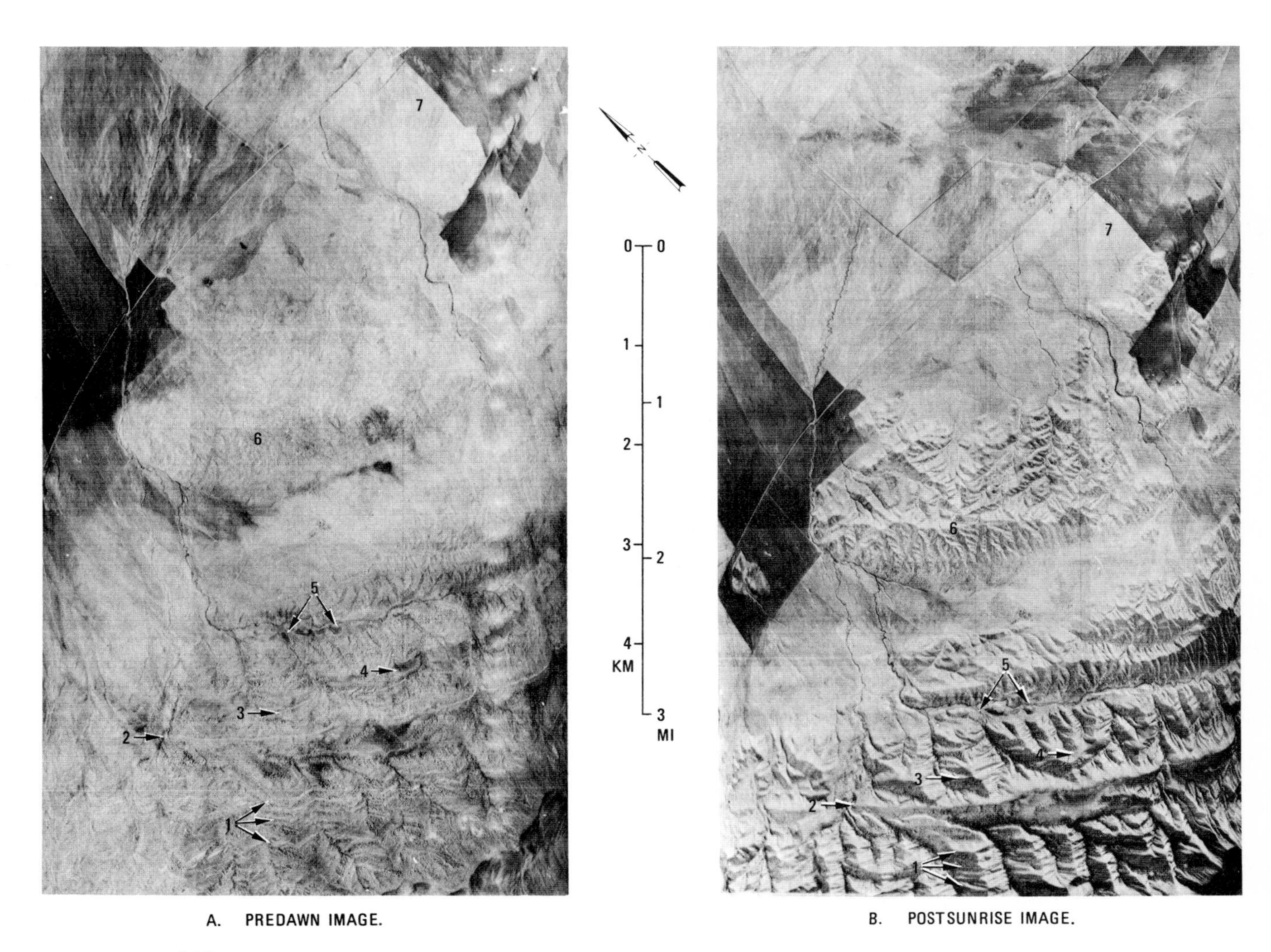

FIGURE 5.33
Thermal IR images (8 to 13 μm) of Caliente Range and Carrizo Plain, California. From Wolfe (1971, Figures 3A and 4A). Courtesy E. W. Wolfe, U.S. Geological Survey.

3. Bright line is the outcrop of hard, light-colored mudstone. Adjacent bedrock is mantled by soil.
4. Dark band is the outcrop of Quatal Formation, a mudstone consisting largely of expanding clay. The overlying and underlying sandstones of the Morales and Caliente formations, respectively, appear brighter.
5. Small dark patches are outcrops of expanding clay in the Morales Formation surrounded by sandy soil and alluvium that appear brighter.
6. Low hill underlain by mudstone, sandstone, and conglomerate of the Morales Formation. Surrounding area is the floor of the alluvium-covered Carrizo Plain. On postsunrise imagery the hill is defined by differential solar heating and shadowing of the slopes.
7. Ground information is not available for agricultural conditions at time the images were acquired. However, experience elsewhere indicates that fields with a cool signature are probably plowed ground and those with a warm signature are covered with growing vegetation or stubble.

Summary of Interpretation

The following points can be summarized from this interpretation:

1. For geologic interpretation, the predawn image is far superior to the postsunrise image, which is dominated by topographic effects caused by differential solar heating and shadowing.
2. On the predawn image, shale and mudstone have cool signatures; basalt, sandstone, and conglomerate have warm signatures. These signatures are reversed on the postdawn image, but topographic effects partially mask the rock signatures.
3. Density of the rocks influences their radiant temperature because thermal inertia increases with increasing density. Rocks with higher density (sandstone, conglomerate, and basalt) have warmer signatures on the predawn image than do rocks with lower density (shale and mudstone). On the postsunrise image, the thermal signatures of different rock types are obscured by topographic effects.

HAWAIIAN VOLCANOES

Kilauea Volcano in Hawaii was the site of the first volcano IR surveys in 1963, which were reported by Fischer and others (1964). Much valuable information on fractures, vents, and buried caldera margins was interpreted from the uncalibrated imagery. In February, 1973 Daedalus Enterprises, under contract to the National Science Foundation, acquired nighttime IR images of Kilauea and Mauna Loa, both on the island of Hawaii, with a quantitative scanner equipped with a magnetic tape recorder. The tapes were used to produce the calibrated images displayed in color in Plate 4.

Mauna Loa Summit

Mauna Loa is the largest volcano on earth, rising approximately 4,200 m above sea level and more than 9,000 m above the adjacent ocean floor. It is a classic shield volcano with a convex-upward profile and a large depression, Mokuaweoweo Caldera, at its summit. The IR image (Plate 4A) and the aerial photomosaic (Figure 5.34A) cover the caldera and part of the Southwest Rift Zone including the pit craters, South Pit, Lua Hohonu, and Lua Hou. These are identified on the geologic map of Figure 5.34B, which also shows the dates of the

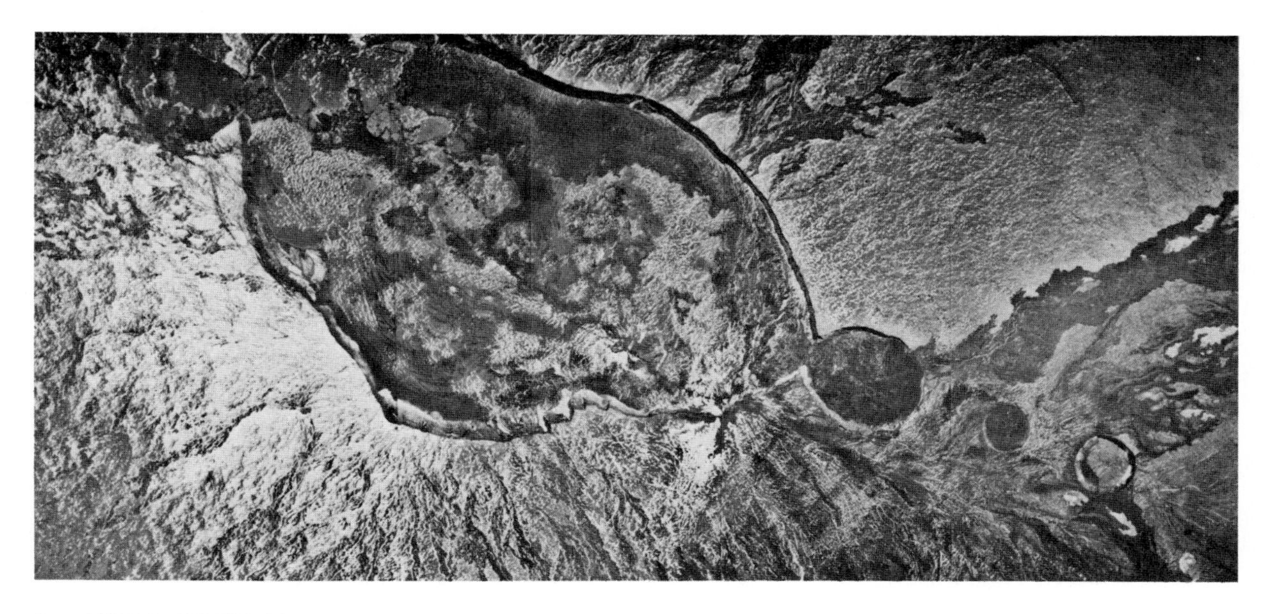

A. AERIAL PHOTOMOSAIC ACQUIRED JANUARY 30, 1965 BY U.S. DEPARTMENT OF AGRICULTURE.

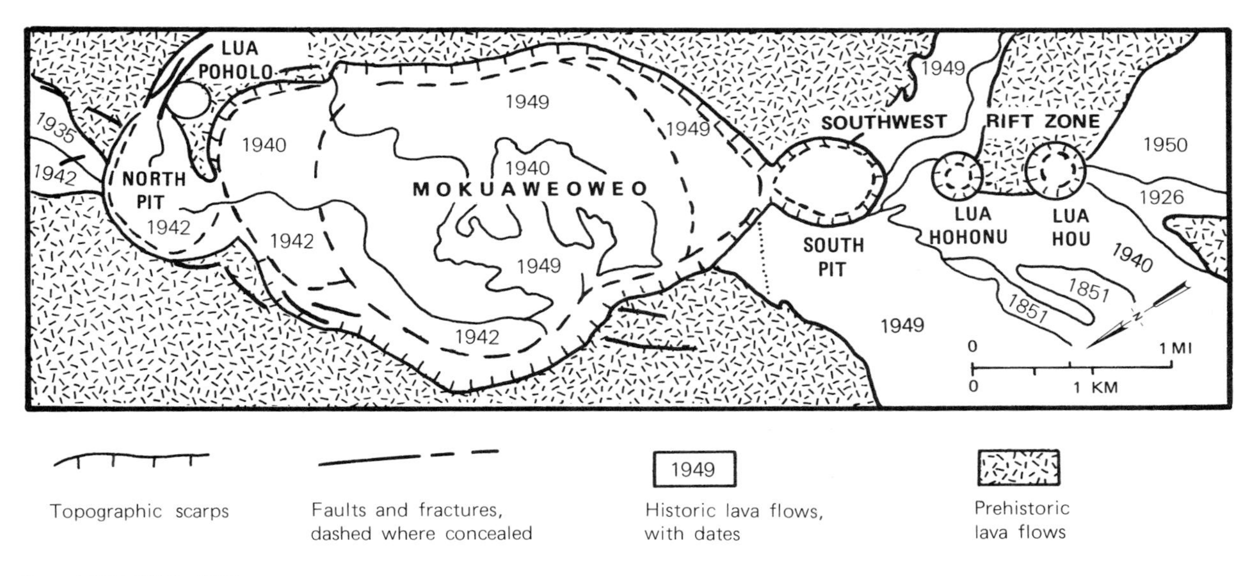

B. GEOLOGIC MAP.

FIGURE 5.34
Summit of Mauna Loa and upper part of Southwest Rift Zone, Hawaii. Map from Macdonald (1971).

historic lava flows. Pit craters form by subsidence of the surface and are not primarily vents for lava. Lua Hohonu is the youngest pit crater, having formed since 1841 (Macdonald and Abbott, 1970, p. 308) and was partially filled with lava in 1940. From rim to floor, these pit craters range in depth from almost 90 m at Lua Hou to about 30 m at Lua Hohonu. The north end of the summit caldera coalesces with another pit crater, North Pit. East of the junction between these is Lua Poholo, a small pit crater that probably formed between 1874 and 1885 (Macdonald, 1971). The cliffs bounding the summit caldera are slightly eroded fault scarps that have a maximum height of 180 m on the west side, but only 60 m on the east. The main faults along which the floors of the caldera and the pits have subsided are covered by young lava flows. The inferred position of these faults is shown by dashed lines in Figure 5.34B. Smaller faults and fractures are common.

The historic lava flows shown on the geologic map are known from contemporary records. The flows originated from fissures on the floor of the summit caldera and in the Southwest Rift Zone. Lava from different flows has spilled into the pit craters. Near the fissures the lava flows are the smooth, ropey pahoehoe type and change character downslope to the rough, fragmented, and clinkery aa type. Mauna Loa had not erupted since June, 1950, but on July 5, 1975 a line of lava fountains erupted across the floor of the summit crater and extended one km down the Southwest Rift Zone (Lockwood and others, 1976). Lava cascaded into the three pit craters along the Southwest Rift Zone. The eruptive fissures extended northeastward across North Pit and into the Northeast Rift Zone. Lava from North Pit cascaded into the Lua Poholo pit crater. The volcanic activity lasted for less than one day and erupted approximately $30 \cdot 10^6$ m^3 of lava that covered 13.5 km^2 of the Mauna Loa summit (Lockwood and others, 1976). These flows are not shown on the geologic map (Figure 5.34A) because they postdate the IR image that was acquired in 1973. Lockwood and others (1976, p. 15) speculate that the 1975 event was the initial phase of an eruptive sequence that may culminate with a future major eruption on the flanks of Mauna Loa.

On the thermal IR image in Plate 4A, the temperature range from –7 to +8°C is digitally subdivided, or sliced, into six intervals. Each interval represents 2.5°C and is shown in a color. Radiant temperatures cooler than –7°C are shown in black; those warmer than 8°C are shown in white. The –7 to +8°C range was selected for color display because preliminary study of the image data indicated that most features of geologic interest had radiant temperatures within this range. The low overall temperature level is caused by the 4,200 m elevation of the summit of Mauna Loa. The individual lava flows cannot be distinguished on the basis of radiant temperature; the flows of the 1940s within the summit caldera have the same temperature ranges as the prehistoric flows on the flanks of the summit. The flows are more readily mapped by tonal differences on the aerial photographs (Figure 5.34A). After the flows have cooled, their radiant temperatures are determined by albedo and thermal properties of the rocks. The aerial photographs and geologic descriptions indicate that all the flows have a similar range of rock properties, which suggests that the thermal properties are also similar. This fact explains the lack of unique thermal signatures for the different flows.

Some of the warmest radiant temperatures occur at the walls of the scarps bounding the summit caldera, North Pit, South Pit, and Lua Hou. The dense interior portions of lava flows that are exposed in the scarps have higher thermal inertias than the porous vesicular surfaces of the flows exposed on the flanks of the volcano and in the floors of the caldera and the pit craters. The dense rocks with higher thermal inertia retain solar heat which is reradiated at night. The reticulate pattern of very warm signatures on the floor of the summit crater appears to be concentrated within the area of the 1940 flow (Figure 5.34A). Fumes of steam and sulfur dioxide issue from cracks in the surface of the 1940 flow (Macdonald, 1971) and probably are responsible, at least in part, for the warm pattern.

The floors of the three pit craters along the Southwest Rift Zone have marked differences in radiant temperature. South Pit is the warmest ($>3°C$), Lua Hohonu the coolest ($< -2°C$), and Lua Hou has intermediate temperatures (-2 to $+0.5°C$). The warmer radiant temperatures are probably caused by higher rates of convective heat transfer at South Pit and Lua Hou. The relative temperatures of the craters was confirmed in early 1975 when the summit of Mauna Loa was blanketed by 2 m of snow. Aerial photographs acquired in April, 1975 (Lockwood and others, 1976, p. 12) revealed that the snow was essentially melted on the floors of South Pit and Lua Hou, while the floor of Lua Hohonu was completely snow covered.

Kilauea Caldera

The IR image in Plate 4B covers a portion of Kilauea Caldera that forms the summit of Kilauea Volcano, a shield volcano east of Mauna Loa. Halemaumau is a collapse crater near the southwest margin of the caldera floor that has been the center of much of the surface activity at Kilauea for the past 150 years. The historic flows within Kilauea Caldera, shown on the geologic map (Figure 5.35B), are mainly pahoehoe lava. The prehistoric lava flows southwest of Kilauea escarpment are mantled by ash deposits up to 2 m thick. The faults in the vicinity of the Crater Rim Road are generally downdropped toward the caldera and form topographic benches. The southwest-trending swarm of cracks lack appreciable displacement although many are open at the surface and range from a few centimeters to 4.5 m in width (Peterson, 1967).

On the nighttime image of Kilauea the temperature range from 8 to 17°C was digitally sliced into six intervals of 1.5°C for the color display of Plate 4B. Temperatures cooler than 8°C are shown in black; those warmer than 17°C are shown in white. The lower elevation of the Kilauea Caldera floor (1,050 m) accounts for the temperatures being warmer than at Mauna Loa. In Plate 4B, the highest temperatures ($>17°C$) occur at three localities:

1. The outer escarpment of Halemaumau is consistently warmer, probably due to the dense interior portions of the lava flows exposed there.
2. Discontinuous arcuate warm zones on the floor of Halemaumau partially coincide with a low interior escarpment. This may represent the convective transfer of heat along an unmapped fracture zone.
3. Irregular reticulate warm zones occur on the floor of Kilauea Caldera. The zone intersecting the north rim of Halemaumau is especially pronounced. Similar warm zones on the 1963 imagery were correlated in part with steaming fissures on the caldera floor (Fischer and others, 1964).

As at Mauna Loa there is no apparent correlation of radiant temperature with the different flows. Between Halemaumau and the Crater Rim Road (Figure 5.35B) the escarpment of Kiluaea Caldera

A. AERIAL PHOTOMOSAIC ACQUIRED MAY 28, 1972 BY R.M. TOWILL CORPORATION, HONOLULU.

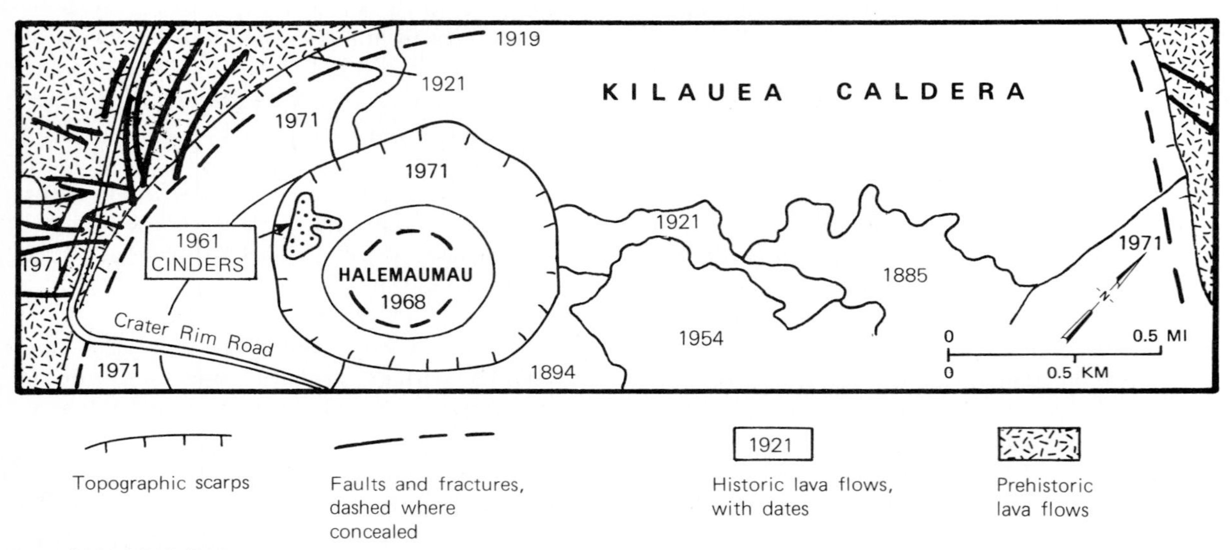

B. GEOLOGIC MAP.

FIGURE 5.35
Halemaumau Crater and portion of Kilauea Caldera, Hawaii. Map from Peterson (1967).

is marked by temperatures 3°C higher than the surroundings. This is due to dense interior portions of flows exposed at the escarpment. The fractures and faults in the vicinity of the Crater Rim Road have distinctly warmer signatures than their surroundings. One possible explanation is that escaping hot volcanic gases could heat the fractures, but no such activity has been reported. A more likely explanation is based on the fact that solar radiation heats the walls of the open fractures during the day. The heat radiated from one wall is absorbed and reradiated by the opposite wall, causing the fractures to appear warm on the nighttime image. Similar fractures in extinct volcanic provinces also have warm signatures, probably caused by this mechanism. Thermal features on the 1963 images interpreted by Fischer and others (1964) are similar to those in the 1973 image shown in Plate 4B. This similarity is remarkable because the eruptions and lava flows of 1968 and 1971 (Figure 5.35B) occurred during the intervening time. These younger flows are not distinguishable on the 1973 image.

The reticulate warm pattern of steaming fissures on the floor of the main crater is similar in both the 1963 and 1973 images. In 1973 a prominent fissure intersected the north rim of Halemaumau, but 10 years earlier the fissure was less extensive. The 1961 vent and spatter cone inside the southwest rim of Halemaumau (Figure 5.34B) was the highest thermal anomaly on the 1963 image, with rock temperatures of 100°C measured 1 m below the surface (Fischer and others, 1964, p. 735). On the 1973 image (Plate 4B) there is no longer a warm anomaly at this vent because of cooling over the intervening 10 years. The warm anomalies at this general location on the 1973 image appear to be steaming fissures on the main caldera floor, outside Halemaumau crater. In August, 1974 lava flows covered the eastern and southeastern portions of Kilauea Caldera. In September, 1974 a flow covered the floor of Halemaumau and a small area of the caldera floor southwest of Halemaumau. On November 29, 1975 a major earthquake triggered a small eruption at Kilauea. There is no apparent relationship of these events to the thermal features on Plate 4B.

HEAT LOSS SURVEYS

Recent shortages and price increases of fuel, especially of natural gas and heating oil, have emphasized the need to conserve energy. A vital aspect of conservation is to eliminate energy waste, particularly in heating of buildings which accounts for a large portion of energy consumption. Major causes of heat loss are:

1. Inadequate or damaged insulation.
2. Worn or damaged roofing materials.
3. Leaks in buried steam lines.

The first step in waste prevention is to identify the localities where heat loss is occurring in order to make the necessary repairs and corrections. Building roofs may be difficult of access and even more difficult to monitor with contact thermometers. Access to buried steam lines requires excavation and the exact location of lines is often uncertain. Thermal IR images acquired at low altitudes have proven valuable in locating sources of heat loss. The tape-recorded data are played back to produce quantitative color images similar to those of the Hawaiian volcanoes (Plate 4). From these images the exact location and the surface temperature of areas of heat loss can be identified. Heat loss surveys are available as routine services and have proven valuable. At one mid-western university a break in a buried steam line that had increased fuel bills by approximately $10,000 per year was located. The heat loss survey cost $6,000 and located other areas for energy conservation.

IR IMAGES FROM SATELLITES

Many thermal IR images with low spatial resolution have been acquired from meteorologic satellites launched by the United States beginning in 1960 with TIROS I (Television and IR Operational Satellite). There were 10 satellites in this series, the last of which was launched in 1965. Nimbus, ESSA (Environmental Science Services Administration), ATS (Applications Technology Satellite), ITOS (Improved TIROS Operational Satellite) and NOAA (National Oceanographic and Atmospheric Administration) are successive generations of meteorological satellites. The NOAA satellites circle the globe at an altitude of 1500 km in near-polar orbits that are programmed to occur at approximately the same time each day on the lighted and dark sides of the earth. Images of equatorial localities are acquired twice daily, once during the day and again at night. The NOAA orbits converge at high latitudes and images of polar localities are acquired 12 to 13 times daily. Two separate imaging systems provide coverage in the visible (0.5 to 0.7 μm) and thermal IR (10.5 to 12.5 μm) bands. The scanning radiometer (SR) has a ground resolution of 4.0 and 7.5 km for the visible and IR bands, respectively. The very high resolution radiometer (VHRR) has a ground resolution of approximately 1 km for both the visible and IR bands.

A cloud-free VHRR image of the southwestern United States, acquired in the evening, is shown in Figure 5.36. Temperature signatures are the reverse of those for aircraft thermal IR images, being dark for warm radiant temperatures and bright for cool temperatures. These signatures are employed so that clouds, which are relatively cool, have light tones on both the IR and visible images acquired by the NOAA satellite. It is instructive to compare Figure 5.36 with the corresponding part of the Landsat mosaic of the United States shown in Chapter 4. Thermal patterns on the NOAA image are clearly related to topography, as indicated by the cool signatures of the Sierra Nevada and the warm signature of the Great Valley to the west. Individual mountains and valleys of the Basin and Range province in Nevada and Utah are defined. The Grand Canyon is particularly well defined. Day and night IR data acquired by Nimbus were used with a thermal model to compute thermal inertia values of rocks in Oman by Pohn, Offield, and Watson (1974).

Channel 13 of the Skylab multispectral scanner (S-192 experiment) acquired daytime IR imagery in the 10.0 to 12.5 μm band with a ground resolution cell that measured 79 by 79 m. An example of this imagery was illustrated in Chapter 3. Landsat 1 and 2 do not record thermal IR radiation, but Landsat 3 will record a 10.0 to 12.5 μm image in addition to the four MSS spectral bands. In contrast to the 1 km ground resolution of NOAA images, the ground resolution cell of the Landsat 3 thermal IR images will be approximately 238 m. The planned NASA Heat-Capacity Mapping Mission satellite is designed to acquire imagery in the 0.5 to 1.1 μm and 10.5 to 12.5 μm bands. The predawn and mid-day thermal IR data together with albedo information will be used to calculate thermal inertia of surface materials using thermal model methods similar to those described earlier in this chapter.

COMMENTS

Thermal IR images record a property that is not visible to the eye, namely, the pattern of heat radiated from materials. Radiant heat of a material

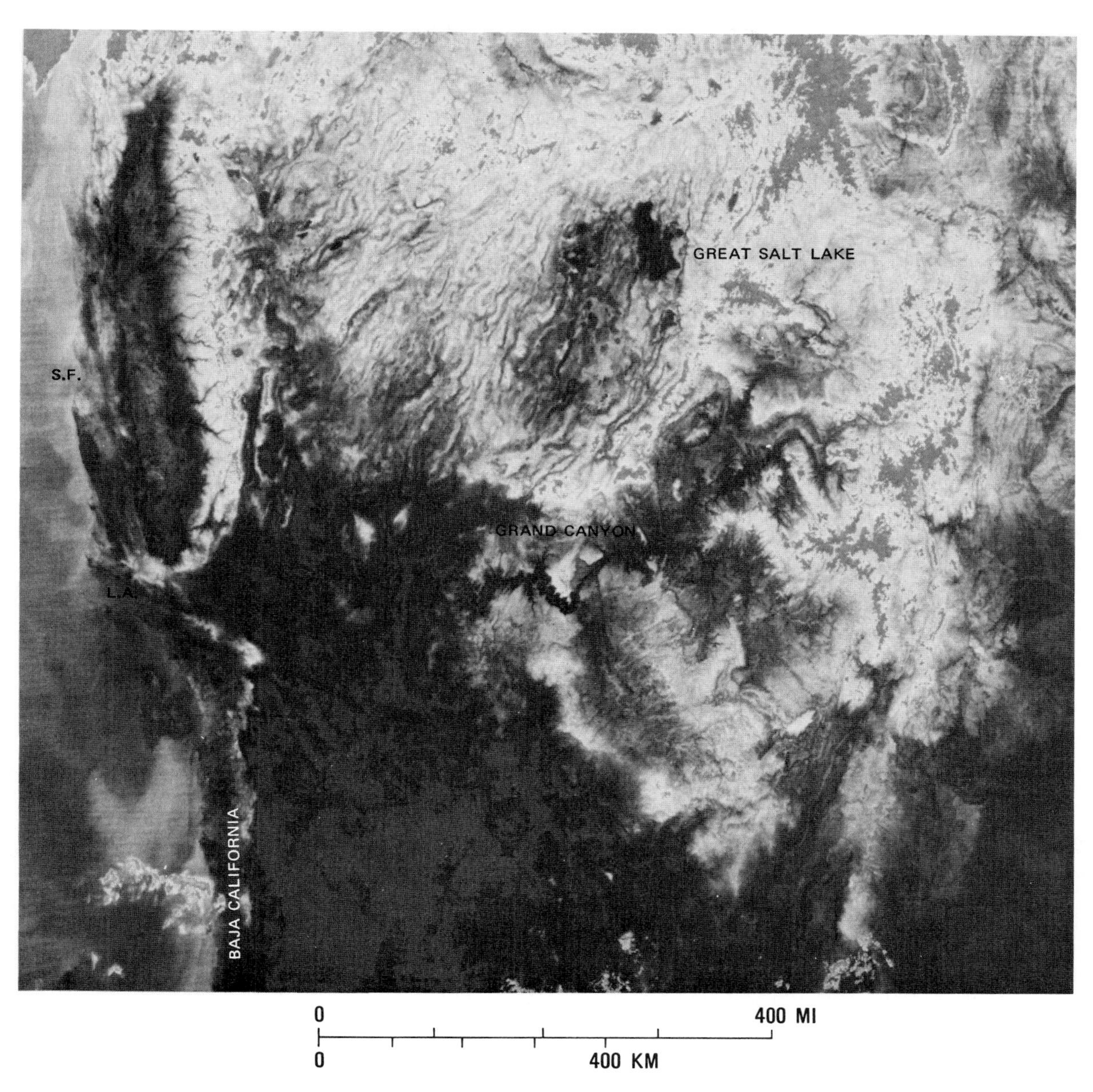

FIGURE 5.36
NOAA-3 satellite thermal IR (10.5 to 12.5 μm) image of southwestern United States acquired August 30, 1975 at 8:26 P.M., MST. Light tones are relatively cool radiant temperatures; dark tones are relatively warm. Courtesy L. C. Breaker, National Environmental Satellite Service.

is determined by the kinetic heat and by the emissivity, which is a measure of the ability to radiate and to absorb thermal energy. Thermal IR images are useful for many applications, including:

1. Differentiation of rock types. Denser rocks, such as basalt and sandstone, have higher thermal inertias than less dense rocks, such as cinders and siltstone. On nighttime thermal IR images the rocks with higher thermal inertias have warmer signatures.
2. Mapping surface moisture. Damp ground has a cool signature on thermal IR images that is caused by evaporative cooling.
3. Mapping geologic structure. Faults may be marked by cool linear anomalies caused by evaporative cooling of moisture concentrated along the fault zone. Folds may be indicated by the thermal pattern caused by the outcrops of different rock types.
4. Surface temperatures in volcanic terrains are measureable on thermal IR images.
5. The ability to distinguish some rock types on thermal IR imagery has an application to uranium exploration, as described later in Chapter 8.
6. Monitoring environmental conditions. Thermal plumes, oil films on water, and underground coal fires can be detected on thermal IR images, as described later in Chapters 9 and 10.

Thermal IR images are affected by environmental factors that must be considered by the interpreter. These include:

1. Clouds and surface winds that produce confusing patterns.
2. Time of day. Daytime images record the differential solar heating and shadowing of topographic features. For geological and other interpretations nighttime imagery is required.
3. Surface moisture effectively masks other features on thermal IR images.

REFERENCES

Allen, C. R., M. Wyss, J. N. Brune, A. Grantz, and R. E. Wallace, 1972, Displacements on the Imperial, Superstition Hills, and San Andreas faults triggered by the Borrego Mountain earthquake *in* The Borrego Mountain earthquake of April 9, 1968: U.S. Geological Survey Professional Paper 787, p. 87–104.

Baker, L. R. and R. M. Scott, 1975, Electro-optical remote sensors with related optical sensors *in* Reeves, R. G., ed., Manual of remote sensing: ch. 7, p. 325–366. American Society of Photogrammetry, Falls Church, Va.

Buettner, K. J. K. and C. D. Kern, 1965, Determination of infrared emissivities of terrestrial surfaces: Journal Geophysical Research, v. 70, p. 1329–1337.

Cummings, J. R., 1964, Coachella Valley investigation: California Department Water Resources Bulletin 108.

Fischer, W. A., R. M. Moxham, F. Polycn, and G. H. Landis, 1964, Infrared surveys of Hawaiian volcanoes: Science, v. 146, n. 3645, p. 733–742.

Janza, F. J., ed., 1975, Interaction mechanisms *in* Reeves, R. G., ed., Manual of remote sensing: ch. 4, p. 75–179, American Society Photogrammetry, Falls Church, Va.

Kahle, A. B., A. R. Gillespie, and A. F. H. Goetz, 1976, Thermal inertia imaging—a new geologic mapping tool: Geophysical Research Letters, v. 3, p. 26–28.

Lockwood, J. P., R. Y. Koyanagi, R. I. Tilling, R. T. Holcomb, and D. W. Peterson, 1976, Mauna Loa threatening: Geotimes, v. 21, p. 12–15.

Lowe, D. S., 1975, Imaging and nonimaging sensors *in* Reeves, R. G., ed., Manual of remote sensing: ch. 8, p. 367–399, American Society of Photogrammetry, Falls Church, Va.
Macdonald, G. A., 1971, Geologic map of the Mauna Loa Quadrangle, Hawaii: U.S. Geological Survey Geologic Quadrangle Map GQ-897.
Macdonald, G. A. and A. T. Abbott, 1970, Volcanoes in the sea, the geology of Hawaii: University of Hawaii Press, Honolulu.
Moxham, R. M., 1971, Thermal surveillance of volcanoes *in* The surveillance and prediction of volcanic activity: p. 103–124, UNESCO, Paris, France.
Offield, T. W., 1975, Thermal-infrared images as a basis for structure mapping, Front Range and adjacent plains in Colorado: Geological Society America Bulletin, v. 86, p. 495–502.
Peterson, D. W., 1967, Geologic map of the Kilauea Crater Quadrangle, Hawaii: U.S. Geological Survey Geologic Quadrangle Map GQ-667.
Pohn, H. A., T. W. Offield, and K. Watson, 1974, Thermal inertia mapping from satellites—discrimination of geologic units in Oman: U.S. Geological Survey Journal of Research, v. 2, p. 147–158.
Quade, J. G., P. E. Chapman, P. A. Brennan, and J. C. Blinn, 1970, Multispectral remote sensing of an exposed volcanic province: Jet Propulsion Laboratory Tech. Memo. 33-453.
Rowan, L. C., T. W. Offield, K. Watson, P. J. Cannon, and R. D. Watson, 1970, Thermal infrared investigations, Arbuckle Mountains, Oklahoma: Geological Society America Bulletin, v. 81, p. 3549–3562.
Sabins, F. F., 1967, Infrared imagery and geologic aspects: Photogrammetric Engineering, v. 29, p. 83–87.
———, 1969, Thermal infrared imagery and its application to structural mapping in southern California: Geological Society America Bulletin, v. 80, p. 397–404.
———, 1973A, Flight planning and navigation for thermal IR surveys: Photogrammetric Engineering, v. 39, p. 49–58.
———, 1973B, Recording and processing thermal IR imagery: Photogrammetric Engineering, v. 39, p. 839–844.
Sawatzky, D. L., R. J. Kline, and K. Watson, 1975, Application of the thermal inertia mapping technique to aircraft and satellite data (abstract): Society of Exploration Geophysicists Annual Meeting, Denver, Col.
Wallace, R. E. and R. M. Moxham, 1967, Use of infrared imagery in study of the San Andreas fault system, California: U.S. Geological Survey Professional Paper 575-D, p. D147–D156.
Watson, K., 1971, Geophysical aspects of remote sensing: Proceedings of the International Workshop on Earth Resources Survey Systems, NASA SP 283, v. 2, p. 409–428.
———, 1975, Geologic applications of thermal infrared images: Proceedings IEEE, v. 63, p. 128–137.
Wolfe, E. W., 1971, Thermal IR for geology: Photogrammetric Engineering, v. 37, p. 43–52.

ADDITIONAL READING

Gillespie, A. R., and A. B. Kahle, 1977, Construction and interpretation of a digital thermal inertia image: Photogrammetric Engineering and Remote Sensing, v. 43, p. 983–1000.
Kappelmeyer, O. and R. Haenel, 1974, Geothermics with special reference to application: Gebruder Borntraeger, Berlin, Germany.
Lee, K., 1976, Ground investigations in support of remote sensing *in* Reeves, R. G., ed., Manual of remote sensing: ch. 13, p. 804–856, American Society of Photogrammetry, Falls Church, Va.
Lowe, D. S., 1976, Nonphotographic optical sensors *in* Lintz, J. and D. S. Simonette, eds., Remote sensing of environment: p. 155–193, Addison-Wesley Publishing Co., Reading, Mass.

6 RADAR IMAGERY

Radar imaging systems provide a source of electromagnetic energy to "illuminate" the terrain. Energy returned from the terrain is detected by the system and recorded as imagery. Radar is called an *active* form of remote sensing in contrast to *passive* forms, such as photographic and thermal IR systems, which detect the available energy reflected or radiated from the terrain. Radar systems can be operated independently of lighting conditions and largely independently of weather. In addition, the terrain can be illuminated in the optimum direction to enhance features of interest.

Radar is the acronym for *Radio Detection and Ranging*, indicating that it operates in the radio and microwave bands of the electromagnetic spectrum ranging from a meter to a few millimeters in wavelength. The reflection of radio waves from solid objects was noted in the late 1800s and early 1900s. The definitive investigations of radar began in the 1920s in the United States and Great Britain for the detection of ships and aircraft. Radar was developed during World War II for navigation and target location, using the familiar rotating antenna and circular cathode-ray tube display. The continuous strip-mapping capability of *Side-Looking Airborne Radar*, or SLAR, was developed in the 1950s to acquire reconnaissance imagery without the necessity of overflying unfriendly regions. A comprehensive history of radar development was prepared by Fischer and others (1975, p. 41–43).

SLAR SYSTEM

Radar imagery is acquired with complex engineering systems that are described in detail by Moore and others (1975). A typical SLAR system is discussed here to aid the reader in understanding the image-forming process.

Radar Components

A typical SLAR antenna and components are illustrated in Figure 6.1. The timing pulses of electromagnetic energy from the pulse-generating device serve two purposes: (1) they control the bursts of energy from the transmitter and (2) they trigger the sweep of the cathode-ray tube (CRT) film-recording device. The bursts of electromagnetic energy from the transmitter are of specific wavelength and duration, or pulse length.

The same antenna transmits the radar pulse and receives the return from the terrain. An electronic

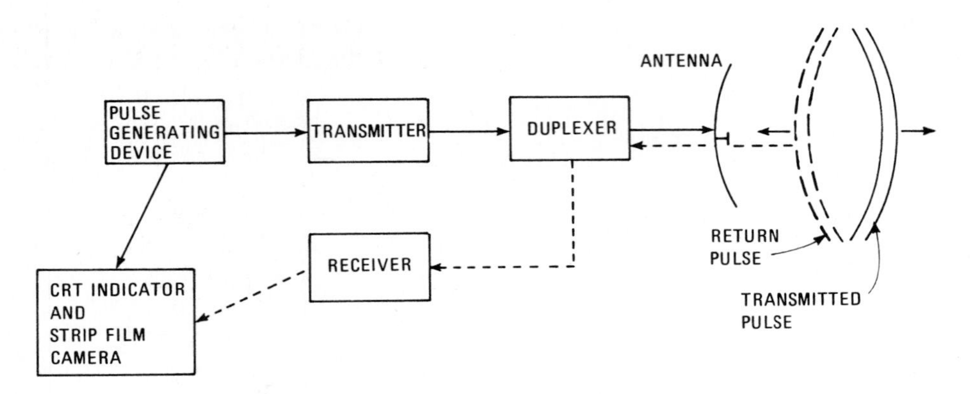

FIGURE 6.1
Block diagram of a side-looking airborne radar system.

switch, or *duplexer*, prevents interference between the transmitted and received pulse by blocking the receiver during transmission and blocking the transmitter during periods of reception. The *antenna* is a reflector that focuses the pulse of radio energy into the desired shape for transmission and also collects the energy returned from the terrain. A *receiver*, similar to a home radio, amplifies the weak energy waves collected by the antenna while preserving the shape of the pulse. This factor is vital because the timing of the returned energy pulse determines the position of terrain features on the image. The return pulse may be displayed as a line sweep on a CRT and recorded on film.

Airborne Imaging System

The SLAR antenna illustrated in Figure 6.2 is 4.9 m long and resembles a cylinder cut in half lengthwise and counted with the long axis parallel with the aircraft fuselage. Pulses of energy transmitted from the antenna illuminate narrow strips of terrain oriented normal to the aircraft flight direction. Figure 6.3 illustrates such a strip of terrain and the shape of the energy pulse that it returns to the antenna. The pulse is displayed as a function of two-way travel time on the horizontal axis with the shortest times at the right, or *near range*, closest to the aircraft flight path. The longest travel times are at the *far range.* To locate positions on the ground, travel time is converted to distance by multiplying the speed of electromagnetic radiation ($3 \cdot 10^8$ m $\cdot$ sec^{-1}). The amplitude, or intensity, of the returned pulse is a complex function of the interaction between the terrain and the transmitted pulse. In Figure 6.3 note that the *azimuth direction* is the aircraft flight path. The *range direction*, normal to the azimuth direction, is the look direction of the radar antenna.

The mountain front facing the antenna in Figure 6.3 has a strong return because of its orientation with respect to the radar antenna. The mountain blocks the transmitted pulse from the terrain immediately downrange, and there is no return from that terrain. The resulting dark signature on the image is called a *radar shadow.* Because of the diverse shapes and orientation, vegetation produces a speckled signature of intermediate intensity. Metallic objects, such as the bridge in Figure 6.3, produce very strong returns and bright signatures because of their geometry and electrical

FIGURE 6.2
Side-looking airborne radar antenna with look direction to the right of the aircraft. This system produced the Ka-band images in this book.

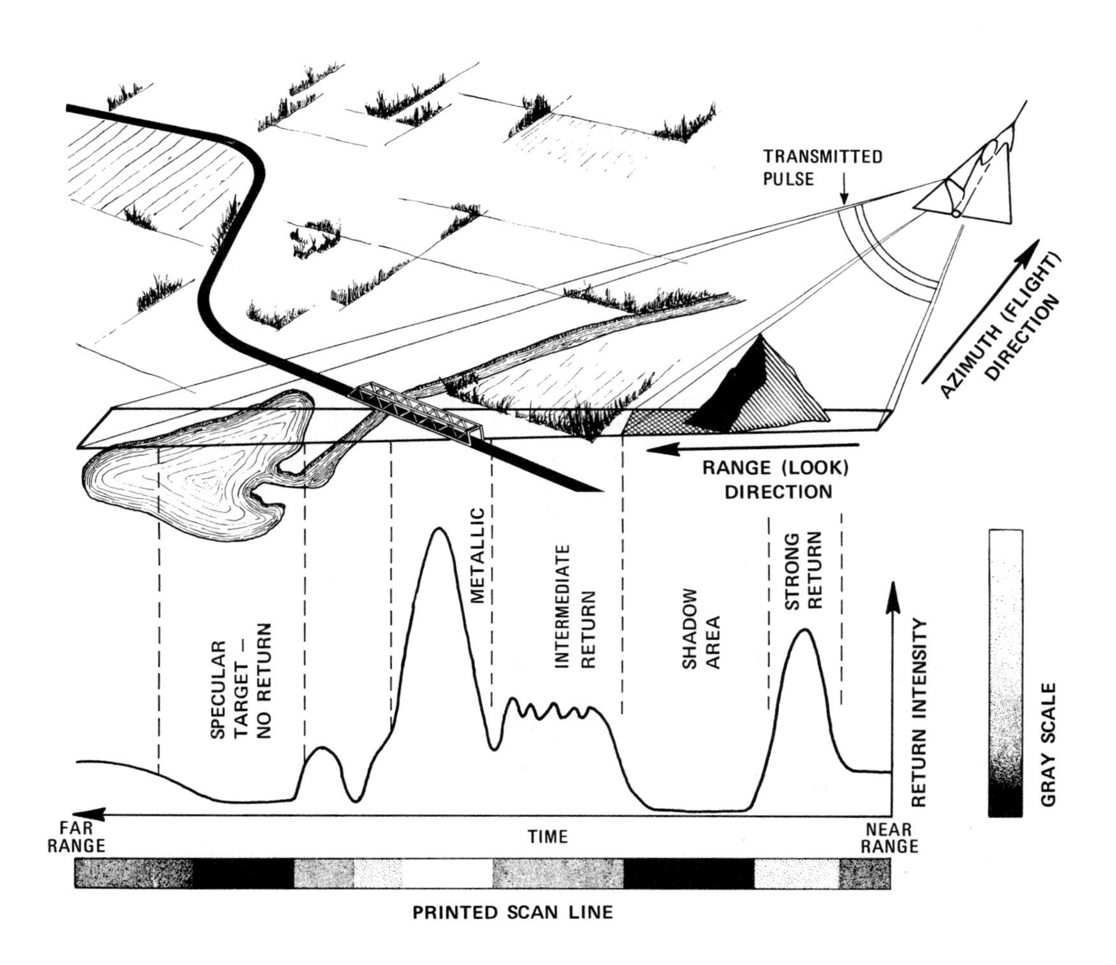

FIGURE 6.3
Ground coverage and signal processing of a pulse of SLAR energy. From Sabins (1973, Figure 7).

A. ROGUE RIVER, OREGON.

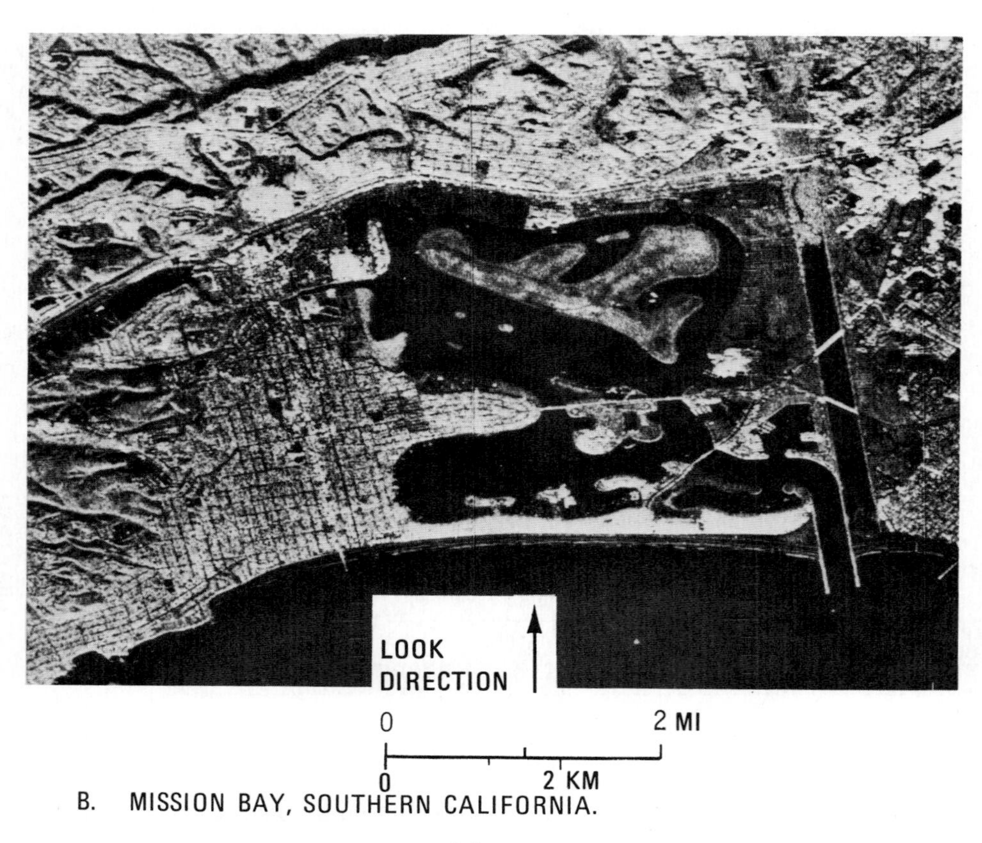

B. MISSION BAY, SOUTHERN CALIFORNIA.

FIGURE 6.4
Ka-band radar images.

properties. Radar energy striking calm water is almost totally reflected with the angle of reflection equal and opposite to the angle of incidence. Very little energy is returned to the antenna and a dark signature results. Smooth surfaces such as calm water, dry lake beds, highways, and airport runways are called *specular targets* because of their mirror-like reflection of radar energy. When these surfaces are located in the extreme near-range position, a bright return can result; at other locations they produce dark signatures.

The return pulse shown in Figure 6.3 is converted to a scan line on a cathode-ray tube by assigning the darkest tones of a gray scale to the lowest intensity returns and the lightest tones to the highest intensity returns. The resulting scan line is recorded on a film strip that moves past the CRT synchronously with the aircraft ground speed. As the aircraft moves forward, successive scan lines are generated to form images such as those in Figure 6.4. In addition to being an active system, radar differs from other remote sensors because data are recorded on the basis of time rather than angular distance, as in the case of cameras and optical-mechanical scanners. Time can be much more precisely measured and recorded than angular distance; hence, radar imagery can be acquired at long ranges with high resolution. Also, atmospheric absorption and scattering are minimal at most microwave wavelengths.

Typical Images

The images in Figure 6.4 illustrate the terrain features and signatures shown diagrammatically in Figure 6.3. On these and other examples the *look direction* of the radar antenna is shown by arrows. Thus the lower edge of each image in Figure 6.4 is the near range closest to the aircraft flight path and the upper edge is the far range. The specular nature of the Pacific Ocean and the Rogue River, Oregon produce very weak returns and the dark signatures seen in Figure 6.4A. Very strong returns come from the bridge over the Rogue River and from sea stacks in the vicinity of the arrow. The mountain ridges produce bright signatures from slopes facing the antenna. The radar shadows become progressively longer farther away from the antenna in the far-range direction.

The bright radar signatures from urban areas are illustrated in the image of Mission Bay, located just north of San Diego (Figure 6.4B). The buildings adjacent to the bay produce strong returns, similar to the bridge in Figure 6.3, and the streets are specular targets with no return. The beach sand is also a specular reflector with a dark signature on the image; it is separated from the calm ocean by the bright lines caused by the rough surf. The large island in Mission Bay was covered with grasses that produced the intermediate signature characteristic of vegetation.

TABLE 6.1
Radar wavelengths and frequencies used in remote sensing

Band designation	λ Wavelength, cm	v Frequency, megahertz (10^6 cycle · sec^{-1})
Ka (0.86 cm*)	0.8 to 1.1	40,000 to 26,500
K	1.1 to 1.7	26,500 to 18,000
Ku	1.7 to 2.4	18,000 to 12,500
X (3 and 3.2 cm*)	2.4 to 3.8	12,500 to 8,000
C	3.8 to 7.5	8,000 to 4,000
S	7.5 to 15.0	4,000 to 2,000
L (25 cm*)	15.0 to 30.0	2,000 to 1,000
P	30.0 to 100.0	1,000 to 300

*Indicates wavelengths commonly used in imaging radars.

Radar Wavelengths

Table 6.1 lists the various radar wavelengths and corresponding frequencies. Although frequency is a more fundamental property, most interpreters can visualize wavelengths more readily; therefore, wavelengths are used here to designate various SLAR systems. Equation (1.1) enables any frequency v to be converted into wavelength λ in the following manner

$$c = \lambda v$$

$$\lambda = \frac{3 \cdot 10^8 \text{ m} \cdot \text{sec}^{-1}}{v}$$

where c is the speed of electromagnetic radiation. The random letter code for frequencies on Table 6.1 was assigned during the early classified stages of radar development to avoid mention of the wavelength regions under investigation. Most of the available images have been acquired by Ka-band or X-band systems, together with some L-band images. The capability to penetrate precipitation and beneath a surface is increased with longer wavelengths.

Polarization

The electrical field vector of the transmitted energy pulse may be *polarized* (or vibrating) in either the vertical or horizontal plane. Upon striking the terrain most of the energy returned to the antenna has the same polarization as the transmitted pulse. This energy is recorded as *parallel-polarized* or like-polarized imagery and is designated HH (horizontal transmit and horizontal return) or VV (vertical transmit and vertical return). A portion of the returned energy has been depolarized by the terrain surface and vibrates in various directions. The mechanism responsible for depolarization is not definitely known, but the most widely accepted theory attributes it to multiple reflections at the surface. This is supported by the fact that depolarization effects are much stronger from vegetation than from bare ground. The leaves, twigs, and branches cause the multiple reflections thought to be responsible for depolarization. Some radar systems have a second antenna element that receives the depolarized energy vibrating at right angles to the polarization direction of the transmitted pulse. The resulting imagery is termed *cross polarized* and may be either HV (horizontal transmit, vertical return) or VH (vertical transmit, horizontal return). Most images are acquired in the HH mode because this produces the strongest return signals.

Figure 6.5 illustrates simultaneously acquired HH and HV Ka-band images of the Caribbean coastal plain of Nicaragua. There are no rock outcrops or bare soil in this heavily vegetated area. The signature differences between the two images are related to the vegetation cover. The interpretation map (Figure 6.6) is based on limited visual observations during the radar flight and by analogy with known radar signatures of similar terrain. On both images the forest has a bright signature and the water has a dark signature. The grasslands in the eastern part of the image have a dark signature on the HH image that contrasts with the bright signature of the forest (Figure 6.5A). On the HV image (Figure 6.5B) the grasslands signature has approximately the same brightness as the forest. The coarser texture and branching pattern of the forested areas makes them stand out slightly on the HV image. The relatively bright HV signature of grasslands indicates that the structure of grass produces a relatively strong vertically polarized radar return. In the western part of the images are areas interpreted as swamps that have a relatively darker signature on the HV image than on the HH image. By analogy with swamps in the southern United States it may be inferred that the Nicaragua swamps are covered by floating plants with flat leaves. These horizontal surfaces would not cause multiple reflections; this factor could account for the weak return from the swamp areas on the HV image.

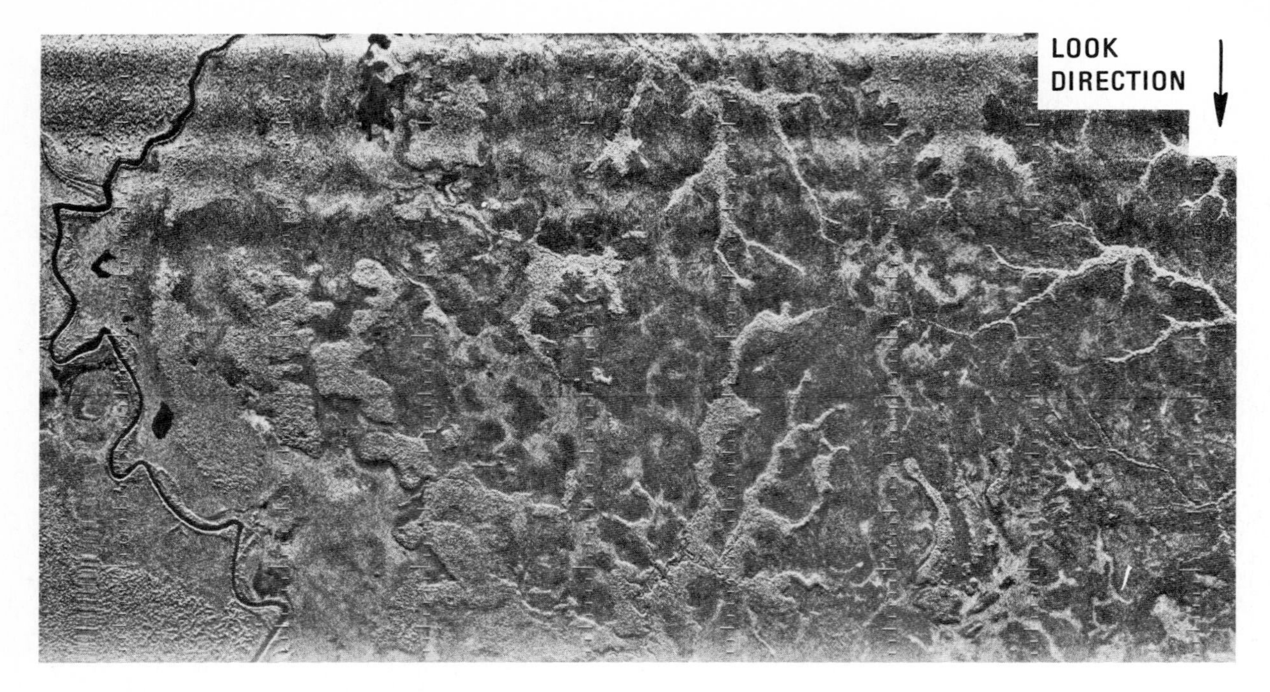

A. PARALLEL POLARIZED (HH) IMAGE.

B. CROSS POLARIZED (HV) IMAGE.

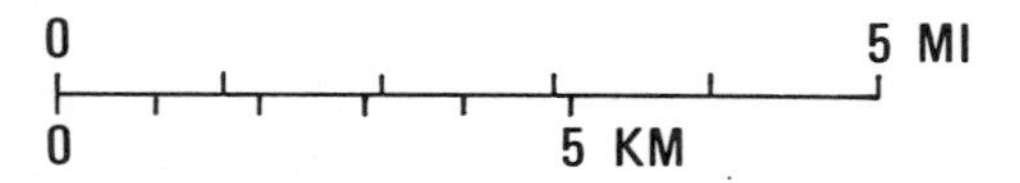

FIGURE 6.5
Parallel-polarized and cross-polarized Ka-band radar images of eastern Nicaragua. Arrows indicate look direction.

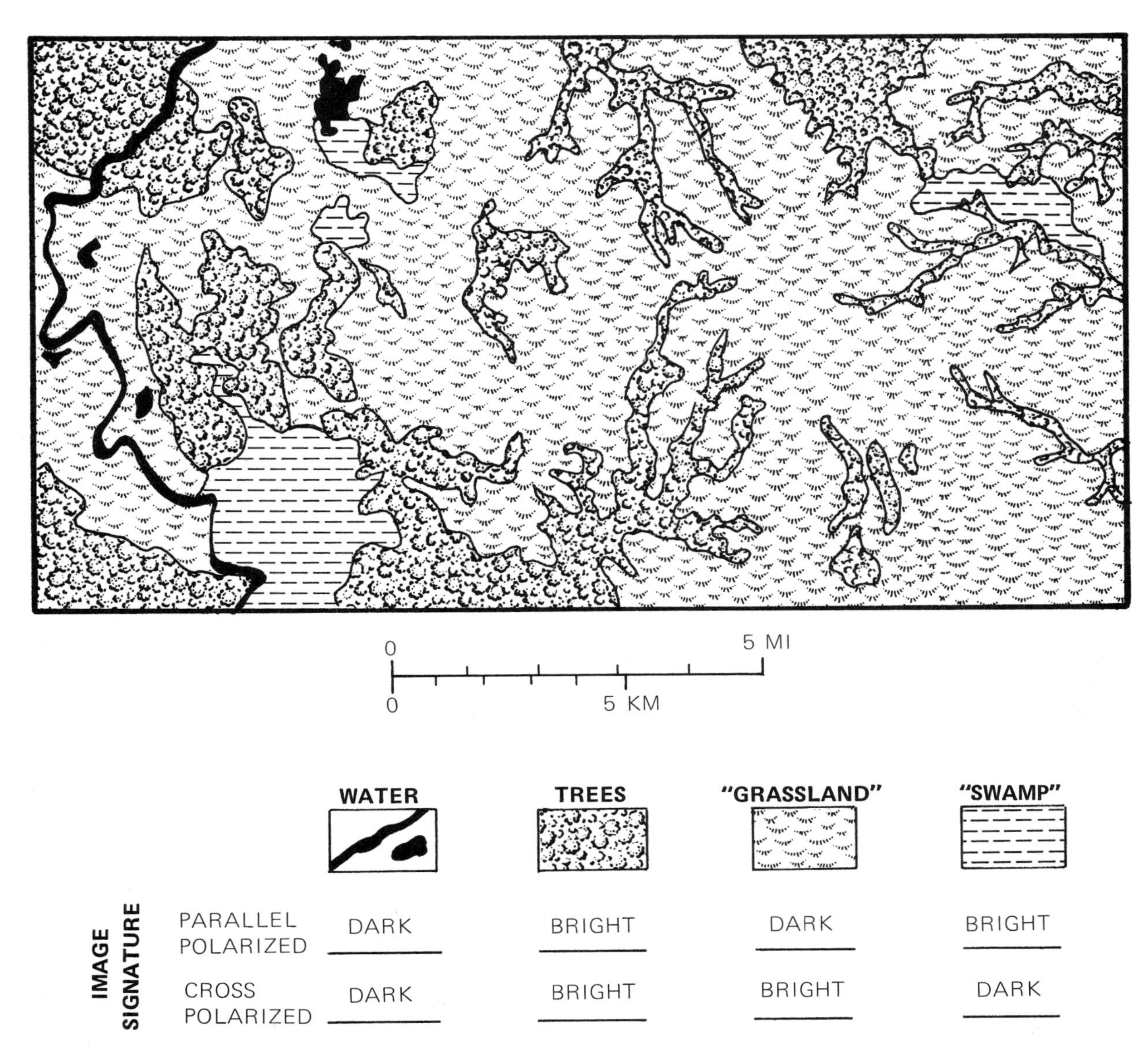

FIGURE 6.6
Interpretation of parallel-polarized and cross-polarized radar images of eastern Nicaragua. Not field checked.

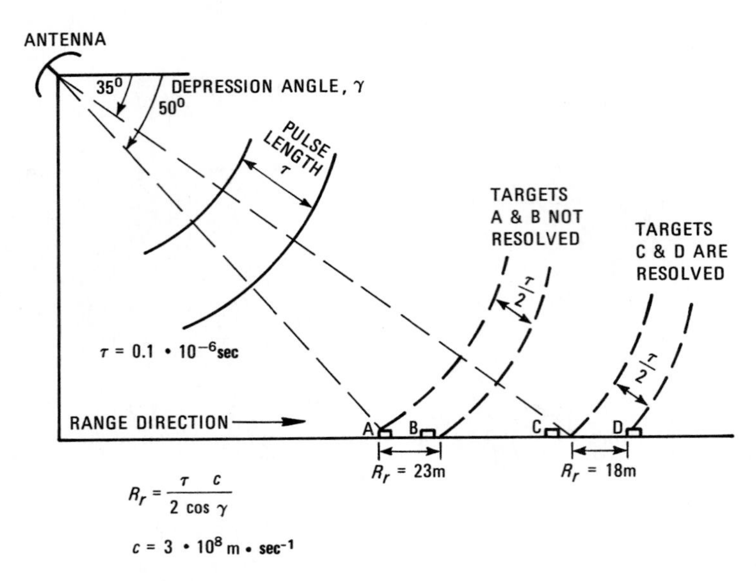

FIGURE 6.7
Radar resolution in the range direction R_r. From Barr (1969).

Spatial Resolution

Resolution in the range and azimuth directions is determined by separate engineering characteristics of the SLAR system. The combination of range resolution and azimuth resolution determines the dimensions of the ground resolution cell.

Range Resolution Spatial resolution in the range direction, R_r, is determined by the duration, or length, of the pulse of transmitted energy and is theoretically equal to one-half the pulse length (Figure 6.7). *Pulse length*, τ, is measured in microseconds, 10^{-6} sec, and is converted from time into distance by multiplying by the speed of electromagnetic radiation ($c = 3 \cdot 10^8$ m $\cdot$ sec^{-1}). The resulting distance is measured in the *slant range*, which is the propagation direction from the antenna to the target. Slant-range distance is divided by the cosine of the depression angle, γ, to convert into ground distance. The *depression angle* (Figure 6.7) is the angle between the horizontal plane and the line connecting the target and the antenna.

The equation for range solution is

$$R_r = \frac{\tau c}{2 \cos \gamma} \tag{6.1}$$

For a depression angle of 50° and pulse length of 0.1 microsecond, range resolution is calculated as

$$R_r = \frac{(0.1 \cdot 10^{-6}\,\text{sec})\ (3 \cdot 10^{8}\,\text{m} \cdot \text{sec}^{-1})}{2 \cos 50^\circ}$$

$$R_r = \frac{0.3 \cdot 10^{2}\,\text{m}}{2 \cdot 0.64}$$

$$R_r = 23\ \text{m}$$

This means that at a depression angle of 50°, targets must be separated by more than 23 m in the range direction to be resolved. In Figure 6.7 targets A and B are separated by less than 23 m and are not resolved on the image, where they will appear as a single signature. At a depression angle of 35° range resolution is 18 m. Targets such as C and D of Figure 6.7, which are separated by more than this distance, appear as separate signatures on the image. One method of improving range resolution is to shorten the pulse length, but this reduces the amount of energy in each pulse transmitted from the antenna. The energy and the pulse length cannot be reduced below the level required for a sufficiently strong return

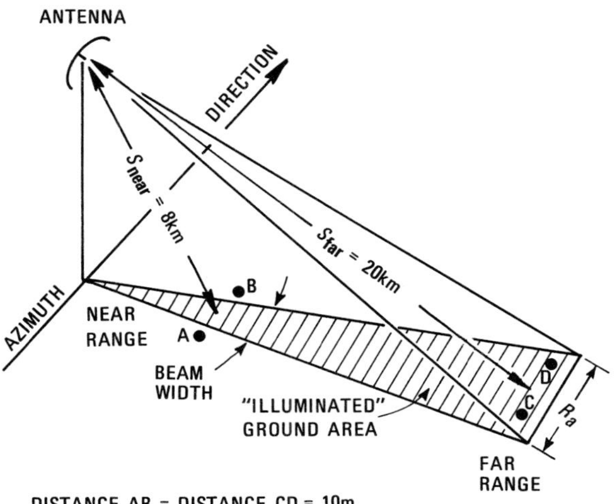

FIGURE 6.8
Radar beam width and resolution in the azimuth direction, R_a, for real-aperture SLAR system. From Barr (1969).

from the terrain. Electronic techniques have been developed for shortening the apparent pulse length while providing adequate signal strength.

Azimuth Resolution Resolution in the azimuth direction, R_a, is determined by the angular width of the terrain strip illuminated by the radar beam. To be resolved, targets must be separated in the azimuth direction by a distance greater than the beam width on the ground. As shown in Figure 6.8, the fan-shaped beam is narrower in the near range than in the far range, causing azimuth resolution to be smaller in the near-range portion of the image. Angular width of the radar beam is directly proportional to wavelength of the transmitted energy, λ; therefore, azimuth resolution is higher for shorter wavelengths, but the short wavelengths lack the desirable weather-penetration capability. Beam width is inversely proportional to antenna length, D; therefore, resolution improves with longer antennas, but there are practical limitations to the maximum antenna length.

The equation for resolution in the azimuth direction, R_a, is

$$R_a = \frac{0.7 S\lambda}{D} \qquad (6.2)$$

where S is the slant range. For a typical Ka-band SLAR system, such as the one shown in Figure 6.2, λ is 0.86 cm and D is 490 cm. At the near-range position, the slant-range distance (S_{near} in Figure 6.8) is 8 km and R_a is calculated from Equation (6.2) as

$$R_a = \frac{0.7 \cdot 8 \text{ km} \cdot 0.86 \text{ cm}}{490 \text{ cm}}$$

$$R_a = 9.8 \text{ m}$$

Therefore targets in the near range, such as A and B in Figure 6.8, must be separated by approximately 10 m in order to be resolved. At the far-range position, the slant-range distance (S_{far} in Figure 6.8) is 20 km; R_a is calculated as 24.6 km, and targets C and D are not resolved.

For most of the SLAR images in this book, azimuth resolution in the middle range and range resolution are both approximately 10 m. The radar return from each 10 by 10 m ground resolution cell is determined by the average radar response of all the terrain features within the cell. Small but strongly reflecting targets may dominate the return. For example, suitably oriented metal fence gates a couple of meters wide may produce strong returns because of their electrical properties. On the image, however, the size of this target cannot be measured below the resolution limit of 10 m.

Real-Aperture and Synthetic-Aperture Systems

The two basic types of SLAR systems are real-aperture radar and synthetic-aperture radar, which differ primarily in the method for achieving resolution in the azimuth direction. The *real aperture*, or "brute force," system uses an antenna of the maximum practical length to produce a narrow angular beam width in the azimuth direction, as illustrated in Figure 6.8.

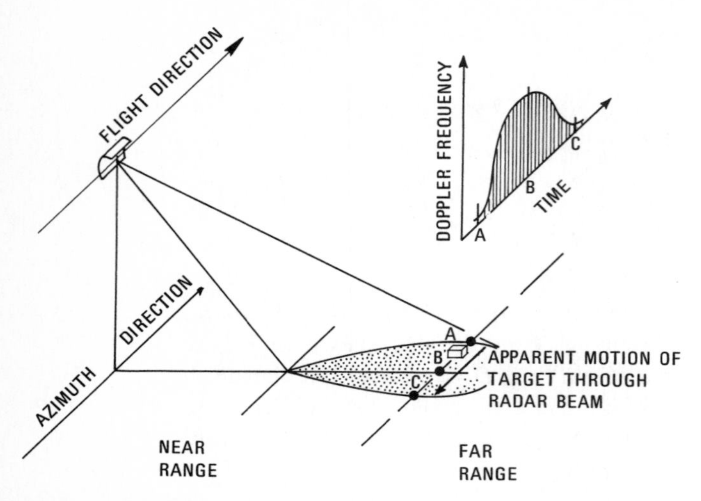

A. DOPPLER FREQUENCY SHIFT DUE TO RELATIVE MOTION OF TARGET THROUGH RADAR BEAM.

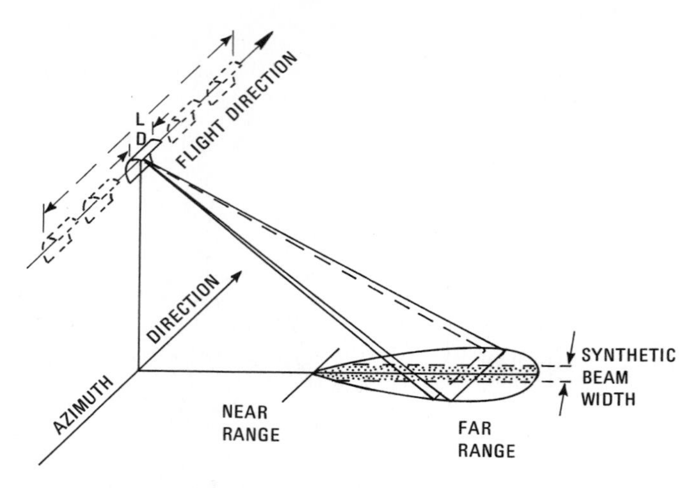

B. RESOLUTION OF SYNTHETIC APERTURE RADAR. NOTE THE SYNTHETICALLY LENGTHENED ANTENNA, L. THE PHYSICAL ANTENNA LENGTH IS D.

FIGURE 6.9
Synthetic aperture radar system. From Craib (1972, Figures 3 and 5).

The *synthetic aperture* system employs a relatively small antenna that transmits a moderately broad beam (Figure 6.9A). The Doppler principle and data processing techniques are employed to synthesize the azimuth resolution of a very narrow beam. Using the familiar example of sound, the *Doppler principle* states that the frequency (pitch) of the sound heard differs from the frequency of the vibrating source whenever the listener or the source are in motion relative to one another. The rise and drop in pitch of the siren as an ambulance approaches and recedes is a familiar example. This principle is applicable to all harmonic wave motion, including the microwaves employed in radar systems.

Figure 6.9A illustrates the apparent motion of a target through the repeated radar beam from points A to C as a consequence of the forward motion of the aircraft. As illustrated by the Doppler frequency diagram, the frequency of the energy pulse returned from the target increases from a minimum at point A to a maximum at point B directly opposite the aircraft. As the target recedes from B to C the frequency decreases. A holographic film record is made of the amplitude and phase history of the returns from each target as the repeated radar beams pass across it from A to C (Figure 6.9A). The holographic film is played back on the ground to produce an image film, as described by Jensen and others (1977). The record of Doppler frequency changes enables the target to be resolved on the image film as though it had been observed with an antenna of length L, as shown in Figure 6.9B. This synthetically lengthened antenna produces the effect of a very narrow beam with constant width in the azimuth direction, depicted by the shaded area in Figure 6.9B. This pattern can be compared with the fan-shaped beam of a real-aperture SLAR system (Figure 6.8). For both real- and synthetic-aperture systems, resolution in the range direction is determined by pulse length (Figure 6.7).

Synthetic-aperture and real-aperture images of the San Francisco Peninsula are compared in Figure 6.10. This is not a completely objective comparison because the images are of different wavelengths and were acquired at different times with opposite look directions. The synthetic-aperture image is a ground-range display and the real-aperture image is a slant-range display, described later in the discussion of image distortion. Contrast appears better on the real-aperture image, but final conclusions should not be drawn from such a limited comparison.

A. Ka–BAND (λ = 0.86 cm) REAL APERTURE IMAGE WITH LOOK DIRECTION TO SOUTHWEST. DISTORTION IN NEAR RANGE IS DUE TO SLANT RANGE DISPLAY.

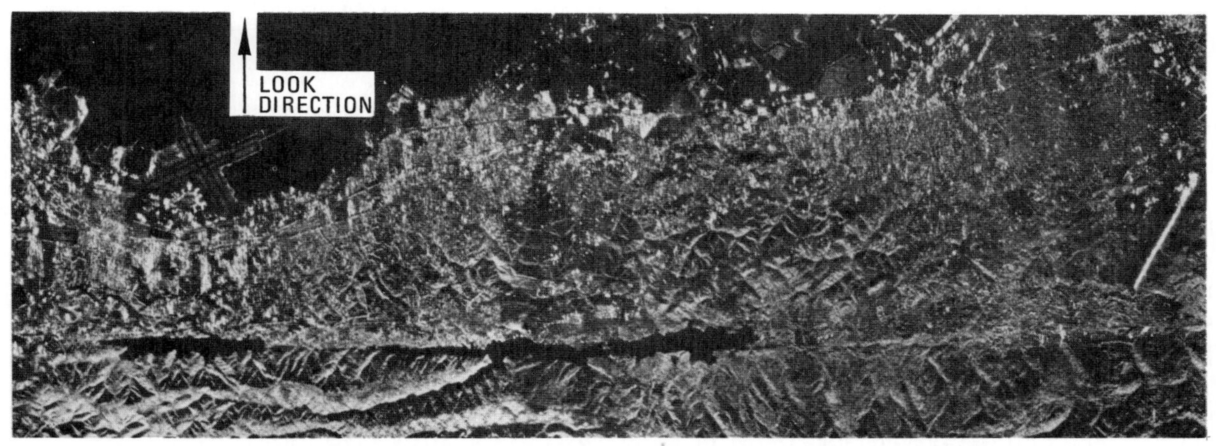

B. X–BAND (λ = 3 cm) SYNTHETIC APERTURE IMAGE WITH LOOK DIRECTION TO NORTHEAST. GROUND RANGE DISPLAY WITH LITTLE DISTORTION.

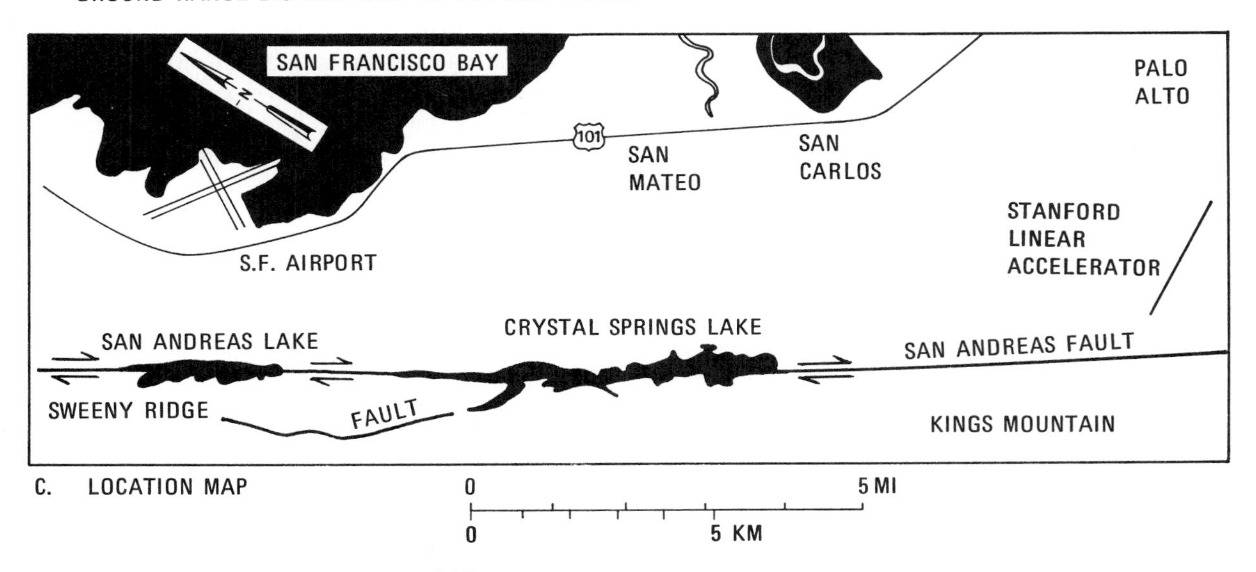

C. LOCATION MAP

FIGURE 6.10
Comparison of radar images, San Francisco Peninsula.

TABLE 6.2
Typical SLAR systems

Operator	Equipment	Remarks
Aero Service Corp.	X-band synthetic aperture.	Radar system built by Goodyear Aerospace Corp.
Environmental Research Institute of Michigan	Simultaneous X-band and L-band synthetic-aperture system.	Parallel- and cross-polarized images acquired at both wavelengths.
Jet Propulsion Laboratory	Simultaneous X-band and L-band synthetic-aperture system.	Parallel- and cross-polarized images acquired at both wavelengths.
University of Kansas	X-band, real aperture.	Compact inexpensive system, fabricated from commercially available components.
Motorola Aerial Remote Sensing, Inc.	X-band, real aperture.	Dual antennas simultaneously acquire images on both sides of flight path.

Real-aperture systems are simple in design and do not require sophisticated data recording and processing. However, coverage in the range direction is relatively limited and only shorter wavelengths can be employed if high resolution is required. Synthetic-aperture systems maintain high azimuth resolution at long distances in the range direction at both long and short wavelengths. The added complexity and cost are the trade off for these advantages.

Available Radar Systems and Imagery

Table 6.2 lists the currently operational nonmilitary radar systems. The Aeroservice–Goodyear system and the Motorola system are available for commercial surveys; the other systems are used for research projects.

Images of most of the United States acquired by the Strategic Air Command have been placed in the public domain through a data bank operated by Goodyear Aerospace Corporation, Department 408A, Litchfield Park, Arizona 85340. Inquiries may be addressed to that facility and should include the latitude and longitude boundaries of the area of interest. The reply will consist of a computer printout listing available coverage with scale and date of acquisition. Resolution of the images ranges from 3 to 15 m. The scale of the original image films ranges from 1:100,000 to 1:600,000 and can be photographically enlarged. A price list and instructions for ordering image reproductions are also provided. The X-band imagery of Death Valley that appears later in this chapter was originally acquired from this source.

RADAR RETURN AND IMAGE SIGNATURES

Energy reflected from the terrain to the radar antenna is called *radar return*. As may be seen diagrammatically in Figure 6.3 and in the images of Figure 6.4, the stronger returns produce brighter signatures on the image. Intensity of the radar return is determined by the following:

1. Radar system properties
   - Polarization
   - Depression angle
   - Wavelength
2. Terrain properties
   - Complex dielectric constant
   - Surface roughness

The influence of polarization has already been discussed. Because the effect of depression angle and wavelength of the radar system are related to surface roughness of the terrain, these properties are discussed together.

Complex Dielectric Constant

The electrical properties of matter influence the interaction with electromagnetic energy and are called the *complex dielectric constant*. At radar wavelengths the dielectric constant of dry rocks and soils ranges from about 3 to 8 and water has a value of 80 (MacDonald and Waite, 1973, p. 149). The dielectric constant of rocks and soils increases almost linearly with increasing moisture content. The depth of penetration of radar waves below the surface varies inversely with the dielectric constant of the material and directly with the radar wavelength. An increase in the dielectric constant increases the reflectivity of a surface; therefore, moist ground reflects radar energy more readily than dry ground. It is difficult to measure the dielectric constant of natural materials and few values have been published. The properties of natural surfaces that control their radar signatures are primarily surface roughness and moisture content. Dielectric properties themselves are rarely considered in image interpretation.

Surface Roughness

Surface roughness of terrain strongly influences the strength of radar returns. Surface roughness is distinct from topographic relief, which is measured in meters and hundreds of meters. Topographic relief features include hills, mountains, valleys, and canyons that are expressed on the imagery by highlights and shadows. Surface roughness is determined by surface textural features such as leaves and twigs of vegetation and sand, gravel, and cobble particles.

The average surface roughness within a ground resolution cell (10 by 10 m for typical SLAR systems) determines the intensity of the return for that cell. Average surface roughness is a composite of the vertical and horizontal dimensions and spacing of the small-scale features, together with the geometry of the individual features (leaves, twigs, particles of sand and gravel). Because of the complex geometry of most natural surfaces it is difficult to characterize them mathematically, particularly for the large area of a resolution cell. For most surfaces the average vertical relief, on the scale of centimeters, is an adequate approximation of surface relief.

Surfaces may be grouped into the following three roughness categories:

1. A *smooth surface* reflects all the incident radar energy with the angle of reflection equal and opposite to the angle of incidence.
2. A *rough surface* diffusely scatters the incident energy at all angles. The rays of scattered energy may be thought of as enclosed within a hemisphere, the center of which is located at the point where the incident wave encounters the surface.
3. A surface of *intermediate roughness* reflects a portion of the incident energy and diffusely scatters a portion.

Strength of the radar return is determined by the relationship of surface roughness to radar wavelength and to the depression angle of the antenna. These relationships are described in the following discussion.

Radar Wavelength and Surface Roughness The relationship of radar wavelength and depression angle to surface roughness may be described by the *Rayleigh criterion* that considers a surface to be smooth if

$$h < \frac{\lambda}{8 \sin \gamma} \qquad (6.3)$$

where

h = the height of surface irregularities, or surface roughness

λ = the radar wavelength

γ = the grazing angle between the terrain and the incident radar wave.

Both h and λ are given in the same units, usually centimeters. For horizontal terrain surfaces (assumed in this discussion), the grazing angle equals the antenna depression angle. For an X-band radar ($\lambda = 3$ cm) at a depression angle of 45°, the roughness value at which the surface will appear smooth is determined by substituting into the Rayleigh criterion (Equation 6.3)

$$h < \frac{3 \text{ cm}}{8 \sin 45°}$$

$$h < \frac{3 \text{ cm}}{8 \cdot 0.71}$$

$$h < 0.53 \text{ cm}$$

This means that a vertical relief of 0.53 cm is the *theoretical boundary* between smooth and rough surfaces for the given wavelength and depression angle. The Rayleigh criterion does not consider the important category of surface relief that is intermediate between definitely smooth and definitely rough surfaces. The Rayleigh criterion was modified by Peake and Oliver (1971) to define the upper and lower values of h for surfaces of intermediate roughness. Their *smooth criterion* considers a surface to be smooth if

$$h < \frac{\lambda}{25 \sin \gamma} \tag{6.4}$$

Using this relationship and substituting for a radar system with λ of 3 cm and γ of 45°, enables the limiting value of surface roughness to be calculated as

$$h < \frac{3 \text{ cm}}{25 \sin 45°}$$

$$h < \frac{3 \text{ cm}}{25 \cdot 0.71}$$

$$h < 0.17 \text{ cm}$$

This means that a vertical relief of 0.17 cm is the boundary between smooth surfaces and surfaces of intermediate roughness for the given wavelength and depression angle.

Peake and Oliver (1971) derived a *rough criterion* that considers a surface to be rough if

$$h > \frac{\lambda}{4.4 \sin \gamma} \tag{6.5}$$

Using this relationship and substituting for the same radar system with λ of 3 cm and γ of 45° enables the limiting value of surface roughness to be calculated as

$$h > \frac{3 \text{ cm}}{4.4 \sin 45°}$$

$$h > \frac{3 \text{ cm}}{4.4 \cdot 0.71}$$

$$h > 0.96 \text{ cm}$$

This means that a vertical relief of 0.96 cm is the boundary between surfaces of intermediate roughness and definitely rough surfaces for the given radar wavelength and depression angle. Note that the value determined earlier from the Rayleigh criterion ($h < 0.53$ cm) is intermediate between those derived for the smooth criterion and the rough criterion. The relationships of radar return intensity to the vertical relief values derived in the preceding expressions are illustrated in Figure 6.11. The smooth surface (Figure 6.11A) specularly reflects all the energy, and there is no energy returned to the antenna at this depression angle. The surface with intermediate roughness specularly reflects part of the energy and scatters the remainder (Figure 6.11B). The waves that are scattered in a direction back toward the antenna—*backscattered*, in radar terminology—produce a return of intermediate brightness on the image.

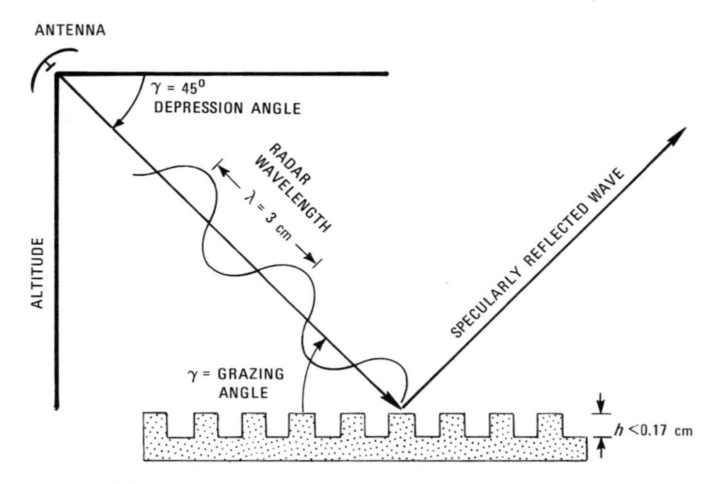

A. SMOOTH SURFACE WITH SPECULAR REFLECTION; NO RETURN.

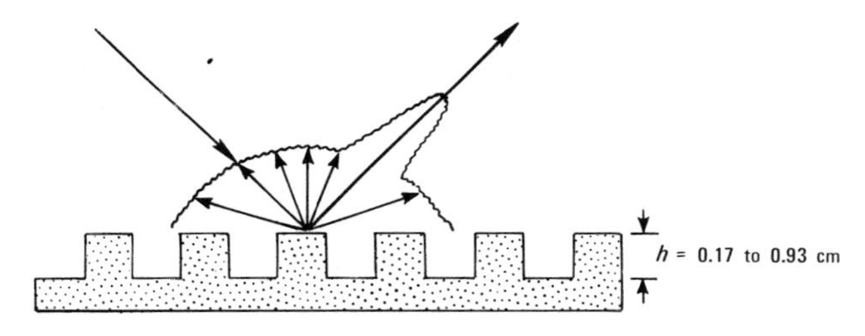

B. INTERMEDIATE SURFACE ROUGHNESS; MODERATE RETURN.

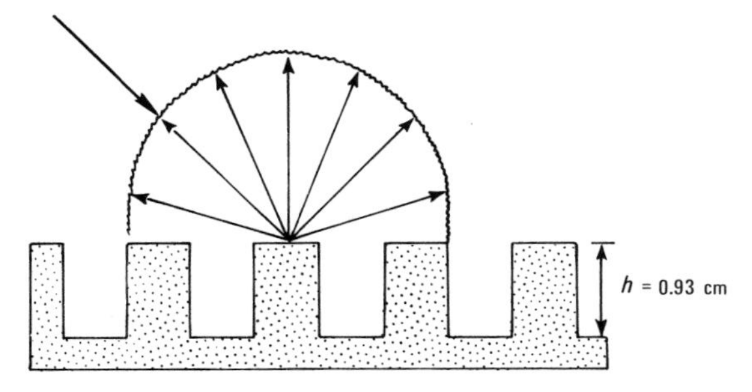

C. ROUGH SURFACE WITH DIFFUSE SCATTERING; STRONG RETURN.

FIGURE 6.11
Models of surface roughness and return intensity for X-band radar (λ = 3 cm).

TABLE 6.3
Limiting values of vertical relief h for surface roughness categories with depression angle γ of 45°

Roughness category	Ka band (λ = 0.86 cm)	X band (λ = 3 cm)	L band (λ = 25 cm)
Smooth	$h < 0.05$ cm	$h < 0.17$ cm	$h < 1.41$ cm
Intermediate	h = 0.05 to 0.28 cm	h = 0.17 to 0.96 cm	h = 1.41 to 8.04 cm
Rough	$h > 0.28$ cm	$h > 0.96$ cm	$h > 8.04$ cm

TABLE 6.4
Response of different values of vertical relief h at different radar wavelengths with depression angle γ of 45°

h (cm)	Ka band (λ = 0.86 cm)	X band (λ = 3 cm)	L band (λ = 25 cm)
0.05	Smooth	Smooth	Smooth
0.10	Intermediate	Smooth	Smooth
0.5	Rough	Intermediate	Smooth
1.5	Rough	Rough	Intermediate
10.0	Rough	Rough	Rough

The rough surface (Figure 6.11C) diffusely scatters all the energy, causing a relatively strong backscattered component that produces a bright signature on the image.

In addition to rough surfaces, bright radar returns are produced by metallic targets and by *corner reflectors* that are formed by three planar surfaces intersecting at right angles. Regardless of the incidence angle at which a radar wave enters the cavity of a corner reflector, it is reflected directly back toward the antenna. Outcrops of regularly jointed and layered volcanic and sedimentary rocks may form natural corner reflectors. Man-made structures, such as buildings and bridges, also form corner reflectors. Corner reflectors are carried in the rigging of boats that are too small to produce a return on the navigation radars of large vessels. The corner reflector produces a return on the radar screens that indicates the presence of the small boat at night and during foggy weather.

Equations (6.4) and (6.5) were used to calculate the limiting values of vertical relief, h, that define smooth, intermediate, and rough surfaces at the wavelengths of three SLAR systems, all at a depression angle of 45°. The resulting values of h are listed in Table 6.3. This information is shown in a different fashion in Table 6.4, where the roughness response for different values of h is given for different radar wavelengths. For example, a surface with vertical relief of 0.5 cm, typical of medium to coarse sand, appears rough on Ka-band images, intermediate on X-band, and smooth on L-band images. The image signature of this surface is light, medium gray, or dark at these three wavelengths. Table 6.4 also illustrates the advantage of acquiring radar images at more than one wavelength for terrain analysis. By comparing the image signatures at two or more wavelengths, the surface relief may be estimated more accurately than from the signature at a single wavelength.

Depression Angle and Surface Roughness In addition to the wavelength of a SLAR system, the depression angle, γ, and incidence angle, θ, are important. Note that $\gamma + \theta = 90°$. Figure 6.12A shows that at low to intermediate depression angles the specular reflection from a smooth surface returns little or no energy to the antenna. At very high depression angles (80 to 90°) however, the specularly reflected wave may be received by the antenna and produce a strong return. A rough surface produces diffuse scattering of relatively uniform intensity regardless of look direction or depression angle of the radar system. As shown in Figure 6.12B, a rough surface produces a strong and nearly uniform return at all depression angles. Figure 6.12C compares the relative return intensity for smooth and rough surfaces at different depression angles. The relatively uniform return from the rough surface decreases somewhat at low depression angles because of the longer two-way travel distance. Smooth surfaces produce strong returns at depression angles near vertical, but little or no return at lower angles.

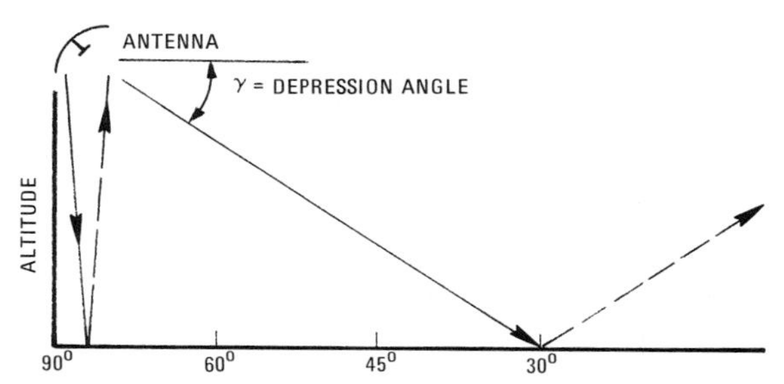

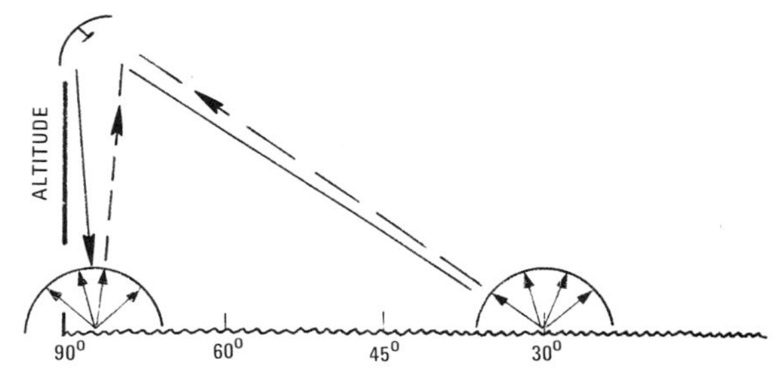

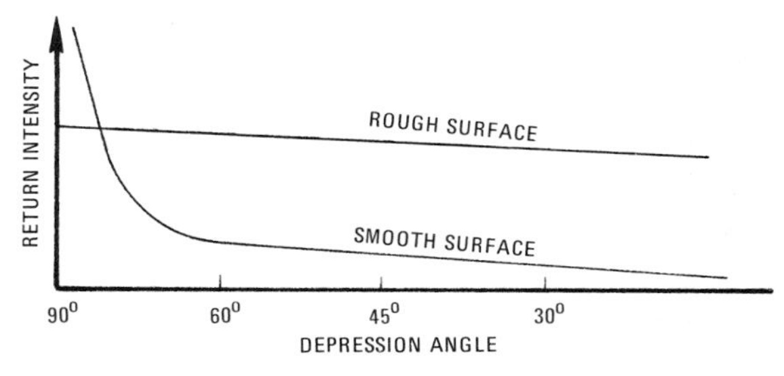

FIGURE 6.12
Radar return from smooth and rough surfaces as a function of depression angle.

Radar Return and Surface Roughness at Cottonball Basin

Previously the relationship between radar backscatter and surface roughness has been based largely on theoretical analysis supplemented by laboratory studies of artificial surfaces. Recently, however, the U.S. Geological Survey and the Jet Propulsion Laboratory have conducted empirical field investigations of radar signatures from terrain in Death Valley, California. The following presentation is summarized from the work of Schaber, Berlin, and Brown (1976) and Schaber, Berlin, and Pitrone (1975). These reports constitute important contributions toward understanding the quantitative relationships between radar signatures and terrain roughness.

Cottonball Basin, at the north end of Death Valley, is one of a number of closed evaporation basins in the Basin and Range physiographic

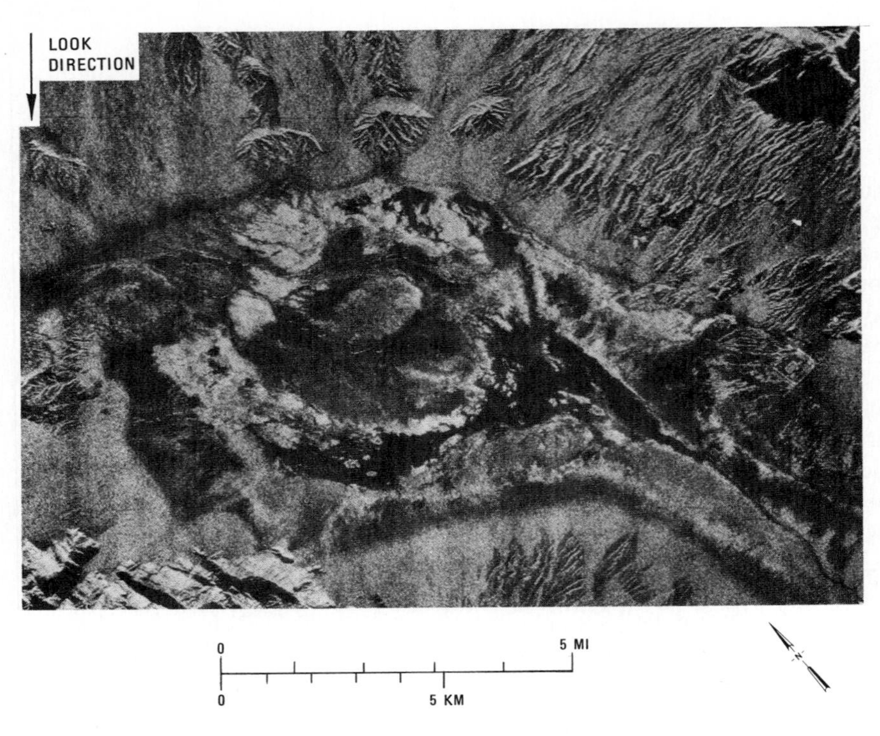

FIGURE 6.13
X-band (λ = 3 cm) radar image of Cottonball Basin, Death Valley, California. Synthetic aperture image acquired in November 1964 by U.S. Strategic Air Command. Depression angle is 23° at center of basin. Look direction toward the southwest. From Schaber, Berlin, and Pitrone (1976, Figure 1). Courtesy G. G. Schaber, U.S. Geological Survey.

province. The floor of the basin is a flat saltpan with a range of average surface roughness from 0.2 to >30 cm caused by periodic solution and re-precipitation of salts. The gravel surfaces of the alluvial fans bordering the basin provide additional degrees and types of surface roughness. The diversity of surface relief and scarcity of vegetation make Cottonball Basin an excellent site for analyzing radar backscatter theory.

Image Interpretation The X-band, synthetic-aperture radar image of Figure 6.13 was acquired in November, 1964 at an altitude of 10,732 m above sea level. Polarization is HH and ground resolution in range and azimuth is approximately 15 m. Look direction is toward the southwest, as shown by the orientation of highlights and shadows from ridges of bedrock around the margins of the basin. Figure 6.14 is a Skylab photograph enlarged to the same scale as the radar image for comparison. The Skylab photograph and aircraft photographs (not illustrated) do not show the variety of signatures within Cottonball Basin that is apparent on the radar image. The various saline deposits of the basin are white to light gray, resulting in very low contrast on the photographs. The deposits, however, are characterized by variations in small-scale surface relief that cause the different signatures on the radar image. On the interpretation map (Figure 6.15) the surface deposits are separated into *radar-rock units* that combine the radar signature and lithologic characteristics of the materials. The radar-rock units are illustrated in Figure 6.16. The vertical relief is the average of many field measurements as reported by Schaber, Berlin, and Brown (1976). The origin and mineralogy of the Death Valley deposits were described by Hunt and Mabey (1966).

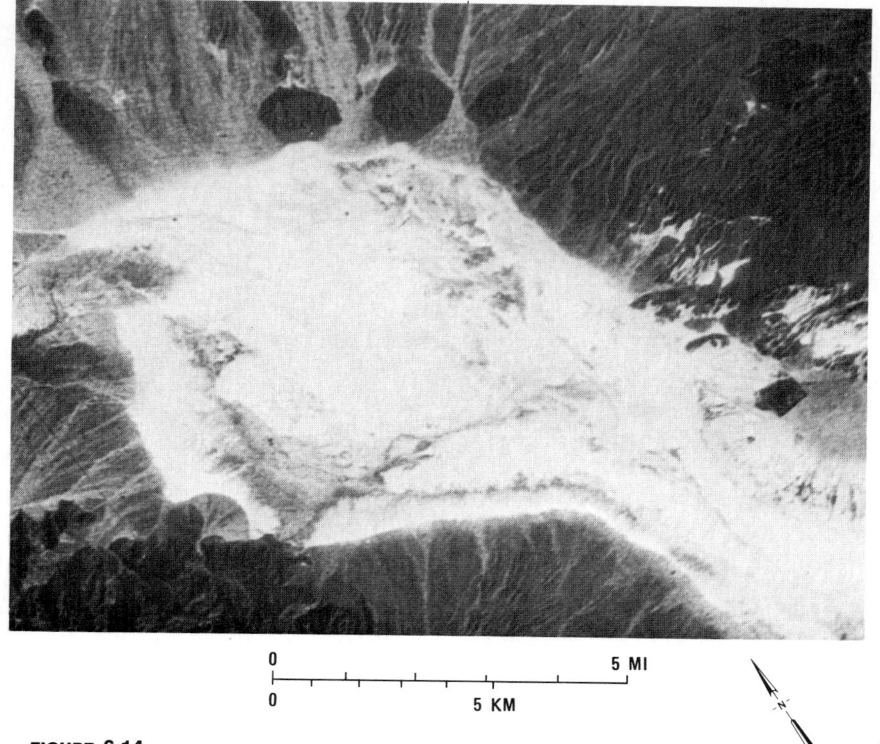

FIGURE 6.14
Skylab photograph of Cottonball Basin. Enlarged from normal color photograph No. 4-94-013. Courtesy G. G. Schaber, U.S. Geological Survey.

FIGURE 6.15
Geologic interpretation of X-band radar image of Cottonball Basin.

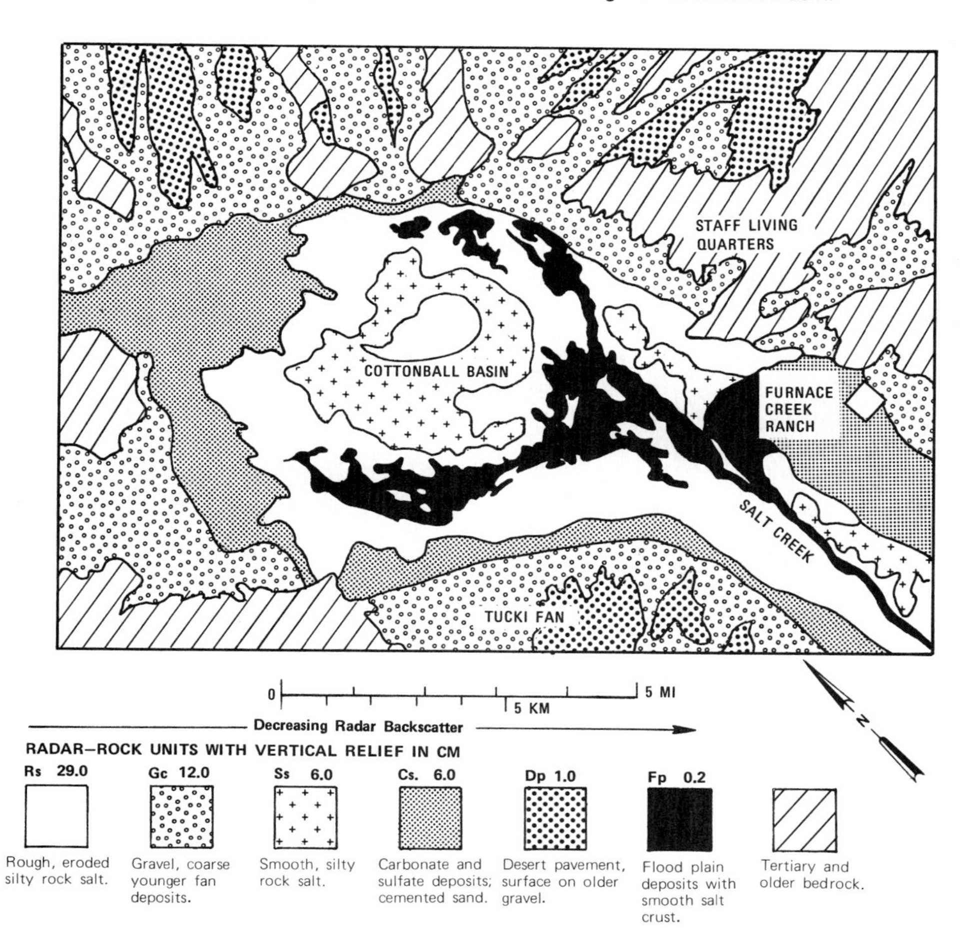

FIGURE 6.16
Oblique ground photographs of radar-rock units at Cottonball Basin and Tucki fan. Numbers indicate average vertical relief.

TABLE 6.5
Surface relief and radar signatures at Cottonball Basin

Surface relief	Limiting value of h	Predicted signature on X band image
Smooth	$h < 0.31$ cm	Dark
Intermediate	$h = 0.31$ to 1.74 cm	Intermediate
Rough	$h > 1.74$ cm	Bright

TABLE 6.6
Observed and predicted signatures for radar-rock units at Cottonball Basin

Radar-rock unit	h, average (measured)	Observed signature	Predicted signature
Flood plain deposits	0.2 cm	Dark	Dark
Desert pavement	1.0 cm	Intermediate	Intermediate
Carbonate-cemented sand	6.0 cm	Intermediate	Bright
Smooth rock salt	6.0 cm	Intermediate	Bright
Coarse gravel	12.0 cm	Bright	Bright
Rough, eroded rock salt	>30.0 cm	Bright	Bright

Comparison of Scattering Theory and Field Data
The theoretical backscattering relationships presented earlier in this chapter may be evaluated by using the radar parameters and the surface relief of the radar-rock units in Cottonball Basin. The X-band image (Figure 6.13) was acquired with an antenna depression angle γ of 23° at the center of the basin. From near range to far range across the basin the antenna angles vary by only 7°; therefore the angle for the center may be used for the entire basin. By substituting these values into Equations (6.5) and (6.6) the limits of h for different surfaces may be calculated and are listed in Table 6.5.

For each limiting value of h the corresponding signature on the X-band image is predicted. Table 6.6 lists the radar-rock units and their vertical relief determined from field measurements, together with their signature on the image (Figure 6.13). The predicted signature for each unit is also listed in Table 6.6 and agrees with the signature on the image for all units, except the carbonate-cemented sand and the smooth rock salt. The vertical relief h of these two units is 6.0 cm, and they should produce the bright signatures characteristic of rough surfaces. The actual image signatures, however, are only intermediate in intensity. Field checking provides a possible explanation. Immediately below the thin surface crust both units are very moist, which increases the dielectric constant and therefore the absorptivity of radar waves. The radar waves may penetrate the thin crust and be partially absorbed by the moist material, which would explain the anomalously low backscatter from these radar-rock units.

Radar Wavelengths and Surface Roughness at Tucki Fan

X-band and L-band SLAR images (Figure 6.17) have been acquired of the Tucki alluvial fan at the southwest margin of Cottonball Basin (Figure 6.15). These images provide the opportunity to compare signatures of radar-rock units at wavelengths of 3 and 25 cm. The aerial photograph and the interpretation map (Figure 6.18) show the distribution of the radar-rock units that were illustrated in Figure 6.16. An additional radar-rock unit appears on the L-band image that is not apparent on the X-band image. This is a belt of fine-grained gravel a few hundred meters wide at the base of Tucki fan that produces a dark (weak

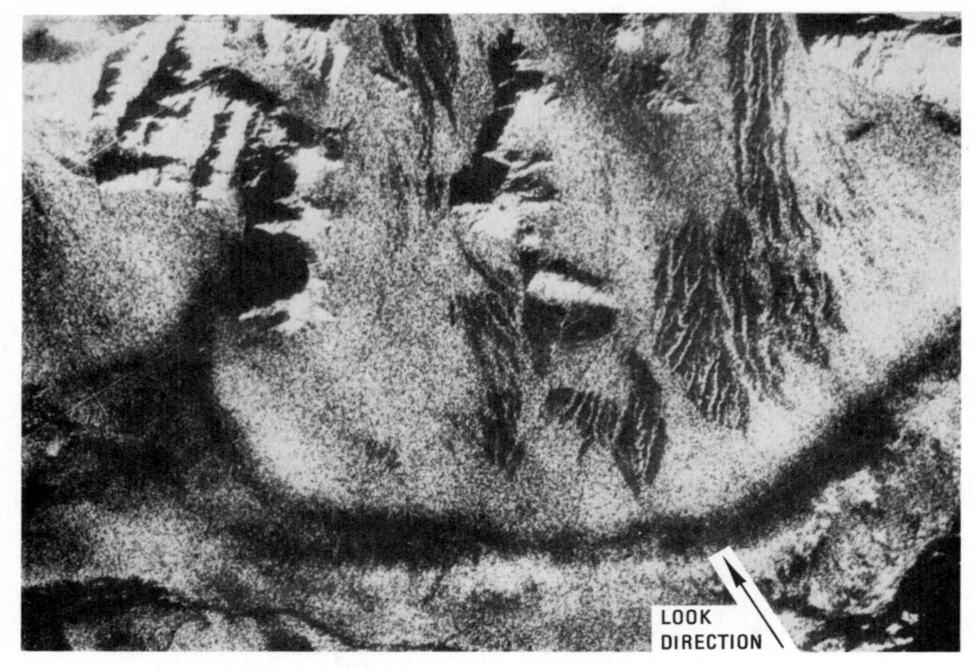

A. X–BAND (λ = 3 cm), IMAGE ACQUIRED BY U.S. STRATEGIC AIR COMMAND. LOOK DIRECTION TO THE SOUTHWEST. DEPRESSION ANGLE AT CENTER OF IMAGE IS 18°.

B. L–BAND (λ = 25 cm) IMAGE ACQUIRED BY JET PROPULSION LABORATORY FOR NASA. LOOK DIRECTION TO THE WEST. DEPRESSION ANGLE AT CENTER OF IMAGE IS 47°.

FIGURE 6.17
Radar images of Tucki fan, Death Valley, California. From Schaber, Berlin, and Brown (1976, Appendix Figure 1, Figure 9). Courtesy G. G. Schaber, U.S. Geological Survey.

A. AERIAL PHOTOGRAPH.

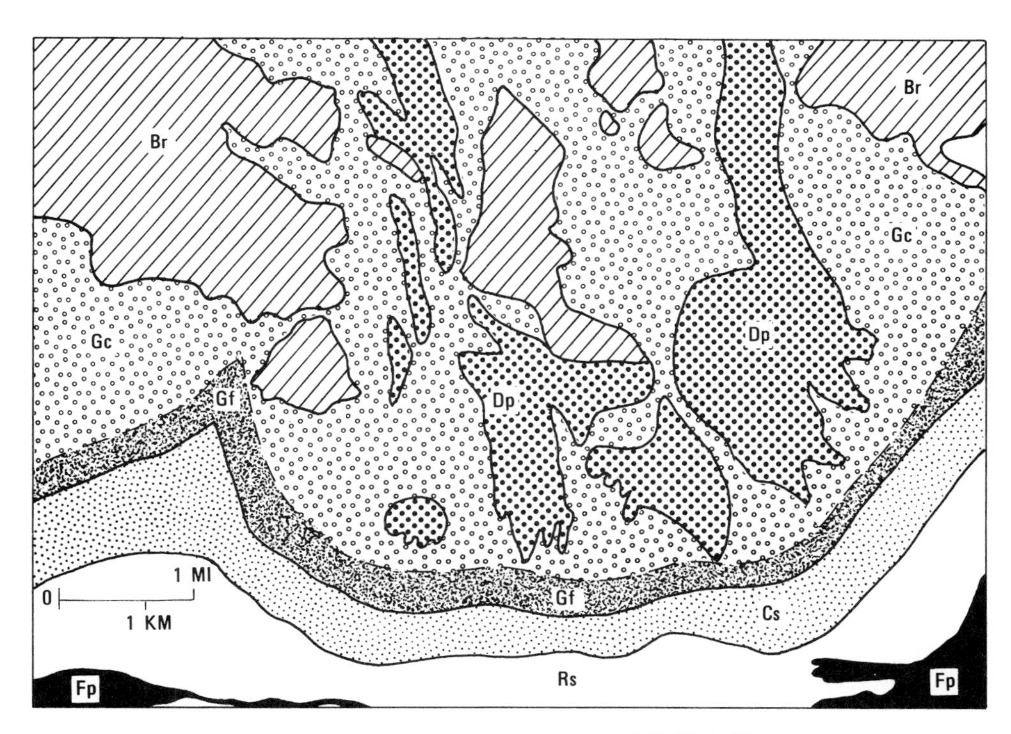

B. RADAR–ROCK UNITS. Br–BEDROCK; Cs–CARBUNATE–CEMENTED SAND; Dp–DESERT PAVEMENT; Fp–FLOOD PLAIN DEPOSITS; Gc–COARSE GRAVEL; Gf–FINE GRAVEL; Rs–ROUGH SALT.

FIGURE 6.18
Aerial photograph and radar-rock units at Tucki fan. Compare with radar images. Photograph from Schaber, Berlin, and Brown (1976, Figure 10A). Courtesy G. G. Schaber, U.S. Geological Survey.

TABLE 6.7
Limiting roughness values calculated for images of Tucki Fan

Surface type	X band image (λ = 3 cm, γ = 18°)	L band image (λ = 25 cm, γ = 47°)
Smooth surface	$h < 0.4$ cm	$h < 1.4$ cm
Intermediate surface	h = 0.4 to 2.2 cm	h = 1.4 to 7.8 cm
Rough surface	$h > 2.2$ cm	$h > 7.8$ cm

backscatter) signature on the L-band image (Figure 6.17B). The fine-grained gravel, which is assigned the symbol Gf, has a sharp contact on the east with carbonate-cemented sand (Cs). On the aerial photograph (Figure 6.18A) the fine-grained gravel has a light gray signature intermediate between the carbonate-cemented sand (bright signature) and the alluvial fan (dark signature). Except for the higher degree of erosion, the desert pavement surface (Dp) is indistinguishable on the photograph from the coarse gravel (Gc). On both the X-band and L-band images, however, the smooth desert pavement produces weaker backscatter (darker signature) than the rough surface of the coarse gravel.

For the same radar-rock units there are marked signature differences on the X-band and L-band images. The first step in understanding these differences is to calculate from Equations (6.4) and (6.5) the limiting values of h for the two SLAR images. By using depression angles γ for the center of the fan, the values listed in Table 6.7 were calculated.

The calculated limiting values of h are shown in Figure 6.19, together with the measured values of h and the signatures of the radar-rock units on the images. The observed signatures agree with the calculated roughness limits for the radar-rock units. The major exception is the intermediate signature of unit Cs on the X-band image that was discussed earlier. Relief of the flood plain deposits (Fp) is below the smooth limit for both wavelengths, which agrees with the dark signature on both images (Figure 6.17). Relief of the eroded rock salt (Rs) and coarse gravel (Gc) exceeds the rough limit for both wavelengths and agrees with the bright signature on both images. The other units have brighter signatures on the X-band image than on the L-band image, consistent with the predicted roughness ranges in Figure 6.19. On the X-band image the intermediate to bright signature of the fine gravel (Gf) grades into the bright signature of the coarse gravel (Gc) and the two are indistinguishable. However, they are readily separated on the L-band image (Figure 6.17B) where Gc has a bright signature and Gf has a dark signature.

IMAGE CHARACTERISTICS

Other factors that contribute to the characteristic appearance of radar images include shadows and highlights, look direction, image geometry, and distortion.

Shadows and Highlights

The oblique illumination of SLAR produces strong returns from the sides of ridges and peaks facing the antenna. These topographic obstructions also create shadows in the look direction, as may be seen on the image of the Santa Rosa Mountains

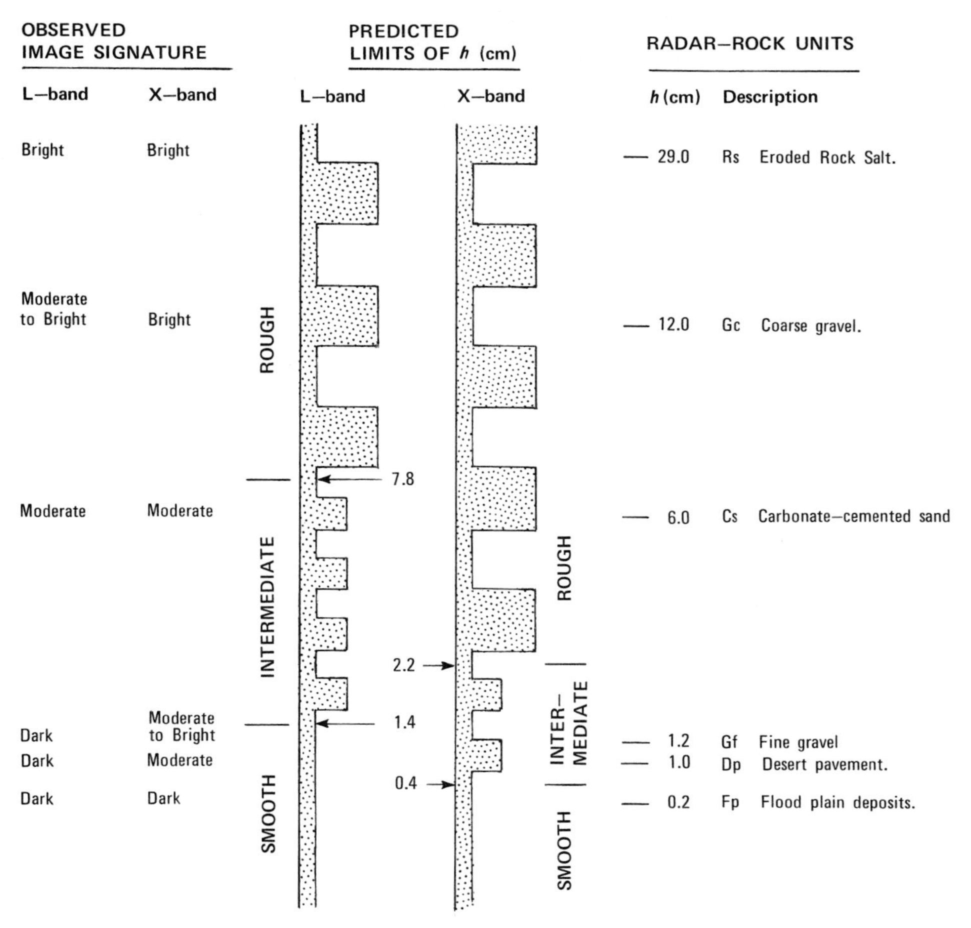

FIGURE 6.19
Comparison of predicted and observed image signatures at L-band and X-band wavelengths for radar-rock units at Tucki fan. Data from Schaber, Berlin, and Brown (1976).

A. Ka–BAND IMAGE WITH LOOK DIRECTION TO THE SOUTHWEST. SCARP OF ACTIVE CLARK FAULT INDICATED AT F.

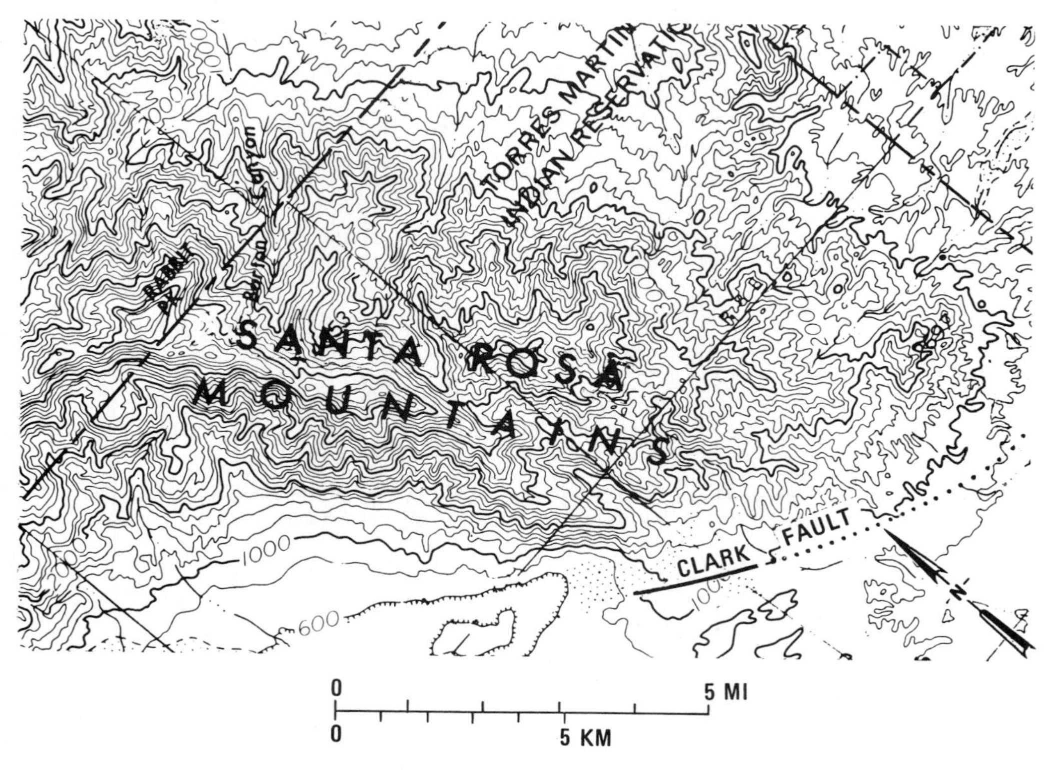

B. TOPOGRAPHIC MAP WITH CONTOUR INTERVAL OF 200 FT (60M). COMPARE PROFILE OF MOUNTAIN CREST WITH RADAR SHADOW.

FIGURE 6.20
Radar shadow of Santa Rosa Mountains, Riverside County, California.

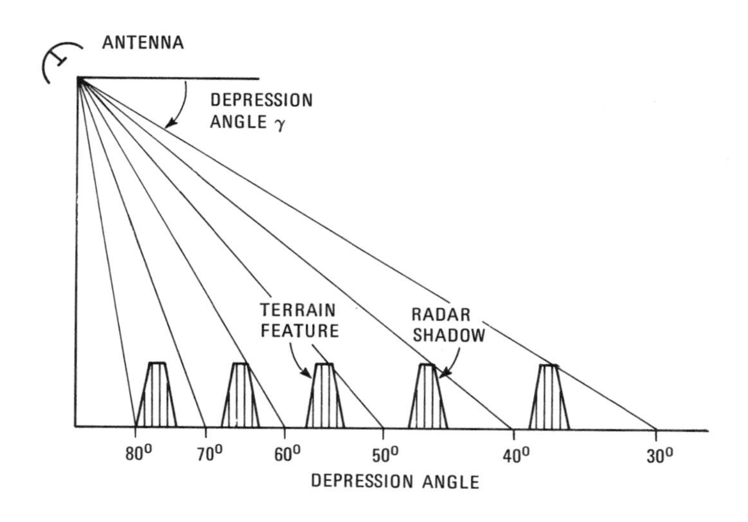

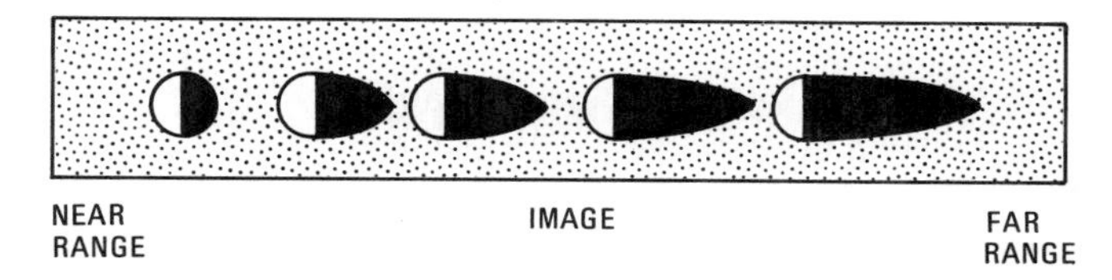

FIGURE 6.21
Radar illumination and shadows at different depression angles.

(Figure 6.20A). The outline of the radar shadow matches the topographic profile of the ridge crest as shown by comparison with the topographic map (Figure 6.20B). The importance of radar highlights is illustrated by the southeast-trending linear feature marked by a very bright highlight at F on the image. This is a topographic scarp along the active Clark fault that is part of the San Jacinto fault zone. Southeastward from this scarp the Clark fault can be traced into the badland topography at the south end of the Santa Rosa Mountains. On the image the fault extension is marked by an aligned series of highlights crossing the badland ridges.

The elevation angle of solar illumination is constant throughout an aerial photograph. On a SLAR image the radar illumination becomes more oblique in the far-range direction and shadows are proportionately longer (Figure 6.21). The importance of the longer shadows is shown on the mosaic of Figure 6.22B, which is compiled from the far-range portions of overlapping image strips. The faults and fractures are much more distinct on this image than on the mosaic compiled from near-range images of the same area (Figure 6.22C). The relatively short shadows of the near-range mosaic are less effective for enhancing faults and fractures.

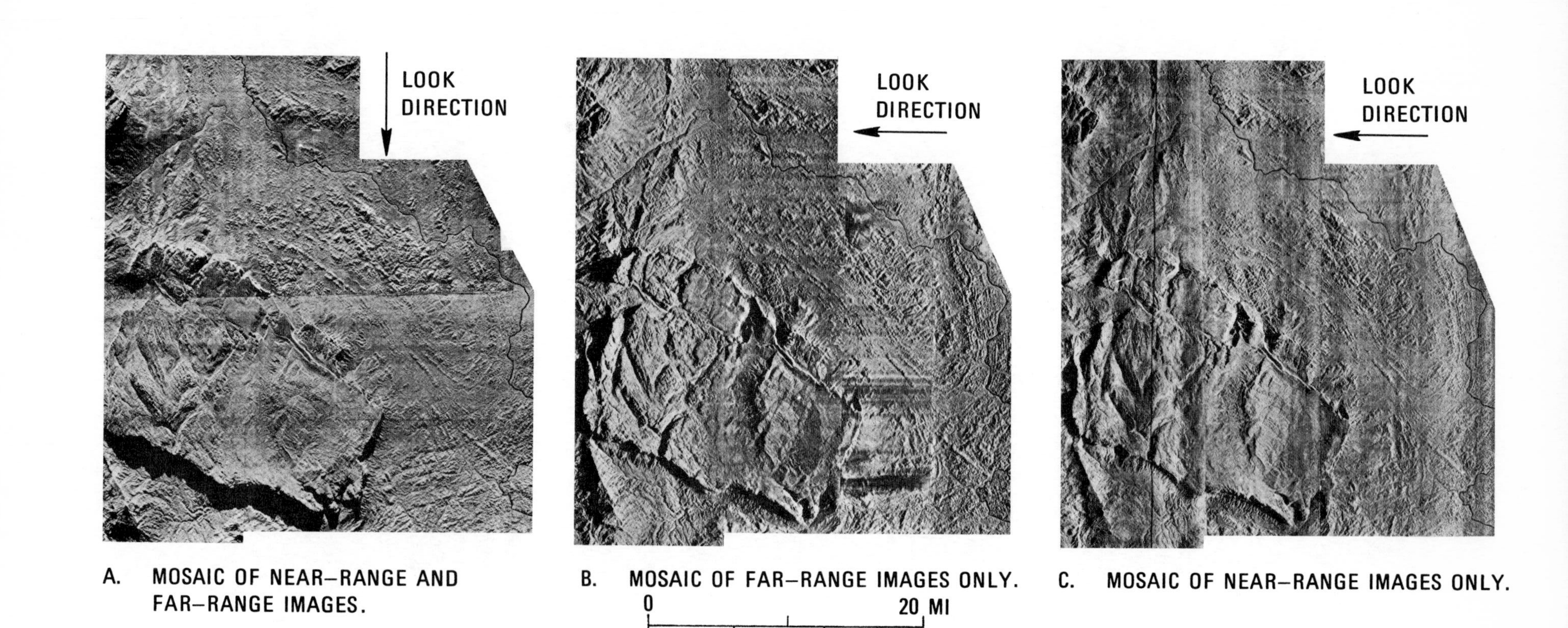

FIGURE 6.22
Influence of look direction and shadow length on X-band radar mosaics of southeastern Venezuela. Courtesy Aero Service Corporation.

Look Direction and Lineament Expression

The relationship between radar look direction and the direction of faults and fractures determines whether these linear features are enhanced or subdued on an image. Shadows and highlights are most pronounced where linear features are oriented normal to the radar look direction. This relationship is illustrated by the fractures on the mesa in Figure 6.22 that trend north and north-northeast. They are relatively subdued in Figure 6.22A with the look direction to the south, which is nearly parallel with the fracture trend. The fractures are more pronounced in Figure 6.22B and 6.22C with the look direction to the west, nearly normal to the fracture trend. The fracture enhancement is particularly evident when the western portions of Figures 6.22A and 6.22B are compared.

Stereo Imagery

Adjacent radar flight lines may be spaced to obtain overlapping images such as the example in Figure 6.23 from Venezuela. These images should be viewed with a pocket stereoscope to obtain the three-dimensional effect. This technique is particularly useful in areas of low relief. Stereo radar coverage may also be obtained by imaging the same area from two different flight altitudes or from two opposing antenna look directions. Stereo models of radar strips are interpreted in the same manner as stereo models of aerial photographs and Landsat images.

Lineament Detection in Forested Terrain

In forested terrain, such as that in Figure 6.22, geologic structure is commonly more apparent on radar images than on aerial photographs. These observations have caused some investigators to report erroneously that Ka-band and X-band radar is "penetrating the vegetation canopy." Radar theory, which was summarized earlier in this chapter, clearly establishes that these short wavelengths must interact with vegetation, for the leaves and branches are approximately the dimension of the radar wavelength. The high moisture content of leaves tends to reflect, rather than transmit, radar energy. The interaction with vegetation is illustrated in images of agricultural areas (Figure 6.24A) where different crop types produce different intensities of backscattering, clearly showing that radar waves do not penetrate crops, let alone dense jungle or rain forest.

The enhancement of structural features in forested terrain appears to be a function of the relatively low spatial resolution of radar images that is typically 10 m. The individual trees are not resolved, which enhances the subtle topographic expression of geologic features that have dimensions of hundreds and thousands of meters. The large ground resolution cell of radar serves as a filter that removes the high-frequency spatial detail of vegetation "noise," thereby enhancing the geologic "signals." The high spatial resolution of an aerial photograph records the vegetation detail that obscures the underlying geologic features.

Recognition of Rock Types

Except for unconsolidated gravel, sand, and clay there is no consistent correlation between surface roughness and rock type. At Cottonball Basin for example, such diverse rock types as smooth silty rock salt (Ss) and carbonate-cemented sand (Cs) have the same vertical relief and the same radar signatures. Therefore identification of rock types on radar images must utilize the geomorphic expression and weathering characteristics of the rocks. In Figure 6.22C the mesa is capped by a

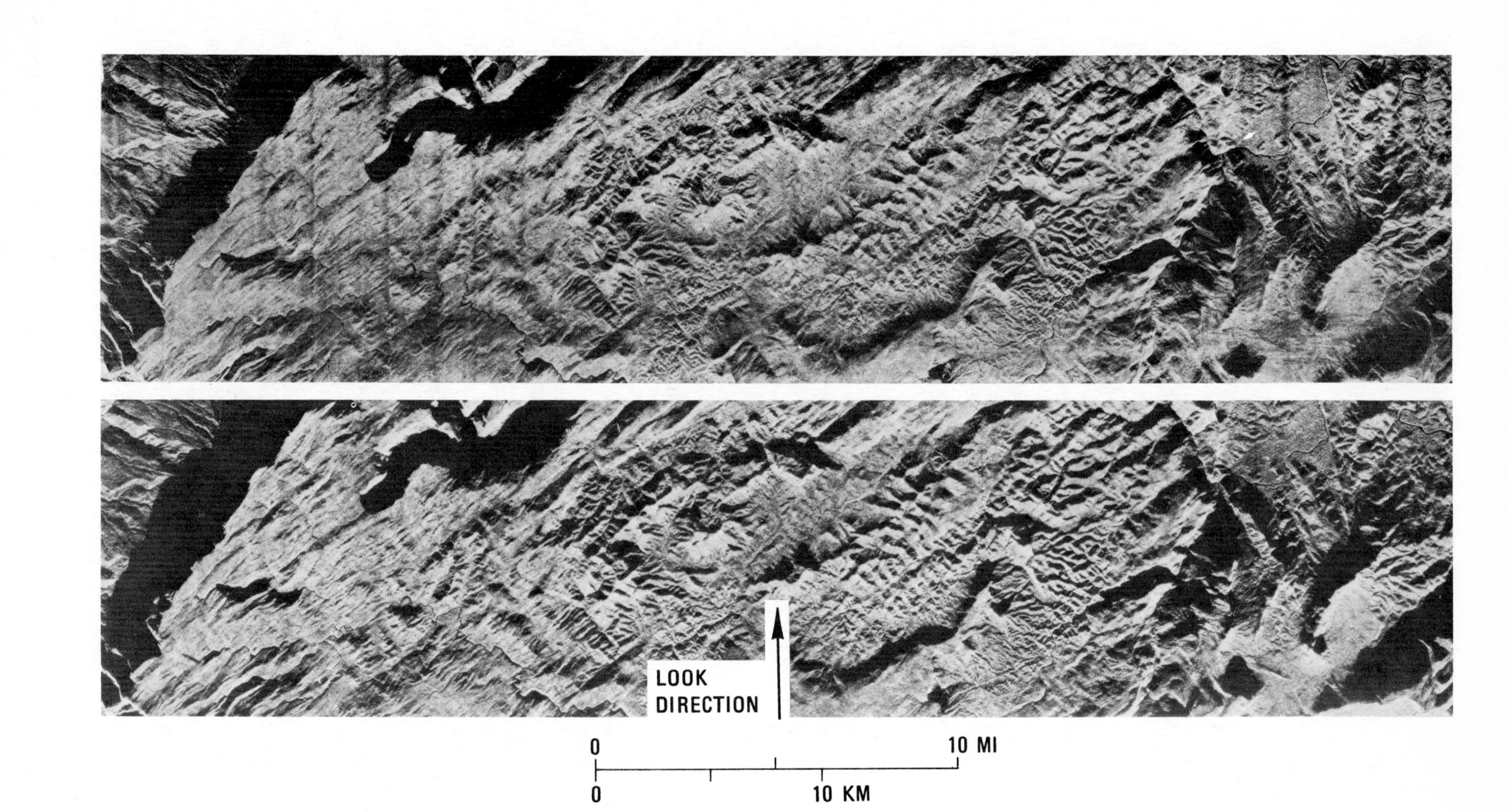

FIGURE 6.23
Stereo pair of X-band radar images for viewing with pocket stereoscope. Area is Cerro Duida in south-central Venezuela. Courtesy Aero Service Corporation.

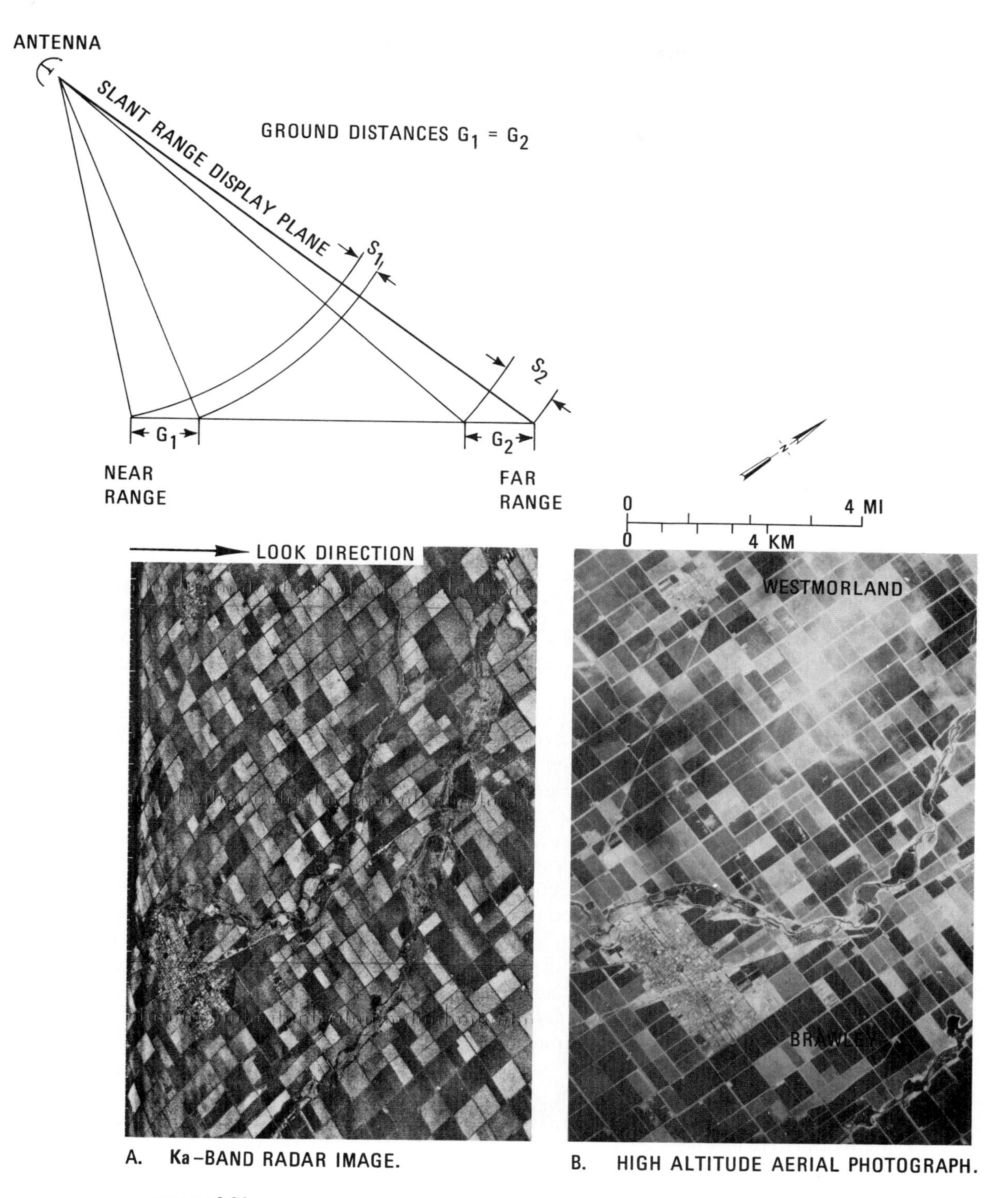

FIGURE 6.24
Distortion of slant-range radar image. Note geometric compression in near-range direction. Imperial Valley, California.

resistant layer and underlain by a nonresistant layer, indicated by the intricately dissected slopes on the south margin. The lack of karst topography in this tropical environment rules out carbonate rocks for the resistant layer, which probably consists of well-cemented sandstone or conglomerate. The low terrain north and east of the mesa appears to be underlain by folded and fractured metamorphic rocks. Young volcanic rocks may be identified on radar images by surface flow features. Plutonic rocks may be recognized by their characteristic fracture patterns, weathering properties, and lack of stratification.

Image Distortion and Irregularities

The geometric distortion and irregularities of radar imagery must be understood and recognized in order to make effective interpretations.

Slant-Range and Ground-Range Images Depending upon design of the system, SLAR images are presented either as slant-range or as ground-range displays. On *slant-range displays*, the scale in the range direction is compressed in the near-range portion of the image (Figure 6.24A) because targets are plotted on the image using the actual two-way travel time. The change in depression angle from near range to far range causes this to be a distorted display, as seen by comparison with an aerial photograph (Figure 6.24B). The square agricultural fields are distorted into rhombic shapes in the near-range part of the image, where compression is most pronounced.

To produce *ground-range images*, which are undistorted, the SLAR system employs a time delay in the plotting circuit that compensates for the travel time differences. This compensation results in approximately orthogonal geometry in which the image scales are equal in the range and azimuth directions. All the Ka-band images in this book are slant-range displays; the other examples are ground-range displays. The geometric differences are shown in Figure 6.10 of San Francisco Peninsula, where the X-band image is a ground-range display and the Ka-band image is a slant-range display.

Image Displacement (Layover) The curvature of the transmitted radar pulse causes the top of a tall vertical target to reflect energy in advance of its base, resulting in displacement of the top toward the near-range direction. This displacement is called *layover*. On the diagram of Figure 6.25A, the curved wavefront encounters the mountain peak at A in advance of the base B. The image of the peak is offset toward the near range relative to the base. The ridges of the Grapevine Mountains are relatively symmetrical, but on the image they appear to be foreshortened, or to "lean," toward the near-range direction because of the layover effect. Comparison with the undistorted Skylab photograph (Figure 6.25B) illustrates the extent of displacement on the radar image. The Jet Propulsion Laboratory SLAR system provides a topographic profile along the aircraft flight path that is shown on the left side of Figure 6.25A.

Side-Lobe Banding A common irregularity in the near-range part of radar images is a pattern of light bands trending parallel with the flight direction, which is called *side-lobe banding* (Figure 6.26A). In addition to the main pulse of energy transmitted from the antenna, subsidiary pulses called side lobes are also generated. The terrain returns from these side-lobe pulses reinforce the main return and cause brighter bands on the image. The lower energy of the side-lobe pulses is largely attenuated in the longer two-way travel to the far range; hence, banding is usually confined to the near-range portion of the image. Cross-polarized terrain returns are very weak and must be electronically amplified, which accentuates the

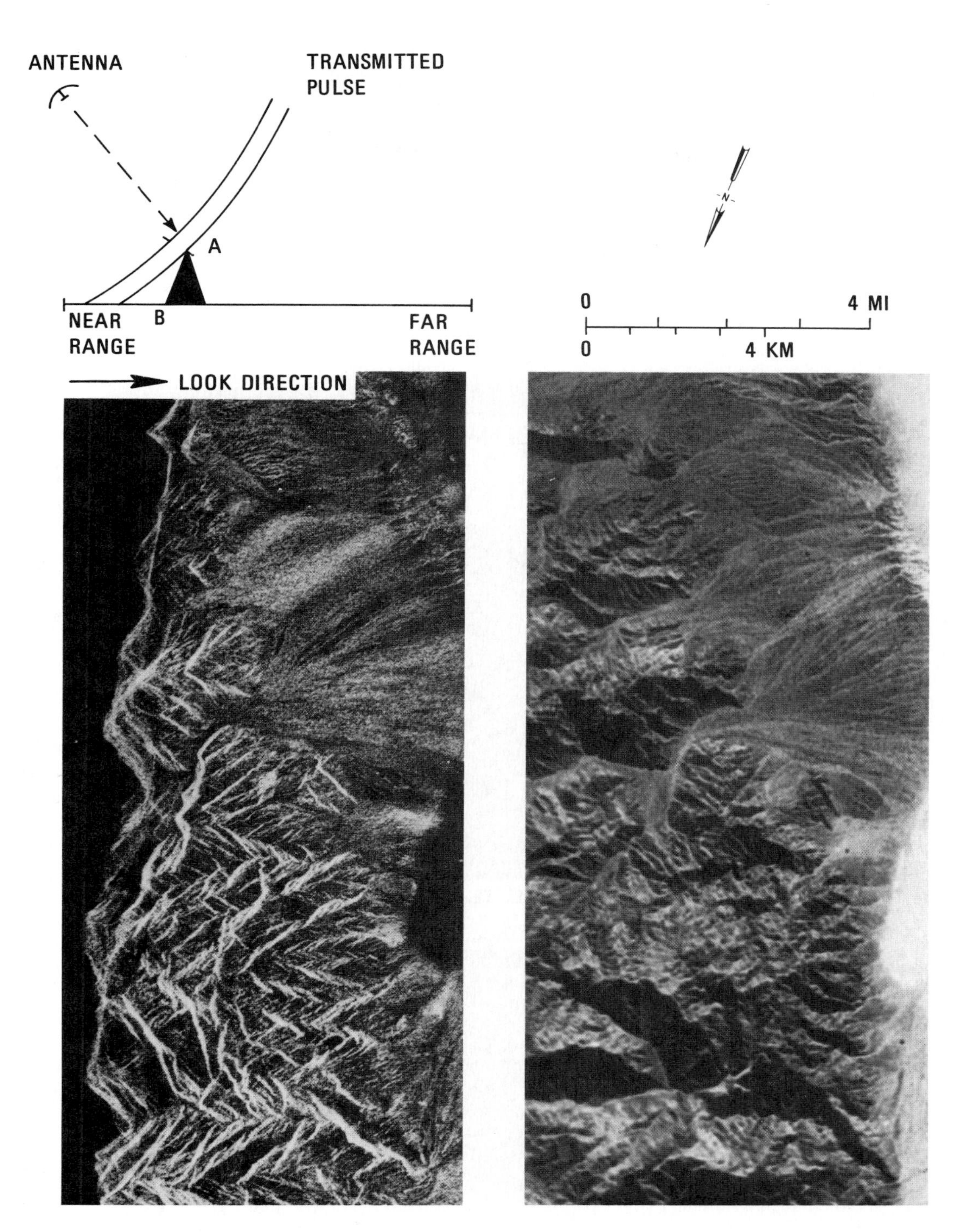

A. L–BAND RADAR IMAGE ACQUIRED BY NASA AND JET PROPULSION LABORATORY, CALIFORNIA INSTITUTE OF TECHNOLOGY.

B. ENLARGED SKYLAB–4 PHOTOGRAPH.

FIGURE 6.25
Displacement, or layover, on radar image. Grapevine Mountains on east side of Death Valley, California. Courtesy G. G. Schaber, U.S. Geological Survey.

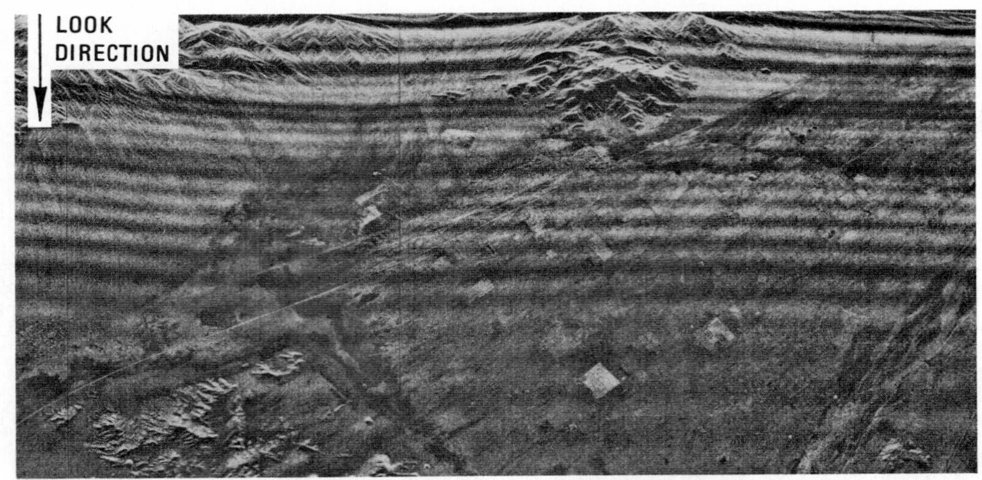

A. SIDE–LOBE BANDING. TROY DRY LAKE AND NEWBERRY MOUNTAINS, CALIFORNIA.

B. RAIN CLOUDS. SOLITARIO UPLIFT, WEST TEXAS.

C. AIRCRAFT ROLL AND YAW. IMPERIAL VALLEY, CALIFORNIA.

FIGURE 6.26
Irregularities on Ka-band radar images. Arrows indicate look direction. Image A is cross polarized; B and C are parallel polarized.

side-lobe banding as well as the terrain returns. For this reason side-lobe banding is more apparent on cross-polarized images, such as Figure 6.26A, than on parallel-polarized images. Pronounced banding is unmistakable, but subtle bands may be confused with true terrain signatures by an inexperienced interpreter.

Backscatter from Precipitation Ka-band and longer wavelengths of microwave energy penetrate clouds that scatter and absorb visible and IR wavelengths. Rain, snow, and sleet, however, cause Ka-band energy to be backscattered and returned to the antenna without reaching the ground. As depicted in Figure 6.26B, there are strong returns from the precipitation and shadows on the image in the down-range direction. This principle is employed by weather radar, which uses short wavelength energy. Dangerous clouds, in which precipitation is occurring can be distinguished (by their strong backscattering) from clouds with no precipitation that do not produce a signature on the image. The 3-cm wavelength energy of X-band SLAR is backscattered only by the heaviest thunderstorms. L-band energy is essentially unaffected by precipitation.

Effects of Aircraft Motion SLAR antennas have stabilization systems to compensate for normal aircraft roll, but strong air turbulence and aircraft maneuvers may exceed the limits of these devices and result in image distortion. In Figure 6.26C the wavy distortion is caused by strong turbulence. The complete loss of image is caused by aircraft turns when the antenna is not directed toward the ground.

Image Scale and Ground Coverage

Image scale and ground coverage are determined by the aircraft altitude and the depression angles from near range to far range. The original scale of the Ka-band images in this book is typically 1:180,000, and ground coverage in the look direction ranges from 13 to 16 km. The original negatives are 7.6-cm wide and can be enlarged considerably. The commercial surveys with synthetic-aperture, X-band SLAR systems are typically flown at an altitude of 12 km and produce original images at a scale of 1:400,000. These are photographically enlarged for mosaic compilation and interpretation. A strip of SLAR imagery may be hundreds of kilometers long in the azimuth direction. The small-scale and continuous coverage are very useful in regional investigations.

Mosaics

Mosaics prepared from parallel strips of radar imagery provide valuable regional coverage. Figure 6.27 of Darien Province, Panama is one of the earliest mosaics and created much interest when it appeared in the late 1960s. Because of difficult terrain and persistent cloud cover, this part of Panama had never previously been adequately mapped or photographed from the air. The mosaic pointed out many errors in existing maps and added much topographic detail. By the standards of the mid-1970s, this is a rather crude mosaic with numerous gaps and duplications of coverage. Side-lobe banding is common. Adjacent strips were flown with opposite aircraft headings causing the shadow directions to be reversed on adjacent strips. The slant-range distortion of the image strips was partially corrected by photographic processing.

Despite these problems, much terrain information can be interpreted from the mosaic (Figure 6.28). Of especial interest are the large *en echelon* anticlines and the strike-slip fault along the Pacific coast. A major drainage anomaly is indicated at locality A in Figure 6.27 by the divergence of two

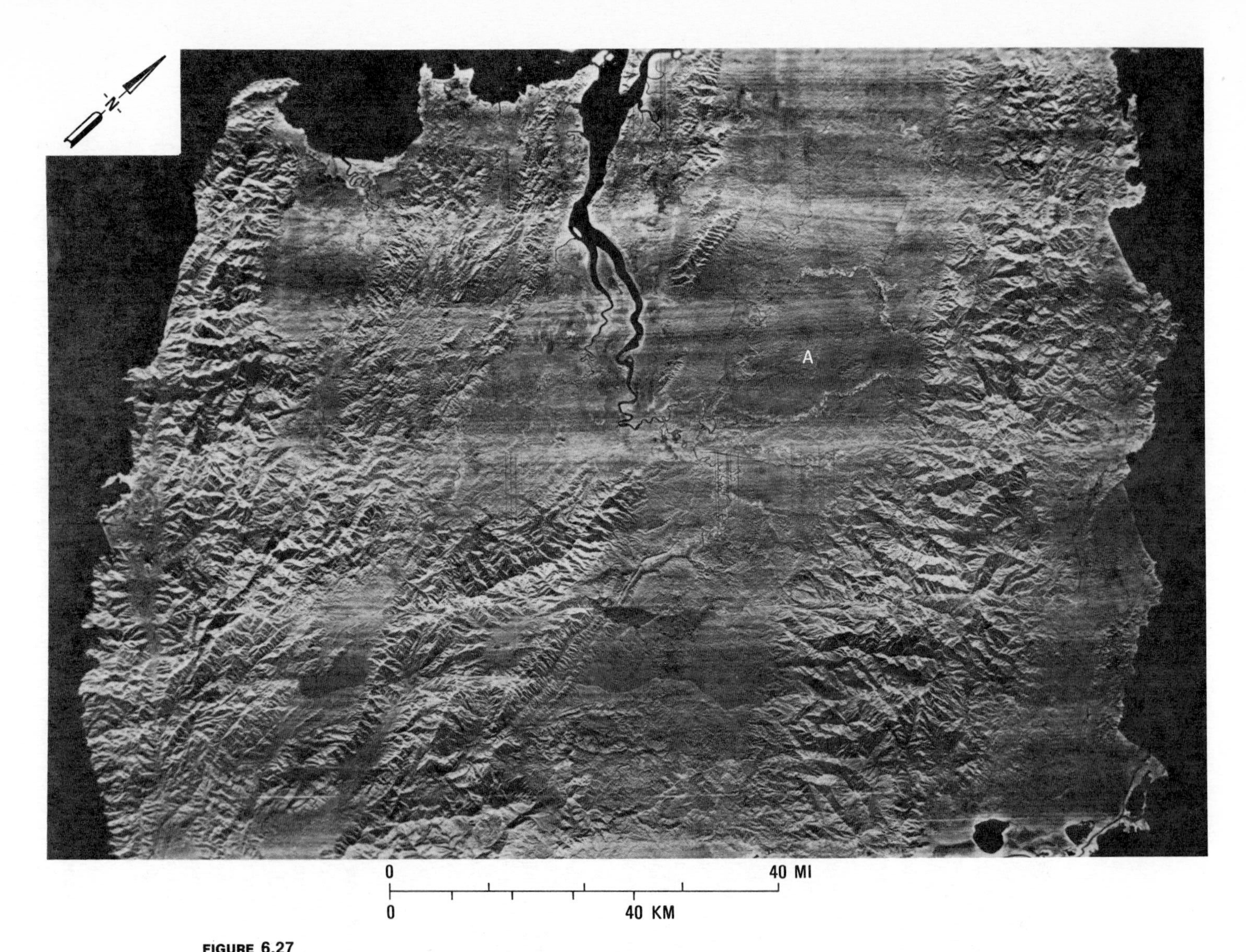

FIGURE 6.27
Ka-band radar mosaic acquired in 1967 of Darien Province, Panama. Note major drainage anomaly at A. Courtesy U.S. Army Engineer Topographic Laboratory.

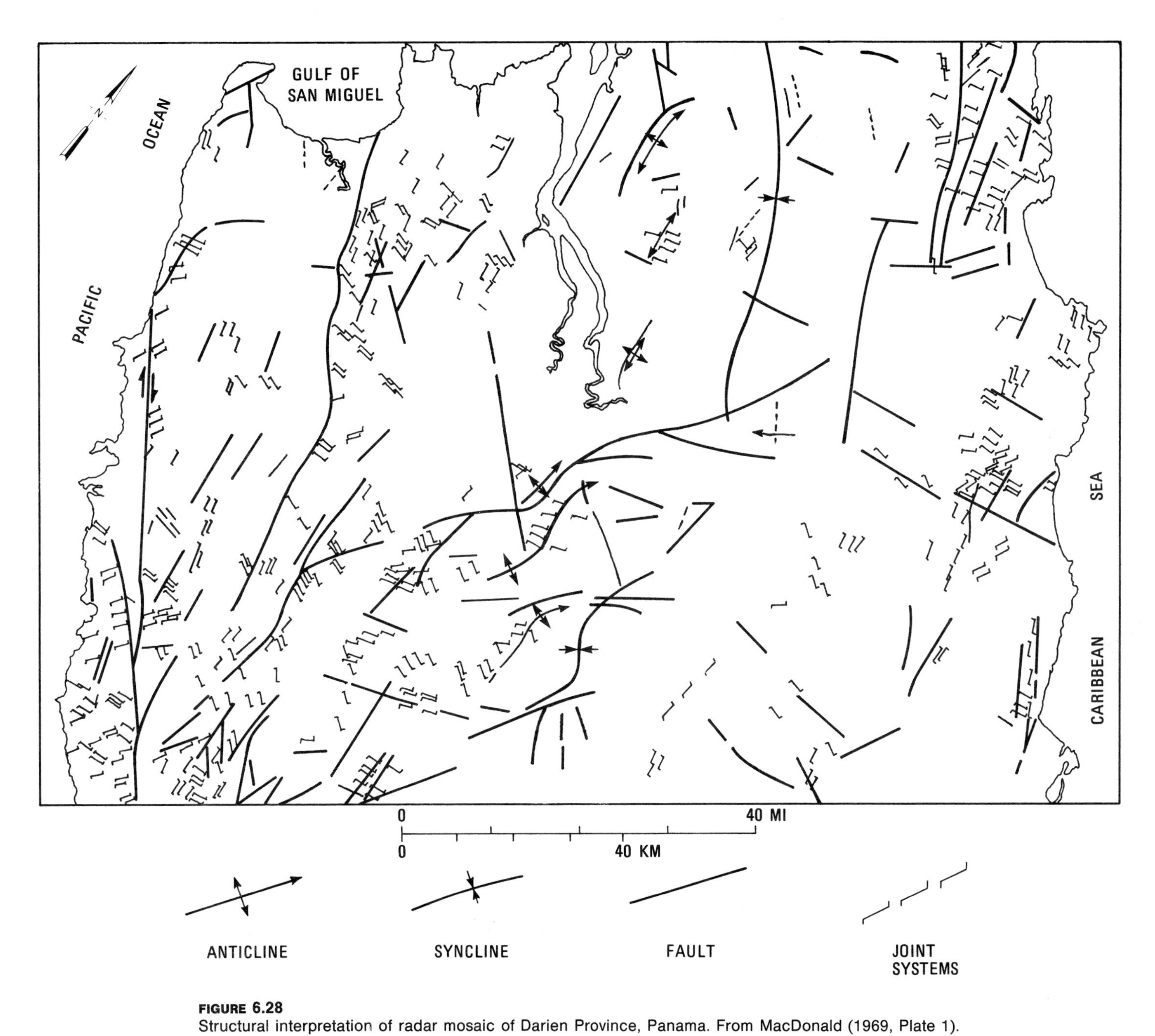

FIGURE 6.28
Structural interpretation of radar mosaic of Darien Province, Panama. From MacDonald (1969, Plate 1).

large streams. The streams may have been deflected around a structural uplift in the interior lowlands.

The X-band mosaic of central Arizona (Figure 6.29) prepared in the early 1970s illustrates the rapid advances in radar technology. All flight lines were made with the same heading so the shadows are all toward the west. Only the far-range half of each image strip was used in the mosaic, resulting in relatively consistent shadow lengths. Side-lobe banding is eliminated and resolution is consistent throughout the mosaic, because a synthetic-aperture system was employed.

Commercial radar mosaics have been prepared for the inland region of Venezuela and most of Brazil. These large, remote areas with poor weather for aerial photography are ideally suited for radar surveys. In Brazil the mosaics are used for preparing base maps, for planning purposes, and for resource exploration.

ADVANTAGES OF RADAR IMAGES

The unique properties of SLAR provide several advantages for interpreters that are summarized in the following section.

Regional Investigations

Individual image strips may be many hundreds of kilometers long and several tens of kilometers wide with minimal geometric distortion. Image strips may be combined into mosaics to provide broader coverage for regional analyses of terrain and geologic structure. The images may be enlarged for detailed interpretation.

Oblique Illumination

The oblique illumination of terrain by the radar beam produces a highlight and shadow effect that enhances faults and fractures, as illustrated in Figure 6.30. The fault that shows so clearly on the radar image is relatively obscure on the vertical aerial photograph. Local geologic maps depict the fault, but do not show it to be as extensive as does the radar image. Orientation of the radar look direction was optimum for enhancing this fault. In areas of high relief, radar shadows obscure part of the terrain and additional flight lines with opposite look direction are needed to acquire images of the shadowed areas.

Suppression of Minor Detail

The advantage of radar images for suppressing details of vegetation cover because of small-scale and low resolution has been discussed. Minor cultural detail is likewise suppressed so the interpreter is not distracted in the search for larger terrain features. This is illustrated in Figure 6.31, where the beach ridges are much more distinct on the radar image than on the aerial photograph, which is "cluttered" with urban features. Oblique illumination of the radar also aids in identifying the beach ridges. The parallel highlights and shadows of the beach ridges could be mistaken for side-lobe banding, but these would also occur on the water in the near-range part of the image.

Detection of Lineaments

The enhancement of linear features results from several characteristics of radar images including the oblique illumination, uninterrupted image coverage, and the suppression of minor details. The interpreter can recognize and integrate subtle changes in tone, texture, and topography that may be aligned on the image to form lineaments with structural significance.

All Weather and Nighttime Capability

Some regions have never been satisfactorily photographed from the air because of prevailing poor weather. SLAR images can be acquired under

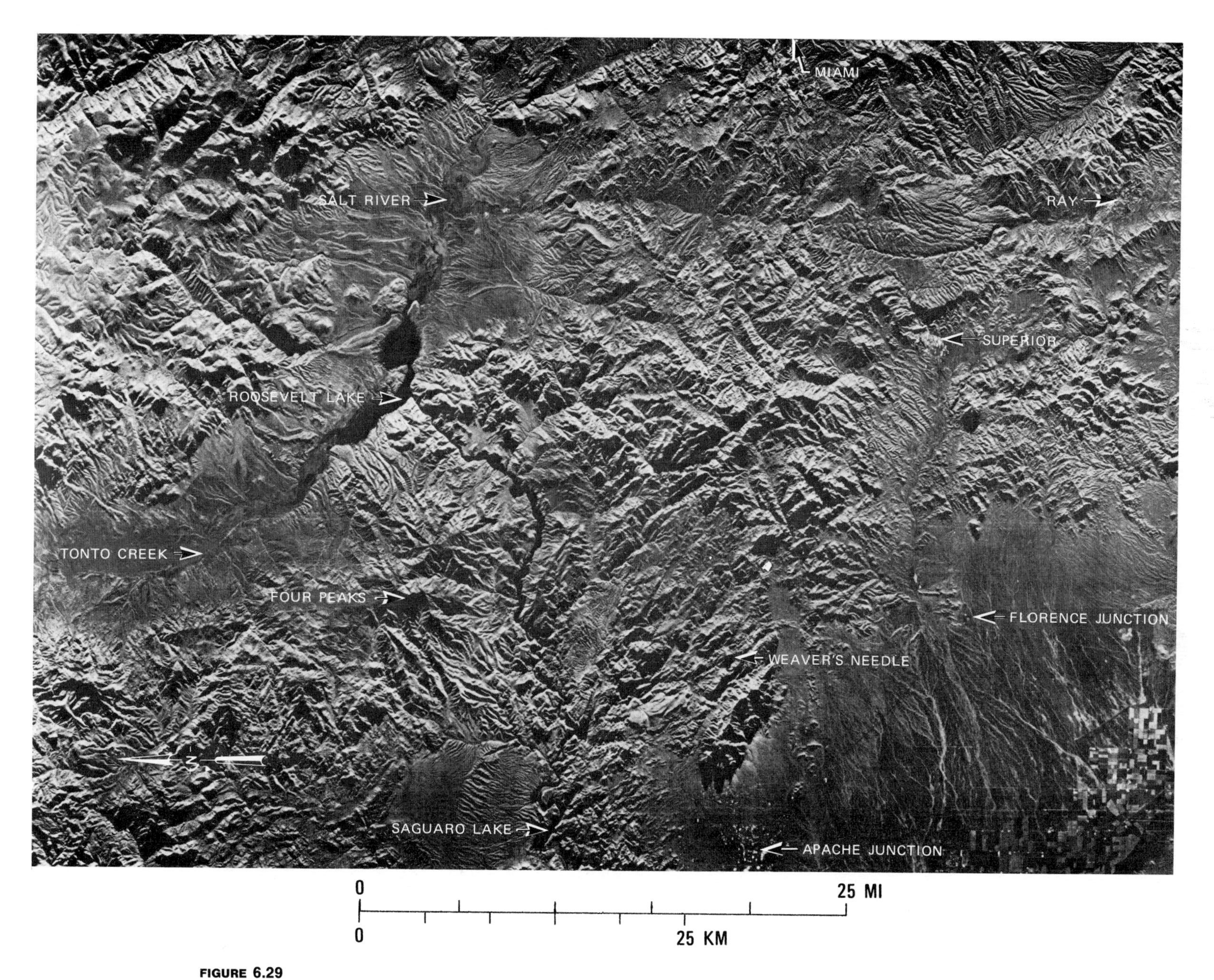

FIGURE 6.29
Radar mosaic of Superior and Miami region, central Arizona. Synthetic aperture, X-band radar images acquired with look direction to the west. Courtesy Aero Service Corporation.

A. AERIAL PHOTOMOSAIC ACQUIRED DECEMBER, 1953. FAULT INDICATED BY ARROWS.

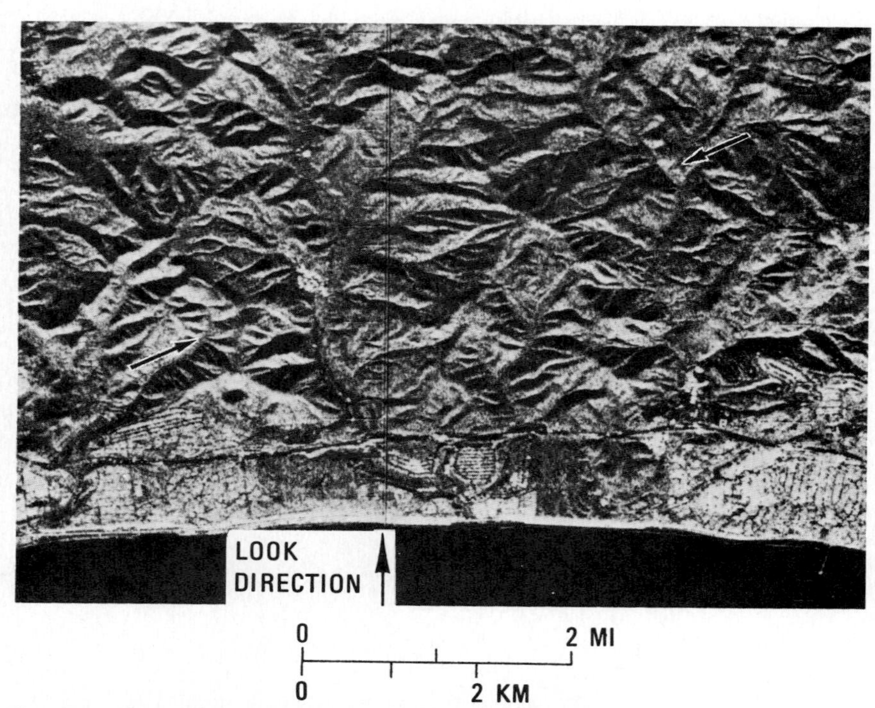

B. Ka–BAND RADAR IMAGE ACQUIRED NOVEMBER, 1965. OBLIQUE RADAR ILLUMINATION CAUSES HIGHLIGHTS AND SHADOWS THAT ENHANCE THE FAULT TRACE.

FIGURE 6.30
Fault near San Clemente, southern California. From Sabins (1973, Figure 8).

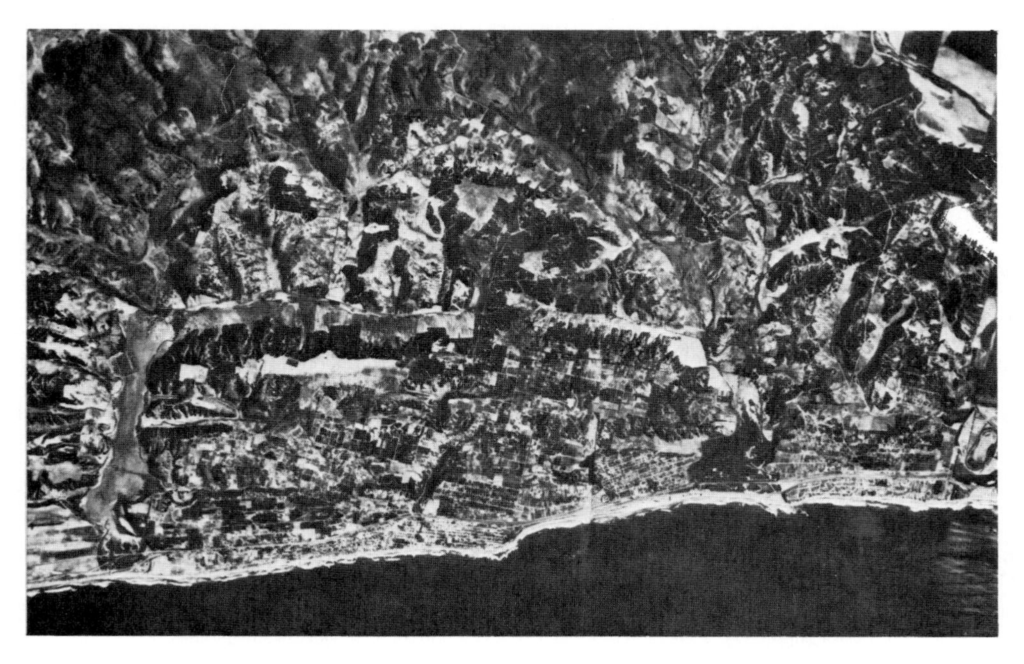

A. AERIAL PHOTOMOSAIC ACQUIRED DECEMBER, 1953. URBAN DEVELOPMENT OBSCURES BEACH RIDGES ALONG THE COAST.

B. Ka-BAND RADAR IMAGE ACQUIRED NOVEMBER, 1965. LOWER RESOLUTION SUPPRESSES URBAN "CLUTTER" AND ENHANCES BEACH RIDGES. OBLIQUE RADAR ILLUMINATION ALSO AIDS.

FIGURE 6.31
Beach ridges near Oceanside, southern California.

TABLE 6.8
Comparison of SLAR and Landsat images

Characteristic	Slar	Landsat
Oblique illumination	Radar depression angle varies from near range to far range across the image.	Sun elevation is constant across the image; varies with the season.
Regional coverage	Typical images are tens of km wide by hundreds of km long.	Each image covers 185 by 185 km.
Image scale	Small to medium scale.	Small scale.
Spatial resolution	Typical ground resolution cell is 10 by 10 m.	Ground resolution cell is 79 by 79 m.
Wavelength	Portion of microwave band (0.86 to 25 cm).	Portions of visible band and reflected IR band (0.5 to 1.1 μm).
Operating mode	Active.	Passive.
Image cost	Dollars per km^2.	Pennies per km^2.

almost any weather conditions, thus providing information on hitherto poorly known regions. Radar is also valuable for monitoring sea ice conditions at high latitudes during periods of winter darkness, as illustrated in Chapter 9.

COMPARISON OF SLAR AND LANDSAT IMAGES

Some significant similarities and differences of SLAR and Landsat images are summarized in Table 6.8. The two forms of imagery are similar in appearance, as shown in Figure 6.32 of the interior of Venezuela. Both images show the topographic and geologic features of this forested region that is underlain by rocks of Precambrian age. The prominent mesas are quartzite and conglomerate beds unconformably overlying deformed crystalline rocks. The Landsat image is one of the few relatively cloud-free images of this region (note the clouds over the mesa). The higher spatial resolution of radar is apparent in the greater terrain detail on the SLAR image.

At 10 A.M. on January 13, 1973, when the Landsat image of Figure 6.32A was acquired, the sun was 46° above the horizon in this North 4° latitude. Shadows and highlights on the Landsat image are subdued in comparison with the SLAR mosaic (Figure 6.32B) that was prepared from the far-range portions of overlapping image strips. The low depression angle of the radar image strongly enhances the lineaments. Streams are much more evident on the band 7 Landsat image because the photographic IR radiation is absorbed by water and reflected by vegetation. The stratified nature of the small mesa in the northwest part of the two scenes is more apparent on the Landsat image than on the SLAR mosaic. The enhancement of stratification on the Landsat image is caused by two factors. (1) The lower spatial resolution of Landsat eliminates much of the finer detail that obscures stratification on the SLAR image; (2) Landsat band 7 records reflectance differences between the stratified rocks in the 0.8 to 1.1 μm spectral region. Radar records differences in surface roughness of the rocks, which apparently do not cause different backscattering at the 3-cm wavelength of the SLAR image. This comparison indicates the complementary nature of the two types of imagery. Because of their low cost, Landsat images can be used economically to supplement an interpretation made from SLAR images.

A. LANDSAT 1174–14091 BAND 7, ACQUIRED JANUARY 13, 1973. SUN ELEVATION 46°.

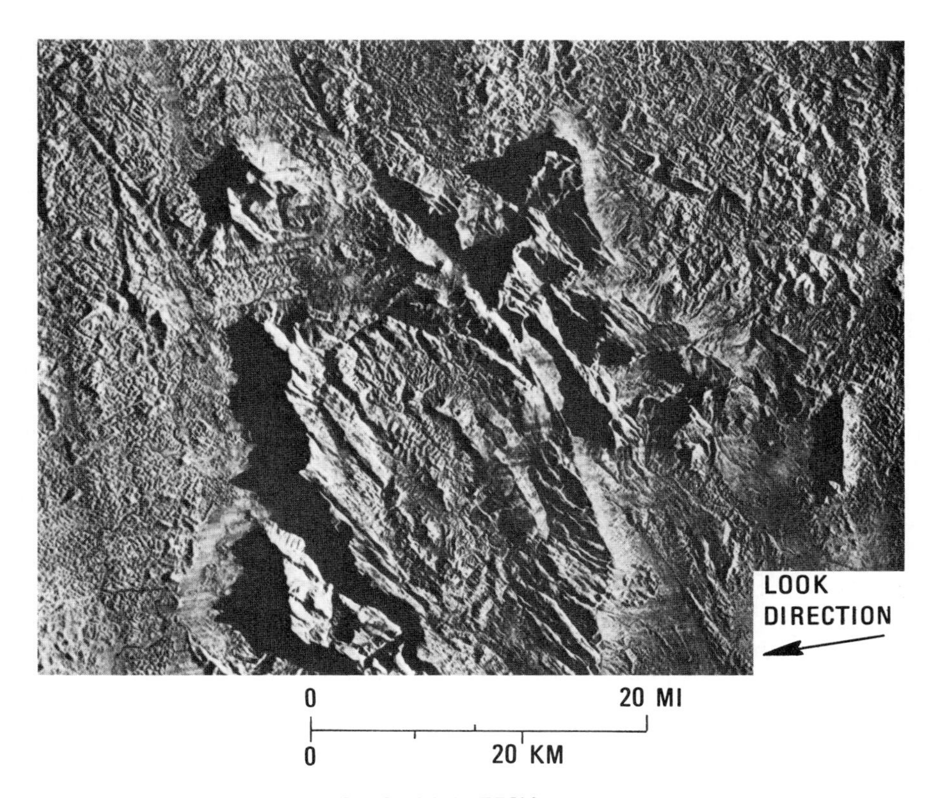

B. X–BAND RADAR MOSAIC, COURTESY AEROSERVICE CORPORATION.

FIGURE 6.32
Comparison of Landsat image and radar mosaic of Cerro Duida, Venezuela.

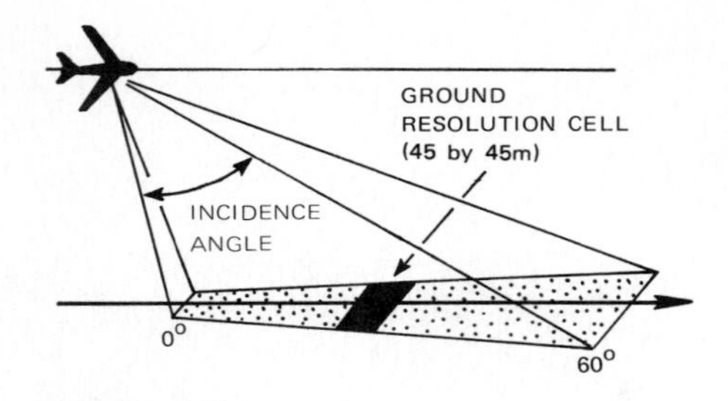

FIGURE 6.33
Diagram of radar scatterometer system.

RADAR SCATTEROMETER

A radar *scatterometer* is a nonimaging airborne system for quantitatively measuring the radar backscatter of terrain as a function of the incidence angle (Moore and others, 1968). Scatterometer data are useful for characterizing surface roughness properties of materials. The technique is particularly useful for identifying types of sea ice.

System Description

In contrast to SLAR systems, a scatterometer transmits a continuous radar signal directly along the flight path. The 3°-wide beam illuminates terrain both ahead and behind the aircraft, but for simplicity only the forward portion is shown in Figure 6.33. The 0° incidence angle is directly beneath the aircraft and the maximum incidence angle recorded is approximately 60°. At a typical altitude of 900 m above terrain the ground resolution cell at a 30° incidence angle is a square measuring 45 m on a side. The return signal is a composite of the backscattering properties of all the terrain features within the cell. The wavelength of the scatterometer system described here is 2.25 cm (X band) and both the transmitted and received energy is vertically polarized. As the aircraft advances along the flight path (Figure 6.33) a resolution cell on the ground is initially illuminated at 60° incidence angle and then at successively decreasing angles. The recorded amplitude and frequency of the successive returns are processed with Doppler techniques to obtain the scattering coefficient at each of several incidence angles.

Data Display

Scatterometer data may be displayed as profiles, scattering coefficient curves, or as deviation curves. *Scatterometer profiles* (Figure 6.34B) display the scattering coefficient of the terrain along the flight line for a particular incidence angle. In this example the incidence angles range from 2.5° (almost directly beneath the aircraft) to 52.0°. *Scattering coefficient curves* (Figure 6.35) display the returns at different incidence angles for an area on the ground. As seen in this illustration, and in Figure 6.36, all scattering coefficient curves have a similar shape with strong returns at the low-incidence angles and weaker returns at the high angles. This appearance results from the greater two-way travel distance for the higher angles. The similarity in shape obscures significant differences in the curves.

The differences between scatterometer curves are emphasized on a display called *deviation curves* (Moore and others, 1968). Figure 6.36 shows the scattering coefficient curves of various terrain types along a flight line across Pisgah Crater and vicinity. In Figure 6.37A the average scattering coefficient curve for the entire flight line is shown as a dashed curve, together with the solid curves for basalt and the dry lake. The vertical lines show the deviation of these individual terrain curves from the average curve at various incidence angles. For the dry lake the scattering is greater than the average at low incidence angles, but is below the average at higher angles. The curve for basalt has the opposite relationship to the regional average. In Figure 6.37B the deviation values are plotted (with an expanded vertical scale) as a function of incidence angle, thereby emphasizing the different scatterometer responses of the terrain

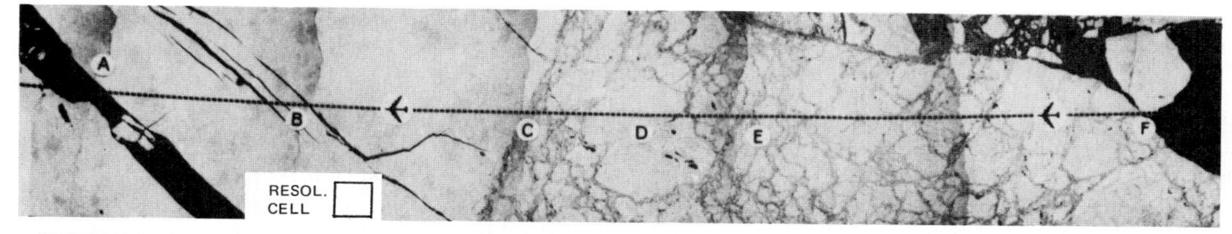

A. PHOTOMOSAIC SHOWING SCATTEROMETER FLIGHT LINE.

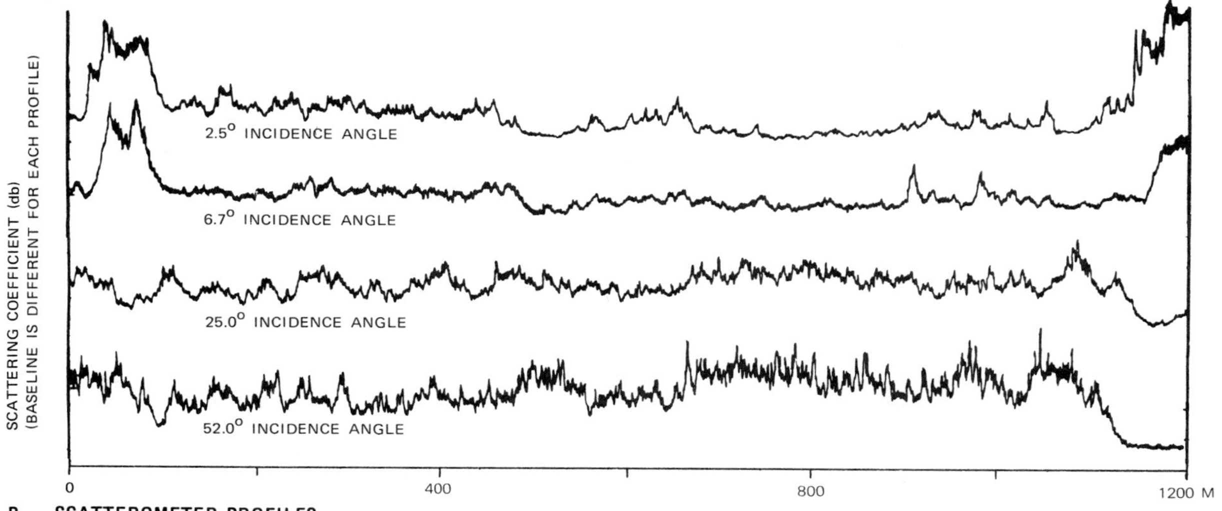

B. SCATTEROMETER PROFILES.

FIGURE 6.34
Scatterometer profiles of sea ice off Point Barrow, Alaska acquired May 12, 1967. From Rouse (1968, Figure 10).

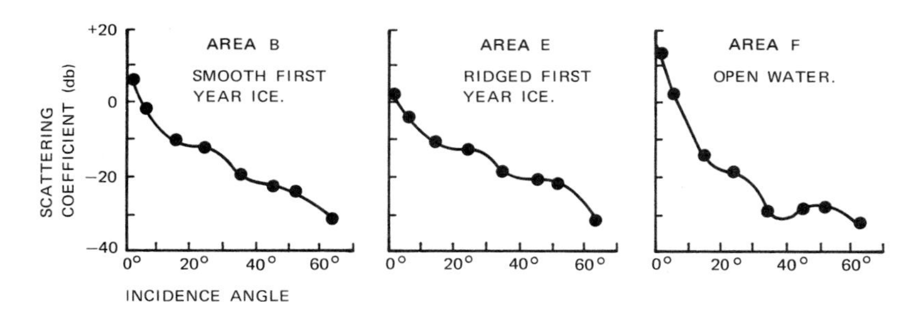

FIGURE 6.35
Scattering coefficient curves for areas along the scatterometer flight line off Point Barrow. See Figure 6.34A for location of areas. From Rouse (1968, Figure 10).

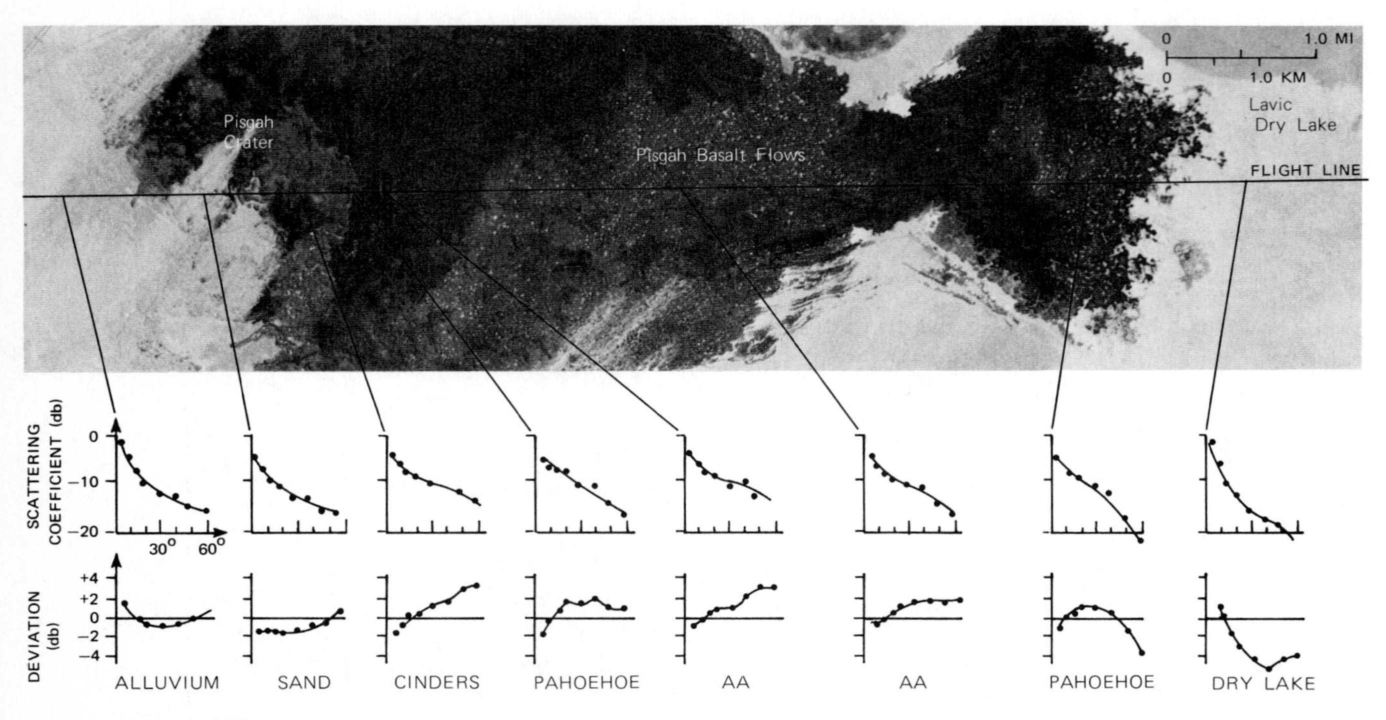

FIGURE 6.36
Scattering coefficient curves and deviation curves for Pisgah Crater and vicinity, California. Scatterometer flight line and data localities are shown on the visible light image. From Moore and others (1968, Figures 5 and 6).

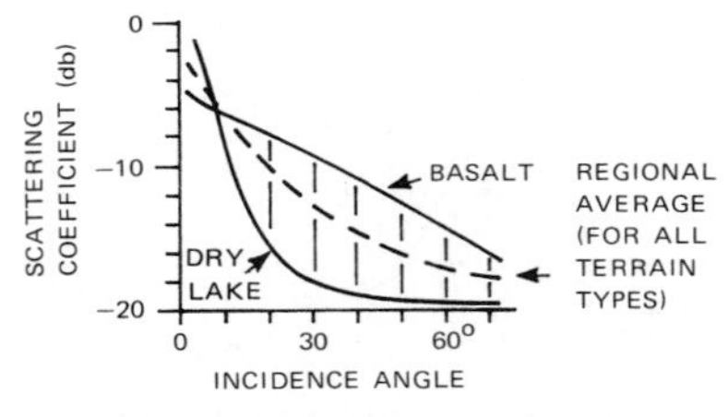

A. DETERMINING DEVIATION FROM REGIONAL AVERAGE.

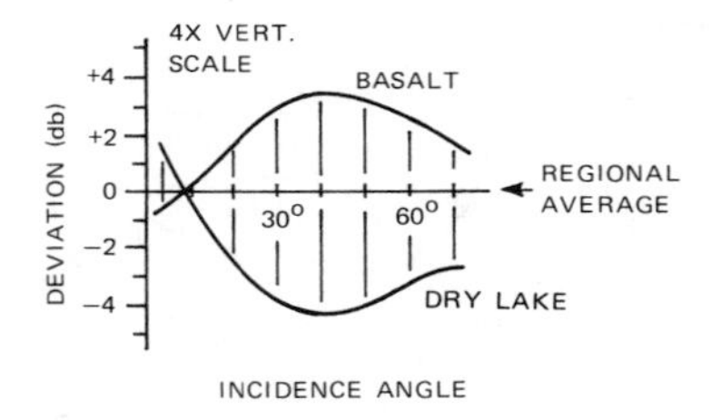

B. DEVIATION CURVES.

FIGURE 6.37
Calculation of deviation curves from scattering coefficient curves.

types. Deviation curves for the Pisgah Crater area are plotted beneath their respective scattering coefficient curves in Figure 6.36. Note that differences among the terrain types are more obvious on the deviation curves than on the scattering coefficient curves.

Arctic Sea Ice Example

Scatterometer profiles and aerial photographs of sea ice off Point Barrow, Alaska were acquired in 1967 and interpreted by Rouse (1968) from which this description was summarized. The photomosaic in Figure 6.34A shows the following features:

A and F	Open water
A to C	Smooth first-year ice
B	Narrow open cracks, or leads
C to F	Mildly ridged first-year ice
C	Major pressure ridge separating the smooth and ridged ice
D	Floe of smooth ice within the ridged ice
B, E, and F	Representative areas of smooth ice, ridged ice, and open water for which scattering coefficient curves are shown (Figure 6.33D).

Characteristics of the various types of sea ice are listed in Table 9.4 of Chapter 9.

The scatterometer profiles (Figure 6.34B) are plotted to the same horizontal scale as the photomosaic. The baselines are different for each profile because the overall return intensity diminishes with increasing incidence angle. At low incidence angles (2.5° and 6.7°) open water has a strong return, as explained earlier in Figure 6.12. Similarly the smooth ice has stronger returns than ridged ice at these low incidence angles. At the higher incidence angles (25.0° and 52.0°), most of the radar energy encountering the water and smooth ice is specularly reflected away from the antenna and produces little return. The backscattering from the rough ridged ice, however, produces a relatively strong and characteristically spiked profile at these incidence angles. Note that the floe of smooth ice at D can be recognized on the 52.0° profile by its reduced return within the strong returns from the ridged ice.

Pisgah Crater Example

The shape of the scattering deviation curves can be used to interpret roughness of the various surface materials of the Pisgah Crater area. Photographs of these materials are illustrated in Chapter 11 on Comparison of Image Types. The deviation curves for alluvium and sand (Figure 6.36) are close to the horizontal axis, indicating that they follow the regional average. There is a tendency toward slightly stronger returns at higher incidence angles. At the Pisgah Crater cinder cone the very strong return at higher incidence angles is caused by two factors: (1) the steep slopes of the cone cause stronger returns at greater incidence angles, and (2) the surface roughness of the volcanic cinders produces strong backscattering at the wavelength of the scatterometer.

The Pisgah basalt flows consist of pahoehoe lava with a smooth, ropey surface and aa lava with a very rough, clinkery surface. The deviation curves of the two aa sites have strong returns at high incidence angles because of the strong backscattering. The pahoehoe site toward the southeast end of the flight line has weak returns at higher incidence angles because of specular reflection from its relatively smooth surface. The second pahoehoe curve near Pisgah Crater is intermediate in character. Field work shows that at this locality there is a mixture of aa and pahoehoe surfaces. The ground resolution cell of the scatterometer integrates returns from both lava types to produce the intermediate deviation curve. The very smooth surface of Lavic dry lake causes a strong return at low incidence angles and a very weak return at high angles. This relationship was explained in Figure 6.12A and is similar to the curve for open water in the Arctic example.

IMAGE INTERPRETATION PROCEDURE

The preceding principles and characteristics of SLAR systems now can be used to interpret some typical images. The Papua example is from an area of dense jungle and poor flying conditions where radar is the only imagery available. The objective is to determine how the imagery can be used in this unmapped area. In contrast, the San Gabriel Mountains example is in a region with good weather that has been thoroughly mapped. The objective is to evaluate the radar interpretation by comparing it with existing maps.

Papua

The image from Papua (Figure 6.38A) is used in a step-by-step demonstration of radar image interpretation procedure.

Step 1 is to orient the image so that topography appears in the correct perspective. For most images the optimum orientation is with the shadows falling toward the interpreter (near range toward the top of the figure). For some interpreters the topography may be inverted (valleys appearing as ridges), which may be corrected by turning the image to reverse the shadow direction.

Step 2 is to trace the drainage on an overlay of drafting film that will serve as a base for the geologic interpretation (Figure 6.38B).

Step 3 is to determine the regional structure. The strong returns in the north part of the image are from escarpments of south-dipping layered rocks that are resistant to erosion. The shadows enhance the triangular flat irons that are typical of tilted layered rocks. Because of geometric compression and layover in this near-range part of the slant-range image, the steepness of dip is exaggerated. South of these flat irons the east-flowing river follows a broad valley formed by rocks that are less resistant to erosion. The valley is bounded on the south by an escarpment that produces strong radar returns. Layered rocks are indicated by the small flat irons on the south flank of the escarpment. The attitude of the flat irons shows that the regional south dip persists in this area. The appropriate dip and strike symbols are added to the overlay. The regional structure has been determined, and the rock units of the area have been differentiated into two resistant stratified units separated by a less-resistant unit.

Step 4 is to determine local structure. The strike valley in the less-resistant belt widens toward the east, suggesting structural complications. Within this broad area are concentric arcuate low ridges, probably formed by thin resistant beds. These are traced onto the overlay and a doubly plunging fold pattern becomes apparent. The asymmetry of the ridges suggests an anticline and the axis is drawn in. Disruption of the ridges near the west plunge suggests a fault, which is drawn with a heavy line.

The straight east-flowing tributary in the center of the image is a linear feature that may be controlled by a fault, which is drawn with the heavy dashed line. Straight aligned notches across the ridges in west-central part of the area are mapped as linear features. In the southwest corner a reversal of dip is indicated by triangular flat irons that

A. Ka–BAND RADAR IMAGE ACQUIRED 1970.

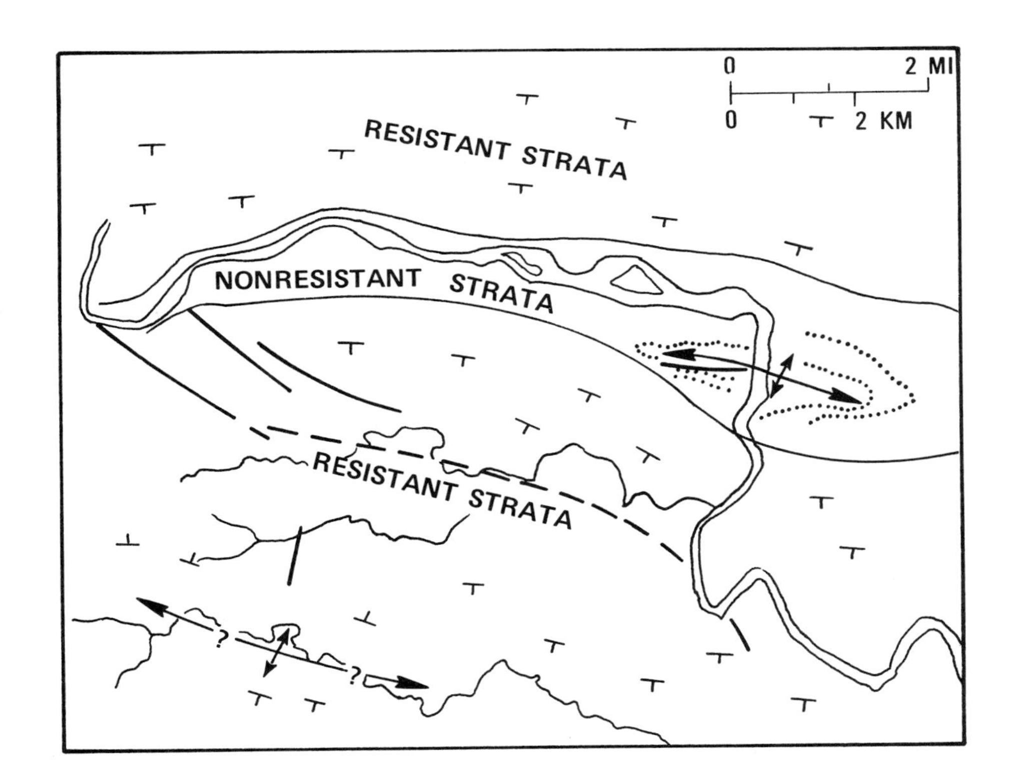

B. GEOLOGIC INTERPRETATION.
SYMBOLS ARE CONVENTIONAL.

FIGURE 6.38
Geologic interpretation of radar image of Papua, New Guinea; not field checked. From Sabins (1973, Figure 9).

dip northward. The curving shadow results from a southwest-facing scarp. This reversal of dip and topography is interpreted as a possible anticline with a meandering stream flowing along the eroded crest.

A reasonable stratigraphic and structural interpretation has been made from the radar image (Figure 6.38B), despite the lack of supporting data. A helicopter field party can quickly check the two possible anticlinal axes.

San Gabriel Mountains, California

The San Gabriel Mountains are part of the Transverse Range province of southern California. Their regional setting is shown on the Landsat image (Plate 2) and interpretation map of southern California (Figure 4.5). The Ka-band image of western San Gabriel Mountains (Figure 6.39A) is part of a continuous strip along the San Andreas fault from San Francisco to San Bernardino acquired in November, 1965 with the look direction to the southwest. This look direction is optimum for depicting the major northwest-trending faults. Faults trending parallel with the look direction are less apparent on the image. There is geometric distortion on this slant-range image in excess of the normal compression in the near range. The scale in the range direction is compressed overall relative to that in the azimuth direction, because the recording film was not advanced at a correct rate relative to the aircraft ground speed. There is no problem, however, in correlating the image with the 1:250,000 scale Los Angeles sheet of the "Geologic Atlas of California." The dark signatures of Fairmont and Bouquet reservoirs are good reference points between the image and the maps.

The interpretation shown in Figure 6.39B followed the method advocated earlier for Landsat images. All linear features on the image were annotated on an overlay without reference to the geologic maps. This is important because it minimizes interpreter bias toward recognizing mapped features and ignoring image features that are not shown on existing maps. The interpretation is then compared with existing maps and the lineaments that correspond to mapped faults are converted from dotted to solid lines. Some of the remaining linear features may be faults that have not previously been recognized. One example is the long lineament between the San Andreas and Liebre faults (Figure 6.39B) that trends parallel with the known faults, cuts across topography without deflection, and has a radar signature similar to that of the Liebre fault. Field checking in this rugged, brush-covered terrain is not easy. The lineament occurs in an area of crystalline rocks where lithologic offsets are difficult to recognize. This example is included to demonstrate that radar provides new geologic perspectives even in such thoroughly mapped areas as southern California.

This example also illustrates the difficulty of mapping lithology from radar images. Granitic, metamorphic, sedimentary, and volcanic rocks crop out in the area, as shown by the geologic map in Figure 6.39B. Comparison with Figure 6.39A demonstrates that these major rock types cannot be distinguished on the image, largely because there are no unique correlations between surface roughness and rock type in this area.

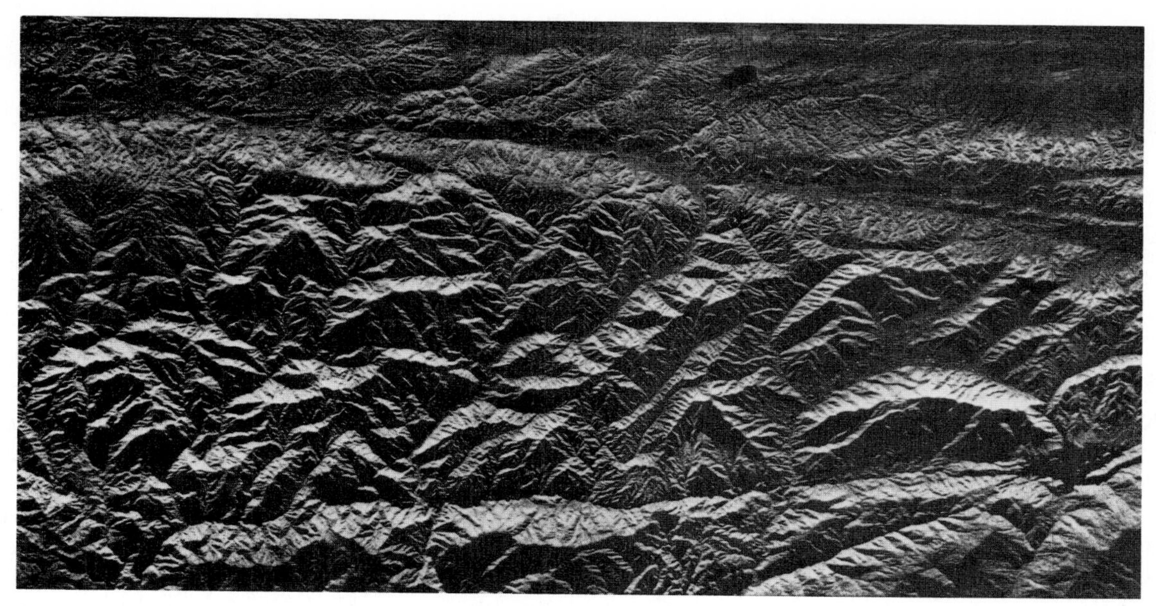

A. Ka-BAND RADAR IMAGE ACQUIRED NOVEMBER, 1965.

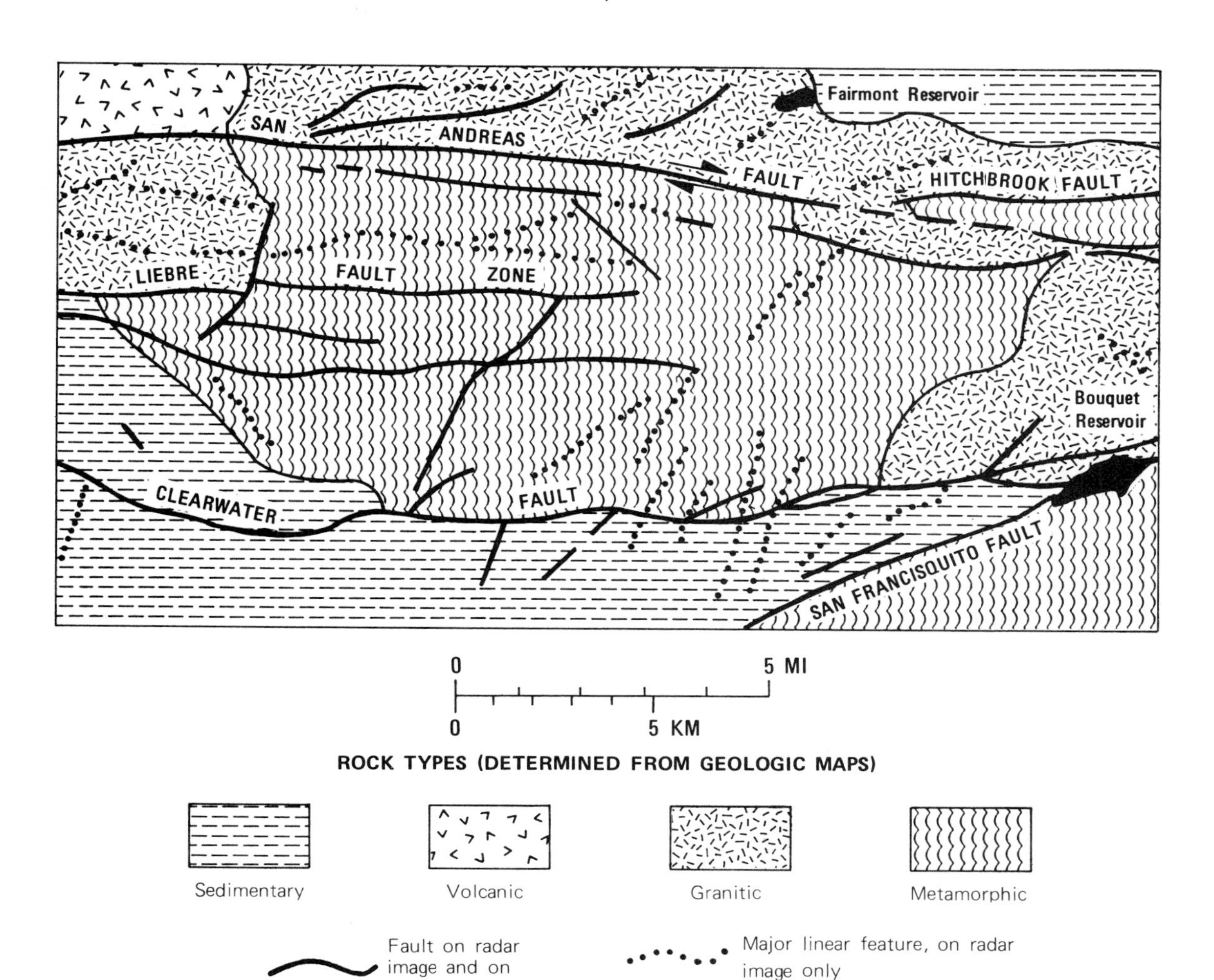

B. IMAGE INTERPRETATION, GEOLOGIC INFORMATION FROM LOS ANGELES SHEET OF "GEOLOGIC ATLAS OF CALIFORNIA."

FIGURE 6.39
Geologic interpretation of radar image of western San Gabriel Mountains, southern California. Not field checked.

RADAR IMAGERY FROM SATELLITES

Until recently radar was not suitable for use on satellites for the following reasons: (1) power requirements for these active systems were excessive; and (2) for real-aperture systems, the azimuth resolution at the long slant ranges of satellites would be too low for imaging purposes. The development of new power systems and radar techniques has overcome the first problem and synthetic-aperture radar systems have overcome the second. Radar images of the lunar surface were acquired during the early Apollo missions and aided in site selection for the moon landings.

Some of the manned Space Shuttle missions scheduled to orbit the earth in the 1980s will include SLAR capabilities. The present plans include synthetic-aperture, X-band and L-band systems designed to record parallel-polarized and cross-polarized images. The Seasat-A satellite, scheduled for 1978, will acquire L-band SLAR imagery primarily of the oceans for determining sea state, but some terrestrial imagery also will be acquired.

COMMENTS

Because radar is an active system that supplies its own illumination, SLAR imagery can be acquired in the dark and through cloud cover. The illumination direction can be oriented to enhance particular lineament trends. Interaction between materials and radar energy is determined by the wavelength and depression angle of the radar beam and by the dielectric constant and surface roughness of the material. Rougher surfaces produce stronger radar reflections, which are recorded as brighter tones on the images. The oblique illumination of SLAR causes highlights and shadows that are well suited for enhancing the appearance of lineaments, faults, and topography.

On SLAR images of heavily forested terrain, lineaments and other surface features are strongly expressed, which leads to the mistaken assumption that radar energy penetrates the tree canopy. This is not the case, however, because radar energy interacts with, and is reflected by, vegetation. Individual trees, with dimensions measured in meters, are not resolved by the relatively large ground resolution cell of radar; this enhances the appearance of lineaments and other terrain features, which are measured in hundreds of meters. Radar imagery of much of the United States is available at reasonable prices, and additional imagery can be acquired by contractors. SLAR images from earth-orbiting satellites will be available from the Seasat and Space Shuttle programs.

REFERENCES

Barr, D. J., 1969, Use of side-looking airborne radar (SLAR) imagery for engineering studies: U.S. Army Engineer Topographic Laboratories, Tech. Report 46-TR, Fort Belvoir, Va.

Craib, K. B., 1972, Synthetic aperture SLAR systems and their application for regional resources analysis *in* Shahrokhi, F. ed., Remote sensing of earth resources. University of Tennessee Space Institute, v. 1, p. 152–178, Tullahoma, Tenn.

Fischer, W. A., ed., 1975, History of remote sensing *in* Reeves, R. G., ed., Manual of remote sensing: ch. 2, p. 27–50, American Society of Photogrammetry, Falls Church, Va.

Hunt, C. B. and D. R. Mabey, 1966, Stratigraphy and structure of Death Valley, California: U.S. Geological Survey Professional Paper 494-A.

Jensen, H., L. C. Graham, L. J. Porcello, and E. N. Leith, 1977, Side-looking airborne radar: Scientific American, v. 237, p. 84–95.

MacDonald, H. C., 1969, Geologic evaluation of radar imagery from Darien Province, Panama: Modern Geology, v. 1, p. 1–63.

———and W. P. Waite, 1973, Imaging radars provide terrain texture and roughness parameters in semi-arid environments: Modern Geology, v. 4, p. 145–158.

Moore, R. K., W. P. Waite, J. R. Lundien, and H. W. Masenthin, 1968, Radar scatterometer data analysis techniques: Proceedings Fifth International Symposium on Remote Sensing of Environment, p. 765–777, University of Michigan, Ann Arbor, Mich.

———, L. J. Chastant, L. J. Porcello, J. Stephenson, and F. T. Ulaby, 1975, Microwave remote sensors *in* Reeves, R. G., ed., Manual of remote sensing: ch. 9, p. 399–537, American Society of Photogrammetry, Falls Church, Va.

Peake, W. H. and T. L. Oliver, 1971, The response of terrestrial surfaces at microwave frequencies: Ohio State University Electroscience Lab., 2440-7, Tech. Rep. AFAL-TR-70-301, Columbus, Ohio.

Rouse, J. W., 1968, Arctic ice type identification by radar: University of Kansas Center for Research, Tech. Report 121-1, Lawrence, Kan.

Sabins, F. F., 1973, Engineering geology applications of remote sensing *in* Moran, D. E., ed., Geology, seismicity, and environmental impact: Association of Engineering Geologists Special Publication, p. 141–155, Los Angeles, Calif.

Schaber, G. G., G. L. Berlin, and W. E. Brown, 1976, Variations in surface roughness within Death Valley, California—geologic evaluation of 25-cm wavelength radar images: Geological Society America Bulletin, v. 87, p. 29–41.

Schaber, G. G., G. L. Berlin, and D. J. Pitrone, 1975, Selection of remote sensing techniques—surface roughness information from 3-cm wavelength SLAR images: Proceedings of Annual Meeting, American Society of Photogrammetry, p. 103–117, Washington, D.C.

ADDITIONAL READING

Long, M. W., 1975, Radar reflectivity of land and sea: D. C. Heath and Co., Lexington, Mass.

Matthews, R. E., ed., 1975, Active microwave workshop report: NASA SP-376, Houston, Tex.

Skolnik, M. I., ed., 1970, Radar handbook: McGraw-Hill Book Co., New York.

Wing, R. S., 1971, Structural analysis from radar imagery: Modern Geology, v. 2, p. 1–21.

7 DIGITAL IMAGE PROCESSING

Remote-sensor images are acquired by technologically advanced systems, but are typically interpreted through the use of simple classical techniques that have progressed little beyond the stereoscopes and magnifiers employed in the early days of aerial photography interpretation. Examination of original images with the unaided eye often constitutes the crucial interpretation phase for data produced by imaging systems that represent millions of dollars in research and development. The photographic methods for contrast enhancement and color compositing described in Chapter 4 enable the interpreter to extract more information from the original images but they lack the advantages of digital techniques. The principal advantages of digital processing methods are their versatility, repeatability, and the preservation of the original data precision.

Computer processing of images is not new; the medical, agricultural, and military intelligence communities have employed digital image processing for a number of years. Geographers were among the first scientists to recognize the potential of digital methods for investigating patterns of land use. The availability of digital multispectral images from the Landsat program accelerated the development and application of digital image processing in the mid-1970s. This field will see major advances in technology and application in the future. Some of the requirements for applying digital image processing to oil exploration have been summarized by Sabins (1974).

IMAGE STRUCTURE

Any image may be thought of as consisting of tiny equal areas, or picture elements, arranged in regular lines and columns. The position of any picture element, or *pixel*, is determined by an X and Y coordinate system with the origin at the upper left corner in the case of Landsat images. The brightness of each pixel has a numerical value ranging from zero for black to some higher number for white. Any image can now be described in strictly numerical terms on a three-coordinate system with X and Y locating each pixel and Z giving the gray-scale intensity value. An image may be recorded originally in this digital format, as in the case of Landsat. An image recorded initially on photographic film may be converted into numerical format by a process known as digitization.

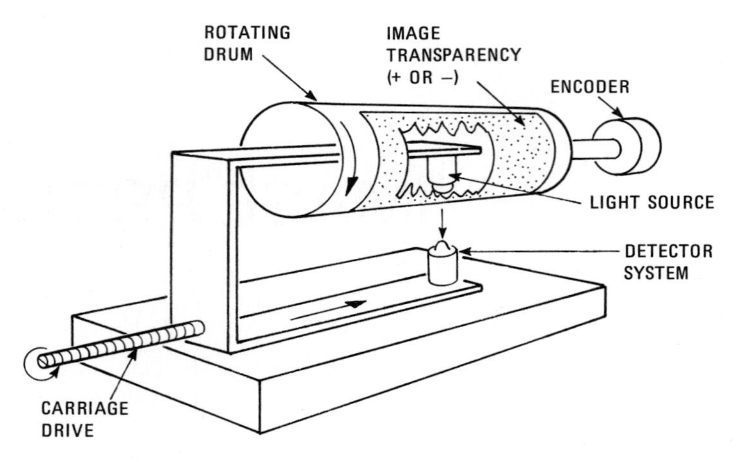

FIGURE 7.1
System for digitizing an image. From Bryant (1974, Figure 2).

Digitization Procedure

Digitization is the process of converting any image recorded on photographic film (radar, thermal IR, aerial photographs) into an ordered array of numbers. Various types of maps may also be digitized. Normally the digitized information is recorded on magnetic tape, but punched paper tapes or cards may be used for small volumes of data. Digitizing systems are grouped into three general categories: drum, flat bed, and flying spot.

A typical *drum digitizing system* (Figure 7.1) consists of a fixed rotating drum and a movable carriage that holds a light source and a detector, similar to a photographic light meter. The positive or negative film is placed over the opening in the drum and the carriage is positioned at a corner of the film. As the drum rotates the detector measures intensity variations of the transmitted light caused by variations in film density. Each revolution of the drum records a scan line of data across the film. At the end of each revolution the encoder signals the drive mechanism, which advances the carriage by the width of one scan line to begin the next revolution.

Line width is determined by the optical aperture of the detector, which typically is a square measuring 50 μm on a side. The analog electrical signal of the detector, which varies with film density changes, is sampled at 50 μm intervals along the scan line, digitized and recorded as a digital value on magnetic tape. Each digital value is recorded as a series of *bits*, which are an ordered sequence of ones and zeroes. Each bit represents an exponent of the base two. An eight-bit series ($2^8 = 256$ levels) is commonly used to represent the gray-scale value, with 0 for black and 255 for white. A group of eight bits is called a *byte*. The film image is thus converted into a regular array of picture elements (pixels) that are referenced by scan line number (Y direction) and pixel count along each line (X direction). A typical 9 by 9 in. (23 by 23 cm) aerial photograph digitized with a 50 μm sampling interval is converted into 30 million pixels.

The intensity of the digitizer light source is calibrated at the beginning of each scan line to compensate for source fluctuations. These drum digitizers are remarkably fast, efficient, and relatively inexpensive. Color photographs may be digitized into the values of the component primary colors on three passes through the system using blue, green, and red filters over the detector. A digital record is produced for each primary color or wavelength band. In the case of *flat-bed digitizers*, the film is mounted on a holder that moves in the X and Y directions between a fixed light source and detector. These devices are usually

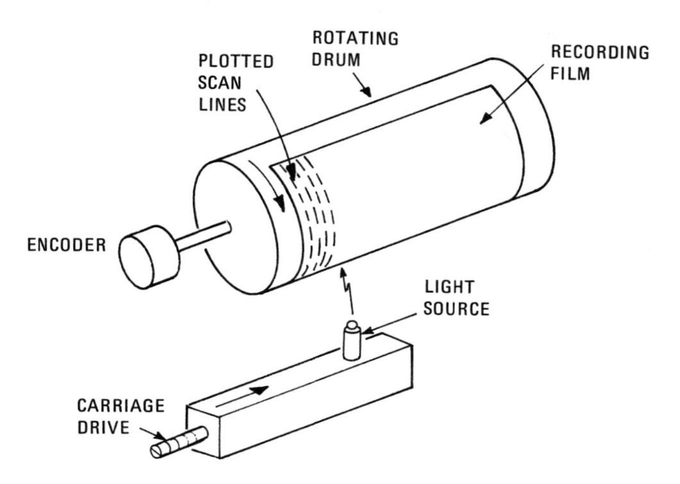

FIGURE 7.2
System for plotting an image from digital data. From Bryant (1974, Figure 3).

slower and more expensive than drum digitizers, but generally are more precise. Some systems employ reflected light in order to digitize opaque paper prints. Another digitizing system is the *flying-spot scanner*, which is essentially a modified television camera, that electronically scans the original image in the X and Y directions. The resulting voltage signal for each line is digitized and recorded.

The digitized image can be read into a computer for various processing operations. The processed data must next be displayed as an image on a viewing device or plotted onto film as described in the following section.

Image Generation Procedure

Digital image data are converted into hardcopy images by film writers (Figure 7.2) that operate in reverse fashion to digitizers. Recording film is mounted on a drum. With each rotation a scan line is exposed on the film by a light source, the intensity of which is modulated by the digital values of the pixels. Upon completion of each scan line, the carriage advances the light source to commence the next line. The exposed film is developed to produce a transparency from which prints and enlargements are made.

Flying-spot plotters produce an image with an electron beam that sweeps across the fixed film in a raster pattern. This system is also employed to display images on a television screen, which enables the interpreter to view the processed image in real time rather than waiting for the processed data to be plotted onto film, developed, and printed.

Landsat Image Format

A major advantage of Landsat multispectral scanner MSS imagery is that it is directly recorded as digital numbers on magnetic tape. This greatly facilitates the application of computer processing programs. The oscillating mirror of the MSS sweeps across the terrain at a right angle to the orbit path (Figure 7.3) and records data only during the eastbound sweep. The 185-km long scan lines form a continuous strip of imagery that is subdivided into lots of 2,340 scan lines to produce individual Landsat scenes (Figure 7.3). The ground resolution cell of each detector is a 79 by 79 m square. Solar energy reflected from this ground area onto the detector generates a response that varies in amplitude proportionally with the intensity of the reflected energy.

A segment of a scan line showing the variation

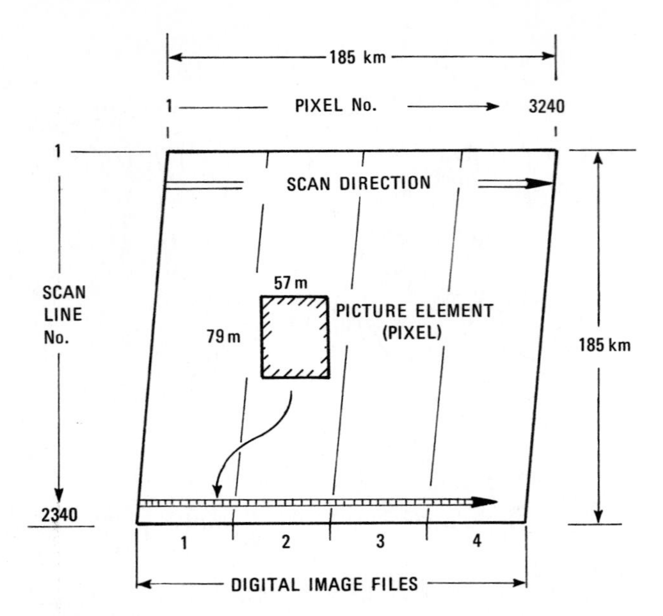

FIGURE 7.3
Reference system of scan lines and pixels for Landsat MSS image. Note location of digital image files on computer compatible tapes.

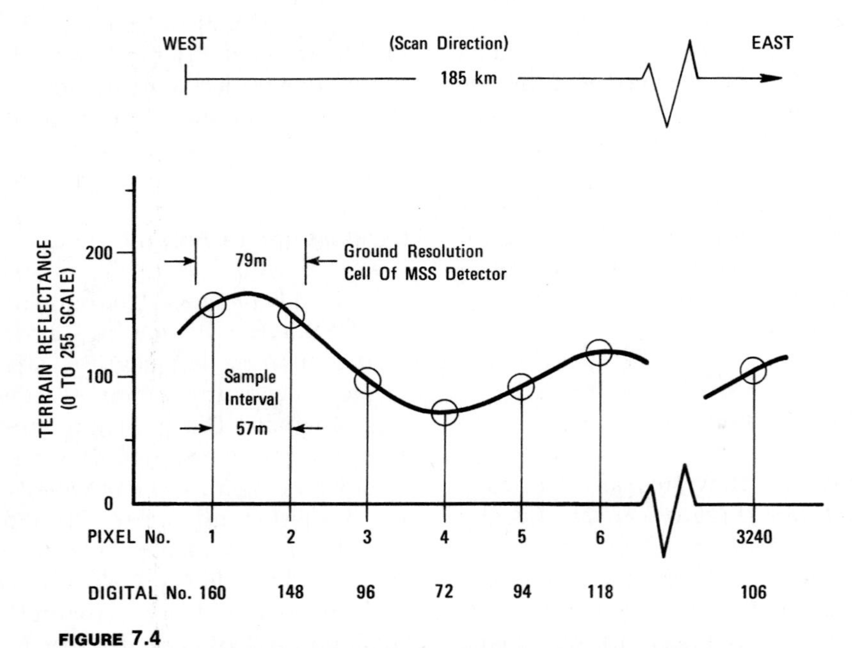

FIGURE 7.4
Plot of terrain reflectance along a Landsat scan line. The 79-m ground resolution cell of each MSS detector produces a reflectance curve that is sampled at intervals of 57 m to generate the digital number for each pixel.

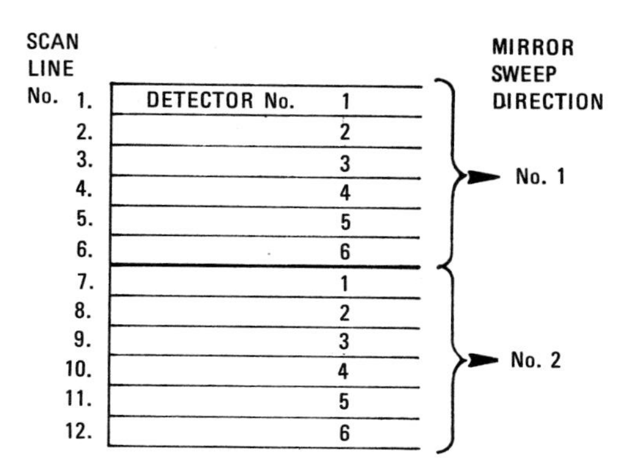

FIGURE 7.5
Detail of detector array for Landsat MSS.

in terrain reflectance is illustrated in Figure 7.4 together with the 79 m dimension of the ground resolution cell. The curve showing reflectance as a function of distance along the scan line is called an *analog display* because the data are shown in directly measurable quantities. Analog displays differ from *digital displays*, which record information in numerical form. Analog data are converted into digital data by sampling the analog display at regular intervals and recording the value at each sample point in digital form. Onboard the Landsat vehicle the analog signals are sampled at 57 m intervals and converted into digital data that are transmitted to earth receiving stations. As shown in Figure 7.4 the sampling of a scan line produces 3,240 pixels, each with a corresponding digital number that represents the reflectance. The analog signal could be sampled at any interval; the 1.4 ratio of the ground resolution cell to the sample interval (79 m/57 m = 1.4) was determined theoretically and experimentally to be adequate for representing the analog signal. A higher ratio (smaller sample interval) would not appreciably improve the digital data; a lower ratio would degrade the quality of the digital data. This information was provided through the courtesy of Virginia Norwood of the Space and Communications Group of Hughes Aircraft Company, which designed and built the MSS. Spatial resolution of the image is ultimately determined by the 79 by 79 m dimensions of the ground resolution cell; the smaller sample interval does not alter this relationship.

Each sweep of the scanner mirror reflects light onto an array of six detectors for each of the spectral bands, simultaneously producing the array of six scan lines shown in Figure 7.5. The advantage of this design is that the number of mirror sweeps and the mirror scan velocity are reduced by a factor of six. A disadvantage is that the response of one detector may differ from that of the other five, causing every sixth scan line to be brighter or darker. Digital methods for correcting this image defect, which is called *sixth-line banding*, are discussed later in this chapter. Each of the four bands of a Landsat image contains $7.6 \cdot 10^6$ pixels (2,340 scan lines, each with 3,240 pixels) that are recorded on computer compatible tapes (CCTs). A tape contains four data files, each of which represents a strip that is 46-km wide in the east-west direction and 185-km long in the north-south direction (Figure 7.3). Each data file contains the brightness values for the four spectral bands interleaved. A fifth file contains data for the image annotation. The CCTs are available from the EROS Data Center at a cost of $200 per Landsat scene. Prior to processing, most laboratories convert the CCT data into four new files, each containing a single spectral band for the entire scene. Details of the tape format are described by Thomas (1975) and in the "Landsat Data Users Handbook" by NASA (1976).

The brightness values for bands 4, 5, and 6 are recorded on CCTs using a seven-bit scale (0 to 127); band 7 is recorded on a six-bit scale (0 to 63). Prior to processing, some digital systems multiply these values by two or four to produce a consistent eight-bit scale for all four bands. In addition to consistency between MSS bands, the uniform eight-bit format minimizes errors introduced by rounding of decimal values. To reduce computer processing time and core usage, integer (whole number) mathematics are used wherever possible. Decimal values are integer truncated (rounded) to the nearest integer. In six-bit format (0 to 63) computed values of 8.6 and 9.3 are both truncated to an integer value of 9; in eight-bit (0 to 255) format the equivalent of a six-bit value of 8.6 is represented by 34 and 9.3 by 37 ($4 \cdot 8.6 = 34.4$; $4 \cdot 9.3 = 37.2$).

The computer printout of pixel values in Figure 7.6A represents a small subarea of a Landsat image. The upper left corner (pixel 1, line 1) coincides with the northwest corner of the image. Each pixel is represented by a *digital number* (DN) with higher values indicating higher reflectance of the terrain. The outlined diagonal strip of pixels with higher values represents the highly reflective concrete surface of the 280 Freeway, which is also shown in the aerial photograph (Figure 7.6D). The printout is geometrically distorted because two spaces are required for each pixel, plus a blank space between adjacent pixels.

The initial step in converting the numerical pixel array into an image is to determine the statistical distribution of the brightness values of the pixels. This is shown in the histogram (Figure 7.6B) where the number of pixels corresponding to each DN value are plotted. The histogram was prepared for a larger scene of which the pixel array in Figure 7.6A is only a subarea; hence the histogram does not exactly match the pixel population of the subarea. The 14 gray-scale symbols of the line printer are then assigned to specific ranges of DN values, with the darker symbols for the lower DNs. Values of 10 and less are plotted with a solid black symbol; values of 35 and higher are plotted with solid white. The 12 intermediate gray-scale symbols are each assigned to two digital values as shown along the abscissa of the histogram in Figure 7.6B. The resulting image is shown in Figure 7.6C, where each pixel is represented by a gray-scale symbol. This portion of the Landsat band 5 image covers the northwest corner of the aerial photograph (Figure 7.6D). The dark tones on the image and photograph are vegetation; the light tones are the 280 Freeway, the Stanford Linear Accelerator, and residential areas. These examples also provide a comparison of spatial resolution of the Landsat image with an aerial photograph at an original scale of 1:60,000.

The MSS image data telemetered from Landsat are recorded on magnetic tape at the Fairbanks, Goldstone, and Goddard receiving stations. At Goddard Space Flight Center (GSFC) these tapes are used to produce a 70-mm archival film for each of the four bands. A second-generation film is sent to EROS Data Center (EDC) which reproduces the images for sale to the public. As described in Chapter 4, this procedure is scheduled for change in late 1977. Most EDC images reproduced from the GSFC films are satisfactory for many users; the major problem has been the high density and low contrast of 70-mm black-and-white positive transparencies, which was discussed in Chapter 4. In addition, some images have defects that can be corrected or improved by digitally processing the data from the CCTs.

IMAGE PROCESSING SYSTEMS

Digital systems for processing Landsat and other images have been developed by a number of universities, government facilities, and commercial organizations. One of these is the Video Image Communication and Retrieval (VICAR) system developed at the Jet Propulsion Laboratory (JPL) which is a NASA facility at Pasadena, California, operated by the California Institute of Technology. Details of the VICAR system are given by Goetz and others (1975). The Laboratory for Applications of Remote Sensing (LARS) at Purdue University has developed the LARS System (LARSYS) for

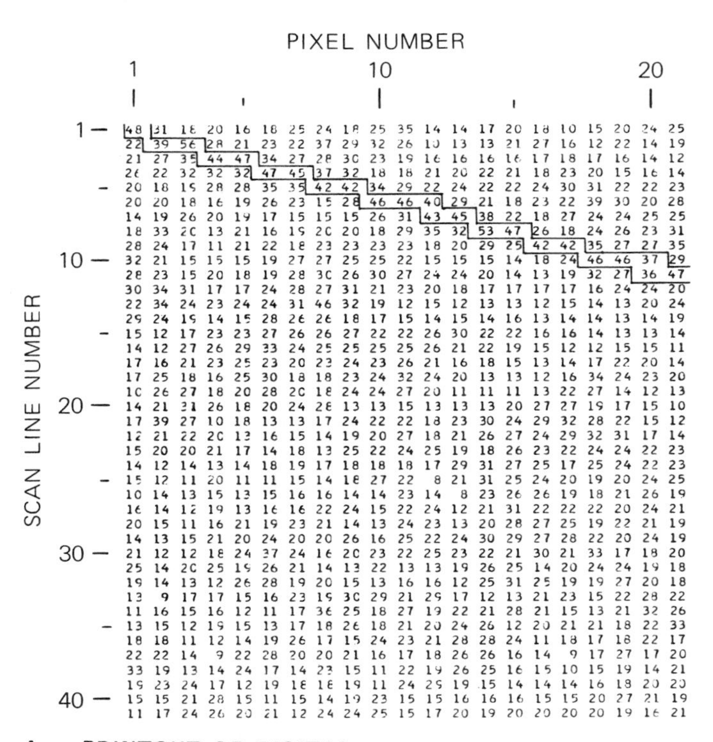

A. PRINTOUT OF DIGITAL VALUES OF PIXELS.

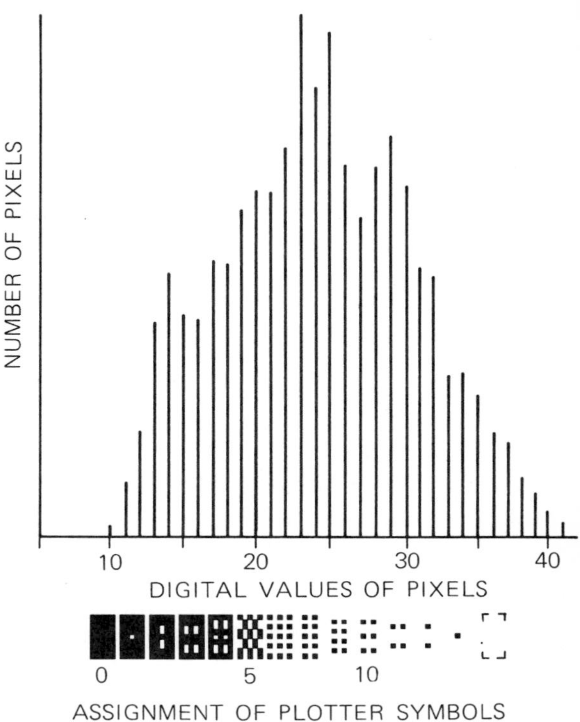

B. HISTOGRAM OF PIXEL VALUES WITH GRAY SCALE ASSIGNMENTS.

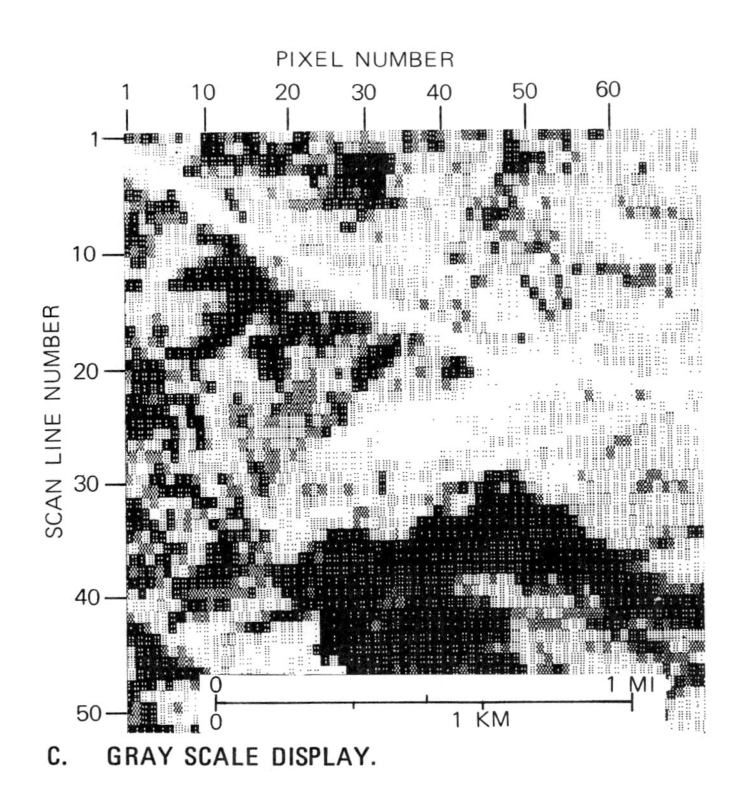

C. GRAY SCALE DISPLAY.

280
LINEAR ACCELERATOR
AREA OF FIGURE C
0
1 MI
0
1 KM

D. AERIAL PHOTOGRAPH.

FIGURE 7.6
Digital structure of Landsat image, showing conversion of pixel values to gray scale display. Area is Palo Alto, California, on Landsat 1525-18145, band 5, acquired December 30, 1973. Data courtesy R. J. P. Lyon, Stanford University.

digitally processing a variety of multispectral data, including Landsat (Landgrebe, and others, 1968). VICAR, LARSYS, and other image processing software systems have been placed in the public domain and are available from the COSMIC Data Center, which is described at the conclusion of this chapter. The digitally processed images illustrated in this chapter were provided by IBM, Goddard Space Flight Center, JPL, Stanford University, TRW, and the U.S. Geological Survey Image Processing Facility at Flagstaff, Arizona.

Image processing methods may be grouped into the functional categories of restoration, enhancement, and information extraction that are listed here, together with specific processing routines:

1. Image restoration
   a. Sixth-line dropouts
   b. Sixth-line banding
   c. Scan-line offsets
   d. Atmospheric correction
   e. Geometric corrections
   f. Synthetic stereo images
2. Image enhancement
   a. Contrast enhancement
   b. Density slicing
   c. Edge enhancement
   d. Spatial and directional filtering
   e. Simulated normal color images
   f. Digital mosaics
3. Information extraction
   a. Band ratio images
   b. Other ratio images
   c. Multispectral classification
   d. Change-detection images

The subsequent discussion concentrates on principles and applications of digital image processing rather than program descriptions. Hardware and software systems are summarized at the end of this chapter, together with a listing of available systems. The mathematical basis for image processing is described in books such as those by Andrews and Hunt (1977), Pratt (1977), and Rosenfeld and Kak (1976).

IMAGE RESTORATION

Image restoration processes recognize and compensate for data errors, noise, and geometric distortion introduced in the scanning and transmission processes. The objective is to make the image resemble the original scene; therefore, some workers refer to these as "cosmetic processes." Image restoration is relatively simple because the pixels from each MSS band are processed separately.

Sixth-Line Dropouts

On some images acquired in the early months of Landsat 1, data from one of the six detectors were not recorded because of a hardware problem. On the CCT every sixth scan line is a string of zeros that plots as a black line or "dropout" on the image (Figure 7.7A). The average DN value per scan line is calculated for the entire scene. The average DN value for each individual scan line is then compared with this scene average. Any scan line deviating from the average by more than a designated threshold value is identified as defective. The correction step is done in the following manner. For each pixel in the dropout line the average value for the corresponding pixel on the preceding and succeeding scan lines is calculated and substituted, as shown on the printout of Figure 7.7B. The resulting image is a major improvement, although every sixth scan line consists of artificial data. This restoration program is equally effective for random-line dropouts that do not follow a systematic pattern.

Line														
1	18	18	20	19	19	19	19	19	19	22	18	18	22	22
	19	21	19	18	18	19	18	13	18	21	19	23	25	19
	16	20	25	24	23	21	21	21	20	21	18	18	18	20
	16	23	31	32	25	23	19	20	20	20	19	19	19	19
	16	17	28	26	22	22	22	23	20	19	20	23	22	23
6	0	0	0	0	0	0	0	0	0	0	0	0	0	0
	14	19	18	18	18	21	20	20	20	20	20	20	20	20
	19	20	18	18	18	20	19	20	22	20	20	20	22	19
	20	21	18	17	17	18	21	24	19	20	20	21	22	18
	19	19	21	19	19	19	19	19	19	19	19	19	19	19
	19	19	19	19	21	19	19	21	19	18	18	18	20	18
12	0	0	0	0	0	0	0	0	0	0	0	0	0	0
	18	18	21	21	18	18	21	17	21	19	14	19	18	18
	18	18	21	20	15	19	20	20	19	16	19	18	19	16
	18	18	18	16	18	16	18	18	17	19	18	16	17	17
	15	19	15	15	18	18	18	18	14	21	17	17	20	19
	18	19	16	16	20	19	15	16	20	16	16	20	16	17
18	0	0	0	0	0	0	0	0	0	0	0	0	0	0
	20	20	20	20	20	20	21	18	19	15	16	16	16	15
	19	18	18	20	18	20	18	19	16	19	18	18	19	16
	16	18	16	18	16	18	18	17	19	18	16	18	16	17
	19	17	18	18	18	18	18	17	19	17	17	19	19	17
	20	18	19	16	16	20	16	19	19	16	19	19	16	17
24	0	0	0	0	0	0	0	0	0	C	0	0	0	0
	19	21	19	18	19	15	16	16	15	20	15	16	15	20
	19	16	19	18	18	18	20	19	16	17	17	17	17	14
	18	18	16	18	18	21	18	19	17	18	18	17	21	21
	18	18	18	18	18	18	15	16	16	16	15	18	17	20
	19	16	20	16	19	18	18	18	20	18	18	18	18	18
30	0	0	0	0	0	0	0	0	0	0	0	0	0	0
	15	16	16	16	16	16	16	14	22	26	25	19	20	21
	19	20	20	21	18	20	18	18	18	20	21	20	25	25
	19	18	18	21	21	18	18	18	21	21	18	18	18	18
	18	18	20	19	20	18	17	19	18	20	20	18	20	18
	18	18	21	18	18	20	18	18	18	20	20	19	16	20

A. BEFORE DIGITAL PROCESSING.

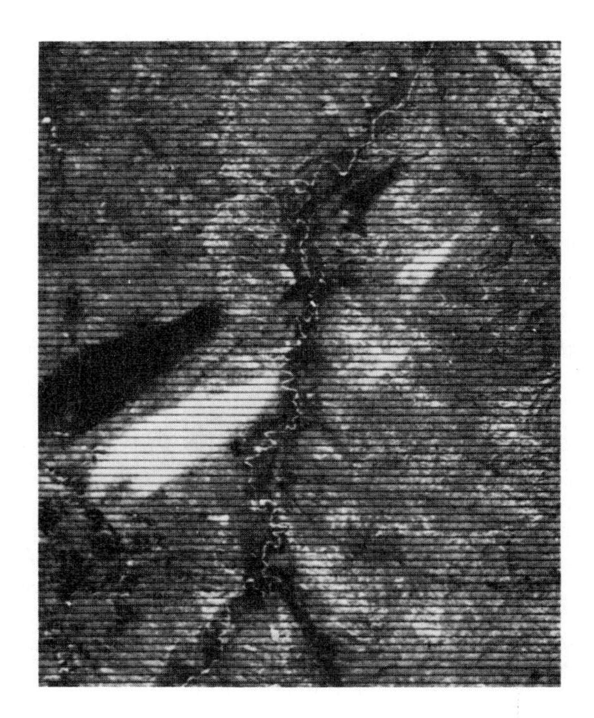

Line														
1	18	18	20	19	19	19	19	19	19	22	18	18	22	22
	19	21	19	18	18	19	18	13	18	21	19	23	25	19
	16	20	25	24	23	21	21	21	20	21	18	18	18	20
	16	23	31	32	25	23	19	20	20	20	19	19	19	19
	16	17	28	26	22	22	22	23	20	19	20	23	22	23
6	15	18	23	22	20	22	21	22	20	20	20	22	21	22
	14	19	18	18	18	21	20	20	20	20	20	20	20	20
	19	20	18	18	18	20	19	20	22	20	20	20	22	19
	20	21	18	17	17	18	21	24	19	20	20	21	22	18
	19	19	21	19	19	19	19	19	19	19	19	19	19	19
	19	19	19	19	21	19	19	21	19	18	18	18	20	18
12	19	19	20	20	20	19	20	19	20	19	16	19	19	18
	18	18	21	21	18	18	21	17	21	19	14	19	18	18
	18	18	21	20	15	19	20	20	19	16	19	18	19	16
	18	18	18	16	18	16	18	18	17	19	18	16	17	17
	15	19	15	15	18	18	18	18	14	21	17	17	20	19
	18	19	16	16	20	19	15	16	20	16	16	20	16	17
18	19	20	18	18	20	20	18	17	20	16	16	18	16	16
	20	20	20	20	20	20	21	18	19	15	16	16	16	15
	19	18	18	20	18	20	18	19	16	19	18	18	19	16
	16	18	16	18	16	18	18	17	19	18	16	18	16	17
	19	17	18	18	18	18	18	17	19	17	17	19	19	17
	20	18	19	16	16	20	15	19	19	16	19	19	16	17
24	20	20	19	17	18	18	16	18	17	18	17	18	16	19
	19	21	19	18	19	15	16	16	15	20	15	16	15	20
	19	16	19	18	18	18	20	19	16	17	17	17	17	14
	18	18	15	18	18	21	18	19	17	18	18	17	21	21
	18	18	18	18	18	18	15	16	16	16	15	18	17	20
	19	16	20	16	19	18	18	18	20	18	18	18	18	18
30	17	16	18	16	18	17	17	16	21	22	22	19	19	20
	15	16	16	16	16	16	16	14	22	26	25	19	20	21
	19	20	20	21	18	20	18	18	18	20	21	20	25	25
	19	18	18	21	21	18	18	18	21	21	18	18	18	18
	18	18	20	19	20	18	17	19	18	20	20	18	20	18
	18	18	21	18	18	20	18	18	18	20	20	19	16	20

B. AFTER DIGITAL PROCESSING.

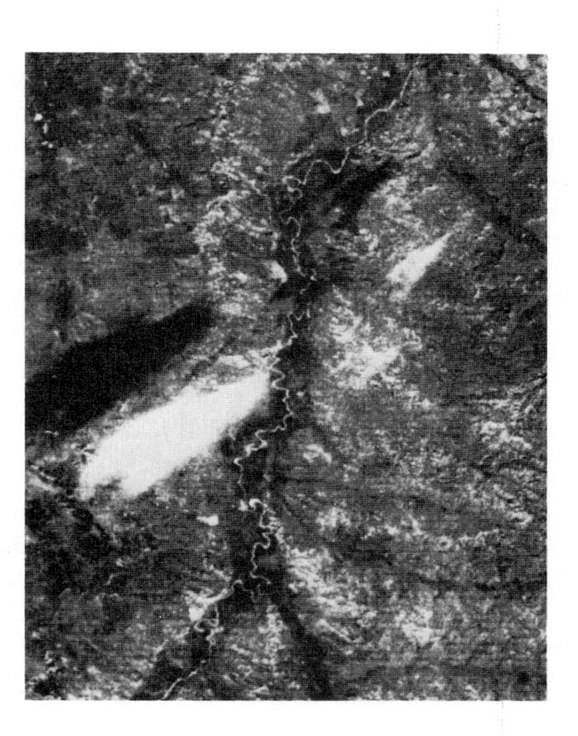

FIGURE 7.7
Correction of sixth-line dropouts by replacing zeros with average pixel values. Printouts on the left represent a portion of each image. Landsat 1084-17233, band 4, southeast Montana. Image processed at Jet Propulsion Laboratory, California Institute of Technology.

Sixth-Line Banding

The six detectors for each spectral band were carefully calibrated and matched before Landsat was launched. With time, however, the response of some detectors has drifted to higher or lower levels resulting in sixth-line banding. This image defect is illustrated in Figure 7.8A, where the gain, or amplitude, of detector number 6 is twice that of the other detectors, causing every sixth scan line to be twice as bright as the scene average. Valid data are present in the defective lines, but must be corrected to match the overall scene. One correction method employs a routine that sweeps across the image normal to the scan lines and determines amplitude of the pixel values. Any periodic deviations from the norm for the scene are recognized and a correction factor is calculated. In Figure 7.8B the amplitude of every sixth line is multiplied by a factor of 0.5 to produce the correct digital values from which the restored image is plotted. Another correction method plots a histogram of DNs for each of the six detectors. Differences in mean and median values for the histograms are used to recognize and determine corrections for detector differences.

Scan-Line Offsets

On some EDC images individual scan lines are offset horizontally from adjacent scan lines in either a random or periodic fashion. This becomes a serious defect when images are enlarged to 1:250,000 scale or larger. The example in Figure 7.9A was enlarged to 1:100,000 for use as a base map, and is significantly degraded by the numerous scan lines that are offset to the right by one kilometer or more. These offsets are particularly evident to the east of the lakes in Figure 7.9A. The corrected image in Figure 7.9B was produced by precisely aligning the start of each scan line.

Atmospheric Correction

As was discussed in Chapter 2, the atmosphere selectively scatters the shorter wavelengths of light. Landsat MSS band 4 (0.5 to 0.6 μm) has the highest component of scattered light, and band 7 (0.8 to 1.1 μm) has the least. Contrast ratio of images is improved by correcting for the effects of this differential scattering. Two techniques for determining the correction factor for atmospheric scatter (or haze) in different MSS bands are shown in Figure 7.10, which is taken from the work of Chavez (1975). Both techniques are based on the fact that band 7 is essentially free of atmospheric effects, which can be verified by examining the DNs corresponding to bodies of clear water and to steep, shadowed slopes. Both the water and shadows have values of either zero or one on band 7.

The first haze correction technique (Figure 7.10A) employs an area within the image that has shadows caused by irregular topography. For each pixel the DN in band 7 is plotted against the DN value in band 4, and a straight line is fitted through the plot, using a least-squares technique. If there were no haze in band 4, the line would pass through the origin, but the intercept is offset along the band 4 axis as shown in Figure 7.10A. The amount of offset is caused by haze (atmospheric scattering), which has an additive effect on scene brightness as shown earlier in Figure 2.3. To correct the haze effect on band 4 the value of the intercept offset is subtracted from the DN value of each band 4 pixel for the entire image. Bands 5 and 6 are also plotted against band 7 and the procedure is repeated. In this technique the data must be passed through the computer several times. Also, materials with widely different values for the two bands cause the cluster to be scattered thus making the least-squares fit difficult.

The second correction technique requires only one pass of data through the computer. In most

1	22	17	20	18	23	26	26	26	26	25	22	27	24	24
	24	24	25	24	26	30	30	26	24	28	24	25	19	19
	23	21	22	32	31	28	27	22	24	23	23	22	23	20
	13	16	24	32	27	23	24	22	18	23	22	23	23	19
	16	21	30	31	25	23	24	23	23	24	23	23	22	22
6	18	20	36	40	26	24	26	22	20	26	24	22	28	22
	16	19	33	31	27	25	25	23	20	23	24	29	25	26
	19	16	20	24	27	27	25	22	25	25	23	31	25	30
	17	18	21	20	22	23	24	23	22	22	23	23	28	23
	23	19	18	20	16	19	24	24	23	23	23	23	24	24
	23	24	18	20	18	18	23	27	24	23	22	22	23	23
12	44	44	40	40	44	44	44	44	48	44	40	44	36	44
	21	21	21	21	24	21	20	27	20	22	19	19	19	22
	24	20	18	23	26	22	24	23	23	24	23	19	20	19
	22	22	23	23	23	22	23	19	23	20	17	23	23	20
	19	17	23	23	17	20	18	24	19	17	19	17	19	18
	18	18	19	17	19	18	20	19	20	18	17	18	16	18
18	36	36	36	36	32	40	32	40	36	44	36	36	36	40
	20	18	17	17	20	18	17	17	20	18	17	19	18	18
	25	24	18	20	20	19	18	25	20	18	18	20	18	17
	23	21	23	23	23	23	22	23	20	18	18	18	19	17
	18	18	24	18	19	20	18	17	17	19	17	19	18	19
	18	18	17	20	20	18	20	19	19	20	17	19	19	17
24	44	36	40	32	36	44	36	44	36	44	36	44	44	36
	22	18	18	20	18	20	19	20	18	17	19	18	19	17
	20	12	20	21	19	20	20	20	19	19	20	18	14	15
	23	24	23	23	16	19	17	19	18	21	18	17	18	18
	23	18	19	18	24	18	19	20	17	16	18	18	19	13
	18	20	18	17	23	22	22	23	18	20	17	17	25	23
30	36	36	44	36	40	32	36	32	36	32	32	40	44	44
	18	18	20	19	19	19	19	19	19	22	18	18	22	22
	19	21	19	18	18	19	18	13	18	21	19	23	25	19
	16	20	25	24	23	21	21	21	20	21	18	18	18	20
	16	23	31	32	25	23	19	20	20	20	19	19	19	19
	16	17	28	26	22	22	22	23	20	19	20	23	22	23

A. BEFORE DIGITAL PROCESSING.

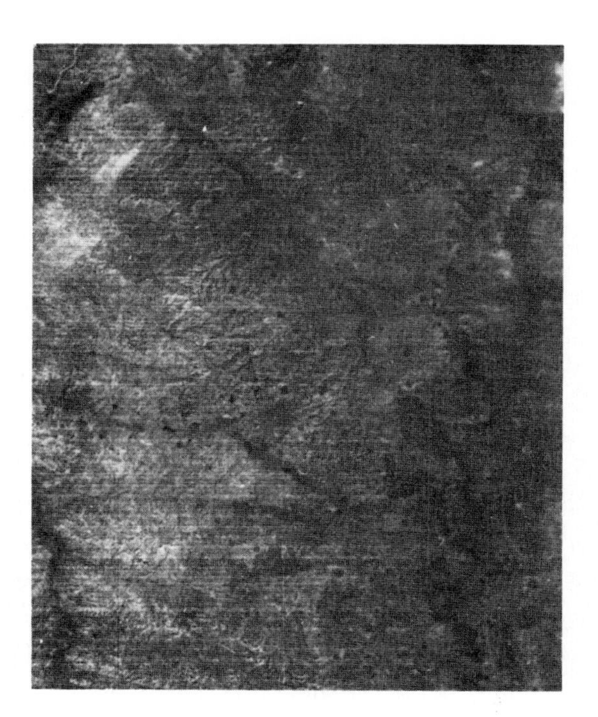

1	22	17	20	18	23	26	26	26	26	25	22	27	24	24
	24	24	25	24	26	30	30	26	24	28	24	25	19	19
	23	21	22	32	31	28	27	22	24	23	23	22	23	20
	13	16	24	32	27	23	24	22	18	23	22	23	23	19
	16	21	30	31	25	23	24	23	23	24	23	23	22	22
6	18	20	36	40	26	24	26	22	20	26	24	22	28	22
	16	19	33	31	27	25	25	23	20	23	24	29	25	26
	19	16	20	24	27	27	25	22	25	25	23	31	25	30
	17	18	21	20	22	23	24	23	22	22	23	23	28	23
	23	19	18	20	16	19	24	24	23	23	23	23	24	24
	23	24	18	20	18	18	23	27	24	23	22	22	23	23
12	22	22	20	20	22	22	22	22	24	22	20	22	18	22
	21	21	21	21	24	21	20	27	20	22	19	19	19	22
	24	20	18	23	26	22	24	23	23	24	23	19	20	19
	22	22	23	23	23	22	23	19	23	20	17	23	23	20
	19	17	23	23	17	20	18	24	19	17	19	17	19	18
	18	18	19	17	19	18	20	19	20	18	17	18	16	18
18	18	18	18	18	16	20	16	20	18	22	18	18	18	20
	20	18	17	17	20	18	17	17	20	18	17	19	18	18
	25	24	18	20	20	19	18	25	20	18	18	20	18	17
	23	21	23	23	23	23	22	23	20	18	18	18	19	17
	18	18	24	18	19	20	18	17	17	19	17	19	18	19
	18	18	17	20	20	18	20	19	19	20	17	19	19	17
24	22	18	20	16	18	22	18	22	18	22	18	22	22	18
	22	18	18	20	18	20	19	20	18	17	19	18	19	17
	20	12	20	21	19	20	20	20	19	19	20	18	14	15
	23	24	23	23	16	19	17	19	18	21	18	17	18	18
	23	18	19	18	24	18	19	20	17	16	18	18	19	13
	18	20	18	17	23	22	22	23	18	20	17	17	25	23
30	18	18	22	18	20	16	18	16	18	16	16	20	22	22
	18	18	20	19	19	19	19	19	19	22	18	18	22	22
	19	21	19	18	18	19	18	13	18	21	19	23	25	19
	16	20	25	24	23	21	21	21	20	21	18	18	18	20
	16	23	31	32	25	23	19	20	20	20	19	19	19	19
	16	17	28	26	22	22	22	23	20	19	20	23	22	23

B. AFTER DIGITAL PROCESSING.

FIGURE 7.8
Removal of sixth-line banding by correcting gain level of detector number 6. Printouts on the left represent a portion of the image. Landsat 1084-17233, band 6, southeast Montana. Image processed at Jet Propulsion Laboratory, California Institute of Technology.

A. BEFORE DIGITAL PROCESSING.

B. AFTER DIGITAL PROCESSING.

FIGURE 7.9
Correction of offset scan lines by digital processing. Landsat 1198-07494, band 7, southern part of Sudan. Image processed at Jet Propulsion Laboratory, California Institute of Technology.

images there are some steep shadowed slopes that are represented by pixels with DN values of zero on band 7. These DNs are shown on the histogram of band 7 (Figure 7.10B). The histogram of band 4 lacks zero values, and the peak is offset toward higher DN values because of the additive effect of haze. Band 4 also shows a characteristic abrupt increase in frequency of pixels on the low (left) side of the histogram. The lack of DNs below this level is attributed to illumination from light scattered by the atmosphere. For band 4 the abrupt increase typically occurs at a DN of 11; this value is subtracted from all of the band 4 pixels to correct for haze. For band 5 the correction is usually a value of 7 and for band 6 a value of 3. There can be a problem with this correction technique if the Landsat scene lacks steep topography and there are no band 7 pixels with DNs of zero.

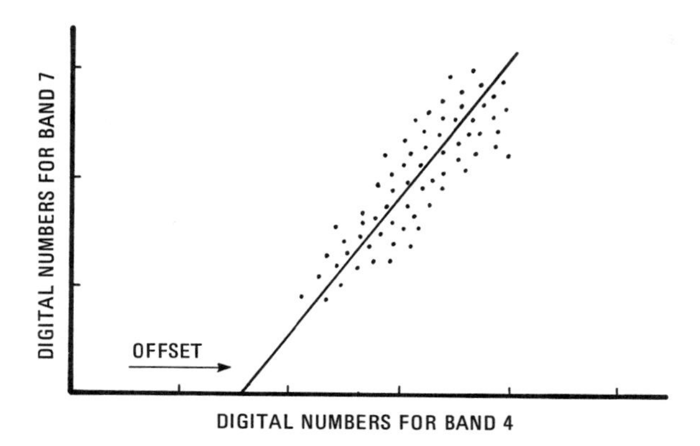

A. PLOT OF BAND 7 VERSUS BAND 4 FOR AN AREA WITHIN THE IMAGE THAT HAS SHADOWS. OFFSET OF THE LINE OF LEAST–SQUARES FIT ALONG THE BAND 4 AXIS IS ATTRIBUTED TO ATMOSPHERIC SCATTERING IN THAT BAND.

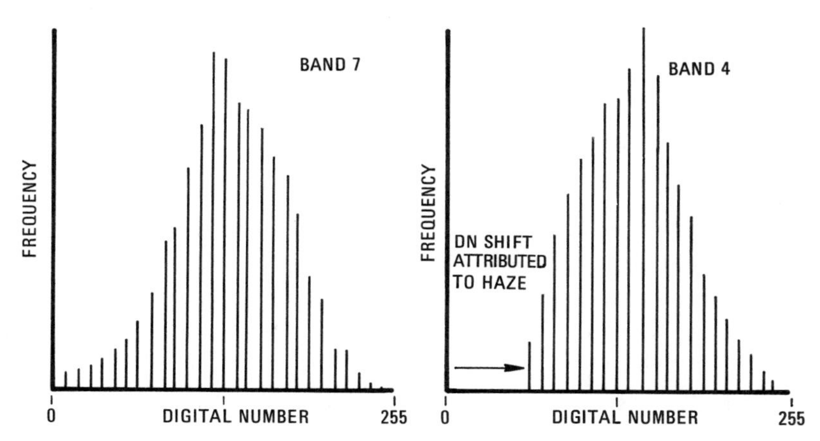

B. HISTOGRAMS FOR BANDS 7 AND 4. THE LACK OF LOW DN'S ON BAND 4 IS CAUSED BY ILLUMINATION FROM LIGHT SCATTERED BY THE ATMOSPHERE (HAZE).

FIGURE 7.10
Methods for determining atmospheric corrections on individual Landsat bands. From Chavez (1975, Figures 2 and 3).

Geometric Corrections

During the scanning process a number of systematic and nonsystematic geometric distortions are introduced into the image data. These distortions are corrected at GSFC during production of the master images, which are remarkably orthogonal. The corrections are not included on the CCTs, however, and images plotted from the tapes must be corrected. In the future, the data will be preprocessed to incorporate some geometric corrections into the CCTs.

Nonsystematic Distortions *Nonsystematic distortions* (Figure 7.11A) are not constant because they result from variations in the spacecraft attitude, velocity, and altitude and therefore are not predictable. These distortions must be evaluated from Landsat tracking data or ground control information. Spacecraft velocity variations cause distortion in the along-track direction only, and are known functions of velocity which can be obtained from tracking data.

The amount of earth rotation during the 28 seconds required to scan an image results in distortion in the scan direction that is a function of spacecraft latitude and orbit. In the correction process, successive groups of six scan lines are offset to compensate for earth rotation, resulting in the parallelogram outline of the image.

Altitude and attitude (roll, pitch, and yaw) variations of the spacecraft result in nonsystematic distortions that must be measured for each image in order to be corrected. The correction process employs detectable and recognizable geographic features on the image, called *ground control points* (GCPs), whose positions are known. Intersections of highways and of airport runways are typical GCPs. Differences between actual and observed GCP locations are used to determine the geometric transformations required to correct the image. The original pixels are resampled to match the correct geometric coordinates. The various resampling methods are described by Rifman (1973), Goetz and others (1975), and Bernstein and Ferneyhough (1975).

Systematic Distortions Geometric distortions, whose effects are constant and can be predicted in advance, are termed *systematic distortions*. Scan skew, scanner distortion, and variations in scanner mirror velocity belong in this category (Figure 7.11B). *Scanner distortion* results from sampling of data along the scan line at regular time intervals that correspond to nominal intervals of 57 m. The length of the ground interval is actually proportional to the tangent of the scan angle and therefore is slightly greater at either margin of the scan line. The resulting marginal compression of Landsat images is identical to that described for thermal IR scanner images; however, the effect is minimal in Landsat images because the scan angle is only 5.8° on either side of vertical in contrast to the 45° to 60° angles of IR scanners. Scanner distortion is also referred to as "panoramic distortion."

Tests made before Landsat is launched determined that the velocity of the MSS mirror is not constant from start to finish of each scan line, resulting in minor systematic distortion along each scan line. The known mirror velocity variations may be used to correct for this effect. *Scan skew* is caused by the forward motion of the spacecraft during the time required for each mirror sweep. The ground swath scanned is not normal to the ground track but is slightly skewed, producing distortion across the scan line. An additional correction is required to compensate for the higher sampling rate in the scan-line direction than in the along-track direction. This correction may be accomplished in the computer by resampling the data or during the film plotting.

Synthetic Stereo Images The Landsat image has a unique cartographic projection that is almost orthogonal and closely matches the Lambert conic conformal map projection. The imagery may be transformed into other cartographic projections

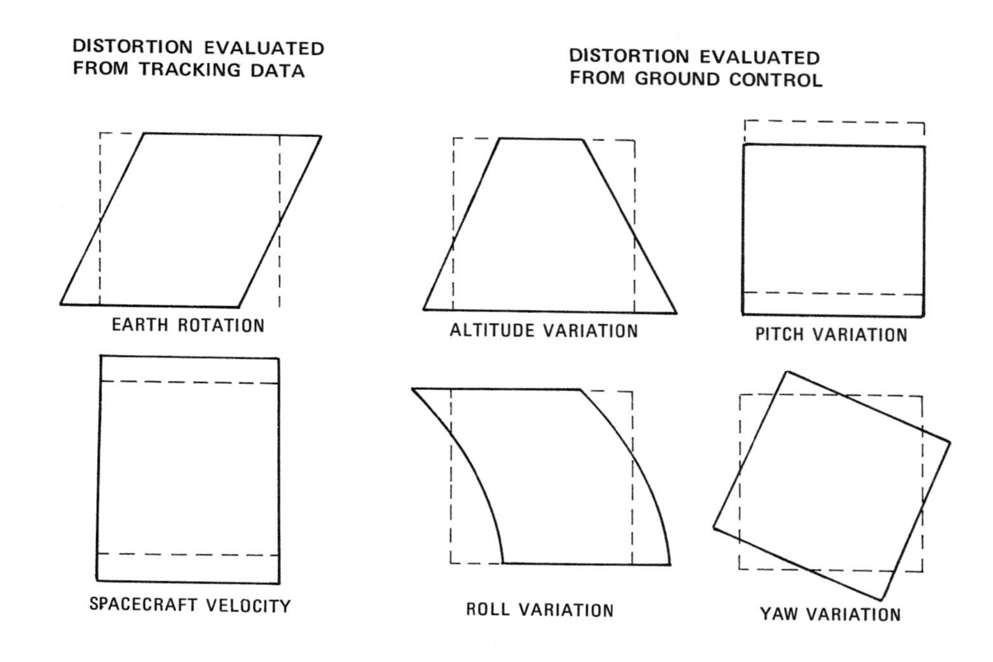

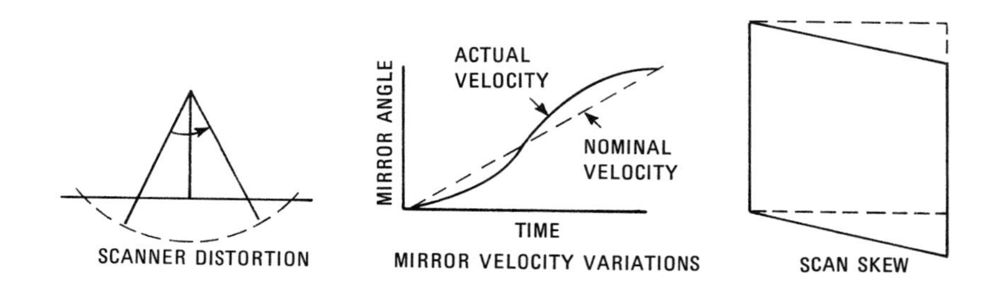

FIGURE 7.11
Geometric distortions of Landsat images. From Bernstein and Ferneyhough (1975, Figure 3).

by use of ground control points on the image.

One such projection technique is to restructure the pixels of a single Landsat scene to produce a synthetic stereo pair (Figure 7.12). This technique was developed at the U.S. Geological Survey Image Processing Facility at Flagstaff, Arizona (Batson, Edwards, and Eliason, 1976). The GCPs of the Landsat scene are first registered to a corresponding topographic map, which provides elevation information. A mathematical model is then used to replot the Landsat pixels into the left and right images of a stereo pair having the desired relief displacement and vertical exaggeration. This technique overcomes the drawbacks, cited in the following list, of using the sidelapping portions of Landsat images for stereo viewing:

1. In low and middle latitudes Landsat orbits do not provide complete sidelap of adjacent images. For instance, there is only 14 percent sidelap at the equator.
2. For images with sidelap, the vertical exaggeration is insufficient for optimum stereo viewing, particularly at higher latitudes where base-height ratio is at a minimum.
3. Because of weather or recording problems, images acquired several months apart often must be used to obtain sidelap. Seasonal and atmospheric differences may obscure the image detail.

To appreciate the synthetic stereo images, Figure 7.12 should be viewed with a pocket lens stereoscope and the resulting stereo model compared with elevation data on the map of Figure 7.13. Processing costs are the main drawback to producing synthetic stereo images of areas where topographic information is available.

IMAGE ENHANCEMENT

Image enhancement is the modification of an image to alter its impact on the viewer. Most enhancement operations distort the original digital values; therefore enhancement usually is not done until all the other processing steps are completed. Like image restoration, enhancement methods are applied separately to each band of a multispectral image.

Contrast Enhancement

The importance of contrast ratio for determining resolving power and detection capability of images has been demonstrated in preceding chapters of this book. For this reason, techniques for modifying contrast are the most commonly applied image enhancement processes. In Chapter 4 it was pointed out that EDC positive transparencies typically lack adequate contrast. Photographic duplication on high-contrast copy film can produce transparencies suitable for making color composites, but this is not an entirely satisfactory remedy for two reasons: (1) each additional reproduction step results in overall loss of information; (2) the limited dynamic range of film, especially the high-contrast type, results in loss of information at the bright and dark extremes of the image gray-scale range. EDC personnel point out that the low contrast is inherent in the original negatives provided by GSFC. Whatever the cause, almost all of the thousands of Landsat black-and-white positive transparencies purchased by Chevron have required contrast enhancement to produce satisfactory color composite images. Beginning in 1976 there has been an improvement in the contrast ratio of EDC transparencies.

Digital methods are more satisfactory than photographic techniques for contrast enhancement because of the precision and wide variety of digital processes. Detectors on the Landsat MSS and other remote-sensing systems must be designed to record a wide range of scene brightness values without becoming saturated. *Saturation* occurs when the sensitivity range of a detector is insufficient to record the full brightness range of a scene. In this case, DN values at the light and dark extremes appear as saturated white or black tones on the image. The sensitivity range of Landsat detectors, for example, must encompass a range of brightness from black basalt outcrops to white sea ice. Few individual scenes have a brightness range that utilizes the full sensitivity range of the Landsat detectors. To produce an image with the optimum contrast ratio it is important to utilize the entire brightness range of the display medium, which is generally film. Figure 7.14A is a portion of band 4 of a Landsat image of the Andes Mountains along the Chile–Bolivia border that was produced from the CCT data with no contrast enhancement.

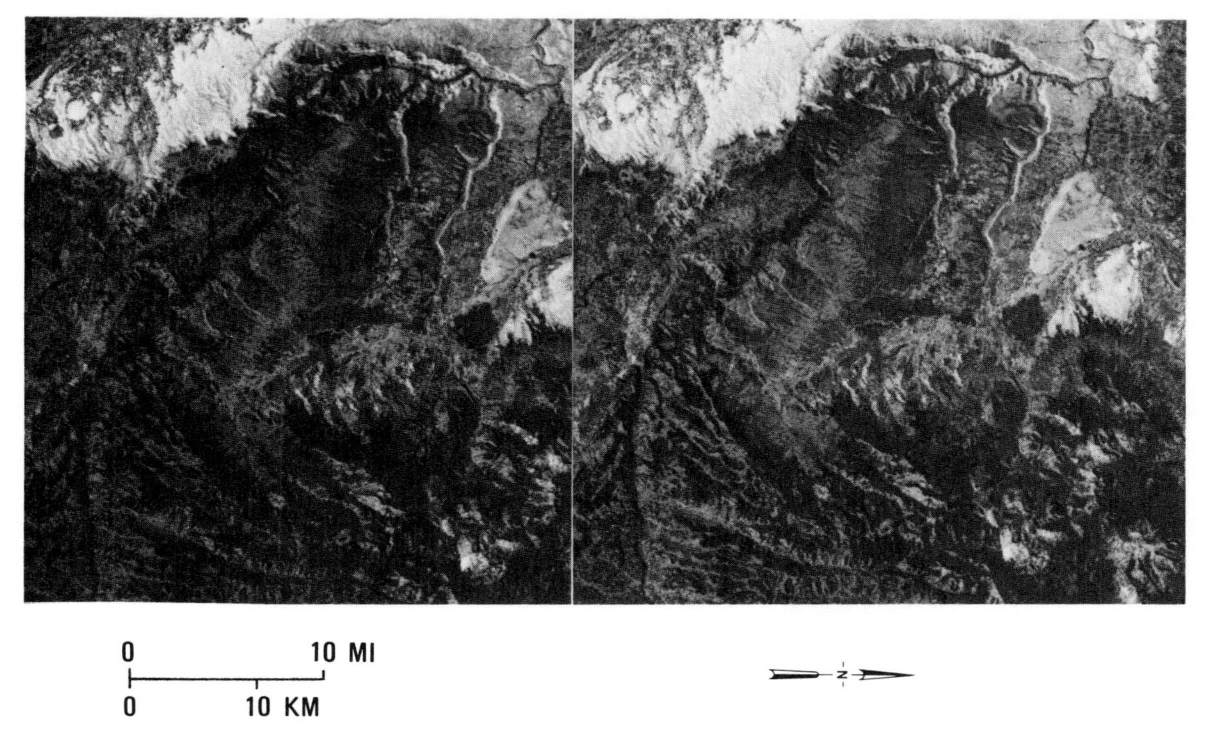

FIGURE 7.12
Synthetic stereo pair of Gunnison River area, west-central Colorado, from Landsat 1407-17190. Through the use of topographic elevation data the pixels were replotted to create relief displacement. Digital processing by U.S. Geological Survey Image Processing Facility. From Batson, Edwards, and Eliason (1976, Plate 1).

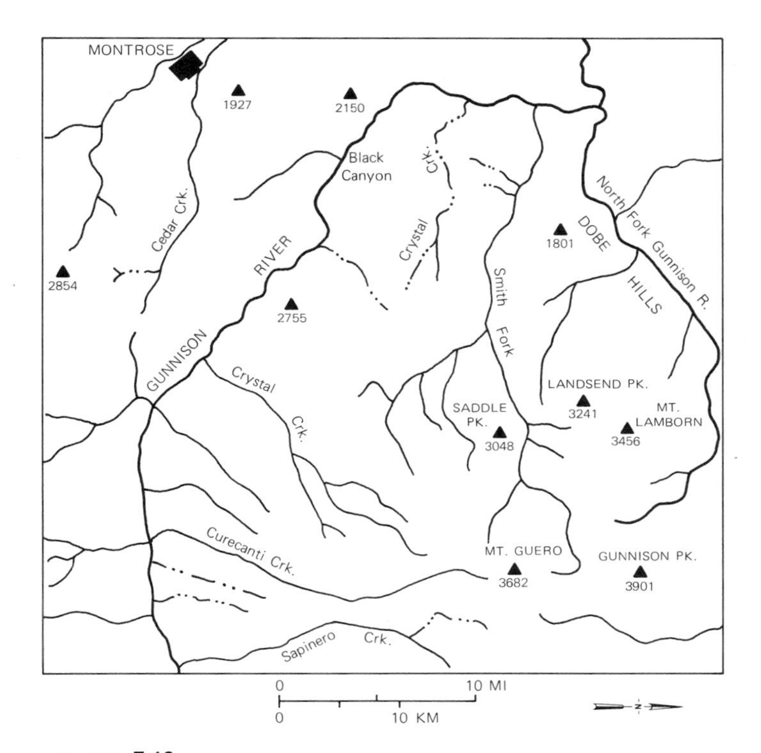

FIGURE 7.13
Geographic map of Gunnison River area, west-central Colorado. Elevations in meters.

No. OF PIXELS

49 106

4% 4%

0 64 128 192 256

A. ORIGINAL IMAGE WITH NO CONTRAST ENHANCEMENT.

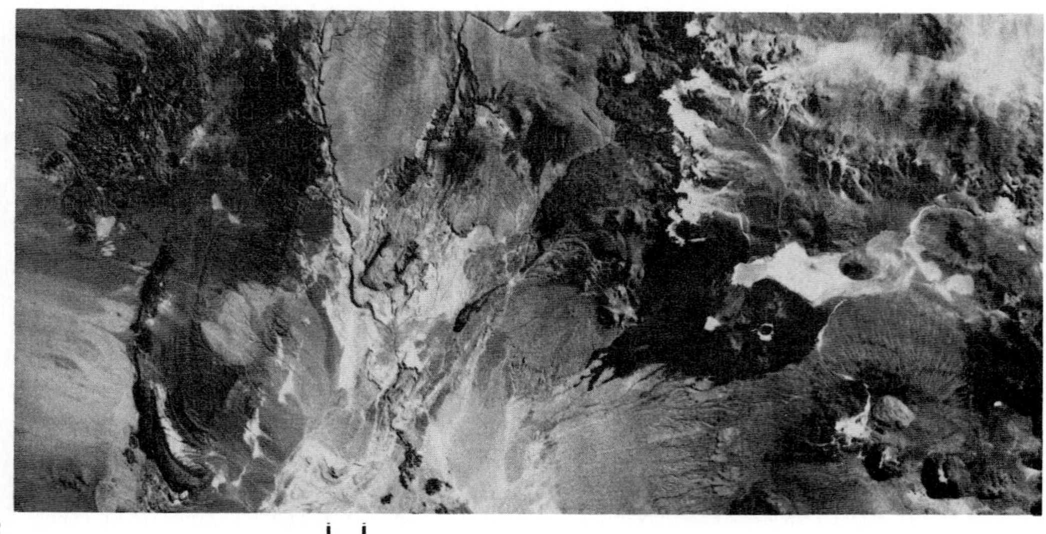

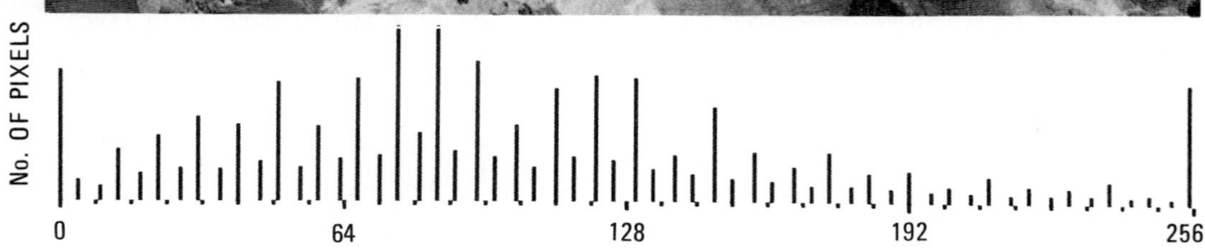

B. LINEAR CONTRAST STRETCH WITH LOWER AND UPPER FOUR PERCENT OF PIXELS SATURATED TO BLACK AND WHITE RESPECTIVELY.

FIGURE 7.14
Contrast enhancement methods. Portion of Landsat 1243-14004, band 4, acquired March 23, 1973 of the border between Chile and Bolivia. See Figure 7.15 for

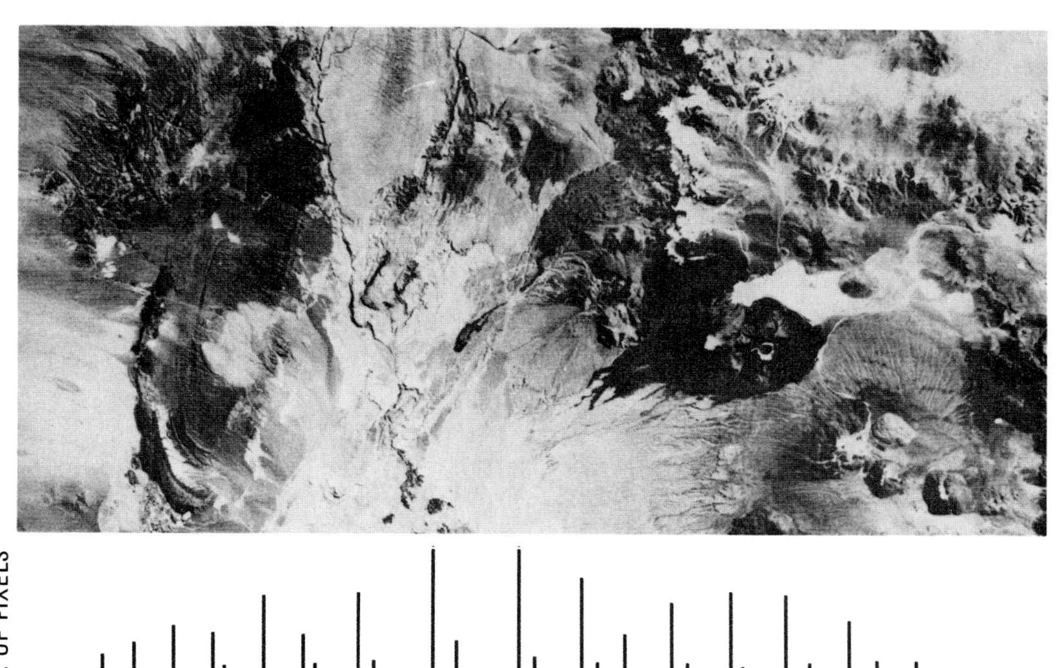

C. UNIFORM DISTRIBUTION STRETCH.

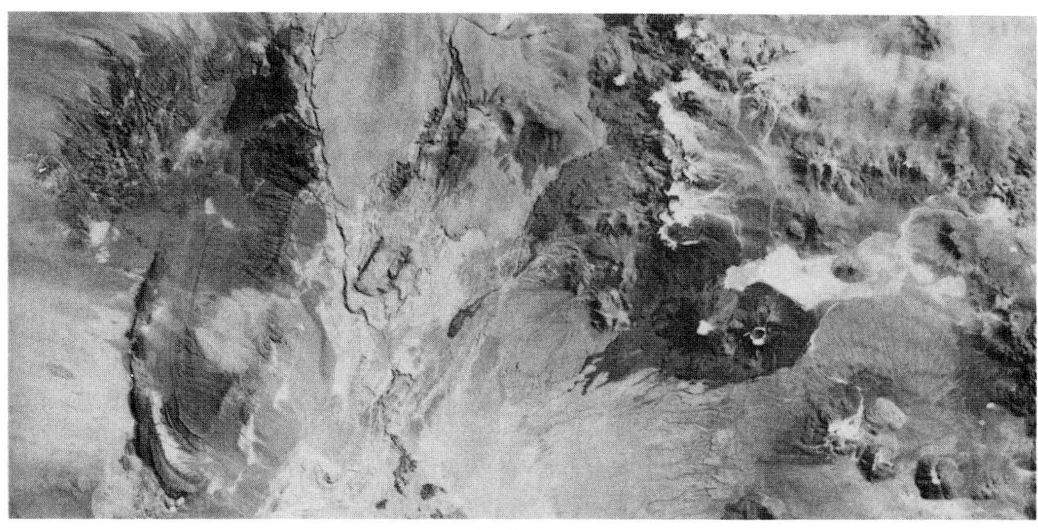

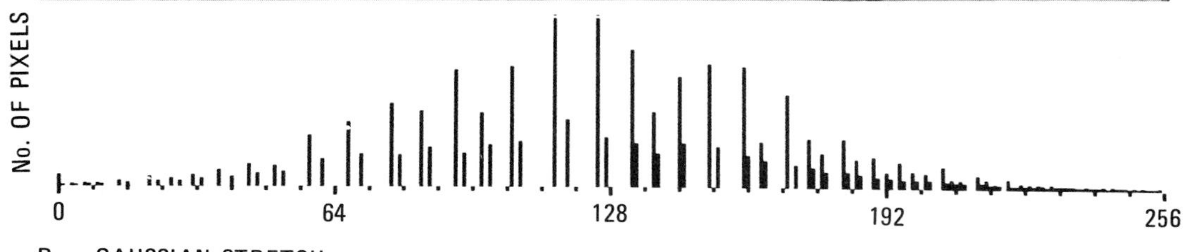

D. GAUSSIAN STRETCH.

location. From Soha and others (1976, Figures 4 and 5). Courtesy J. M. Soha, Jet Propulsion Laboratory, California Institute of Technology.

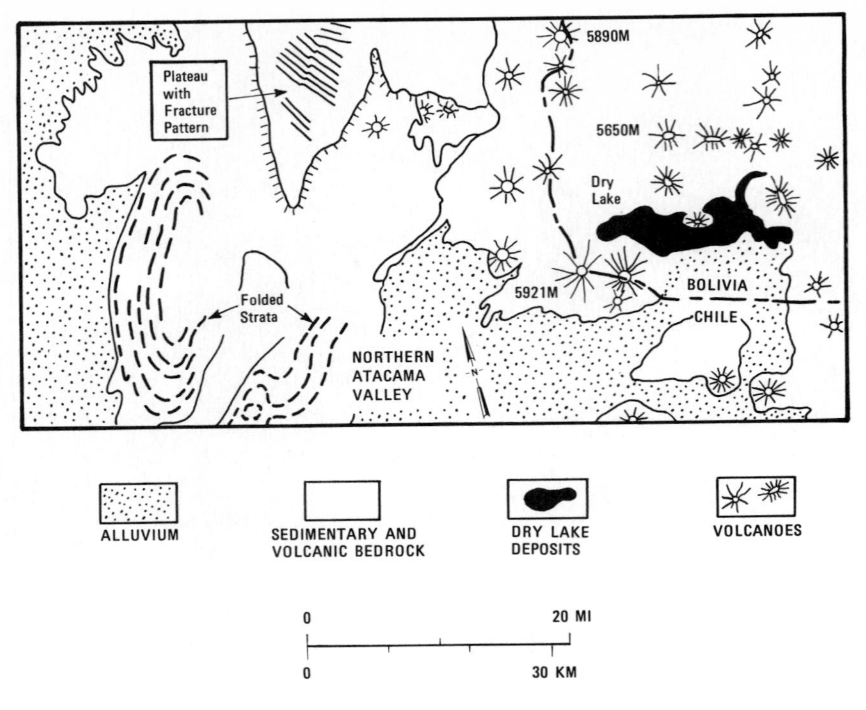

FIGURE 7.15
Location map for Landsat image of the Andes along the border between Chile and Bolivia.

The accompanying histogram shows the number of pixels that correspond to the DN brightness values. The central 92 percent of the histogram encompasses a range of DN values from 49 to 106, which utilizes only 23 percent of the available brightness range [(106 − 49)/256 = 23 percent]. This limited range of brightness values accounts for the low contrast ratio of the original image.

The simplest contrast enhancement is called a *linear contrast stretch*, and is illustrated in Figure 7.14B. A DN value in the low range of the original histogram is assigned to extreme black, and a value at the high end is assigned to extreme white. In this example, the lower four percent of the pixels (DN < 49) are assigned to black, or 0, and the upper four percent (DN > 106) are assigned to white, or 255. The remaining pixel values are distributed linearly between these extremes, as shown in the histogram and the enhanced image of Figure 7.14B. The map of Figure 7.15 is useful in locating features for comparison on the images. The improved contrast ratio of the image with linear contrast stretch is particularly evident in the enhancement of individual rock units in the area of folded strata east of the Atacama Valley. The northwest-trending fracture system in the plateau west of the Chile–Bolivia border is clearly expressed on the stretched image but is obscure on the original image.

In exchange for the greatly enhanced contrast of most original brightness values, there is a trade off in the loss of contrast at the extreme high and low DN values. On the original image (Figure 7.14A) note that the lower elevation limits of the snow caps along the Chile–Bolivia border are clearly defined, but on the stretched image the white tone includes both the snow and the alluvium lower on the flanks of the mountains. In the small dry lake north of the border, patterns that are visible on the original image are absent on the stretched image. The brightness differences within the dry lake and between the snow and alluvium were in the range of DN values > 106. On the stretched image these are now saturated white, as shown on the histogram of Figure 7.14B by the spike at a DN of 255. A similar trade off occurs at the low end of the DN range. On the original image some dark lava flows are differentiated in the vicinity of the 5921 m volcano (see Figure 7.15 for location), but these are all black on the stretched image because of saturation of DNs <49. In comparison to the overall contrast improvement in Figure 7.14B these contrast losses at the brightness extremes are an acceptable trade off unless one were specifically interested in these elements of the scene. Because of the flexibility of digital methods an investigator could, for example, cause all DNs <106 to be saturated black and then linearly stretch the remaining high DN values through a range from 1 through 255. This extreme stretch would enhance the contrast differences within the bright elements at the expense of the remainder of the scene. It should be noted that saturation of the upper and lower four percent of DNs happened to be optimum for this example; for the other MSS bands of this scene and for other images different degrees of saturation may be optimum, and can be determined by inspection of the original histograms. The color Landsat image of Plate 2 was composited from black-and-white transparencies of bands 4, 5, and 7 that were processed with a linear contrast stretch.

Nonlinear contrast enhancement is also possible. Figure 7.14C illustrates a *uniform distribution stretch* in which the original histogram has been redistributed to produce a uniform population density of pixels along the DN axis. This redistribution results in the greatest contrast enhancement being applied to the most populated range of brightness values in the original image. In Figure 7.14C the middle range of brightness values are preferentially stretched, which results in maximum contrast in the alluvial deposits around the flanks of the mountains. The northwest-trending fracture pattern is more strongly enhanced by this stretch. As shown by the histogram and image, the uniform distribution stretch causes severe compression of brightness values at the sparsely populated light and dark tails of the original histogram. The resulting loss of contrast in the light and dark ranges is similar to that in the linear contrast stretch.

A nonlinear stretch that enhances contrast within the tails of the histogram is a *Gaussian stretch* that fits the original histogram to a normal distribution curve between the 0 and 255 limits. As seen in Figure 7.14D, this stretch improves the contrast in the light and dark ranges of the image. The different lava flows are distinguished and some details within the dry lake are emphasized. This enhancement occurs at the expense of contrast in the middle-gray range; the fracture pattern and some of the folds are suppressed in this image.

The examples in Figure 7.14 represent the wide variety of processes that are available for contrast enhancement. A sinusoidal stretch developed at the U.S. Geological Survey Image Processing Facility (Berlin and others, 1976) divides the original DN values into several intervals, each of which is then linearly stretched from 0 to 255. This stretch enhances subtle spectral differences not usually enhanced by other methods. Differences among many types of materials may be enhanced at the same time.

As Soha and others (1976, p. 3) point out, the most effective contrast enhancement depends on the nature of the original histogram and the elements of the scene that are of greatest interest to the user. Through inspection of the original histogram the experienced user can determine the optimum stretch for his needs. For some scenes a variety of stretched images are required to display the original data fully. In all contrast enhancement processes the user should be aware of the inevitable trade offs that are involved. It also bears repeating that contrast enhancement should not be done until other processing is completed, because the stretching distorts the original values of the pixels.

Density Slicing

Density slicing is a process that converts the continuous gray tone of an image into a series of density intervals, or slices, each corresponding to a specified digital range. Each digital slice may be displayed in a separate color, line printer symbol,

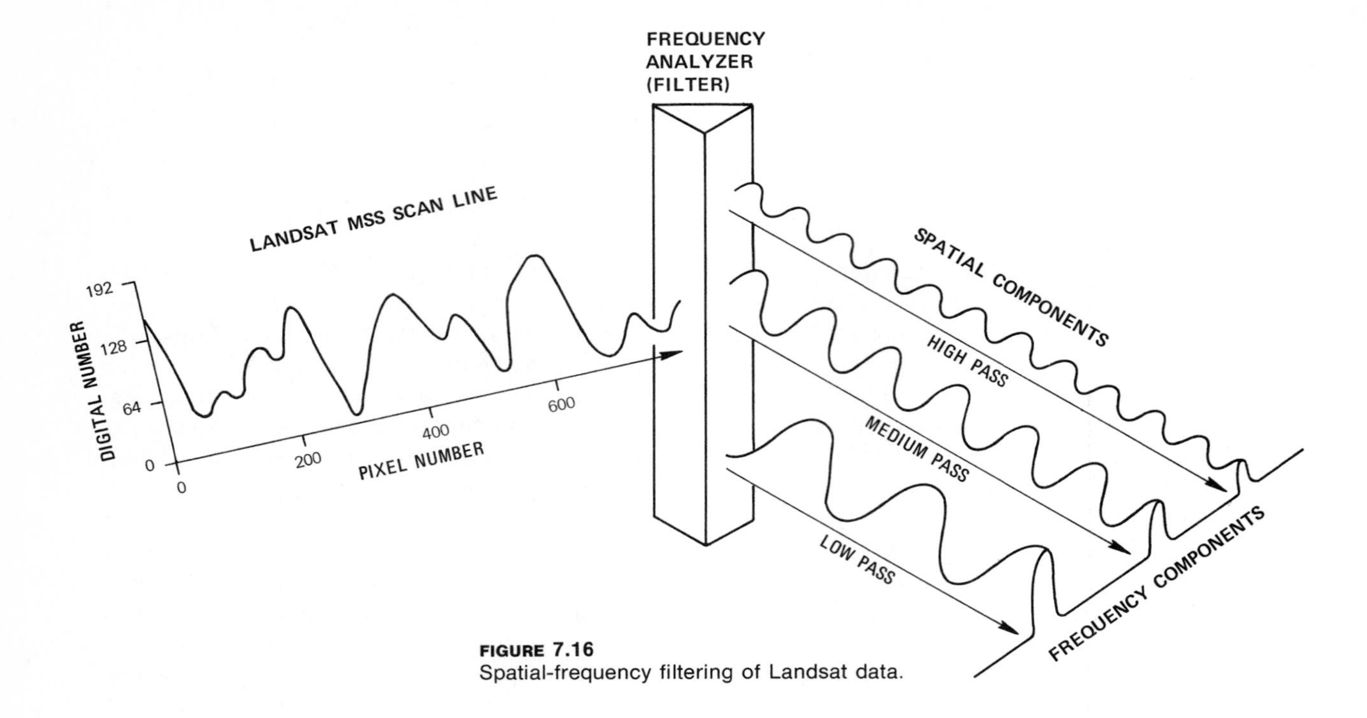

FIGURE 7.16
Spatial-frequency filtering of Landsat data.

or bounded by contour lines. This technique emphasizes subtle gray-scale differences that may be imperceptible to the viewer. In Chapter 5 digital density slicing was illustrated with the examples of calibrated thermal IR images.

Edge Enhancement

Edge enhancement is another technique for emphasizing subtle gray-scale variations. Where the values of adjacent pixels vary by more than a predetermined threshold value, the interface is marked on the image display with a contour, a step in the gray scale, or a color change.

Spatial and Directional Filtering

The regularly spaced pixels of a Landsat scan line may be plotted along the abscissa of a graph with the corresponding DNs as the ordinate. The resulting curve, shown on the left side of Figure 7.16, is a complex wave made up of many simple curves, each with its own constant wavelength. The complex wave of a beam of white light may be separated into its component wavelengths (colors) by a prism. In analogous fashion the complex wave of a Landsat scan line can be separated into its component wavelengths by a mathematical process known as *spatial filtering* (Figure 7.16). Frequency filters are designated high pass, intermediate pass, or low pass depending on the spatial frequency that is transmitted through the filter.

The distances between major terrain features, such as between valleys and mountains, are measured in tens of kilometers; these features have low spatial frequency, with very long wavelengths. Anticlines and synclines have an intermediate spatial frequency with wavelengths measured in kilometers. Joints, fractures, and faults are linear features with high spatial frequency and wavelengths measured in tens to hundreds of meters. High-pass filters may be used to enhance these geologic features. The direction in which the filter is passed across the image determines which linear features are enhanced; those trending normal to the filter direction are enhanced while linear features parallel with the filter direction are suppressed.

An image of the Chaco slope and north flank of the Zuni uplift, New Mexico (Figure 7.17) is used to demonstrate the application of directional high-pass filters. Outcrops are nonmarine sandstones and shales of Mesozoic age. Structural dips are gentle on the Chaco slope, which has relatively low topographic relief interrupted by incised canyons. On the north flank of the Zuni uplift, erosion of the north-dipping strata has produced strike ridges and valleys on the resistant and nonresistant strata. Interstate Highway 40 follows one of the strike valleys. Kelley and Clinton (1960) mapped the strongly developed joints and faults using conventional aerial photographs (Figure 7.17B). The principal fracture directions trend east-northeast and north-northeast; the northeast and northwest trends are less prominent.

The linear high-pass filter was passed across the digital image data in a direction normal to the east-west scan direction (Figure 7.18A) and parallel with the scan direction (Figure 7.18B). In addition to faults and joints, the filter also enhances suitably oriented topographic and cultural features. The directional nature of the filters is demonstrated by the highway and railroad that trend parallel with the scan direction in the southwest part of the image. These features are strongly enhanced by the filter oriented normal to the scan direction (Figure 7.18A) but are suppressed by the filter parallel with the scan direction (Figure 7.18B). Where the road and railroad trend southeast, at an acute angle to both filter directions, they are equally enhanced on both filtered images. Faults and joints are likewise preferentially enhanced with the east-northeast trends being prominent on Figure 7.18A and the north-northeast trends prominent on Figure 7.18B. The northeast and northwest trends are enhanced on both images, in the same fashion as the southeast-trending segment of highway and railroad.

A wide variety of filters are possible, but they must be carefully designed for the problem at hand. The filtering operation itself may produce artificial linear features that must be distinguished from real features.

Simulated Normal Color Images

The lack of a blue band on the MSS system precludes the preparation of normal color composite images from Landsat. A method for computing a synthetic image in the blue spectral region (0.4 to 0.5 μm) was described and illustrated by Eliason, Chavez, and Soderblom (1974). For each pixel the band 5 to band 6 ratio value is calculated using the method given later in this chapter in the section on Information Extraction. Each pixel is classified on the basis of its ratio value as follows: pixels with low values are classed as vegetation; those with intermediate values are classed as soils and rocks; high values are classed as water. For each of these classes an empirical spectral reflectance curve enables the computer to extrapolate a blue value for each pixel. These values constitute a synthetic blue image. The blue, green (band 4), and red (band 5) images are then registered and projected with the corresponding lights to form a *simulated normal color image.* Plate 5 shows a simulated color image prepared in this fashion with an IR color composite image of the same Landsat scene for comparison. This is the same area along the Gunnison River, Colorado shown earlier as a synthetic stereo pair (Figure 7.12) and on the map of Figure 7.13. The change from red to green for vegetation is the most conspicuous difference between the images, but there are also marked color differences in the rock outcrops. The better contrast and spatial resolution of the simulated normal color image results from digital contrast stretching. The IR color image was prepared from transparencies without the benefit of additional digital enhancement.

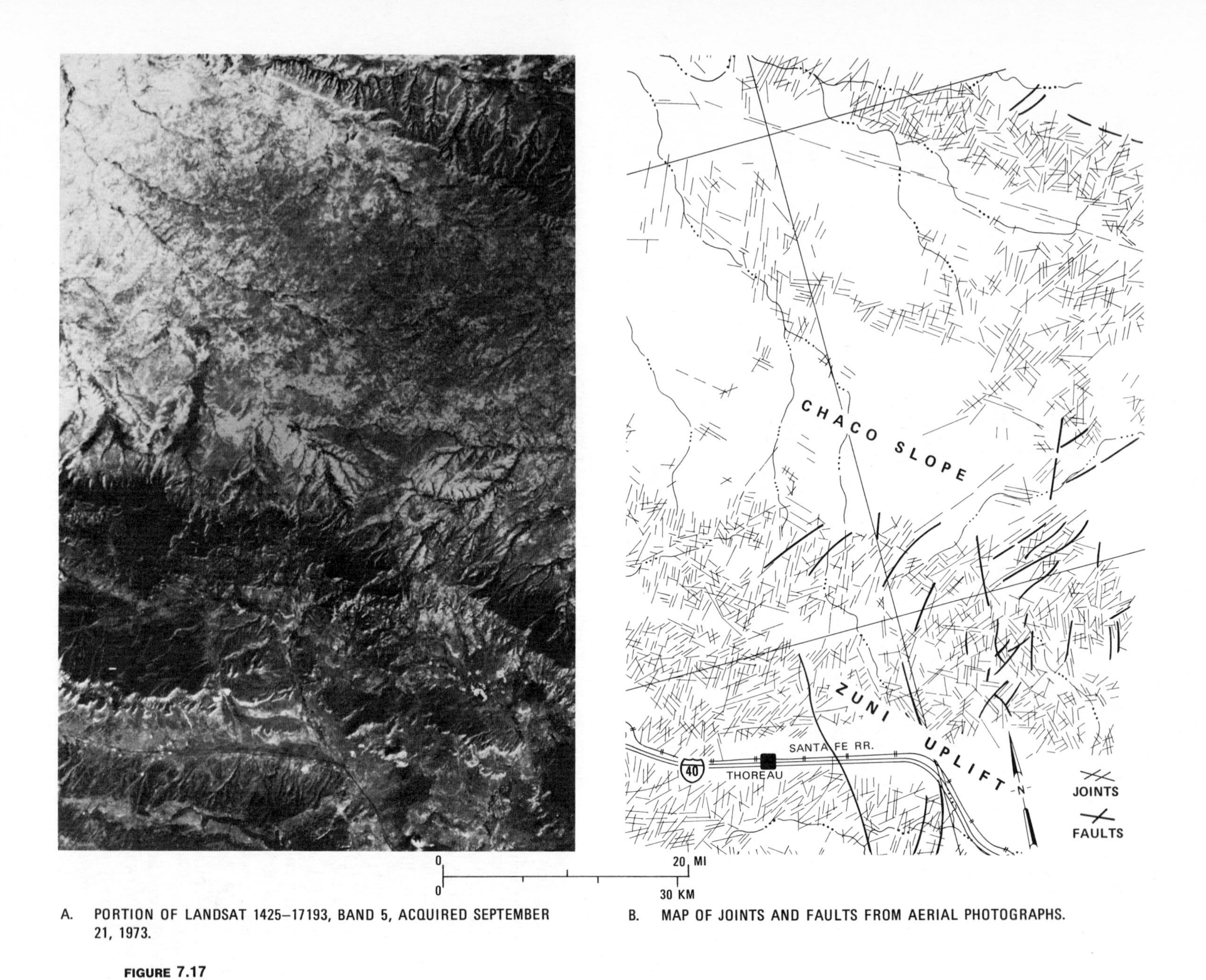

A. PORTION OF LANDSAT 1425–17193, BAND 5, ACQUIRED SEPTEMBER 21, 1973.

B. MAP OF JOINTS AND FAULTS FROM AERIAL PHOTOGRAPHS.

FIGURE 7.17
Landsat image and structure map of Chaco slope and north flank of Zuni uplift, northwest New Mexico. Image processed at NASA Goddard Space Flight Center and provided courtesy M. H. Podwysocki, NASA Goddard Space Flight Center. Image from Podwysocki, Moik, and Shoup (1975, Figure 9A). Map interpreted from aerial photographs by Kelley and Clinton (1960, Figure 2).

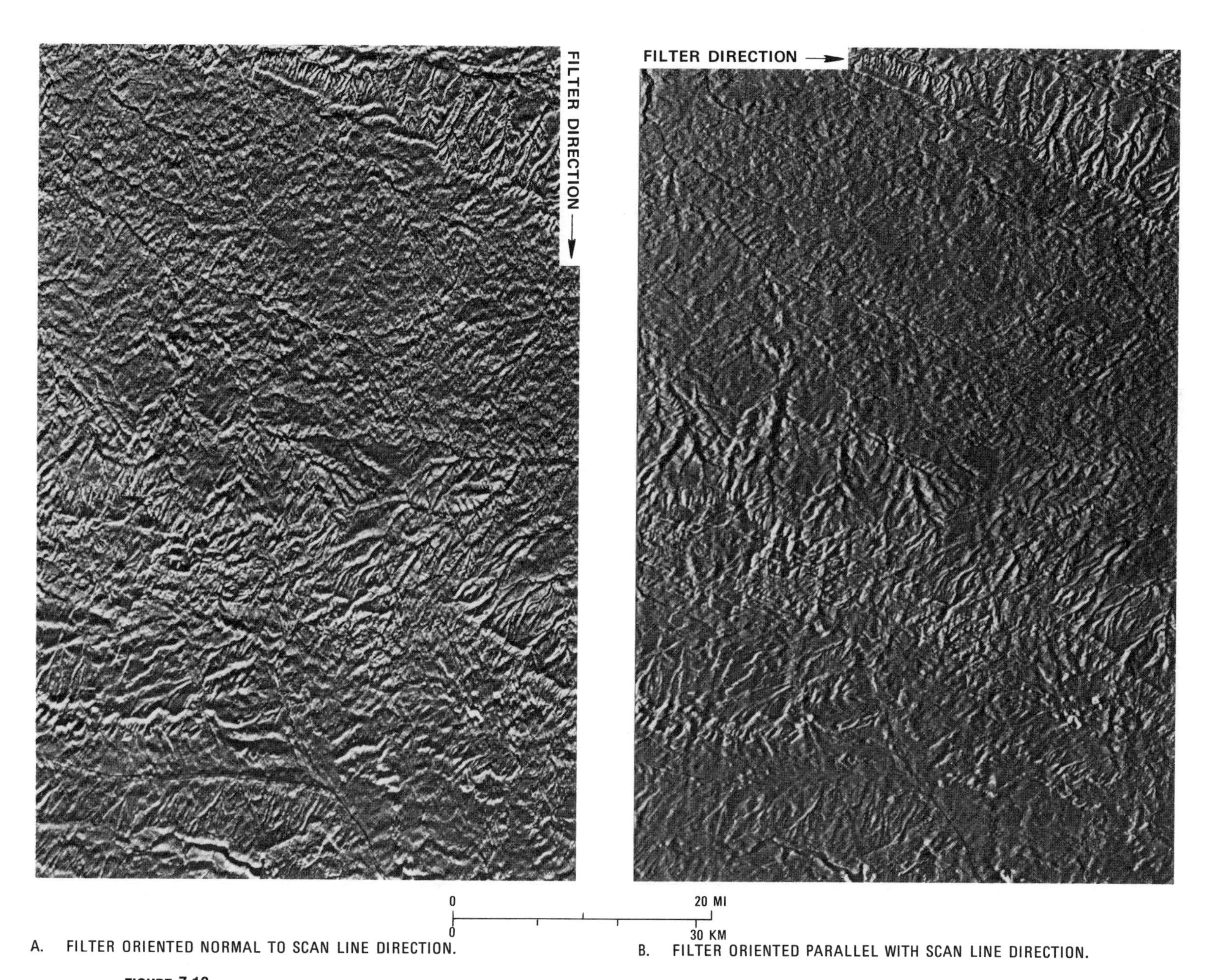

FIGURE 7.18
Landsat image processed with high-pass directional digital filters to enhance linear features. Compare with Figure 7.17. From Podwysocki, Moik, and Shoup (1975, Figures 9C and 9D). Images processed at NASA Goddard Space Flight Center and provided courtesy M. H. Podwysocki, NASA Goddard Space Flight Center.

Digital Mosaics

Conventional Landsat mosaics are prepared by matching and splicing together photographic prints of individual images, as described and illustrated in Chapter 4. Differences in contrast and tone between adjacent images cause the checkerboard pattern that is common on many mosaics. These problems can be largely eliminated by the enhancement technique for preparing mosaics digitally.

Bernstein and Ferneyhough (1975) described a technique whereby the digital data for overlapping Landsat images are merged into one file which can then be plotted on a single piece of film. Adjacent images are geometrically registered to each other by recognizing ground control points in the regions of overlap. The pixels are then geometrically adjusted to match the desired map projection. The next step is to eliminate from the digital file the duplicate pixels within the areas of overlapping images. An optimum contrast stretch is then applied to all the pixels, producing a uniform appearance throughout the mosaic. Figure 7.19A is a digital mosaic prepared in this manner from band 7 of eight images covering the portions of southeastern Montana and adjacent Wyoming, shown on the location map (Figure 7.19B). On the mosaic note the excellent geometric agreement between adjacent images. The eastern strip of imagery was acquired on July 30 and the western strip on July 31, 1973. The tonal differences between strips result from changes in cloud pattern and radiance levels on the two days. Bernstein and Ferneyhough (1975) repeated the registration procedure for bands 4 and 5 of the eight images and composited them with band 7 to produce an IR color mosaic of the area (not shown).

INFORMATION EXTRACTION

Image restoration and enhancement processes utilize computers to provide corrected and improved images for viewing by human interpreters. The computer makes no decisions in these processes. *Information extraction processes*, however, utilize the decision-making capability of computers to identify and extract specific pieces of information. The human operator must provide training data and instructions for the computer and must evaluate the significance of the extracted information. Restoration and enhancement processes operate separately on each of the four individual MSS bands. Information extraction methods, however, simultaneously process corresponding pixels from two or more bands. This processing is facilitated in Landsat data because the pixels in the four MSS bands are spatially registered to each other, having been detected and recorded by a single scanner system.

Ratio Images

Ratio images are prepared by dividing the DN in one band by the corresponding DN in another band for each pixel. The Rocky Point area in southeastern Montana is used to demonstrate this process. Linearly stretched MSS images of Rocky Point are shown in Figure 7.20, but original unstretched DNs were used to compute the ratio images.

Processing Method The four MSS bands were combined to produce the six ratio images shown in Figure 7.21, which were linearly stretched after ratioing. The six inverse ratios are not shown. On a

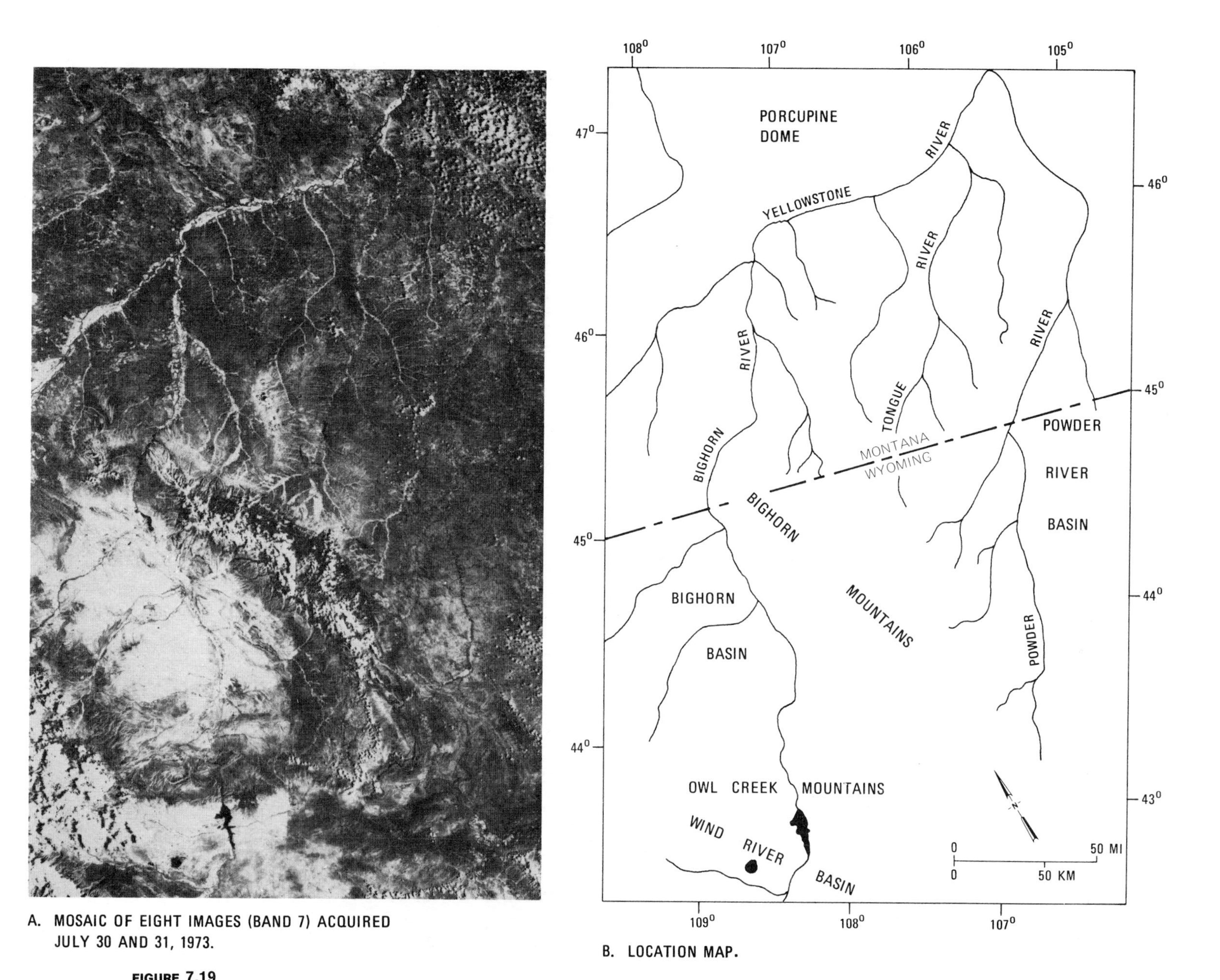

A. MOSAIC OF EIGHT IMAGES (BAND 7) ACQUIRED JULY 30 AND 31, 1973.

B. LOCATION MAP.

FIGURE 7.19
Digitally composited and plotted mosaic of Landsat images of southeast Montana and northeast Wyoming. From Bernstein and Ferneyhough (1975). Imagery processed by IBM and provided courtesy R. Bernstein, IBM.

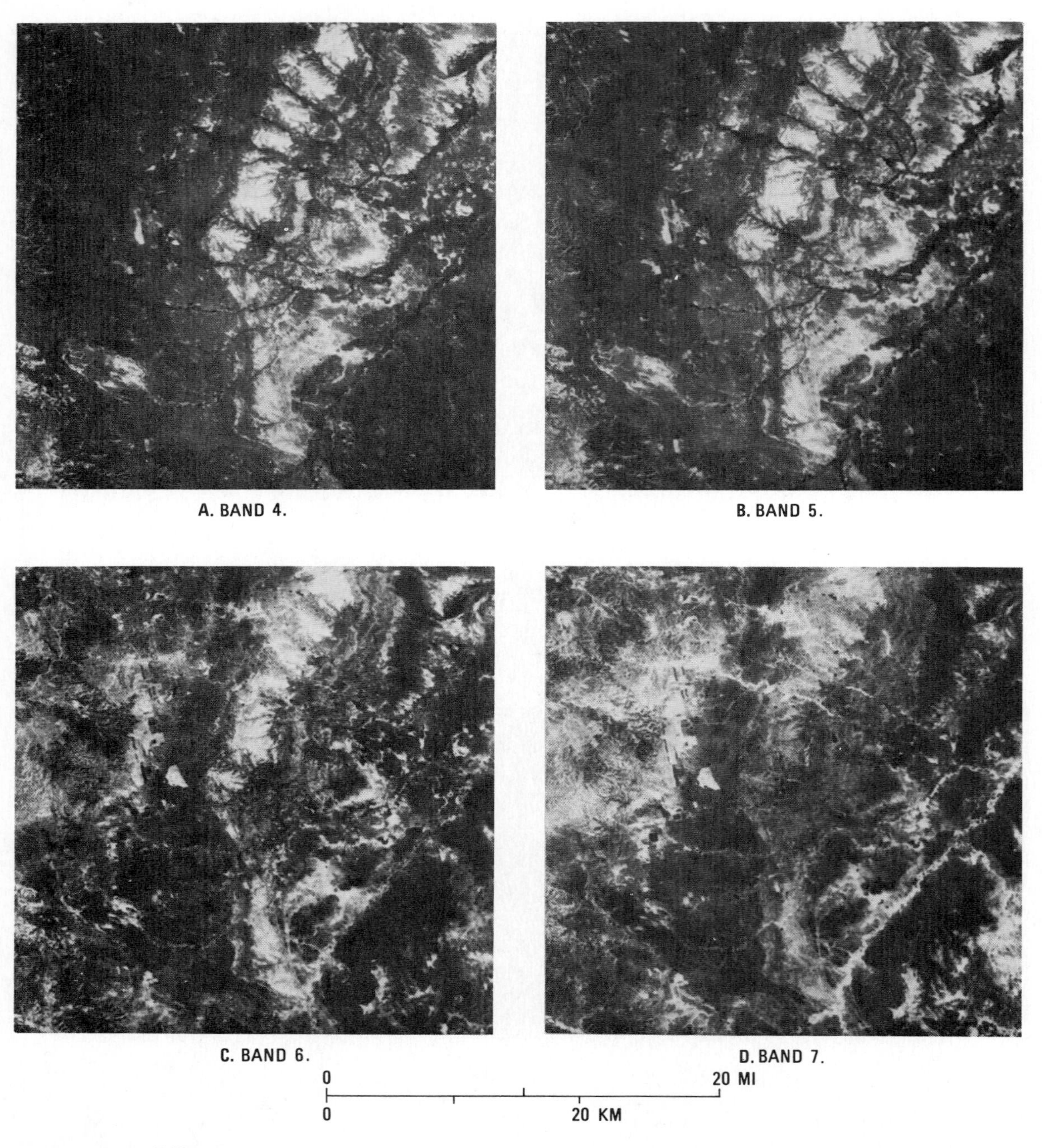

FIGURE 7.20
Contrast-stretched MSS images of Rocky Point area, Montana and Wyoming. Landsat 1353-17181 acquired July 11, 1973. Images processed at Jet Propulsion Laboratory, California Institute of Technology.

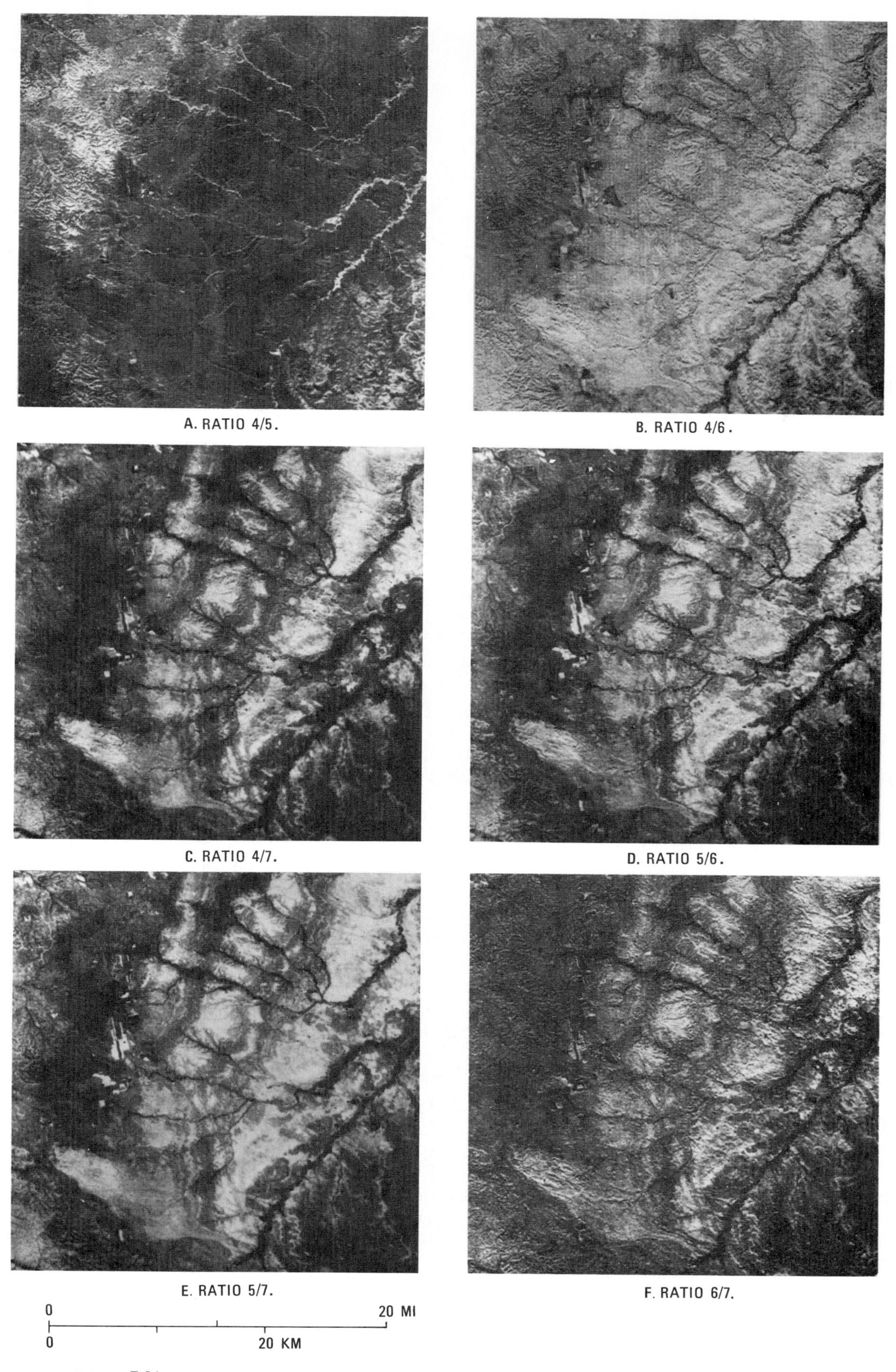

FIGURE 7.21
Contrast-stretched ratio images of Rocky Point area, Montana and Wyoming. Landsat 1353-17181. Images processed at Jet Propulsion Laboratory, California Institute of Technology.

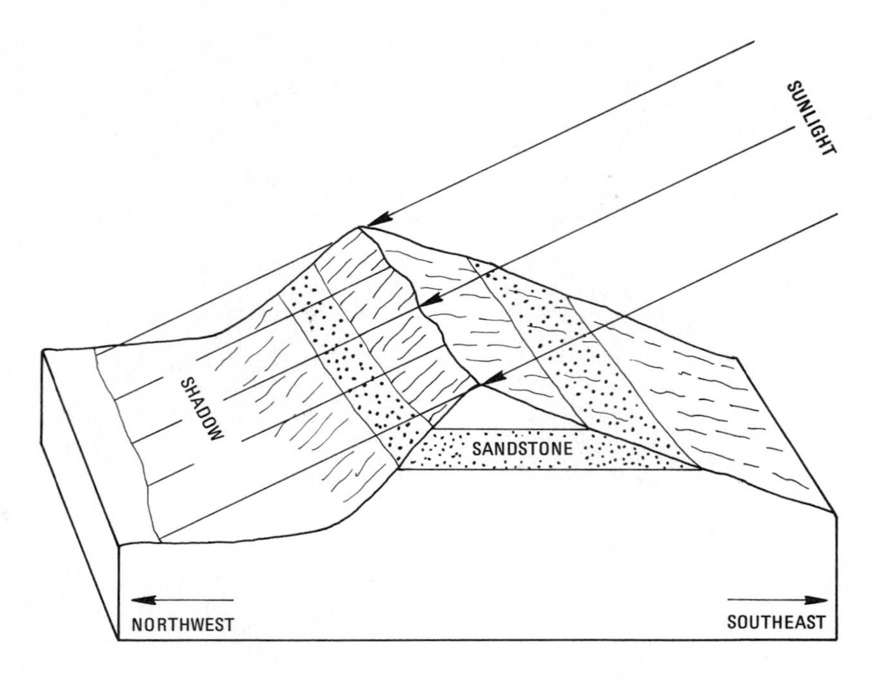

SANDSTONE REFLECTANCE

ILLUMINATION	BAND 4	BAND 5	RATIO 4/5
Sunlight	28	42	0.66
Shadow	22	34	0.65

FIGURE 7.22
Removal of illumination differences on a ratio image.

ratio image the extreme black-and-white tones of the gray scale represent the maximum difference in spectral reflectivity between the two MSS bands. The darkest tones are targets for which the denominator of the ratio is greater than the numerator. Conversely the numerator is greater than the denominator for the lightest tones. For example, the highest reflectance for vegetation occurs in the photographic IR bands 6 and 7. Thus vegetation, which occurs largely along the stream channels at Rocky Point, has a dark signature on ratios with band 6 or 7 in the denominator. Vegetation has a bright signature on the 4/5 image (Figure 7.21) because reflectance is higher for band 4 (green) than for band 5 (red).

One advantage of ratio images is that a material has the same ratio value, regardless of variations in illumination. This factor is illustrated in Figure 7.22 where a sandstone bed crops out on both the sunlit and shadowed sides of a ridge. In the individual Landsat bands the reflectance values of the sandstone are lower in the shadowed area and it would be difficult to match this outcrop with the sunlit outcrop. The ratio values, however, are nearly identical in the shadowed and sunlit areas (Figure 7.22) and the sandstone outcrops would have similar signatures on ratio images. This removal of illumination differences also eliminates the expression of topography on ratio images.

Ratio images show the variations in slopes of the spectral-reflectivity curves between the two wavelength bands. These variations are useful for distinguishing among rock types because the main spectral differences in the visible and photographic IR regions occur in the slopes of the reflectivity curves. A disadvantage of ratio images is that differences in albedo are suppressed. Thus dissimilar materials with different albedos, but similar slopes on their spectral-reflectivity curves, may be inseparable on ratio images.

Rocky Point, Montana Example Outcrops in the Rocky Point area consist of Cretaceous and Tertiary beds of the Western Interior stratigraphic sequence shown on the geologic map of Figure 7.23. The strata dip gently northwestward, reflecting the northwest plunge of the Black Hills uplift farther to the southeast. The only closed structure is the Rocky Point anticline in the southwest part of the area, which has been tested unsuccessfully by wildcat wells. The Bell Creek field in the northwest part of the area is a major oil field that produces from stratigraphic traps in the Muddy Sandstone of the Fall River Formation. Soil cover and vegetation are sparse in this semiarid region. There are concentrations of cottonwood trees along the stream courses.

To evaluate the ratio images of Rocky Point, they should be carefully compared with the geologic map (Figure 7.23) and the original MSS images (Figure 7.20). The first impression is that a remarkable amount of stratigraphic information is present on both sets of images. The upper Pierre Shale outcrops in the core of the Rocky Point anticline are more distinct on several ratio images than on the individual bands. Trends of the reflectance units on the ratio images are parallel with those of the geologic map. Close inspection shows that boundaries of the well-defined units on the ratio images do not match formations on the geologic map. One reason for this mismatch is that geologists are restricted to the visible spectral region in their field mapping and do not have access to photographic IR information. Also, other rock properties in addition to spectral reflectance are used to define geologic formations. These properties include topographic expression, bedding characteristics, and geologic age, none of which are recorded on ratio images. Several reflectance units are recognizable within the upper Pierre Shale interval on ratios 5/6 and 5/7. Field checking is required to determine the geologic significance of these and other units on the ratio images.

Other Ratio Images In addition to ratios of individual bands, a number of other ratios are possible. An individual band may be divided by the average for all the bands, resulting in normalized ratio images. Another combination is to divide the difference between two bands by their sum: $(4-5)/(4+5)$. Little has been reported to date on the applications of these ratio combinations.

Any three ratio images may be made into color composites by projecting the registered black-and-white transparencies with blue, green, and red light sources. This technique has proven valuable for recognizing mineral alteration zones at Goldfield, Nevada (see Chapter 8). The U.S. Geological Survey Image Processing Facility prepares *hybrid-color ratio composites* by projecting two ratio images in color and using the third primary color to project an individual band such as 5 or 7. The individual band restores the topographic detail that is removed from the two ratio images and makes a very effective display for the interpreter. The hybrid-color, ratio-composite image also retains some of the differences in albedo that are lost in pure ratio-composite images.

Multispectral Classification

For each pixel in a Landsat image, the spectral reflectance is recorded at four different wavelength bands. A pixel may be characterized by its spectral signature, which is determined by the relative reflectance in the different wavelength bands. *Multispectral classification* is an information extraction process that analyzes the spectral signa-

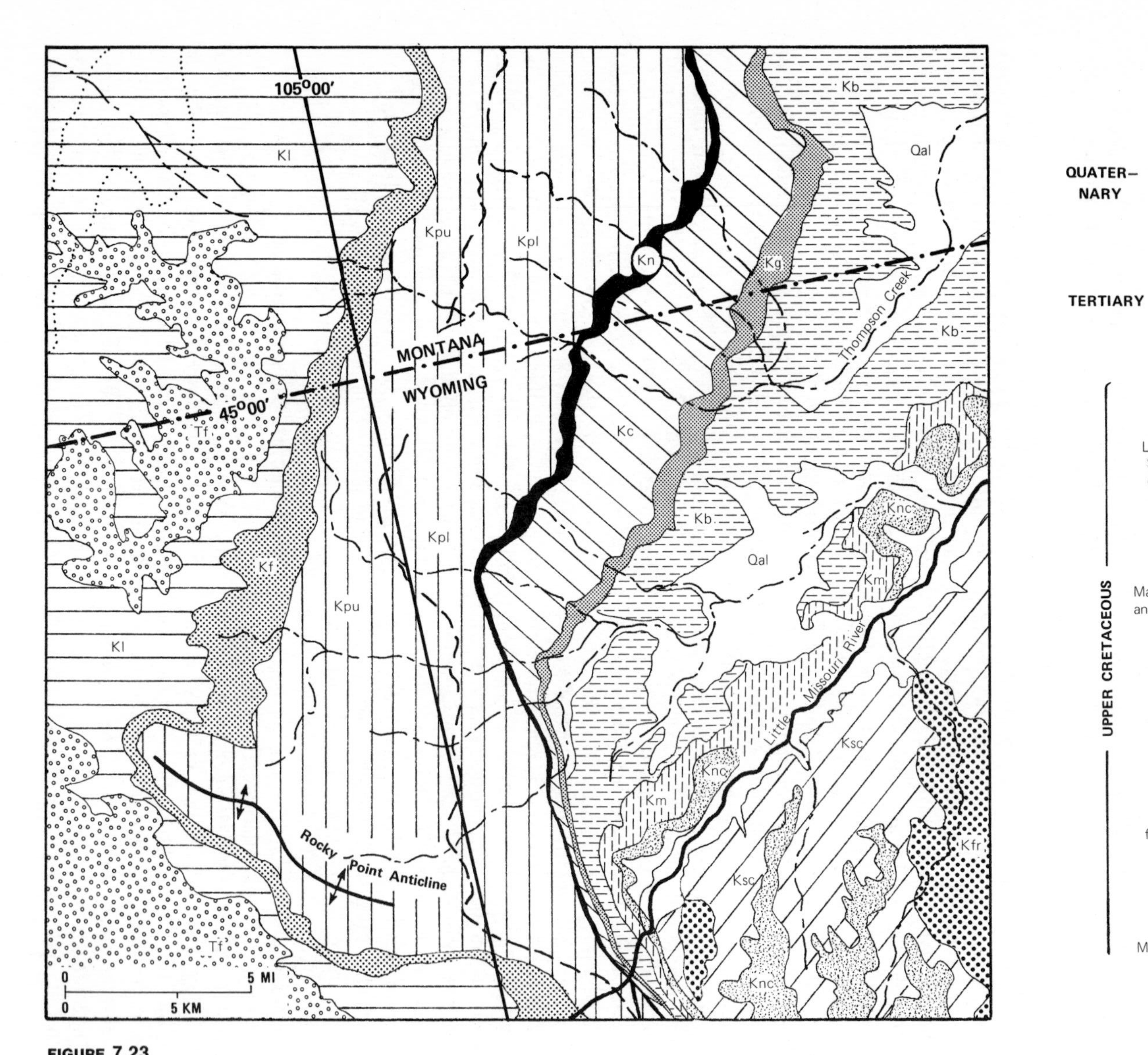

FIGURE 7.23
Geologic map of Rocky Point area, Montana and Wyoming. From Robinson, Mapel, and Bergendahl (1964, Plate 1).

tures and then assigns pixels to categories based on similar signatures. The two general types of classification schemes are supervised and unsupervised classification, which have the following distinctions:

1. *Supervised classification* uses independent information to define training data that are used to establish classification categories. The independent information may be spectral reflectance data for the classification categories. as in the following example for the Salton Sea and Imperial Valley. Another source of information is knowledge of the location of areas within the image that typify each of the desired classification categories. Such localities are known as training areas, and are illustrated by the Pakistan example in Chapter 8. Supervised classification is the most widely used of the two classification schemes.
2. *Unsupervised classification* uses only the statistical properties of the image data as a basis for classification. The computer alone defines the classification categories. This method is potentially useful for classifying images where the analyst has no independent information about the scene.

Each of these classification schemes can be further subdivided into (1) *parametric classification*, which assumes certain mathematical models; and (2) *nonparametric classification*, which assumes no model restraints. Because nonparametric classification makes the fewest assumptions, it is considered to be the more powerful technique.

A simplified example of supervised multispectral classification uses a Landsat image covering the Salton Sea, Imperial Valley, and adjacent mountains and desert areas of southern California (Figure 7.24A). Training data are the generalized reflectance spectra for water, agriculture, desert, and mountains shown in Figure 7.25A. The data points are plotted at the center of the spectral range of each of the four Landsat MSS bands. In Figure 7.25B the reflectance spectra are plotted in three-dimensional space using the values for bands 4, 5, and 6 as coordinates. The solid dots are loci of the spectra shown in Figure 7.25A. Bayesian classification methods assume that the samples belonging to each class form a *cluster* within the decision space. The clusters shown diagrammatically in Figure 7.25B represent ellipsoids whose dimensions are a function of the scatter of the samples within the cluster. The surface of the ellipsoid forms a decision boundary within which all points of that class occur.

During classification of a Landsat CCT, the computer retrieves the four spectral values for each pixel and determines the position of the pixel in the classification space. Should the pixel fall within one of the decision boundaries, or clusters, it is classified accordingly. Pixels that do not fall within a decision boundary are considered *unclassified.* In practice, the computer calculates the mathematical probability that a pixel belongs to a class; if the probability exceeds a designated threshold the pixel is assigned to that class. Applying this method to the digital data of the Salton Sea and Imperial Valley scene produces the classification map of Figure 7.24B. In this example the volume of data is reduced by averaging groups of adjacent pixels and using this value in classification. Note that the unclassified category, indicated by the blank pattern, occurs at the boundaries between classes where the pixels include more than one terrain type. The computer classification of Landsat data actually operates in four-dimensional space using all four spectral bands, but this cannot be shown on the three-dimensional plot of Figure 7.25B.

A. LANDSAT 1052–17495, BAND 5. ACQUIRED SEPTEMBER 13, 1972.

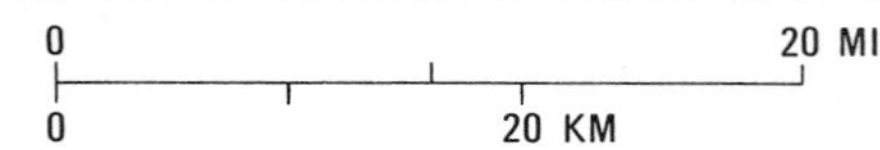


```
WWWWWWWWWWWWWW   DDDDDDDDDDDDDDDDDWDDDDDDDDDDDDD MMMMMM     M  DDDDDDDDDDDDDD
WWWWWWWWWWWWWWWW   DDDDDDDDDDDDDDDDWWDDDDDDDDDDDD MMMMMMMMMMMMMM DDDDDDDDDDDDDD
WWWWWWWWWWWWWWWWWWWW   DDDDDDDDDDDDDDWDDDDDDDDDDDD MMMMMMMMMMMMM DDDDDDDDDDDDDD
 WWWWWWWWWWWWWWWWWWWWWWWW   DDDDWW DDDDWWDDDDDDDDDD    MMMMMMMMM DDDDDDDDDDDDD
  WWWWWWWWWWWWWWWWWWWWWWWWWW  AAWW A DDDDWWDDDDDDDDDDDD  MMMMMMMMM DDDDDDDDDD
DD  WWWWWWWWWWWWWWWWWWWWWWWWW  AW AAA  DDDDWWDDDDDDDDDDDD MMMMMMMMM    DDD D M
DDDD  WWWWWWWWWWWWWWWWWWWWWWW  AAAAAAAA  DDDDWWDDDDDDDDDDD MMMMMMMMMMMM   M MM
DDDDDD  WWWWWWWWWWWWWWWWWWWWWW  AA WW AAA  DDDDWWDDDDDDDDDD MMMMMMMMMMMMMMMMMM
DDDDDDD  WWWWWWWWWWWWWWWWWWWWWWW  AAWWW AAAA  DDDDWWDDDDDDDDDD MMMMMMMMMMMMMMMMM
DDDDDDDD  WWWWWWWWWWWWWWWWWWWWWWW  AAAAAAAAAAA  DDDDWDDDDDDDDD  MMMMMMMMMMMMMMMM
DDDDDDDDD  WWWWWWWWWWWWWWWWWWWWWWWW AAAAAAAAAAAAA DDDDWDDDDDDDDD MMMMMMMMMMMMMMMM
DDDDDDDDDD  WWWWWWWWWWWWWWWWWWWWWW WWAAAAAAAAAAAAA DDDDWWDDDDDDDD   MMMMMMMMMMMMM
DDDDDDDDDDD  WWWWWWWWWWWWWWWWWWWWWW AAWWAAAAAAAAAAAA DDDDDWWDDDDDDD MMMMMMMMMMMM
DDDDDDDDDDDD WWWWWWWWWWWWWWWWWWWW AAAAAWAAAAAAAAAAAA  DDDDDWDDDDDD MMMMMMMMMMMM
DDDDDDDDDDDD  WWWWWWWWWWWWWWWW  AAAAAAWAAAAAAAAAAAAAA DDDDDWDDDDDD MMMMMMMMMMM
DDDDDDDDDDDDD     WWWWWWWWWWW  AAAAAAAWAAAAAAAAAAAAAAA  DDDDWDDDDDDD MM  M M MMM
DDDDDDDDDDDD AAAAA  WWWWWWWW AAAAAAAAWWAAAAAAAAAAAAAA DDDDWDDDDDDD  DD    MMM
DDDDDDDDDDDD AAAAAA  WWWWWW  AAAAAAAAAAWAAAAAAAAAAAAAAAA  DDDWWWDDDDDDDDDD M
DDDDDDDDDDDDDD AAAAAAAA  WWW WWWAAAAAAAAAAAAWWWWAAAAAAAAAAAAA DDDDDDWDDDDDDDDDD  D
DDDDDDDDDDDDDDD     AAAAAAAAAAAWWWWWWWWWWAAWWWAAAAAAAAAAAA DDDDDDDWDDDDDDDDDDDD
DDDDDDDDDDDDDDDDDDDD  AAAAAAAAAAAAAAAAAAWAAWWAAAAAAAAAAAAAAA DDDDDDDWDDDDDDDDDD
DDDDDDDDDDDDDDDDDDDDD  AAAAAAAAAAAAAAAAAAWAAWWWAAAAAAAAAAAAAA DDDDDDDDWDDDDDDDD
DDDDDDDDDDDDDDDDDDDDDD  AAAAAAAAAAAAAAAAAAWAAAAWAAAAAAAAAAAAA DDDDDDDDWWDDDDDDD
DDDDDDDDDDDDDDDDDDDDDDD  AAAAAAAAAAAAAAAAAAAWAAAWAAAAAAAAAAAAAAA DDDDDDDDDWDDDDDD
DDDDDDDDDDDDDDDDDDDDDDDDD AAAAAAAAAAAAAAAAAWAAAAAWAAAAAAAAAAAAAAA DDDDDDDDDDWDDDDD
 DDDDDDDDDDDDDDDDDDDDDDDD  AAAAAAAAAAAAAWWA  AAAWAAAAAAAAAAAAAAAA DDDDDDDDDDWWDDD
M DDDDDDDDDDDDDDDDDDDDDDDD  AAAAAAAAAAAAW   AAAAWAAAAAAAAAAAAAA  DDDDDDDDDDDDWDD
MMM DDDDDDD      CDDDDDDDDDD AAAAAAAAAAAWA   AAAAWAAAAAAAAAAAAAAAA DDDDDDDDDDDDDWD
MMMM DDDDD MMMMMM  DDDDDDDDD  AAAAAAAAWAAAAAAAAAAAAWAAAAAAAAAAAAAA DDDDDDDDDDDDDW
MMMMM DDDD MMMMMMMMM  DDDDDDD  AAAAAAWWAAAAAAAAAAAAAAWAAAAAAAAAAAAA DDDDDDDDDDDDDD
MMMM DDDDDDD   MMMMMM DDDDDDDD  AAAAAWAAAAAAAAAAAAAAAWAAAAAAAAAAAAAA  DDDDDDDDDDDDD
MMM DDDDDDDDDDDD MM  DDDDDDDDDD  AAAAWAAAAAAAAAAAAAAAWAAAAAAAAAAAAAA DDDDDDDDDDDDD
M  DDDDDDDDDDDDDDDDDDDDDDDDD  A  AAAWAAAAAAAAAAAAAAAWAAAAAAAAAAAAAA DDDDDDDDDDDDD
 DDDDDDDDDDDDDDDDDDDDDDDDDDDDD AAAAAAAWAAAAAAAAAAAAAAAAWAAAAAAAAAAAAA  DDDDDDDDDDDD
M   DDDDDDDDDDDDDDDDDDDDDDD   AAAAAAWWAAAAAAAAAAAAAAAAAAWAAAAAAAAAAAAA  DDDDDDDDDDDD
MMMM  DDDDDDDDDDDDDDDDDD   AAAAAWWAAAAAAAAAAAAAAAAAAAAAWAAAAAAAAAAAAA DDDDDDDDDDDD
MMMMMM DDDDDDDDDDDDDDDD  AAAAAAWAAAAAAAAAAAAAAAAAAAAAAAAAWAAAAAAAAAAA DDDDDDDDDDDD
MMMMMMM  DDDDDDDDDDDDD AAAAAAWAAAAAAAAAAAAAAAAAAAAAAAAAAAAWAAAAAAAAAAA  DDDDDDDDDDD
MMMMMMMM   DDDDDDDDDD  AAAAAAWAAAAAAAAAAAAAAAAAAAAAAAAAAAAAAWAAAAAAAAAAA  DDDDDDDDDD
MMMMMMMMMMM DDDDDDDDD AAAAAAAWAAAAAAAAAAAAAAAAAAAAAAAAAAAAAAWAAAAAAAAAAA  DDDDDDDDDD
```

B. CLASSIFICATION USING THE CLUSTER DIAGRAM OF FIGURE 7.25B.
A=AGRICULTURE, D=DESERT, M=MOUNTAINS, W=WATER, BLANK=UNCLASSIFIED.

FIGURE 7.24
Digital classification of Landst CCT data for Salton Sea and Imperial Valley, California.

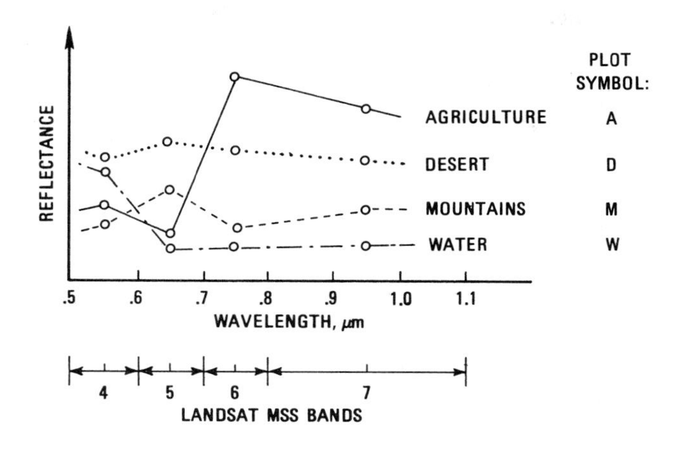

A. SPECTRAL REFLECTANCE CURVES FOR MAJOR TERRAIN TYPES.

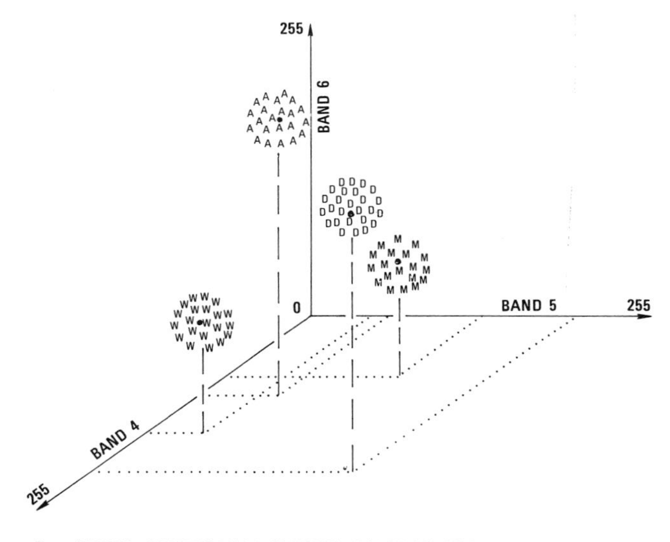

B. THREE–DIMENSIONAL CLUSTER DIAGRAM FOR CLASSIFICATION.

FIGURE 7.25
Spectral reflectance curves and cluster diagram for Landsat CCT data of Salton Sea and Imperial Valley.

For demonstration purposes the classification categories mapped in Figure 7.24B are very broad. For detailed work, the individual crop types in the Imperial Valley can be classified by their spectral properties. The different types of rock and detrital material in the mountains and desert can be separated in similar fashion. The example from Pakistan in Chapter 8 illustrates the use of more restricted classification categories to recognize potential copper deposits.

Change-Detection Images

Temporal information about an area may be extracted by comparing two or more images of the area that were acquired at different times. The first step is geographic registration of the images. Corresponding ground control points are recognized on the different images and registered, using the techniques described in the discussion of geometric corrections. The intervening pixels between GCPs are then redistributed to complete the registration process. A precision of one-half pixel is claimed for some registration techniques. Repeated Landsat images of the same ground area must be registered in this manner because of differences in spacecraft position, altitude, and attitude. After two images have been geometrically registered, the DNs of one image may be subtracted from those of the corresponding band of an image acquired earlier or later. The resulting values for each pixel are positive, negative, or zero; the latter indicates no change. These values are then plotted as a *change-detection image* with a neutral gray tone representing zero and saturated black and white representing the maximum negative and positive differences, respectively. Contrast stretching is employed to emphasize the differences. Because of the subtractive process, change detection images are also referred to as *difference images.*

The change-detection process is illustrated in Figure 7.26, where image B is subtracted from image A, which was acquired 72 days later. On the resulting change-detection image (Figure 7.26C), the neutral gray tones representing areas of little change are concentrated in the upper-left and lower-right regions and correspond to forested terrain. Note that this terrain has a similar dark signature on both original images. Some patches within ephemeral Goose Lake have similar light signatures on image A and image B, resulting in a neutral gray tone on the difference image. The clouds and shadows on image B produce dark and light tones, respectively, on the difference image. The agricultural practice of seasonally alternating between cultivated and fallow fields is clearly shown by the light and dark tones on the difference image. On the original images, the fields with light tones have crops or stubble and the fields with dark tones are bare earth.

Change-detection processing may also be applied to images other than Landsat. Thermal IR images acquired in the day and night may be registered and a change-detection image prepared that shows temperature differences.

STRATEGY AND HARDWARE FOR IMAGE PROCESSING

A few comments on implementation methods and equipment are an appropriate conclusion to this discussion of image processing.

Batch and Interactive Methods of Processing

The two basic methods of image processing, and data processing generally, are batch processing and interactive processing. In *batch processing* of a Landsat image, the CCTs are read into computer memory and the desired processing programs are specified. The computer implements the programs to process the image and produce a tape of the new data which is then plotted as an image on film or line printer display. The operator has no control over the processing and sees no results

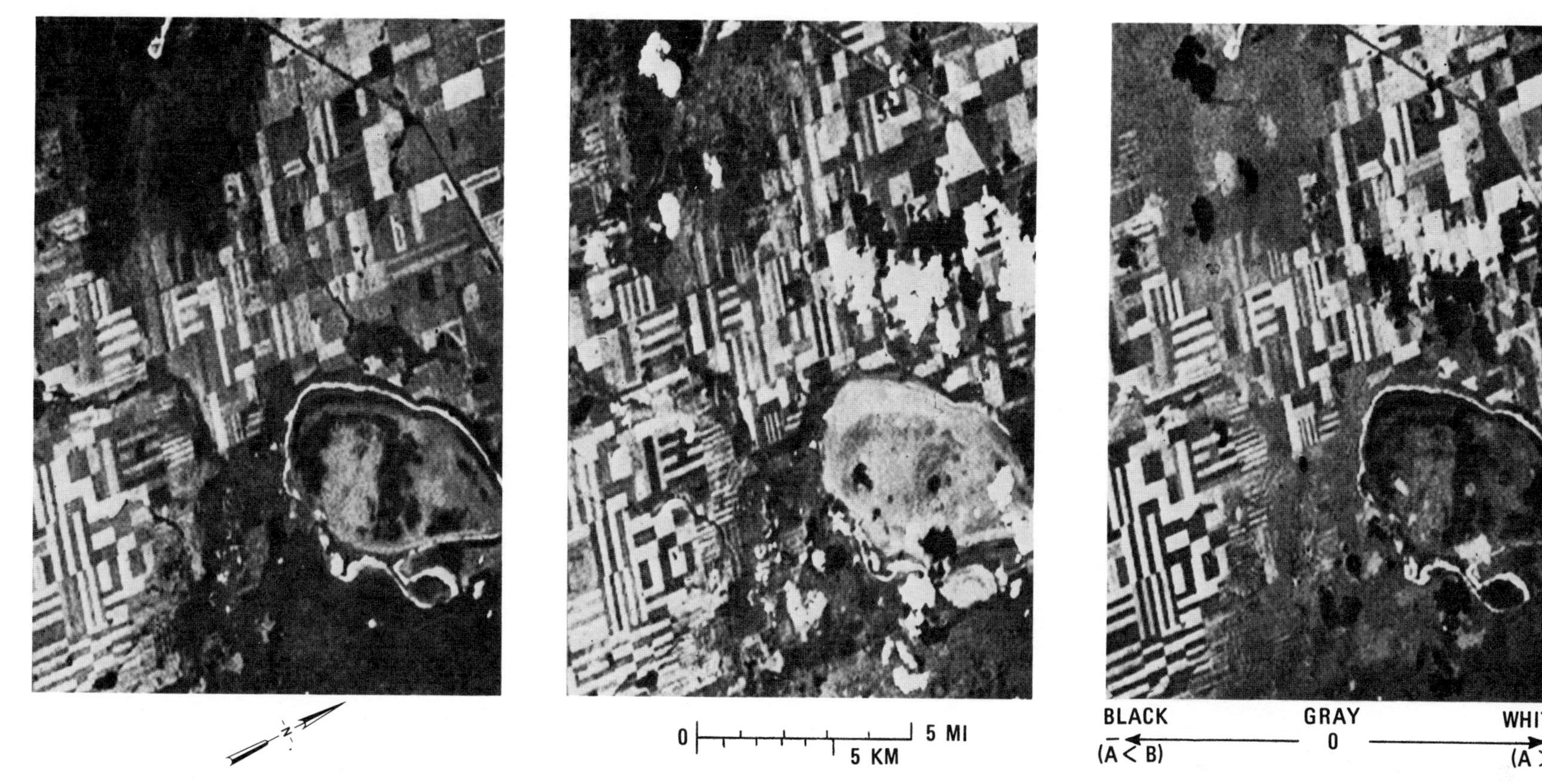

FIGURE 7.26
Procedure for producing a change-detection image. Goose Lake area, 64 km southwest of Saskatoon, Saskatchewan, Canada. From Rifman and others (1975, Figures 2-14, 15, and 17). Images processed at TRW, Inc. and provided courtesy S. S. Rifman, TRW.

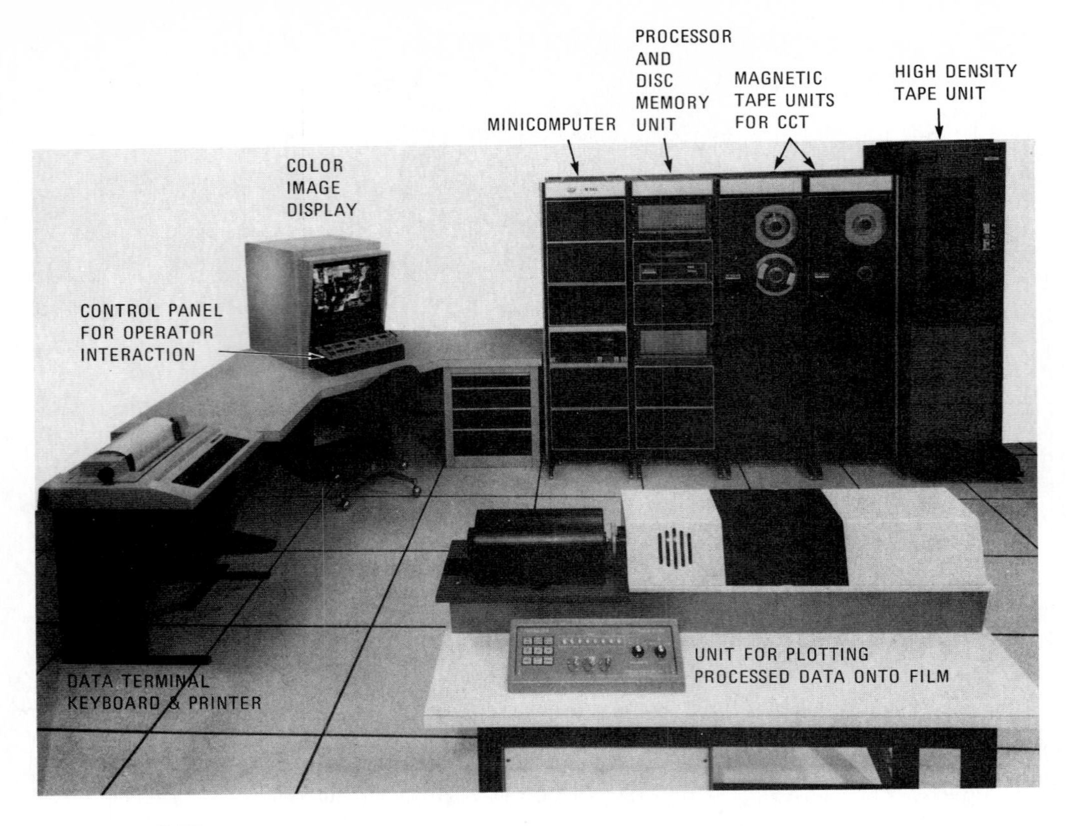

FIGURE 7.27
Typical interactive digital image processing system using a special purpose minicomputer. Courtesy J. B. McKeon, Bendix Aerospace Systems Division.

until the processing is completed. Batch methods are useful for routine image restoration and enhancement processes where an operator is not needed.

A typical *interactive image processing system* is shown in Figure 7.27. The Landsat CCTs are loaded in one of the magnetic tape units. The operator uses the data terminal keyboard to initiate processing which begins by transferring image data from the CCTs to the disc memory unit. The image data are stored in the disc memory unit in a format that is readily accessible to the minicomputer that carries out the computations. Once the data are stored on disc, the operator is notified by a message typed on the printer. The printer then displays a list of program options from which the operator selects one by typing the appropriate message.

Typically the next step is to display the unprocessed data on the color image display (Figure 7.27), which is a version of a color television screen. The blue, green, and red color guns of the display unit are used to display MSS bands 4, 5, and 7 respectively to produce a color composite image on the screen. The screen provides 512 horizontal lines, each of which consists of 512 pixels. This is referred to as a *raster pattern*. Because the pixels of the raster pattern are square, the 57 by 79 m Landsat pixels must be resampled to produce square pixels, typically 79 by 79 m. In order to display an entire Landsat scene, which consists of 2340 scan lines, every fifth scan line and every fifth resampled pixel along the scan line may be shown on the screen. This process of reducing the number of data points is called decimation, although in this case only four percent (one-twenty fifth) of the pixels are displayed. At the small scale of the 30 by 30 cm screen, the decimated image display is remarkably detailed. With an image displayed on the screen, the operator now moves to the control panel of the image display to begin an interactive process, such as classification. The operator interacts with the image display through a keyboard and a track ball or "joy stick" that controls a cursor on the screen. The cursor is a distinctive bright spot on the screen that may be positioned with the track ball control to designate specific pixels in the display. The cursor may also be used to outline specific areas within the image display.

To perform a multispectral classification, the cursor is used to outline a 40 by 40 km subarea on the decimated display of the full scene. The subarea is then displayed at full resolution, with each resampled Landsat pixel of the subarea represented by a pixel on the screen. In supervised classification the operator next uses the cursor to outline training sites for which the identity is known. An agronomist might designate areas of different crop types; a geologist might designate outcrops of different rock types; a geographer might designate different types of land use. The computer uses the data from the training sites to calculate the cluster, or decision space, for each class. Each pixel of the subarea is then assigned to the class to which it most closely corresponds. Each class is then assigned a color and displayed on the screen. The operator may interact additionally with the classification process in the following manner. The statistics for each training site may be displayed on the screen, either as a table or as histograms. Using the keyboard or the cursor, the operator modifies the statistics of the training site to improve the accuracy of the classification. When the classification is complete, the processed data are recorded on a computer tape. An image plotting system, such as the one shown in the foreground of Figure 7.27, is used to produce a film copy of the results. Once the operator is satisfied with the classification results of the subarea, the system may then be instructed to classify the entire Landsat scene, without further operator interaction. The interactive processing feature is essential when the operator must provide instructions and information to the computer system at various stages in the processing cycle. The real-time image display capability enables the operator to view the results of his instructions, which can then be modified to achieve the desired classification.

The interactive mode has obvious advantages over the batch mode where the operator must wait until the image is plotted to view results and modify the processing programs.

General Purpose and Special Purpose Computers

General purpose computers are controlled by programs (software) and can perform a wide variety of data processing tasks. General purpose computers are available in a wide range of capacities from giant "number crunchers" costing millions of dollars to "minicomputers", such as the model shown in Figure 7.27, that cost less than fifty thousand dollars. On general purpose computers, new programs may be added and existing programs modified, providing maximum flexibility. The Computer Software Management and Information Center (COSMIC) at the University of Georgia, Athens, Georgia was established in 1966 under contract to NASA. COSMIC is responsible for making available to the American public computer programs developed by NASA and other organizations under U.S. government sponsorship. Table 7.1 lists and describes seven software systems available from COSMIC that are designed for processing Landsat and other image data. Cost of the programs depends upon their size and complexity. For example, SMIPS is a large and complex system that costs approximately $2,200, which is only a fraction of the cost to duplicate this programming effort. Note that programs are written for specific computer systems; rewriting and modifications are generally required to implement a program on a different system. Additional details on the programs in Table 7.1 may be obtained from COSMIC.

Special purpose computers are designed to perform specific tasks that can be executed rapidly because software commands are not required. A

TABLE 7.1
Software for image processing available through COSMIC

Program	Implemented on	Summary
ASTEP—Algorithm Simulation Test and Evaluation Program	UNIVAC 1100, EXEC 8, batch or interactive mode.	Determines the statistical properties of multispectral data. Evaluates performance of various classification techniques.
DAM—Detection and Location of Surface Water	UNIVAC 1100, EXEC 8, batch or interactive mode.	Identifies and maps surface water using Landsat CCT data.
ELLTAB—Elliptical Table Lookup Algorithm	UNIVAC 1108, EXEC 8, batch or interactive mode.	Classifies multispectral data using an advanced table lookup approach.
LACIE—Large Area Crop Inventory Experiment System	IBM 360 series, batch or interactive mode.	Performs multispectral image analysis and image registration for identifying crop types and estimating yield.
LARSYS—Laboratory for Applied Remote Sensing System, Purdue University	IBM 360 system, batch or interactive mode.	Performs supervised multispectral classification using data from training sites, and evaluates the results.
QUIKLOOK—Landsat	IBM 360/91, batch mode.	Image-enhancement program for quickly performing approximate radiometric and geometric corrections of Landsat data.
SMIPS—Small Interactive Image Processing System	IBM 360/370 series with interactive display on IBM 2250 device.	Modification of VICAR with additional routines for fast access and display of image data. Designed for use by personnel who are not expert programmers.
VICAR—Video Image Communication and Retrieval	IBM 360/44, batch mode.	A very extensive system for image restoration, enhancement, and information extraction.

Source: From McRae (1976).

special purpose computer may be preprogrammed, or "hard wired", to perform various image-processing functions such as ratios, differences, and classification. These preprogrammed functions cannot be modified easily. CCTs for an image are read into the computer memory and the operator activates a switch for the desired function. Special-purpose computers are commonly integrated into complete interactive image-processing systems that include tape-reading devices, memory, display units, and control panels and terminals (Figure 7.27). Representative commercially available interactive systems are listed in Table 7.2 together with the names and addresses of the manufacturers. These systems cost up to several hundred thousand dollars with the exact price depending upon the choice of peripheral equipment. The NASA Earth Resources Laboratory (ESL) at Bay Saint Louis, Mississippi recognized that many potential users can benefit from digital processing of Landsat data, but cannot afford the expensive systems. Many of these users have access to components such as small general-purpose computers, tape drive units, display terminals, and computer line printers. Whitley (1975) described several low-cost systems for image processing that utilize common components that may be available at a user's facility. Whitley also described the software that has been written for such systems and is available through COSMIC.

TABLE 7.2
Commercial interactive image-processing systems

Processing system	Manufacturer	Address
MDAS	Bendix Aerospace Systems Division	3621 South State Road, Ann Arbor, Mi. 48107
Series 9	Comtal Corp.	169 North Halstead, Pasadena, Ca. 91107
IDIMS	ESL Inc.	495 Java Drive, Sunnyvale, Ca. 94086
IMAGE 100	General Electric Co., Space Division	P. O. Box 2500, Daytona Beach, Fla. 32015
System 101	Stanford Technology Corp.	650 N. Mary Avenue, Sunnyvale, Ca. 94086

COMMENTS

Digital image processing has been demonstrated here using examples of Landsat images which are available in digital form. It should be emphasized, however, that any image can be converted into digital format and processed in similar fashion. The three major functional categories of image processing are:

1. Image restoration—to compensate for data errors, noise, and geometric distortions introduced during the scanning, recording, and playback operations.
2. Image enhancement—to alter the visual impact of the image on the interpreter, in a fashion that improves the information content.
3. Information extraction—to utilize the decision-making capability of the computer to recognize and classify pixels on the basis of their digital signatures.

In all of these operations the user should be aware of the trade offs involved, as demonstrated in the discussion of contrast stretching.

A common question is whether the benefits of image processing are commensurate with the cost. This is a difficult question that can be answered only in the context of the individual user. If digital filtering, for example, reveals a previously unrecognized fracture system that leads to the discovery of major ore deposits, the cost benefits are obvious. On the other hand, it is difficult to state the cost benefits of improving the accuracy of geologic and other maps through digital processing of remote-sensor data. It also should be noted that technical advances in software and hardware are steadily increasing the volume and complexity of processing that can be performed, often at a reduced unit cost.

In 1977 EDC first offered for public sale digitally enhanced Landsat images as a special order product. EDC applies the following image restoration procedures to the CCT data: remove line dropouts, correct sixth-line banding, perform geometric corrections. The following image enhancement procedures are employed: contrast enhancement, edge enhancement. The MSS bands are processed individually and made into color composite transparencies from which enlarged color reproductions are made. The digitally processed images are a major improvement over the standard EDC color products. EDC does not provide digital information extraction procedures, such as ratio images and multispectral classification. In late 1977 EDC charged $1000 for digitally processing a Landsat scene, plus a charge for reproductions that was triple the rate given in Table 4.2. These charges are subject to change and should be confirmed with EDC before an order is placed. In addition to EDC, several commercial firms offer digital processing services for Landsat data, including information extraction.

REFERENCES

Andrews, H. C. and B. R. Hunt, 1977, Digital image restoration: Prentice-Hall, Englewood Cliffs, N.J.

Batson, R. M., K. Edwards, and E. M. Eliason, 1976, Synthetic stereo and Landsat pictures: Photogrammetric Engineering, v. 42, p. 1279–1284.

Berlin, G. L., P. S. Chavez, T. E. Grow, and L. A. Soderblom, 1976, Preliminary geologic analysis of southwest Jordan from computer enhanced Landsat 1 image data: American Society of Photogrammetry, Proceedings of Annual Meeting, p. 545–563, Washington, D.C.

Bernstein, R. and D. G. Ferneyhough, 1975, Digital image processing: Photogrammetric Engineering, v. 41, p. 1465–1476.

Bryant, M., 1974, Digital image processing: Optronics International Inc., Publication No. 146, Chelmsford, Mass.

Chavez, P. S., 1975, Atmospheric, solar, and MTF corrections for ERTS digital imagery: American Society of Photogrammetry, Proceedings of Phoenix meeting.

Eliason, E. M., P. S. Chavez, and L. A. Soderblom, 1974, Simulated "true color" images from ERTS data: Geology, v. 2, p. 231–234.

Goetz, A. F. H. and others, 1975, Application of ERTS images and image processing to regional geologic problems and geologic mapping in northern Arizona: Jet Propulsion Laboratory Technical Report 32–1597.

Kelly, V. C. and N. J. Clinton, 1960, Fracture systems and tectonic elements of Colorado Plateau: University of New Mexico Publications in Geology, no. 6.

Landgrebe, D. A. and others, 1968, LARSYAA, a processing system for airborne earth resources data: Purdue University, Laboratory for Applications of Remote Sensing, Information Note 091968.

McRae, W. B., 1976, Image processing software available through COSMIC: Proceedings Caltech/JPL Conference on Image Processing Technology, Data Sources, and Software for Commercial and Scientific Applications. Jet Propulsion Laboratory Report SP 43-30, p. 17-1 to 17-4.

NASA, 1976, Landsat data users handbook: Goddard Space Flight Center, Document no. 76SDS-4258, Greenbelt, Md.

Podwysocki, M. H., J. G. Moik, and W. C. Shoup, 1975, Quantification of geologic lineaments by manual and machine processing techniques: NASA Earth Resources Survey Symposium, NASA TM X-58168, v. 1, p. 885–903.

Pratt, W. K., 1977, Digital image processing: John Wiley and Sons, New York.

Rifman, S. S., 1973, Digital rectification of ERTS multispectral imagery: Symposium on Significant Results Obtained from ERTS-1, NASA SP-327, p. 1131–1142.

Rifman, S. S. and others, 1975, Experimental study of application of digital image processing techniques to Landsat data: TRW Systems Group Report 26232-6004-TU-00 for NASA Goddard Space Flight Center, Greenbelt, Md.

Robinson, C. S., W. J. Mapel, and M. H. Bergendahl, 1964, Stratigraphy and structure of the northern and western flanks of the Black Hills uplift, Wyoming, Montana, and South Dakota: U.S. Geological Survey Professional Paper 404.

Rosenfeld, A. and A. C. Kak, 1976, Picture processing by computer: Academic Press, New York.

Sabins, F. F., 1974, Oil exploration needs for digital processing of imagery: Photogrammetric Engineering, v. 40, p. 1197–1200.

Soha, J. M., A. R. Gillespie, M. J. Abrams, and D. P. Madura, 1976, Computer techniques for geological applications: Proceedings Caltech/JPL Conference on Image Processing Technology, Data Sources and Software for Commercial and Scientific Applications, Jet Propulsion Lab SP 43-30, p. 4-1 to 4-21.

Thomas, V. L., 1975, Generation and physical characteristics of the Landsat 1 and 2 MSS computer compatible tapes: NASA Goddard Space Flight Center Document X-563-75-233, Greenbelt, Md.

Whitley, S. L., 1975, Low-cost data analysis systems for processing multispectral scanner data: NASA Earth Resources Survey Symposium, NASA TM X-58168, p. 1355–1375.

ADDITIONAL READING

Anuta, P. E., 1977, Computer assisted analysis techniques for remote sensing data interpretation: Geophysics, v. 42, p. 468–481.

Bernstein, R., 1976, Digital image processing of earth observation sensor data: IBM Journal of Research and Development, v. 20, p. 40–57.

8
RESOURCE EXPLORATION

Present and impending shortages of mineral and energy resources are well known and will not be reiterated here. Romote-sensing methods have great promise as reconnaissance and detailed exploration techniques for these resources. Several examples are given here as guides for future applications.

MINERAL EXPLORATION

Landsat images have proven valuable for mineral exploration in three ways:

1. Mapping of regional and local fracture systems that controlled ore deposits.
2. Detection of surface alteration effects associated with ore deposits.
3. Providing basic data for geologic mapping.

Mapping Lineaments and Fracture Systems

Prospectors and mining geologists have long realized that in many mineral provinces. mining districts occur along linear trends scores to hundreds of kilometers in length. These are referred to as *mineralized belts* or zones and many mines have been found by exploring along the projections of such trends. Within the mineralized belts, individual mining districts are commonly localized by intersecting fracture systems. Landsat imagery is useful for mapping both the regional lineaments and local fractures.

Regional Lineaments of Nevada In Chapter 4 the seven major lineaments on the Landsat mosaic of Nevada were mapped, and are shown here in Figure 8.1A. These lineaments are the traces of faults and zones of weakness that have been periodically reactivated. The mining districts of Nevada, ranked by dollar value of production, are plotted in Figure 8.1B. One linear belt of mining districts coincides with the northeast-trending Midas Trench lineament. The districts in the southwestern part of the state occur in a broad belt parallel with the northwest-trending Walker Lane lineament.

To aid in evaluating the relationship of mining districts to lineaments, Rowan and Wetlaufer (1975) gridded the mine map, counted the number of districts in each grid square, and contoured this value (Figure 8.1C). The concentration of districts along the Midas Trench lineament is emphasized by these contours. The high concentration is inter-

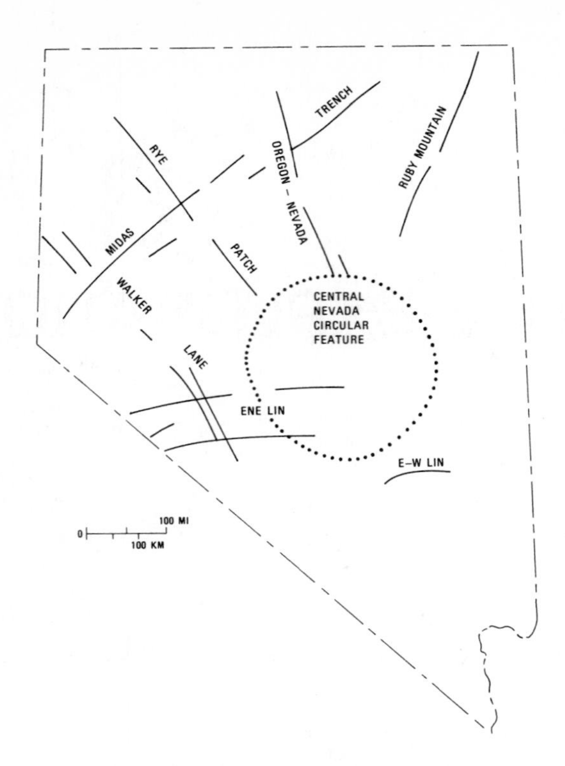

A. MAJOR LINEAMENTS INTERPRETED FROM LANDSAT MOSAIC.

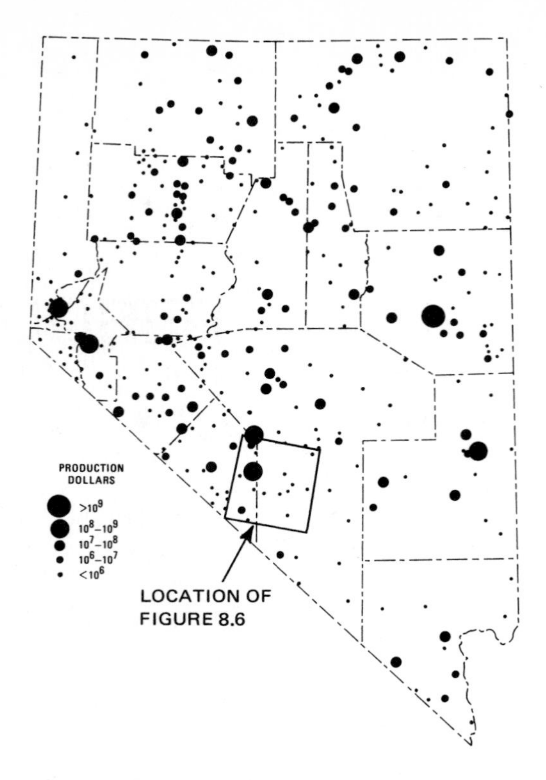

B. METAL MINING DISTRICTS OF NEVADA.

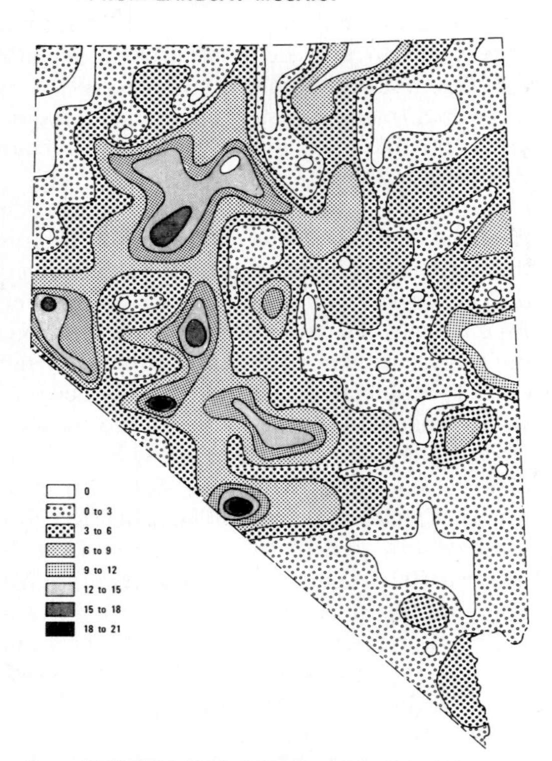

C. CONTOUR MAP OF THE DISTRIBUTION OF METAL MINING DISTRICTS.

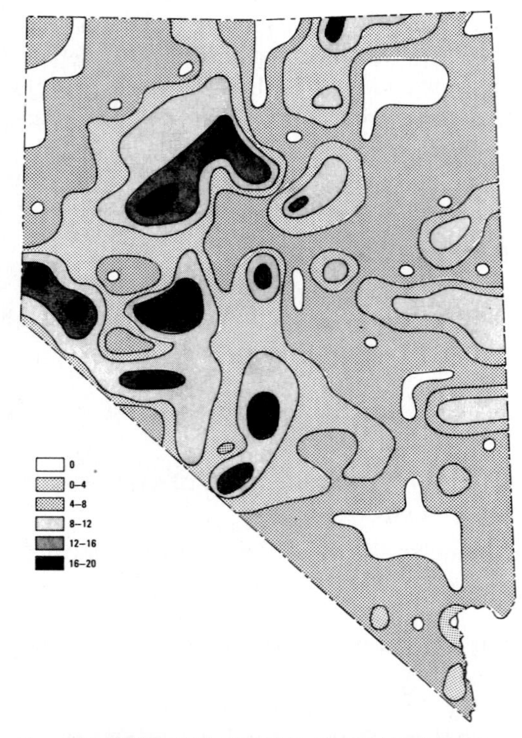

D. CONTOUR MAP OF THE DISTRIBUTION OF METAL–MINING DISTRICTS IN NEVADA, WEIGHTED ACCORDING TO DOLLAR VALUE.

FIGURE 8.1
Landsat lineaments compared with mining districts of Nevada.
Maps from Rowan and Wetlaufer (1975). Mining data from Horton (1964).

rupted by an area of low mining density at the intersection with the Oregon–Nevada lineament. The lack of mining districts along the Oregon–Nevada lineament may be caused by the extensive cover of Tertiary volcanic rocks that masks any underlying deposits. The greatest density of ore deposits along the Midas Trench occurs at the intersection with the Rye Patch lineament. In south-central Nevada the East-Northeast (ENE) lineament system coincides with two of the three east-trending belts of high mining density on Figure 8.1C. The most southerly of the three zones is aligned with the western extension of the East-West (E-W) lineament. Ore deposits along the Walker Lane lineament are generally concentrated at the intersections with east-trending lineaments. The conspicuous lack of ore deposits in southern Nevada coincides with the lack of major lineament systems there.

Figure 8.1D was prepared by gridding and contouring the weighted dollar value of production for the districts. These value trends closely resemble the trends of mining density. The influence of the Walker Lane lineament is more pronounced on the value map, although the high concentrations are still related to the intersection with west-trending lineaments.

Concentrations of mining districts and values around the margin of the central Nevada circular feature correspond to the intersections of lineaments with the circular feature. The lack of mining districts within the circular feature is caused either by nondeposition of ore or by burial of ore deposits beneath the cover of volcanic rocks of mid-Tertiary age.

Fracture Patterns of Central Colorado Regional lineaments are efficiently mapped on Landsat mosaics, as in Nevada. On individual images the fractures and fracture zones, which were conduits for ore-forming solutions, may be mapped. Areas of maximum fracture intensity and fracture intersections are good prospecting targets.

The relationship between Landsat fracture patterns and ore deposits is illustrated in the example from central Colorado, which is summarized from the work of Nicolais (1974). A winter image (Figure 8.2) was used for interpretation because the snow cover and low sun elevation enhance the expression of linear features. On the interpretation map (Figure 8.3) linear, curvilinear, and circular features are plotted together with location of major mining districts. Ten target areas were selected where there are concentrations of intersections of lineaments and intersections of lineaments with curvilinear or circular features. Five of the ten target areas coincide with, or are immediately adjacent to, nine major mining districts. The Landsat interpretation reduced the original 33,500 km^2 image to 10 target areas, each 165 km^2 in area. The five target areas without known ore deposits may be the site of undiscovered deposits.

Mapping of Hydrothermal Alteration

Many ore bodies are deposited by hot watery fluids called *hydrothermal solutions* that invade the host rock, or *country rock.* During formation of the ore minerals these solutions also interact chemically with the country rock to alter the mineral composition for considerable distances beyond the site of ore deposition. This *hydrothermal alteration* is marked by distinctive assemblages of secondary minerals that commonly are laterally and

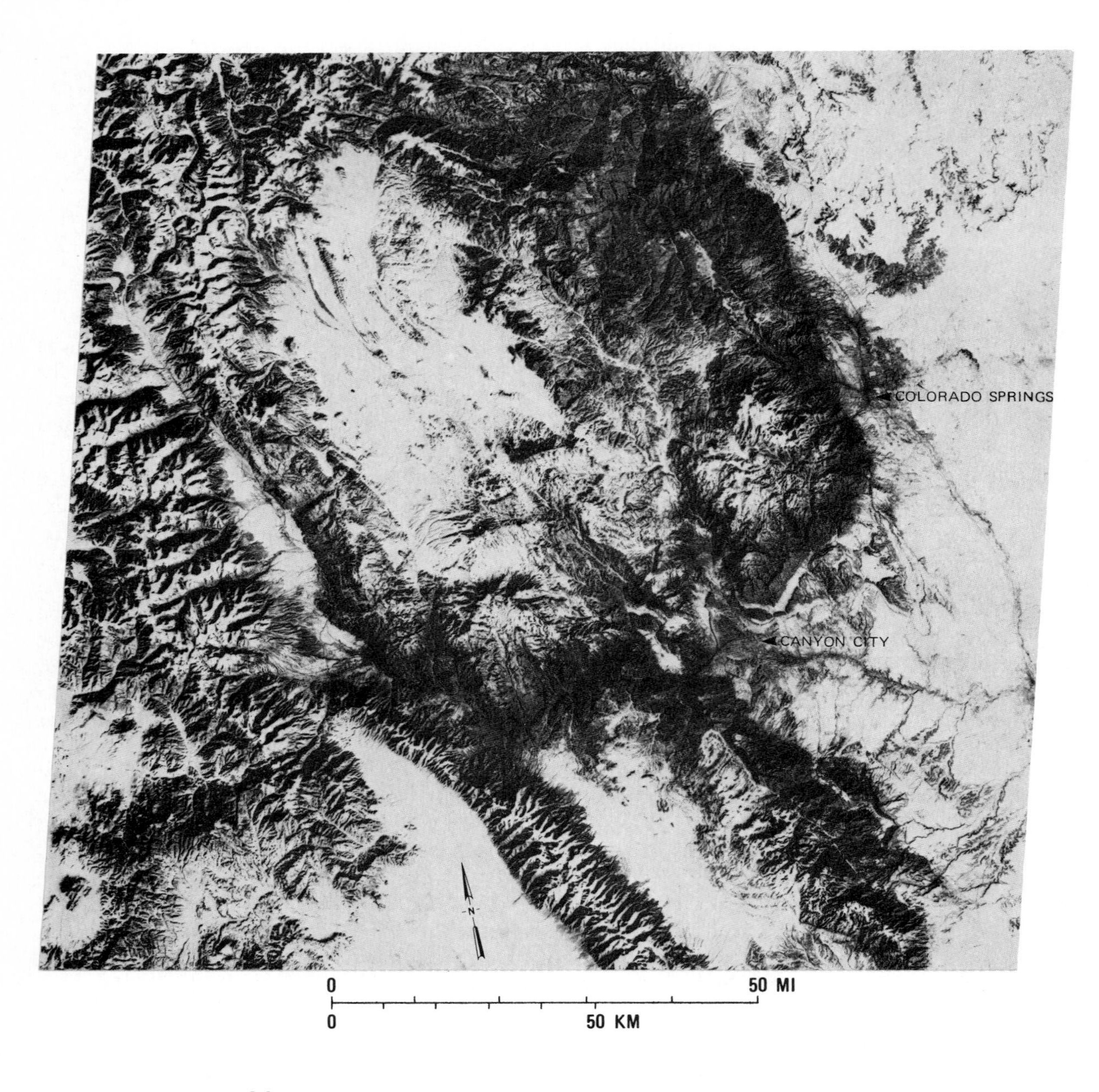

FIGURE 8.2
Landsat image of central Colorado acquired January 11, 1973 at sun elevation of 23°
Landsat 1172-17141, band 7.

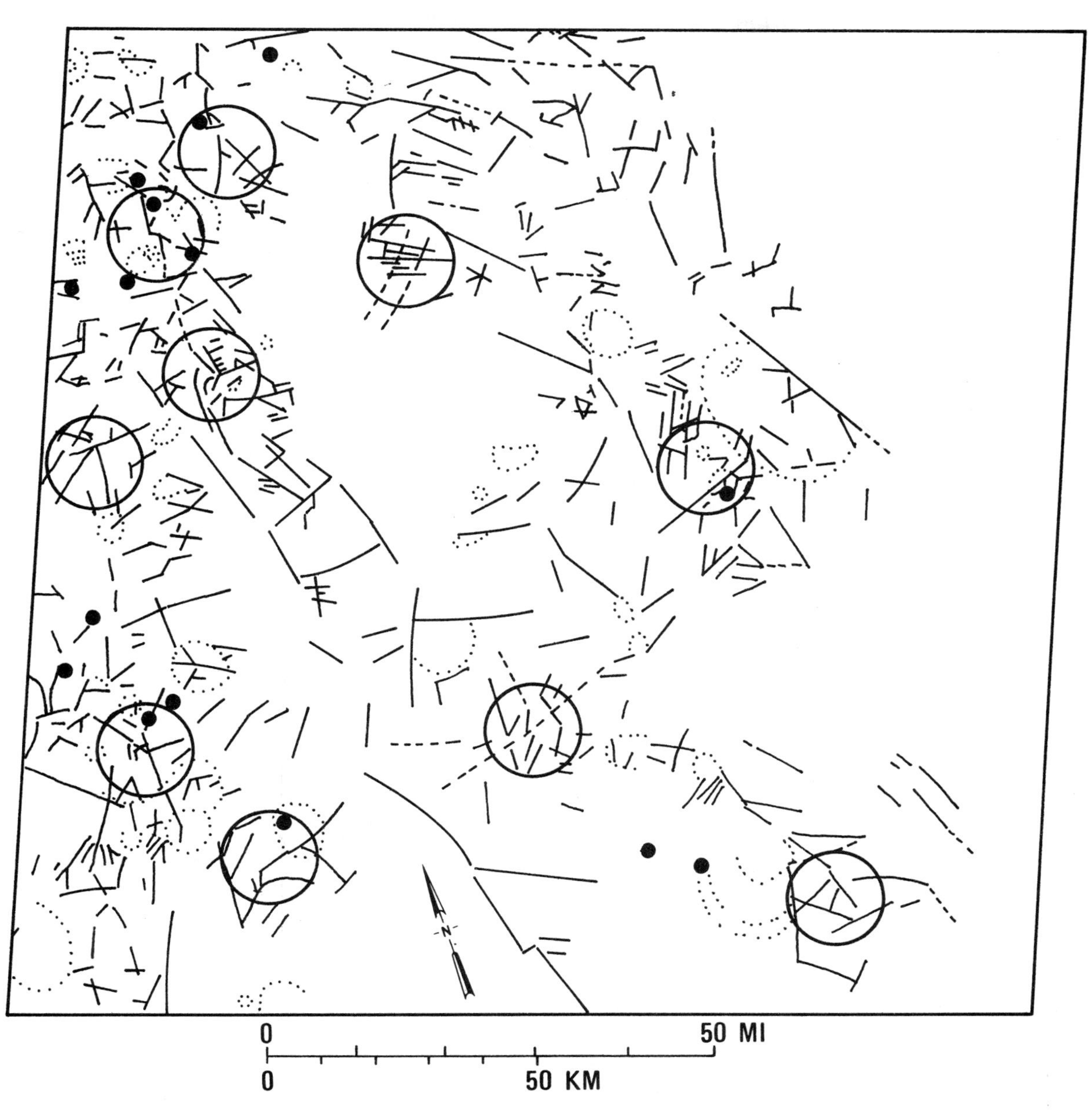

FIGURE 8.3
Interpretation of Landsat 1172-17141 of central Colorado. Solid lines are distinct lineaments; dashed lines are possible lineaments; dotted lines are curvilinear features. Large circles are selected target areas for exploration. Solid dots indicate major mining districts. From Nicolais (1974, Figure 3).

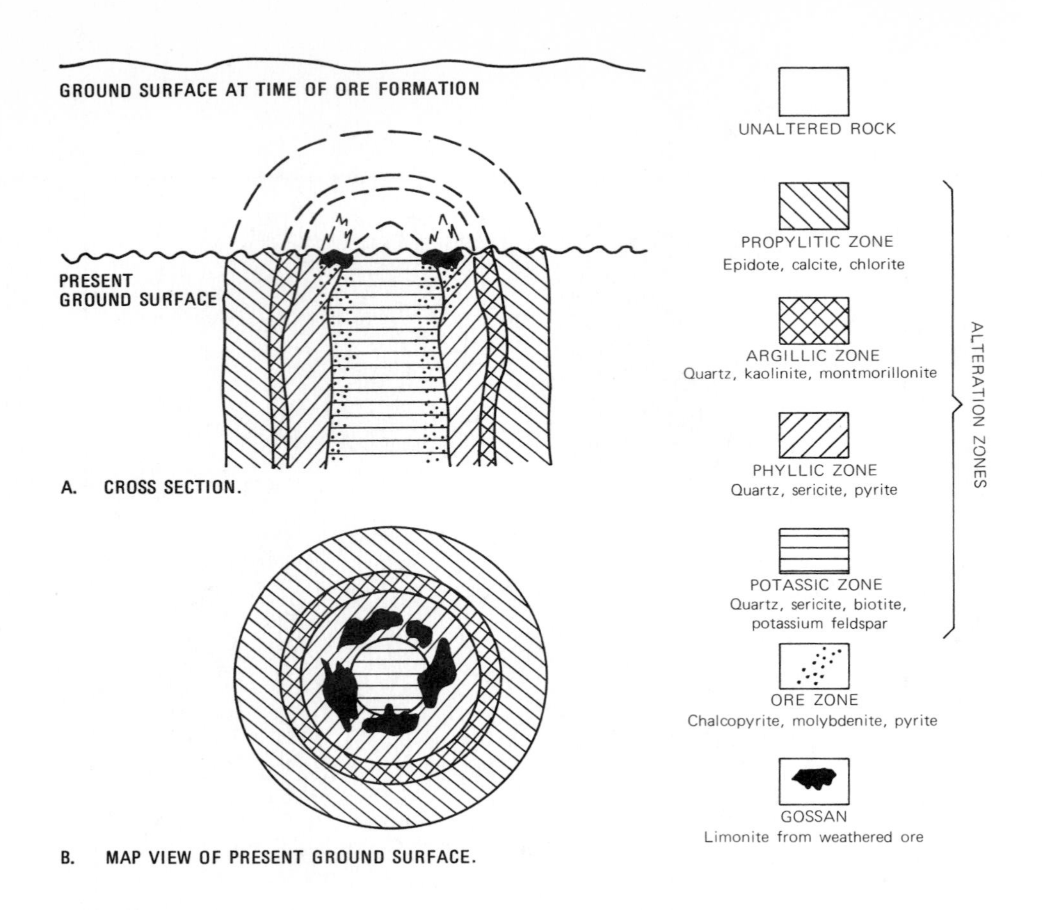

FIGURE 8.4
Model of hydrothermal alteration zones associated with porphyry copper deposits. From Lowell and Guilbert (1970, Figure 3).

vertically zoned with respect to the ore body (Figure 8.4). The zoning is due to changes in temperature, pressure, and chemistry of the hydrothermal solution at progressively greater distances from the ore body. Typical alteration zones and their mineral assemblages are shown in Figure 8.4. At the time of ore deposition, alteration of the country rock may not extend to the surface of the ground. Later uplift and erosion expose successively deeper alteration zones and eventually the ore body itself. Weathering of the metallic sulfide ore minerals forms an oxidized surface deposit called a *gossan* that is red to brown because of the presence of limonite and other iron oxide minerals. The origin of hydrothermal alteration zones is more extensively discussed by Park and MacDiarmid (1975), from which this review was summarized.

Gossans can be valuable indicators of mineral deposits that are concealed beneath the weathered surface, although not all gossans are associated with ore bodies. The colors of gossans contrast with those of adjacent country rocks, but most gossans are smaller than the Landsat pixel size and are difficult to detect on Landsat images. Hydrothermal alteration zones are more areally extensive and generally less conspicuous than gossans. Not all hydrothermal alteration is associated with ore bodies, and not all ore bodies are marked by alteration zones, but these zones are valuable indicators of possible deposits. Field work, laboratory analysis of rock samples, and interpretation of aerial photographs are standard exploration methods for hydrothermal alteration zones. Much research is currently being directed toward using remote-sensing systems for this purpose.

Laboratory reflectance spectra of some silicate minerals that characterize the alteration zones are shown in Figure 8.5A. Kaolinite, montmorillonite, and sericite are generally white or light shades of green, brown, or gray; their presence imparts a light color to the altered rocks. Secondary quartz is also light in color. Epidote and chlorite, which characterize the propylitic zone of Figure 8.4, impart a green color to the altered rocks. Pyrite (iron sulfide) is a common alteration mineral around sulfide ore bodies and may cause pronounced color changes in the country rock. Weathering of pyrite-enriched altered rocks produces the iron oxide minerals, limonite and goethite, that impart conspicuous red and red-brown colors to the outcrops (Figure 8.5B). Limonite and goethite are major components of gossans.

Rowan, Goetz, and Ashley (1977) used a portable spectrometer in the field to measure reflectance of altered and unaltered rocks at the Goldfield mining district, Nevada. Figure 8.5C shows the mean spectra for 284 altered rocks and 342 unaltered rocks that include volcanic rocks, shale, and dry lake deposits. The gaps in the field spectra from 1.3 to 1.5 μm and from 1.8 to 2.0 μm are caused by the water vapor absorption bands shown in Figure 1.3. The differences between the spectra in Figure 8.5C may be summarized as follows:

1. The altered rocks have an overall higher reflectance than the unaltered rocks because of the high reflectance of the quartz, alunite, and clay minerals formed by the hydrothermal processes.
2. The spectral reflectance curve of the altered rocks has considerable variation (peaks and valleys) in contrast to the featureless (flat) shape of the curve for unaltered rocks. The strong reflectance minimum of 2.2 μm is caused primarily by clay minerals and alunite in the altered rocks.
3. A. F. H. Goetz (1976, p. 3) pointed out that a

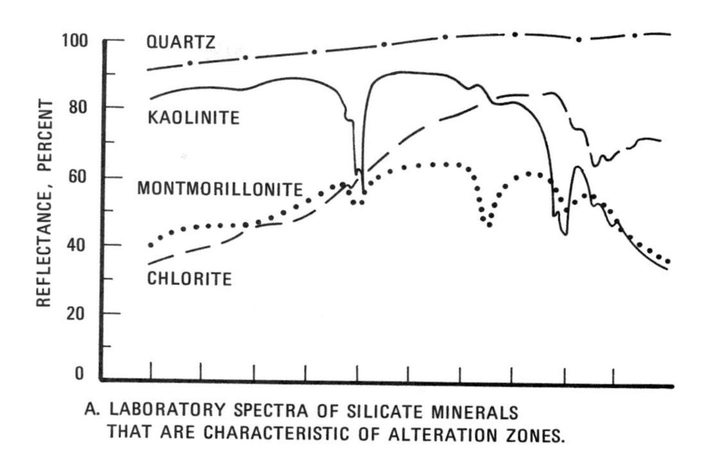

A. LABORATORY SPECTRA OF SILICATE MINERALS THAT ARE CHARACTERISTIC OF ALTERATION ZONES.

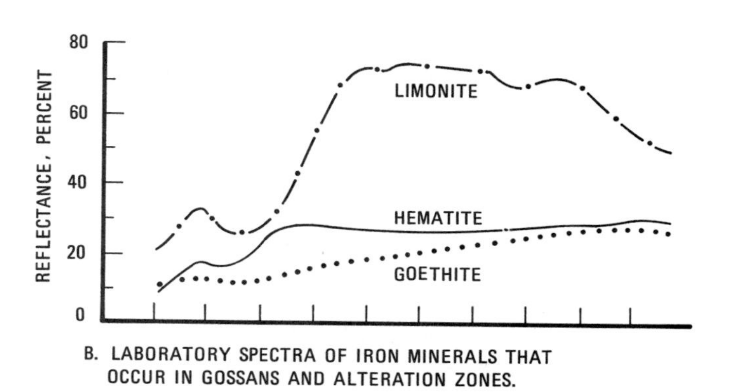

B. LABORATORY SPECTRA OF IRON MINERALS THAT OCCUR IN GOSSANS AND ALTERATION ZONES.

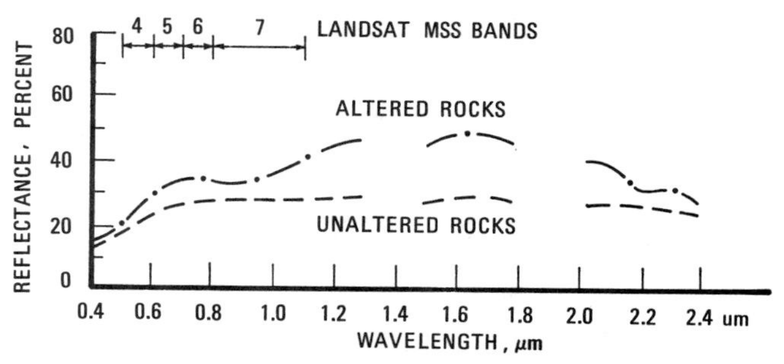

C. FIELD SPECTRA OF ALTERED ROCKS (AVERAGE OF 284 MEASUREMENTS) AND UNALTERED ROCKS (342 MEASUREMENTS) AT GOLDFIELD, NEVADA.

FIGURE 8.5
Reflectance spectra of minerals and rocks that are characteristic of alteration zones. Part A from Hunt and Salisbury (1970, Figures S-7B, 10A, 11B, 18A); part B from Hunt, Salisbury, and Lenhoff (1971, Figure 0-8, 9, 11); part C courtesy L. C. Rowan, U.S. Geological Survey and A. F. H. Goetz, Jet Propulsion Laboratory.

ratio of reflectance values between the broad bands centered at 1.6 and at 2.2 μm (1.6μm/2.2μm) provides maximum discrimination between altered and unaltered rocks. The 1.6 and 2.2 μm bands occur at wavelengths that are beyond the sensitivity range of human vision, IR film, and Landsat. This factor suggests that, even in areas that have been thoroughly explored, there may be alteration zones that have not been detected by conventional prospecting methods.

Abrams and others (1977, p. 2) note that in the spectral range of the Landsat MSS (0.5 to 1.1 μm) limonite-bearing altered rocks have distinctive reflectance spectra because of the intense ferric-iron absorption bands (Figure 8.5B). This relationship was used in the following example from Goldfield to map areas of limonitic altered rocks from ratio images of Landsat bands. Not all altered rocks contain limonite, however, and much limonite occurs in rocks that are not altered, such as red sandstones, shales, and volcanic rocks. In future Landsat systems, the addition of spectral bands in the 1.6 and 2.2 μm regions should be valuable for mapping hydrothermal alteration zones that are not associated with limonite.

Ratio Images of Goldfield, Nevada Digitally contrast-stretched MSS images of part of a Landsat frame covering the Goldfield, Nevada mining district are shown in Figure 8.6. Location of the area is indicated on the mining district map of Figure 8.1B. The Goldfield district is located 10 km south of Mud Lake, which is the bright circular feature in the northwest part of the images in Figure 8.6. The known areas of hydrothermal alteration are not apparent on these bands nor on color composite images made from them. The six primary ratio images were computed and linearly stretched for optimum contrast (Figure 8.7). The alteration areas are inconspicuous on the individual ratio images.

Rowan and others (1974) prepared the color composite image shown in Plate 6 by combining the ratio images in the following manner: 4/5 = blue; 5/6 = yellow; 6/7 = magenta. The resulting color image has the following signatures: blue = dry lakes; white = mafic rocks and clouds; pink = felsic igneous rocks; orange = vigorous vegetation; dark pinkish brown = cloud shadows. For mineral exploration the green and brown colors in Plate 6 are most significant because they correlate with areas of hydrothermal alteration that are characterized by secondary clay, silica, and limonite. These altered areas are inconspicuous on the individual spectral bands, the conventional color composite image, and on individual ratio images. Areas of hydrothermal alteration are recognizable, however, on the color-ratio composite image.

Distribution of the green, brown, and red-brown colors on the color-ratio composite image of Plate 6 are shown in Figure 8.8 together with the mines and prospects in the area. In the Goldfield mining district the green signature corresponds to predominantly limonitic areas that are weathered hydrothermally altered rocks, with two exceptions: (1) the outcrop of slightly ferruginous sandstone and siltstone in the west-central and southwest part of Plate 6, and (2) a mafic intrusive body approximately 20 km northwest of Mud Lake. Red-brown colors coincide with silica-rich, light-colored volcanic rocks. The brown signatures generally represent light-colored volcanic rocks with hydrothermal alteration in at least one district. Although some mines lack anomalous colors

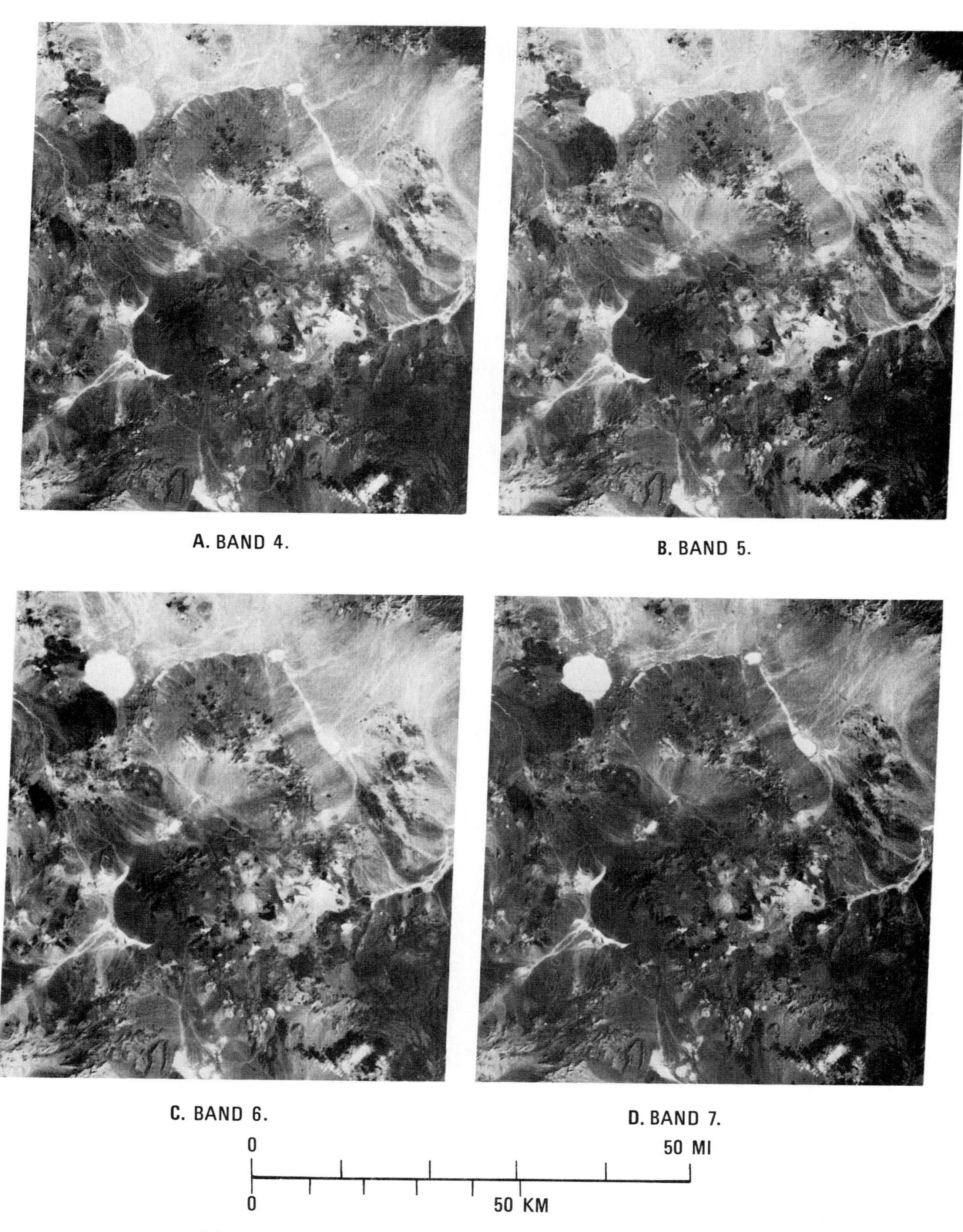

FIGURE 8.6
Stretched MSS images of Goldfield, Nevada, area from Landsat 1072-18001. The images were digitally processed at Jet Propulsion Laboratory. From Rowan and others (1974, Figure 10). Courtesy L. C. Rowan, U.S. Geological Survey.

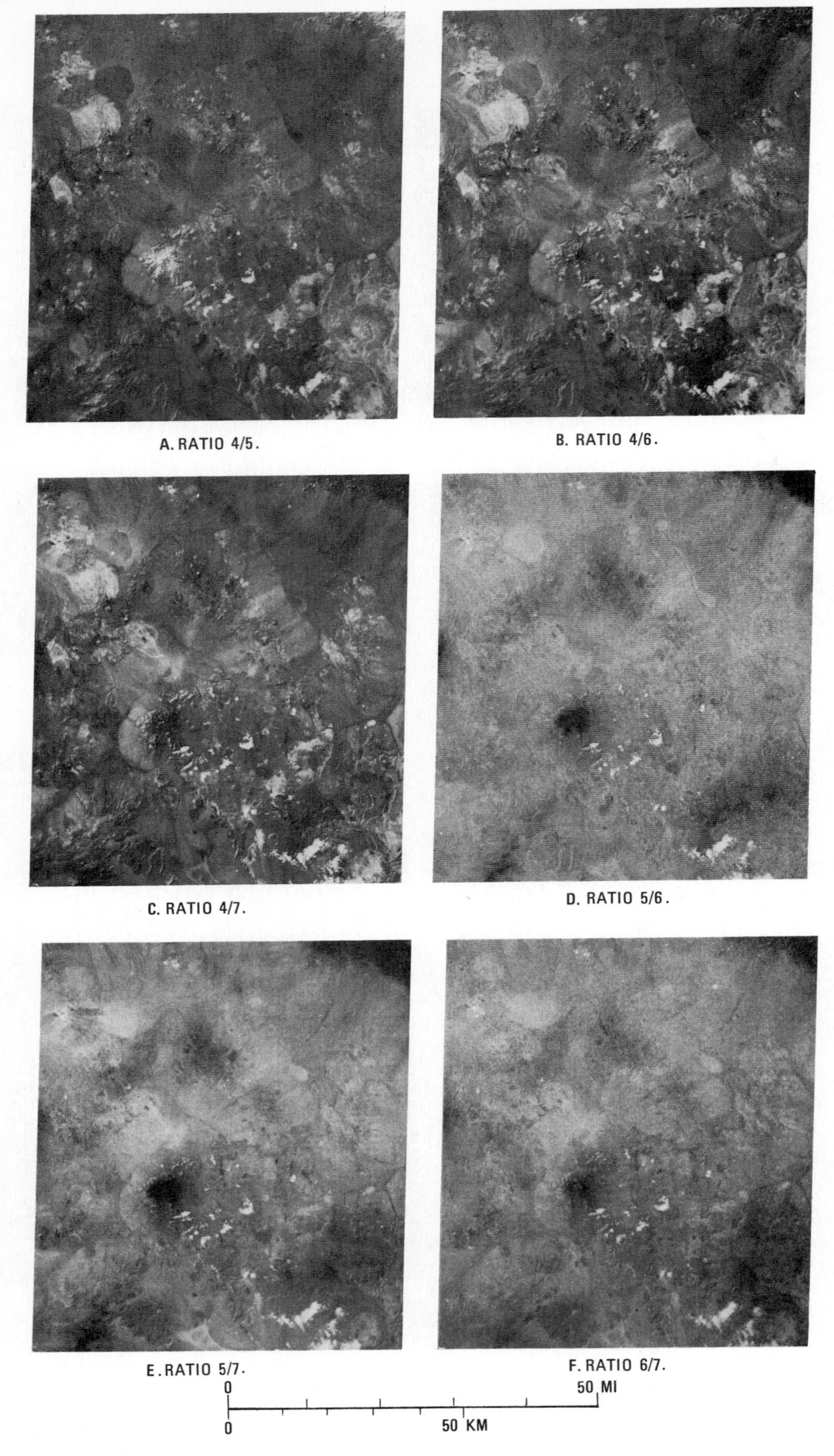

FIGURE 8.7
Stretched ratio images of MSS bands of Goldfield, Nevada, area from Landsat 1072-18001. Images were digitally processed at Jet Propulsion Laboratory. From Rowan and others (1974, Figure 16). Courtesy L. C. Rowan, U.S. Geological Survey.

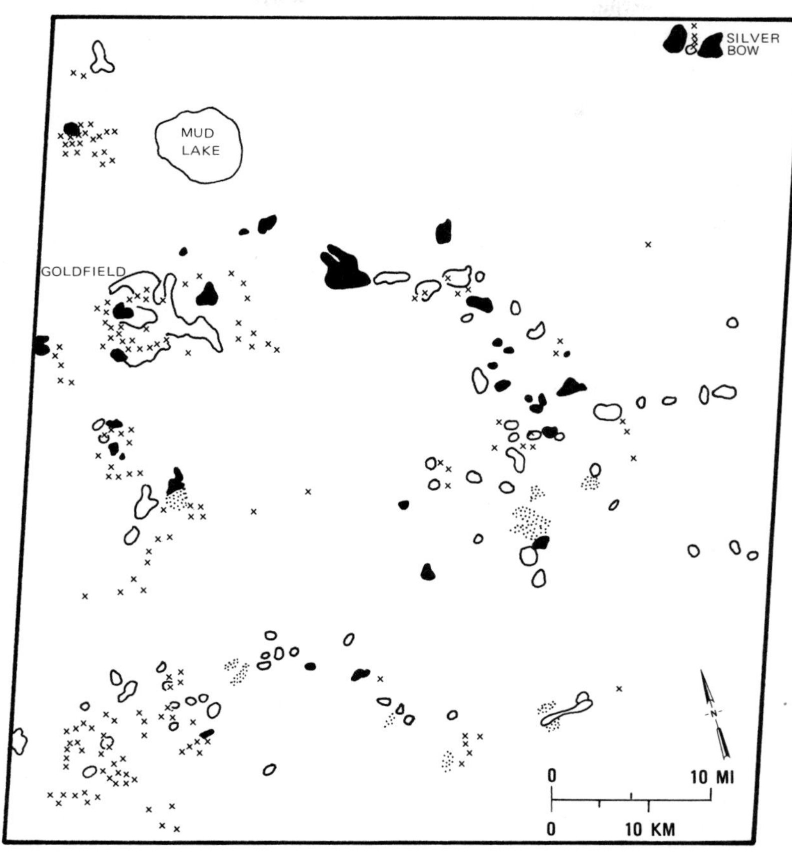

ANOMALOUS PATTERNS FROM COLOR RATIO COMPOSITE:

– GREEN. HYDROTHERMALLY ALTERED LIMONITIC AREAS

– RED BROWN. SILICA–RICH, LIGHT–COLORED VOLCANIC ROCKS

– BROWN. LIGHT–COLORED VOLCANIC ROCKS, WITH SOME HYDROTHERMAL ALTERATION

x – APPROXIMATE LOCATION OF MINES AND PROSPECTS

FIGURE 8.8
Interpretation of hydrothermally altered areas from Landsat color-ratio composite image, correlated with mining districts. From Rowan and others (1974, Figures 18 and 19).

and some color anomalies lack corresponding mineral deposits, the Goldfield investigation demonstrates the potential of ratio images for mineral exploration.

H. A. Pohn of the U.S. Geological Survey (personal communication) reports that the same ratio images and color combinations employed at Goldfield (Plate 6) were applied to a Landsat image of Iran. Surface alteration at the Sarchesma porphyry copper deposit produced an anomalous green signature on the color-ratio composite image. Other green signatures were checked in the field and found to mark ancient mining areas and areas of altered ground worthy of additional exploration.

Classification of Copper Prospects in Pakistan

An early application of digital classification of Landsat data to mineral exploration was by

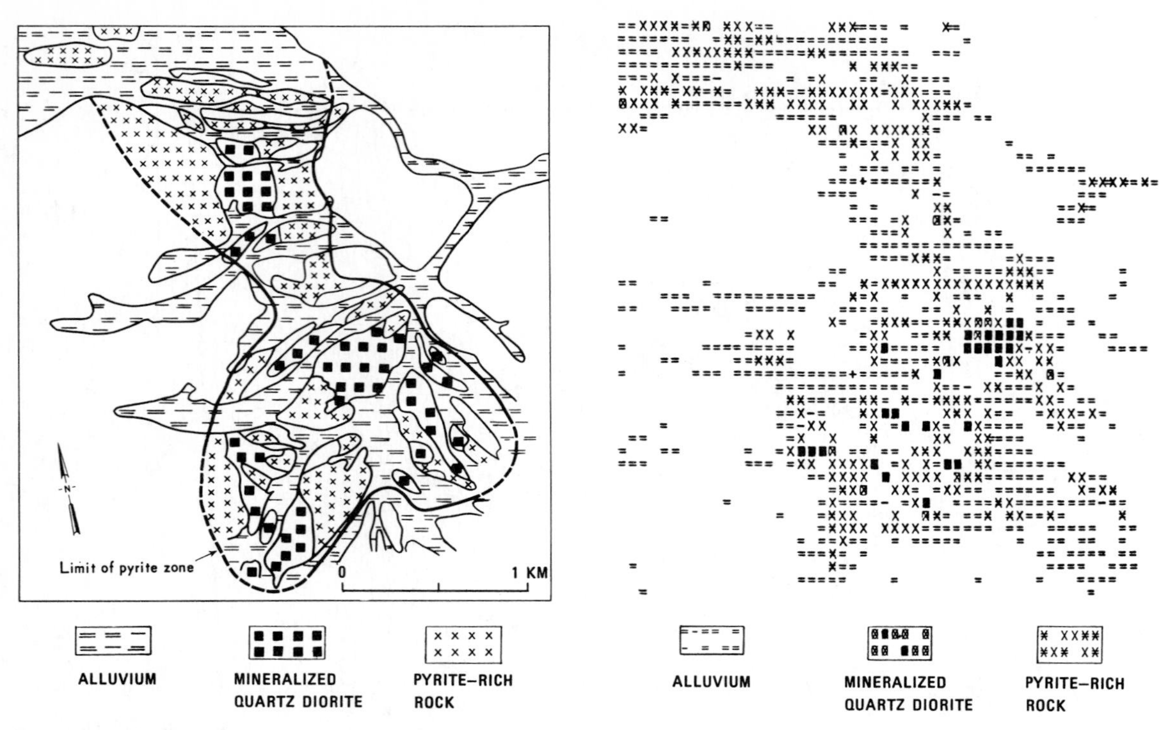

A. GEOLOGY MAPPED IN THE FIELD.

B. DIGITAL CLASSIFICATION MAP.

FIGURE 8.9
Training area for digital classification of Landsat CCT data. Saindak copper deposit, Pakistan. Part A from Khan (1972). Part B from Schmidt, Clark, and Bernstein (1975, Figure 5). Courtesy R. Schmidt, U.S. Geological Survey.

Schmidt, Clark, and Bernstein (1975), who investigated a portion of the Chagai district in western Pakistan. A known porphyry copper deposit at Saindak (Figure 8.9A) was used as a training area to identify the spectral characteristics of copper-bearing rocks and associated rock types. The digital number (DN) values for the pixels belonging to each category are shown in Table 8.1, together with the corresponding computer printout symbols. For each rock type the range of DNs in each MSS spectral band defines the four-dimensional decision space, or cluster, for that rock type. In the vicinity of the training site the classification categories have a high reliability. At localities beyond the training site, however, pixels may be classified near the boundaries of a decision space, or cluster. These classifications are less certain and are shown with a symbol for low reliability in Table 8.1 Recognition of rock types at the Saindak training area is shown on the classification map (Figure 8.9B). Mineralized quartz diorite and pyritic rock are the key categories for recognizing potential copper deposits elsewhere in this region.

The training categories of Table 8.1 were used to classify the Landsat digital tapes covering an area of 2,100 km^2, located east of the Saindak training site. Based on the occurrence of mineralized rock categories on the classification map, 23 exploration targets were selected. Of the 19 targets checked in the field, 5 are prospects with outcrops of hydrothermally altered rock that contain 5 to 10 percent pyrite. Figure 8.10 is a portion of the classification map that covers part of the eroded core of a large volcano and includes four of the five prospects. Sites 5-D, 6-D, and 6-E may belong to a single large mineralized body that is partly covered by alluvium. According to Schmidt, Clark, and Bernstein (1975) their site 6-D, an area of 0.8 km^2, is the most mineralized area found in the field investigation, with some of the unleached rock containing 0.3 percent copper. This is the most promising area for further investigation.

TABLE 8.1
Digital classification table for Saindak training area

Rock type	Classification reliability	Computer symbol
Mineralized quartz diorite	High	0
	Low	⊠
Mineralized pyritic rock	High	⁎
	Low	X
Dry wash alluvium	High	=
	Low	–
Boulder fan	High	+
Eolian sand	High	•
	Low	,
Dark rock outcrops, desert-varnished lag gravels, and black sand		1
		#
		H

Source: From Schmidt, Clark, and Bernstein (1975, Table 1).

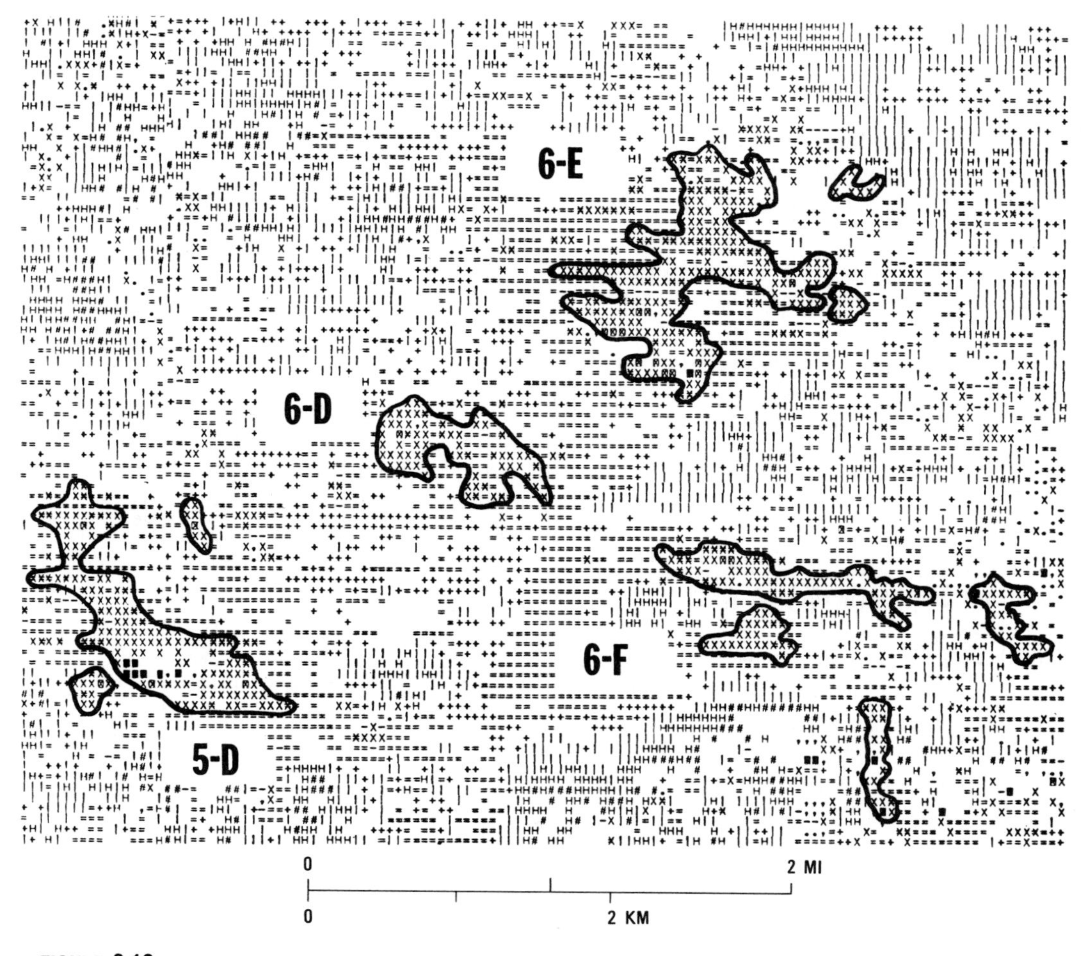

FIGURE 8.10
Copper prospects identified in western Pakistan. The categories developed at Saindak training area were used in this classification. Field work indicates area 6-D is the most promising prospect. From Schmidt, Clark, and Bernstein (1975, Figure 6). Courtesy R. Schmidt, U.S. Geological Survey.

Core drilling is necessary to determine if exploitable copper deposits exist beneath the leached surface rocks.

Photogeologic analyses of enhanced Landsat images did not reveal any of the prospects located on the classification map. None of the prospects on the classification map have more than trace amounts of copper in the weathered outcrops, but the hydrothermal alteration is similar to that of known copper deposits in the region. The absence of copper minerals at the surface may result from three causes:

1. Intense weathering and leaching has removed copper at the surface, offering the possibility of secondary enriched zones at depth.
2. The prospects may be within the alteration zone overlying a deposit that has not been exposed by erosion.
3. The prospects may be areas of altered rock with no associated mineralization.

This investigation illustrates the potential of digital classification for evaluating large areas and locating prospect areas. The time and cost of the classifications are less than for a conventional field reconnaissance.

Basic Data for Geologic Mapping

In addition to locating specific mineral target areas of fracture intersections or rock alteration, remote sensing provides data for preparing and improving geologic maps, which are the fundamental exploration tool. Geologic maps, even at reconnaissance scales, are not available for large areas of the earth. For example, approximately two-thirds of southern Africa lacks published geologic maps at scales of 1:500,000 or larger. This situation can be improved by use of Landsat images. A previously unknown major fault was discovered on a Landsat image of the southern part of South West Africa and the Cape Province of South Africa by Viljoen and others (1975, Figure 3) who named it the Tantalite Valley fault zone. The fault zone appears to have right-lateral, strike-slip displacement and has been mapped for 450 km along the strike. A number of large mafic intrusives have been emplaced along the Tantalite Valley fault zone and are recognized on Landsat images. On a Landsat color mosaic of the northwestern Cape Province of South Africa, Viljoen and others (1975, Figures 11 and 12) mapped a spectacular structural discontinuity, called the Brakbos fault zone, which separates the Kaapvaal craton on the east from the Bushmanland metamorphic complex on the west. The contact between these structural provinces is obscure in the field and had previously been drawn approximately 30 km to the east of the Brakbos fault zone, which is also defined on gravity maps. In Chapter 4 Landsat images were used to map rock types in the Transvaal Basin.

In the Nabesna quadrangle of east-central Alaska, Albert (1975) combined lineament analysis and digital image processing of Landsat data to evaluate known and potential mineral deposits. A preliminary analysis indicates that 56 percent of the known mineral deposits occur within 1.6 km of Landsat lineaments. Color anomalies on the enhanced images coincide with 72 percent of the known mineral occurrences. Of the remaining color anomalies, some coincide with areas of known rock alteration and others constitute potential exploration targets.

URANIUM EXPLORATION

Most uranium deposits in the United States occur in nonmarine fluvial sandstone and conglomerate beds of the Colorado Plateau, Rocky Mountain basins, and south Texas. These host rocks were altered by the solutions that deposited the uranium, resulting in color changes. Black-and-white and color aerial photographs have been used extensively in exploring for these areas of altered rock. Digital processing of Landsat CCTs has great potential for recognizing subtle alteration

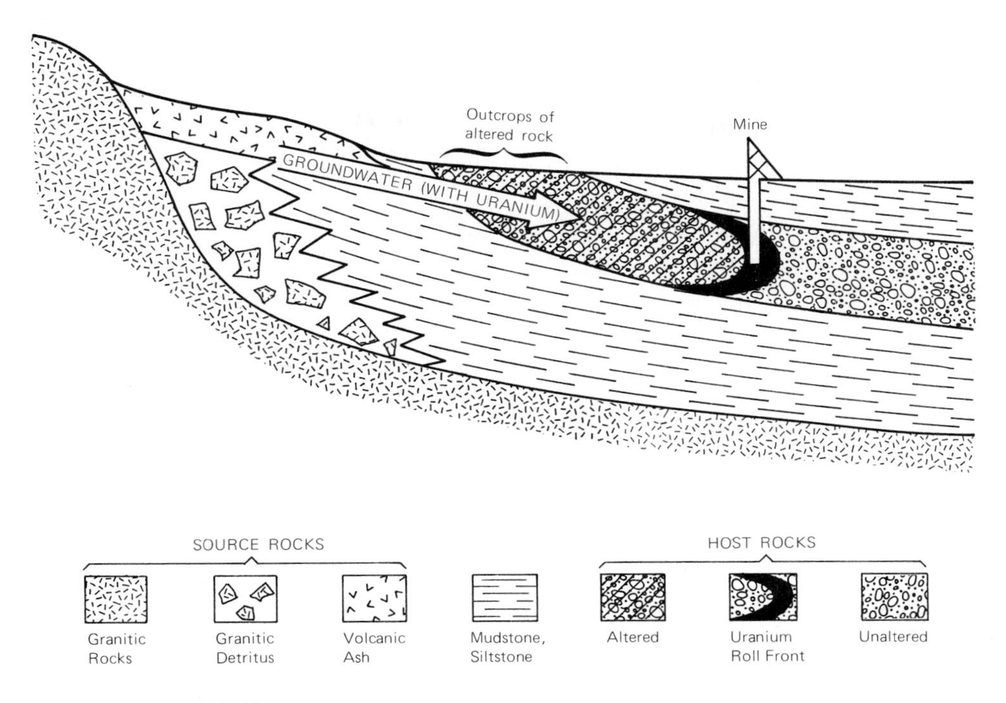

FIGURE 8.11
Model cross section showing origin of sedimentary uranium deposits.

effects that may not be obvious to the eye or readily seen on aerial photographs. A brief description of typical sedimentary uranium deposits is an aid in understanding the subsequent examples of Landsat applications.

Origin of Sedimentary Uranium Deposits

The model shown in Figure 8.11 for the formation of sedimentary uranium deposits is widely accepted, although there is debate about the origin of the uranium and the chemistry of the transporting solutions. Granite, granitic detritus, and silicic volcanic ash and flows are source rocks that contain disseminated uranium in concentrations up to 10 parts per million. Some of this uranium was leached from the source rocks and carried in solution by oxygen-rich ground water that then migrated into porous sandstone and conglomerate beds. Within these eventual host rocks the migrating water encountered reducing conditions caused by the presence of organic material, natural gas, or hydrogen sulfide and pyrite of biogenic origin. The change from oxidizing to reducing conditions caused the uranium to precipitate as oxide minerals, primarily uraninite, which coat sand grains and fill pore spaces in the host rock. The ore deposits typically contain from 0.1 to 0.5 percent U_3O_8 and occur as tabular layers or as arcuate bodies called *roll fronts* (Figure 8.11). Later geologic uplift and erosion may expose the ore to secondary oxidation and migration.

The outcrops of host rock that were altered by the migrating solutions are important for exploration because the altered surface rocks may indicate proximity to ore deposits in the subsurface. The unaltered host rocks are typically drab in color and contain organic carbon and pyrite. Oxidation by the migrating solutions destroys the carbon and converts the dark pyrite to iron oxide minerals that impart characteristic yellow, red, and brown colors to the altered rocks. The application of remote sensing to uranium exploration is described in this chapter for the following examples: (1) the Crooks Gap district in Wyoming is typical of deposits in the Rocky Mountain basins; (2) the Cameron district is an example of Colorado Plateau deposits; (3) the Freer-Three Rivers district is typical of the south Texas uranium belt.

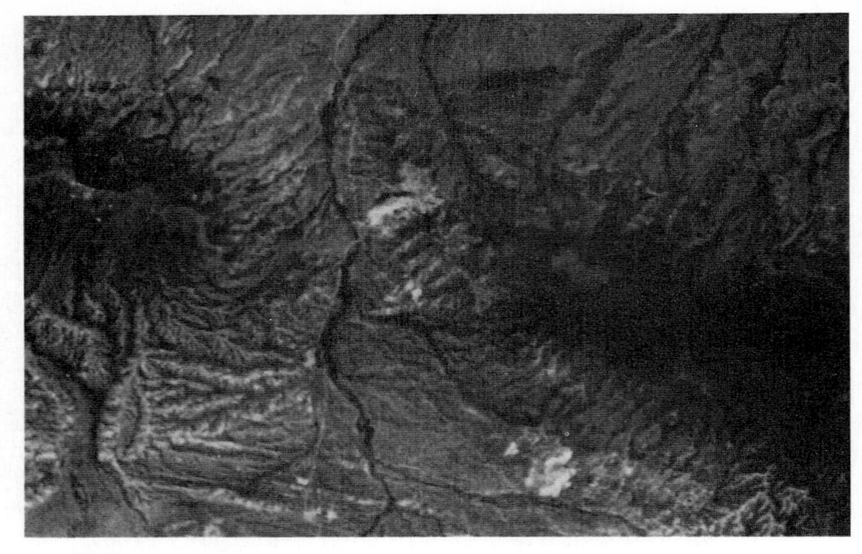

A. BAND 5.

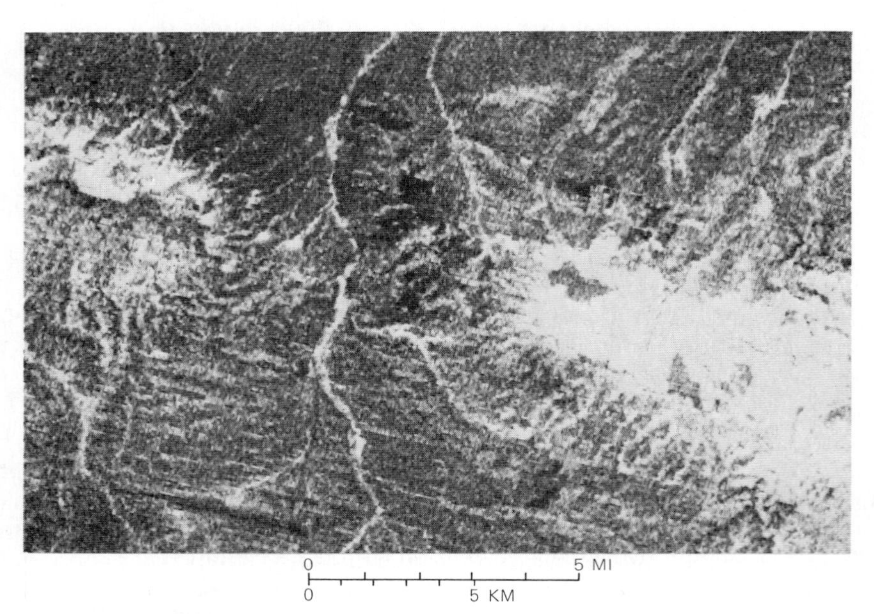

B. RATIO BAND 4/BAND 5. MOST OF THE DARK SIGNATURES MARK ALTERATION ZONES ASSOCIATED WITH URANIUM MINERALIZATION.

FIGURE 8.12
Crooks Gap uranium district, Wyoming. Landsat 1409-17291, acquired September 5, 1973. From Offield (1976, Figures 1 and 2). Courtesy T. W. Offield, U.S. Geological Survey.

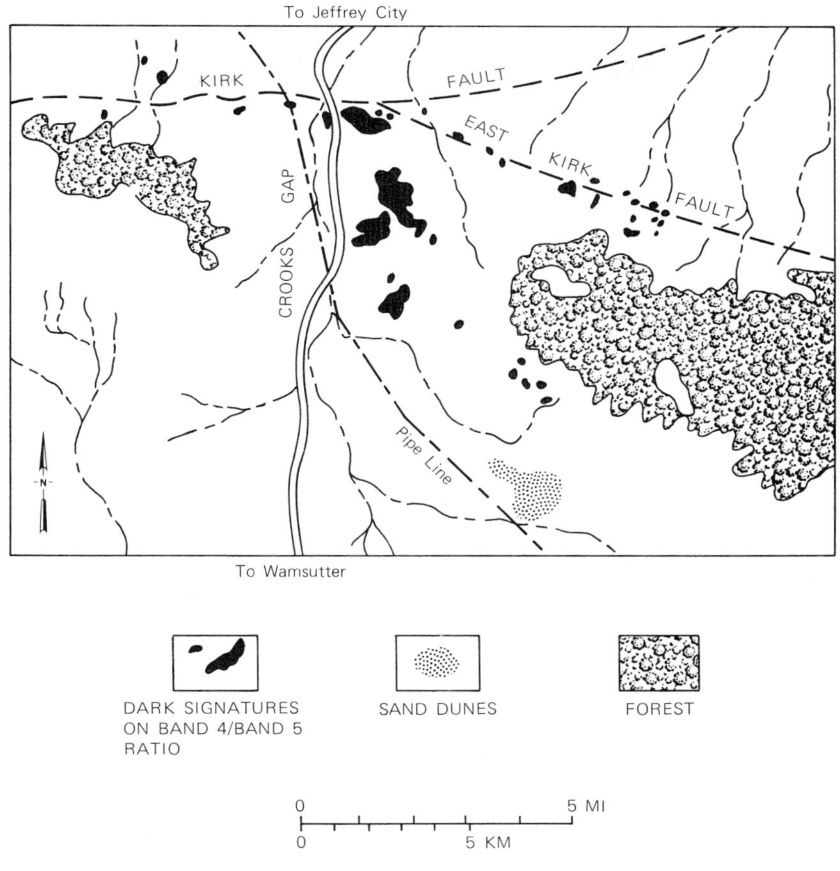

FIGURE 8.13
Interpretation map of Landsat ratio image of Crooks Gap uranium district, Wyoming. After Offield (1976).

Crooks Gap District, Wyoming

The Crooks Gap district, at the north edge of the Great Divide Basin in south-central Wyoming, is typical of the roll-front uranium deposits in the Rocky Mountain basins. Host rocks are sandstone and conglomerate beds of the Battle Spring Formation of Eocene age (Figure 8.12). The normal appearance of the Battle Spring outcrops is drab white to tan and no uranium deposits have been reported from these unaltered rocks. Where mineralizing solutions have passed through the rock, it is altered very subtly with local patches of red where the iron chemistry was suitable. The red areas stand out as color anomalies within the light-colored Battle Spring outcrops. This description is summarized from Offield (1976), who also reported the following Landsat investigation.

Landsat CCTs of the Crooks Gap district were digitally processed by the U.S. Geological Survey Petrophysics and Remote Sensing Branch. On the band 5 image (Figure 8.12A) some altered areas have a relatively bright signature, but cannot be discriminated from outcrops of Battle Spring and older formations that also have bright signatures. Other areas of known alteration lack a bright signature. On the ratio image of band 4/band 5 (Figure 8.12B) the reddish altered areas are readily discriminated by their dark signature. This dark ratio signature is due to the low reflectance on band 4 (green) and high reflectance on band 5 (red) of the altered rocks. The interpretation map (Figure 8.13) shows the distribution of the dark ratio signatures and related features. The dark signatures east of Crooks Gap correspond to altered areas around the two main open-pit uranium mines. The alignment of small dark ratio signatures along the East Kirk fault is caused by a combination of alteration along the fault and reddish outcrops, but this alteration is not related to uranium mineralization. Southeast of the southern mine area a dark signature marks a red alteration area of proven reserves scheduled for open-pit mining. At the north end of Crooks Gap, and west along the Kirk fault, are obscure dark signatures that mark areas of uneconomic mineralization with inconspicuous surface alteration. Terrain areas of very high reflectance, such as the sand dunes in Figure 8.13, cleared mining areas, and tailings piles produce dark ratio signatures that are indistinguishable from those of the reddish altered ground. These ambiguities may be resolved on suitable color-ratio composite images that differentiate altered rocks from bright areas (Offield, 1976).

Cameron District, Arizona

The Cameron district, in north-central Arizona, is currently inactive because of uneconomic mining depths and the low ore grade. New extraction techniques and increased uranium prices have revived interest in the area. The uranium host rocks are conglomerates, sandstones, and siltstones interbedded with mudstone and limestone of the Chinle Formation of upper Triassic age (Figure 8.14). The original pyrite, calcite, and aluminous minerals of the host rocks were altered by the mineralizing solutions to limonite, alunite, gypsum, and jarosite. The resulting light brownish yellow color contrasts with the typical purple to gray color of the unmineralized parts of the Chinle Formation. The outcrops of altered rock form elongate halos up to 400 m long surrounding the ore deposits. The alteration colors are valuable guides to the ore deposits but are not uniquely related to ore deposits for the following reasons:

1. The normal color of some unmineralized parts of the Chinle Formation resembles that of the altered zones.
2. Uranium may have been remobilized and removed after the alteration occurred.
3. Alteration may have occurred without any ore being deposited.

Formation of the uranium deposits at Cameron was more complex than shown in the model of Figure 8.11 because fluids and gases from volcanic centers may have contributed uranium. Nevertheless the deposits are associated with rocks that were altered during uranium deposition. This geologic description is summarized from the work of Spirakis and Condit (1975) of the U.S. Geological Survey, who also reported the Landsat interpretation summarized below.

The digitally processed images of the Cameron district (Plate 7) were prepared by the U.S. Geological Survey facility in Flagstaff, Arizona using methods described in Chapter 7. Grabens, faults, collapse structures, volcanic cones, basalt flows, and sedimentary rock formations are recognizable in Plate 7A. On this simulated normal color image, the light gray and light brown altered rocks cannot be distinguished from the surrounding unaltered rocks of the Chinle Formation, which have similar colors. Contrast-stretched ratio images of the MSS bands were prepared and examined statistically to identify the ratio images that provided maximum discrimination of known alteration zones. The color-ratio composite image of Plate 7B was then prepared by projecting the ratio 4/7 in blue, 6/4 in green, and 7/4 in red. A close correspondence between the distinctive blue color on this image and alteration zones was determined by

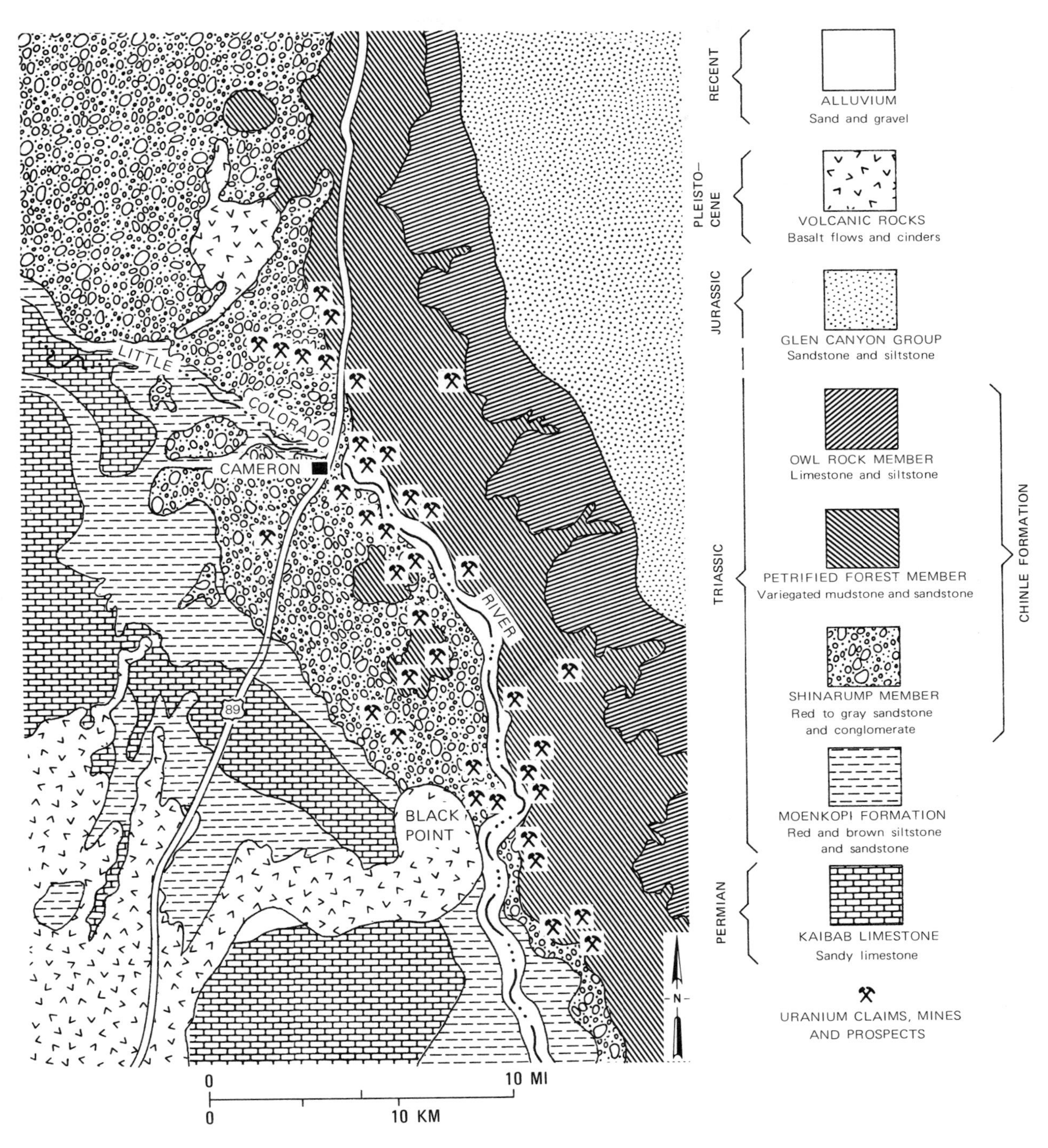

FIGURE 8.14
Geologic map of Cameron uranium district, north-central Arizona. From Chenoweth and Magleby (1971).

aerial reconnaissance of the Cameron district (Spirakis and Condit, 1975). The geologic map (Figure 8.14) does not show the alteration zones, but most of the uranium mines and claims are located at the margins of altered areas. The blue color in Plate 7B closely matches the pattern of altered rocks as indicated by the mines and claims.

The blue signature is also associated with the outcrop of a bluish gray mudstone unit near the base of the Petrified Forest Member. This "false alarm" can be recognized by its association with a particular stratigraphic unit. Additional image processing is being done in an attempt to discriminate the mudstone from altered rocks (C. D. Condit, personal communication).

Similar false alarms were noted on the color-ratio composite image of Goldfield, Nevada and probably occur on most Landsat images that are digitally processed for recognition of alteration zones. As shown in Figure 8.5, there is much overlap between the spectral-reflectance curves of altered and unaltered rocks in the wavelength region recorded on the Landsat MSS bands. The criteria for success in this application of digital image processing are not the absence of false alarms nor a 100 percent correlation between known altered zones and image signatures. The following results are the real criteria of success:

1. The anomalous signatures occupy a small proportion of the image area.
2. A high percentage of the areas of known alteration are marked by distinctive signatures on the processed images.
3. Most of the false alarm signatures can be explained by their similarity to the alteration zones.

Freer–Three Rivers District, Texas

In this typical south Texas uranium district the host rocks are channels filled with sandstone or conglomerate in the Catahoula Tuff (Miocene age). Some of the channel deposits are cut by normal faults. Geologic mapping in this area of low relief is hampered by lack of outcrops, nondistinctive rock types, a partial cover of younger gravel, and restricted land access. Digital processing of Landsat images of the district has not been successful in locating outcrops of host rocks because of heavy vegetation cover.

Thermal IR images were acquired by the U.S. Geological Survey in November, 1974 after a week of heavy rain, and were interpreted by Offield (1976). The high moisture content greatly reduced the thermal contrast between different rock types. A daytime scanner image in the visible region is shown in Figure 8.15A for reference. Despite the poor conditions, a predawn thermal IR image (Figure 8.15B) clearly shows the conglomeratic channel-fill uranium host rocks with warm signatures that contrast with the cooler appearance of the clayey tuff units. These nighttime signatures are similar to those of the conglomeratic sandstone (warm) and siltstone (cool) described earlier from the Indio Hills in Chapter 5. The warm signature of the conglomeratic host rock is caused by the thermal properties of the rock, not by heat from decay of radioactive elements. Calculations have shown that radiogenic heat produced by typical sedimentary uranium deposits is insufficient to produce a detectable thermal anomaly (Kappelmeyer and Haenel, 1974, p. 170). Possible faults (F on Figure 8.15B) are indicated by distinct warm linear features cutting across the drainage pattern.

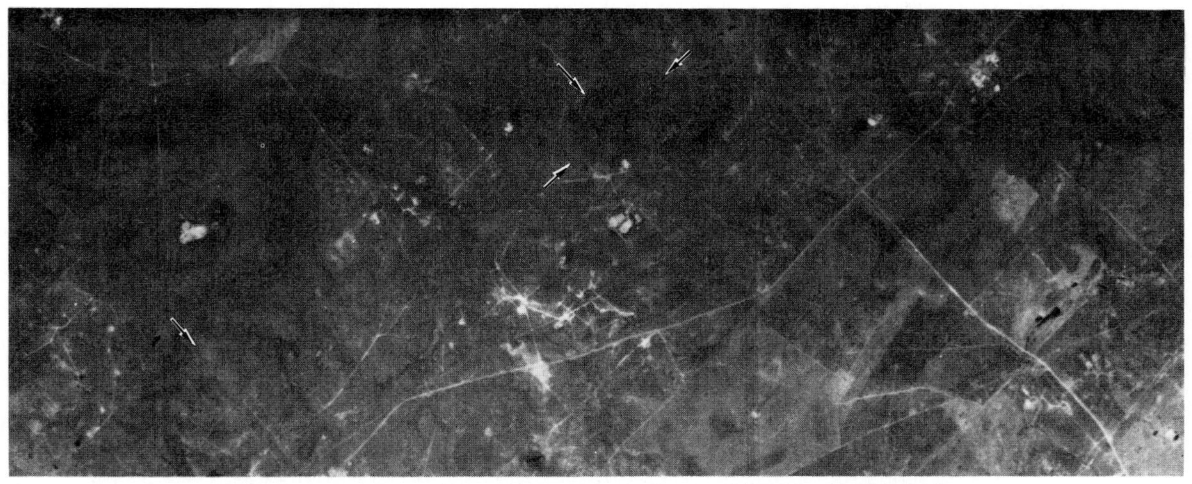

A. SCANNER IMAGE ACQUIRED AT MIDDAY IN THE VISIBLE SPECTRAL REGION. ARROWS MARK CHANNEL–FILL CONGLOMERATES WITH TOPOGRAPHIC EXPRESSION.

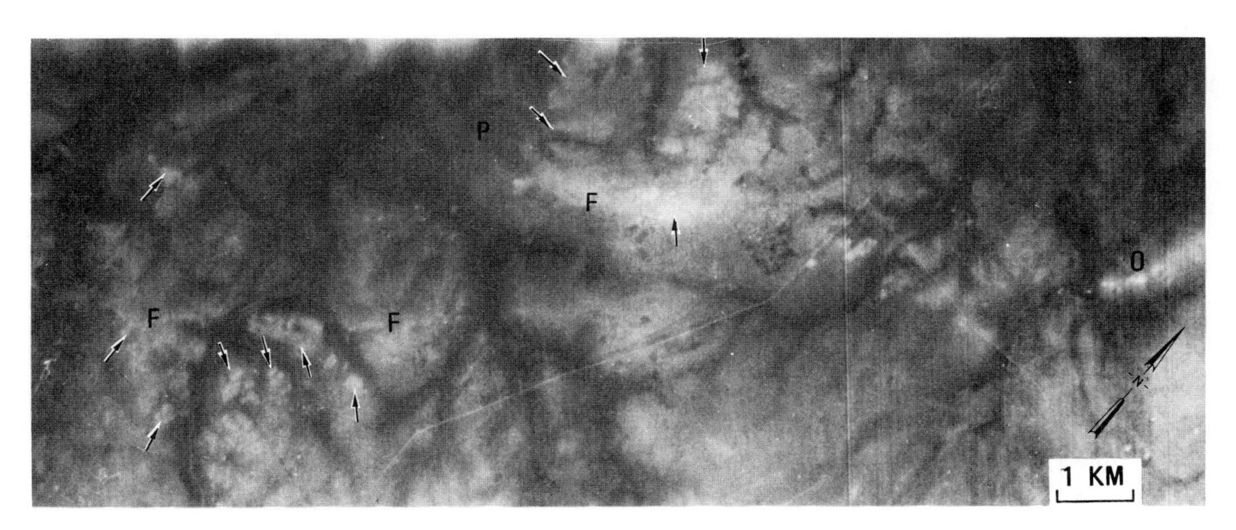

B. PREDAWN THERMAL IR IMAGE. ARROWS MARK CHANNEL–FILL CONGLOMERATES THAT ARE WARMER (BRIGHTER SIGNATURE) THAN THE SURROUNDING CATAHOULA TUFF.
O = OUTLIERS OF OAKVILLE SANDSTONE; F = FAULT TRACE;
P = PIPELINE.

FIGURE 8.15
Freer-Three Rivers uranium district, south Texas. Images acquired November, 1974. From Offield (1976, Figure 4). Courtesy T. W. Offield, U.S. Geological Survey.

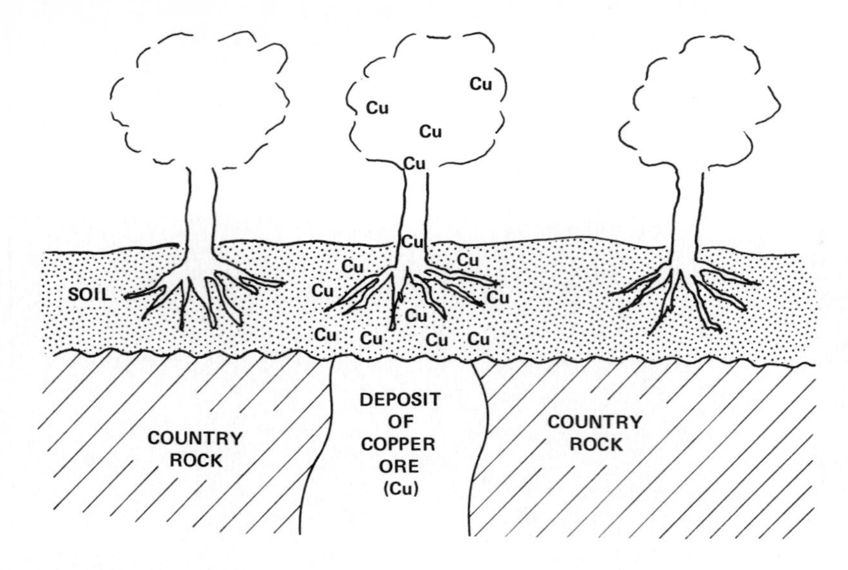

FIGURE 8.16
Copper enrichment of vegetation and soil overlying a concealed copper deposit.

MINERAL EXPLORATION IN COVERED TERRAIN

The examples of mineral exploration in this book, except for the Three Rivers–Freer deposit, are in arid to semi-arid terrain with extensive exposures of bedrock and little vegetation cover. Remote-sensor images from these and similar areas can be analyzed for exposed gossans and alteration zones that are surface indications of mineral deposits. In most of the temperate and humid climate zones of the world, however, mineral deposits are concealed beneath a cover of soil and vegetation. The composition of residual soil reflects the composition of the underlying bedrock from which the soil was derived by weathering processes. Figure 8.16 illustrates the copper enrichment of soil overlying a copper deposit in the bedrock. *Geochemical exploration techniques* are based on collecting water and soil samples and analyzing their metal content. Areas with high metal concentrations are then tested by core drilling. These techniques, and those based on vegetation analyses, are not applicable in areas where the soil has been transported rather than formed in place. Alluvial and glacial soils are examples of transported soils. Figure 8.16 illustrates that vegetation growing in mineralized soil may have a higher metal content in its tissue than vegetation in normal, or background, soil. This concentration of metals in vegetation is the basis for biogeochemical and geobotanical prospecting methods.

Biogeochemical and Geobotanical Exploration

Biogeochemical exploration consists of sampling and analyzing vegetation for anomalously high metal concentrations that indicate a concealed ore deposit.

Geobotanical prospecting searches for anomalous vegetation conditions that may be caused by high metal concentrations in the soil. Sampling and chemical analysis are not required, but skill and experience are needed to recognize the more subtle vegetation anomalies. Remote-sensing techniques are being investigated as possible geobotanical exploration methods. The major geobotanical criteria for recognizing concealed ore deposits are listed here:

1. *Lack of vegetation*—may be caused by concentrations of metals in the soil that are toxic to plants. These anomalous areas are sometimes called "copper barrens". Various types of aerial

photographs can detect this condition, which may result from causes other than mineralization.

2. *Indicator plants*—are species that grow preferentially on outcrops and soils enriched in certain elements. Cannon (1971) prepared an extensive list of indicator plants. For example, in Katanga a small blue-flowered mint, *Acrocephalus robertii*, is restricted entirely to copper-bearing rock outcrops.
3. *Physiological changes*—high metal concentrations in the soil may cause abnormal size, shape, and spectral reflectance characteristics of leaves, flowers, fruit, or entire plants. A relationship between spectral reflectance of plants and the metal content of soils could be the basis for remote sensing of mineral deposits in vegetated terrain.

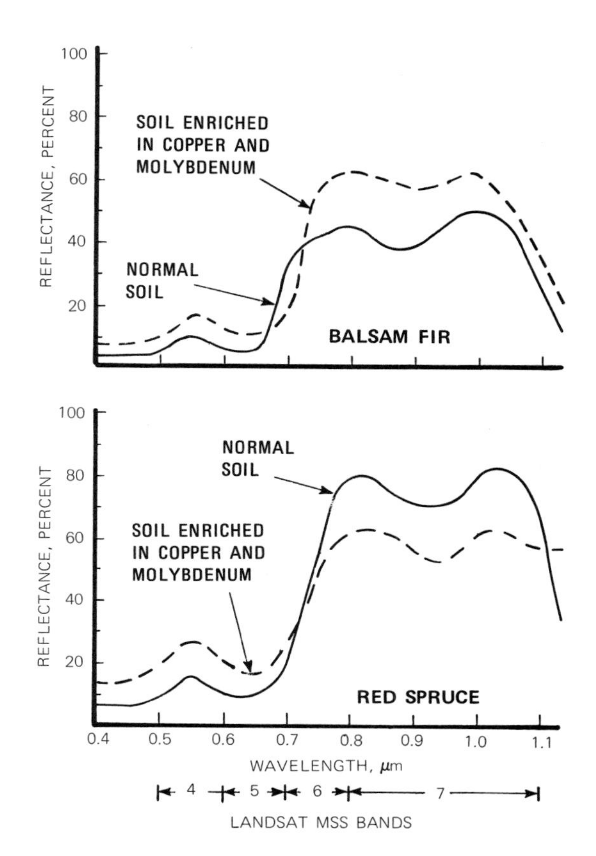

FIGURE 8.17
Reflectance spectra of balsam fir and red spruce growing in normal soil and soil enriched in copper and molybdenum. From Yost and Wenderoth (1971, Figures 5 and 6).

Remote Sensing for Minerals in Vegetated Terrain

Chlorosis, or yellowing of leaves, is a spectral change visible to the eye. Chlorosis results from an upset of the iron metabolism of plants caused by an excess concentration of chlorine, copper, zinc, manganese, or other elements. Relatively high metal concentrations are required to produce chlorosis, and this geobotanical effect has been used for prospecting. High metal content is not consistently indicated by chlorosis, however, for many areas of known mineralized soil support apparently healthy plants with no visible toxic symptoms.

The U.S. Geological Survey and other groups have used the low-grade, copper-molybdenum deposit at Catheart Mountain, Maine as a test site (Canney, 1970). Field spectrometers were used by Yost and Wenderoth (1971) to measure reflectance of trees growing in normal soil and in mineralized soil overlying the deposit (Figure 8.17). Red spruce and balsam fir growing in the mineralized soil both have higher metal concentrations than trees in unmineralized soil. In the reflected IR spectral region the mineralized balsam fir have a higher reflectance than the normal trees, whereas mineralized red spruce have a lower reflectance than the normal trees (Figure 8.17). In the green spectral region the mineralized trees of both species have a higher reflectance. The relationship of anomalous spectral reflectance to mineral content is encouraging for remote-sensing applications. On Landsat images of mixed stands of trees, however, the increased and decreased IR reflectance signatures probably would be cancelled because the signatures of many trees are integrated in the ground resolution cell of the MSS. The ideal situation would be to work in terrain covered with a single species or with species in which the reflectance anomalies caused by mineralized soil are uniformly higher or lower than normal. The problem also would be simplified by using aircraft multispectral scanners with a smaller ground resolution cell. There are other complications to prospecting with remotely sensed spectral reflectance data. Disease and insect attacks can alter spectral reflectance properties of vegetation. Spectral reflectance also varies as a function of the annual plant growth cycle, which is different for different species. Despite these problems the technique has promise and research is continuing. It has been informally reported, but not documented, that two ore bodies have been discovered in the eastern United States through mapping vegetation anomalies on Landsat images.

The problem of recognizing mineral deposits in heavily vegetated areas has been investigated by Lyon (1975), who digitally processed Landsat CCTs of mineralized areas in Norway, Nevada, and Papua. At the Karasjok area of northern Norway, copper sulfide deposits in the schist bedrock are overlain by glacial till that supports a dense growth of birch trees. Springs of copper-rich groundwater issue from the deposits and poison the ground so that birch trees are absent over an area of approximately 100 m by 1 km, which is covered by a sparse growth of grass. On the digitally processed image of band 7, the area of grass has a darker tone than the surrounding area covered by normal birch trees. On the band 5 image, the anomalous area is brighter than the surroundings. A band 7/band 5 ratio image has low values for the grass-covered area (Lyon, 1976, personal communication). This is an encouraging example of mineral exploration in a covered area.

At the Oktedi area in central Papua, copper prospects in intrusive porphyries are covered by soil and jungle. Landsat CCTs were digitally processed to produce images of the individual bands, ratios, and classification maps, but no reflectance anomalies were noted (Lyon, 1975, p. 20). Possible explanations for the lack of vegetation reflectance anomalies include: (1) any anomalies over the mineral deposits may be too small for detection by the 79 by 79 m ground resolution cell of the Landsat MSS system; (2) the vegetation assemblage may have adapted to the copper-rich soil and may not be stressed relative to surrounding vegetation.

The work of Lyon and others has demonstrated both the potential and the problems of remote sensing of mineral deposits in soil-covered and vegetated terrain. Additional research must be done, however, before this method can be used in routine exploration.

OIL EXPLORATION

Under ideal circumstances onshore oil exploration begins with regional reconnaissance and progresses to more detailed and expensive exploration methods. Prior to Landsat, aerial photography was used for regional reconnaissance. In many foreign areas, however, photography is lacking and it may be difficult, expensive, and time consuming to acquire aerial photographs. Coverage of large exploration areas requires tens of thousands of photographs that require many months to interpret. The same area can be covered with a few Landsat images and interpreted in a few days or weeks. Sedimentary basins, regional structural trends, and local structures as small as 1 km in extent can be recognized using the interpretation methods described in Chapter 4. For areas where base maps are unreliable or nonexistent, enlarged Landsat images provide valuable geographic information.

After Landsat studies have defined areas of exploration interest, the next phase of exploration begins. Aerial photographs and radar images may be obtained for more detailed investigation of the areas of interest. Field parties are sent to map and sample outcrops located from Landsat images. The first geophysical work is typically an airborne magnetic survey. Landsat images can aid in orienting flight lines in the optimum direction to evaluate the regional structural trends. Gravity surveys are then made on the ground. Seismic surveys constitute the most expensive operation and should be planned to take advantage of all the information acquired during the earlier reconnaissance studies. In poorly known areas remote-sensing images are useful for determining terrain trafficability and locating access routes.

In actual operations this ideal exploration sequence is difficult to follow because of the pressure of competition and deadlines for decisions. For example, prior to the competitive bidding for state leases on the North Slope of Alaska in 1969, many oil companies were concurrently carrying out all the exploration surveys mentioned here. This was not the most efficient use of funds, but was necessary to acquire data in time for the lease sale. The following Landsat interpretations from Kenya and Egypt outline the reconnaissance use of remote sensing early in the exploration cycle. The Colorado example illustrates the appearance of oil-producing structures on Landsat images.

Kenya

Chevron Overseas Petroleum Incorporated acquired an exploration license in eastern Kenya and completed a photogeologic and field study in 1972. Landsat 1 images of the area became available in 1973 and were compiled into the mosaic depicted in Figure 8.18 that was interpreted by Miller (1975). The drainage patterns on the interpretation map (Figure 8.19) provide valuable geographic reference in this region where base maps are generalized.

Much geologic information is present on the Landsat images, despite the relatively featureless nature of the terrain. Lineaments are a major feature in Kenya, as shown in the interpretation map (Figure 8.19) and in the image covering the northwest part of the Chevron license area (Figure 8.20A). The lineaments and other features are more apparent on the color composite images prepared by Chevron than on the black-and-white versions shown here. The Lagh Bogal lineament, that trends northwest across Figure 8.20A, is particularly significant because it marks the northeast boundary of a sedimentary basin that was confirmed by later geophysical surveys. The major lineaments extend beyond the limits of the mosaic. Some large volcanoes occur along the extensions of several lineaments beyond the Chevron license area.

Young volcanic flows form the dark signature in the west-central part of Figure 8.20A. In the southeast part large arcuate tonal bands may represent depositional patterns in clastic beds. A dark lobe extends southeast from the northwest part of the image and marks the outcrop of crystalline basement rocks with discernable foliation trends. Figure 8.20B covers the southwest part of the license area where dark basement rocks with north-trending foliation patterns crop out in the southwest part of the image. East of the basement outcrops, and probably in fault contact, sands and clays of Pliocene age form a triangular light-colored outcrop though which flows the Tana River with its associated vegetation. North of the river the Pliocene strata are capped by a dark duricrust layer. In the southeastern part of Figure 8.20B a thin wedge of strata with a gray tone occurs between the duricrust and the underlying light-toned strata. The eastward expansion of this wedge represents basinward thickening of the sedimentary section that was first observed on the Landsat image.

Results of gravity, magnetic, and seismic surveys are summarized in Figure 8.21. The Lagh Bogal lineament coincides in part with a subsurface fault, independently interpreted from the geophysical surveys, that forms the northeast boundary of the basin. Geophysical data indicate that the west border of the basin is not a single major fault zone but a combination of faulting and tilting. The northeast and northwest trends of the Landsat lineaments are parallel with the gravity and magnetic positive trends shown in Figure 8.21. Based largely on the Landsat interpretations, Chevron acquired a second exploration license area adjoining the original area on the northwest (Figure 8.21). This is an example of an economic exploration decision based at least in part on remote-sensing information.

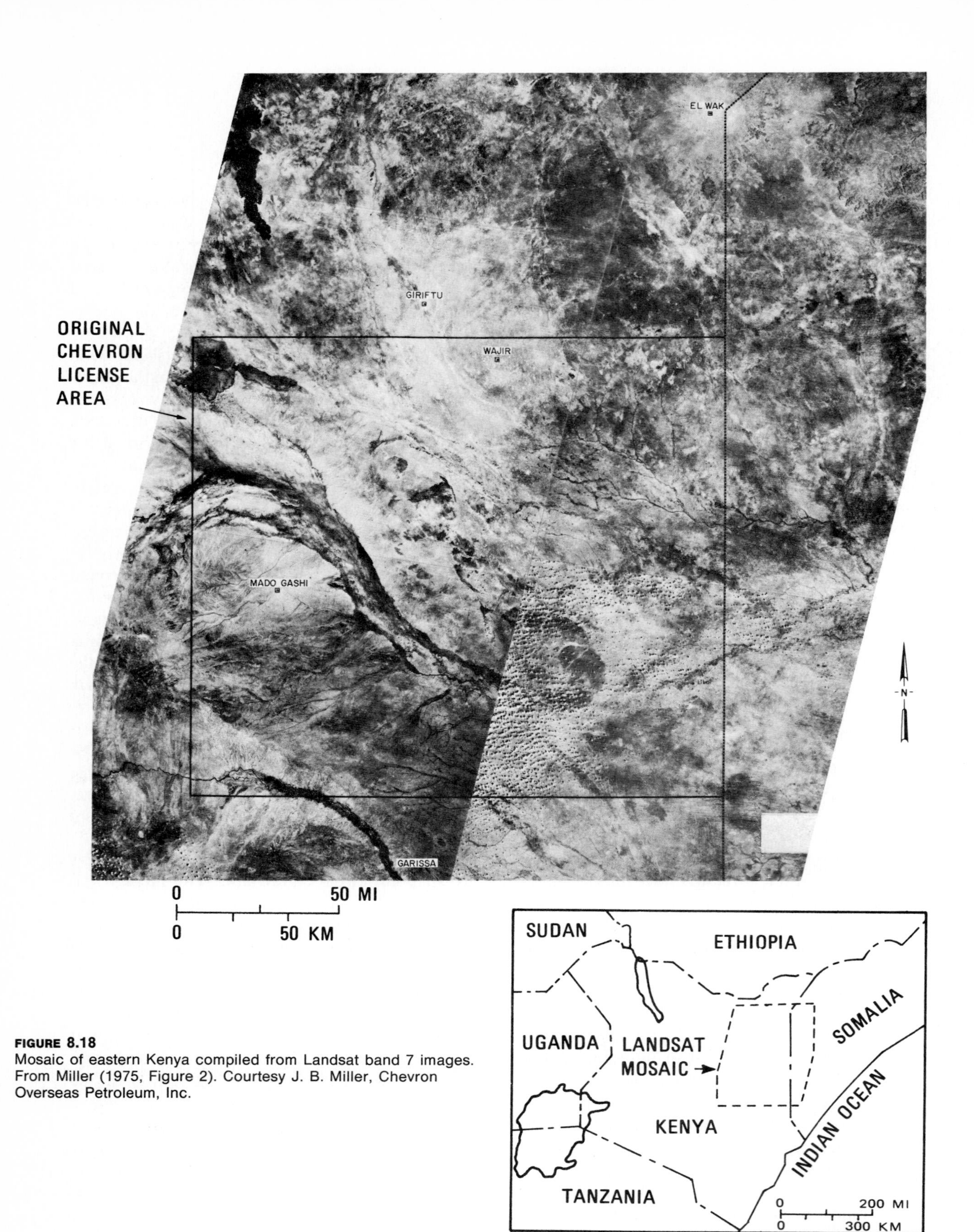

FIGURE 8.18
Mosaic of eastern Kenya compiled from Landsat band 7 images. From Miller (1975, Figure 2). Courtesy J. B. Miller, Chevron Overseas Petroleum, Inc.

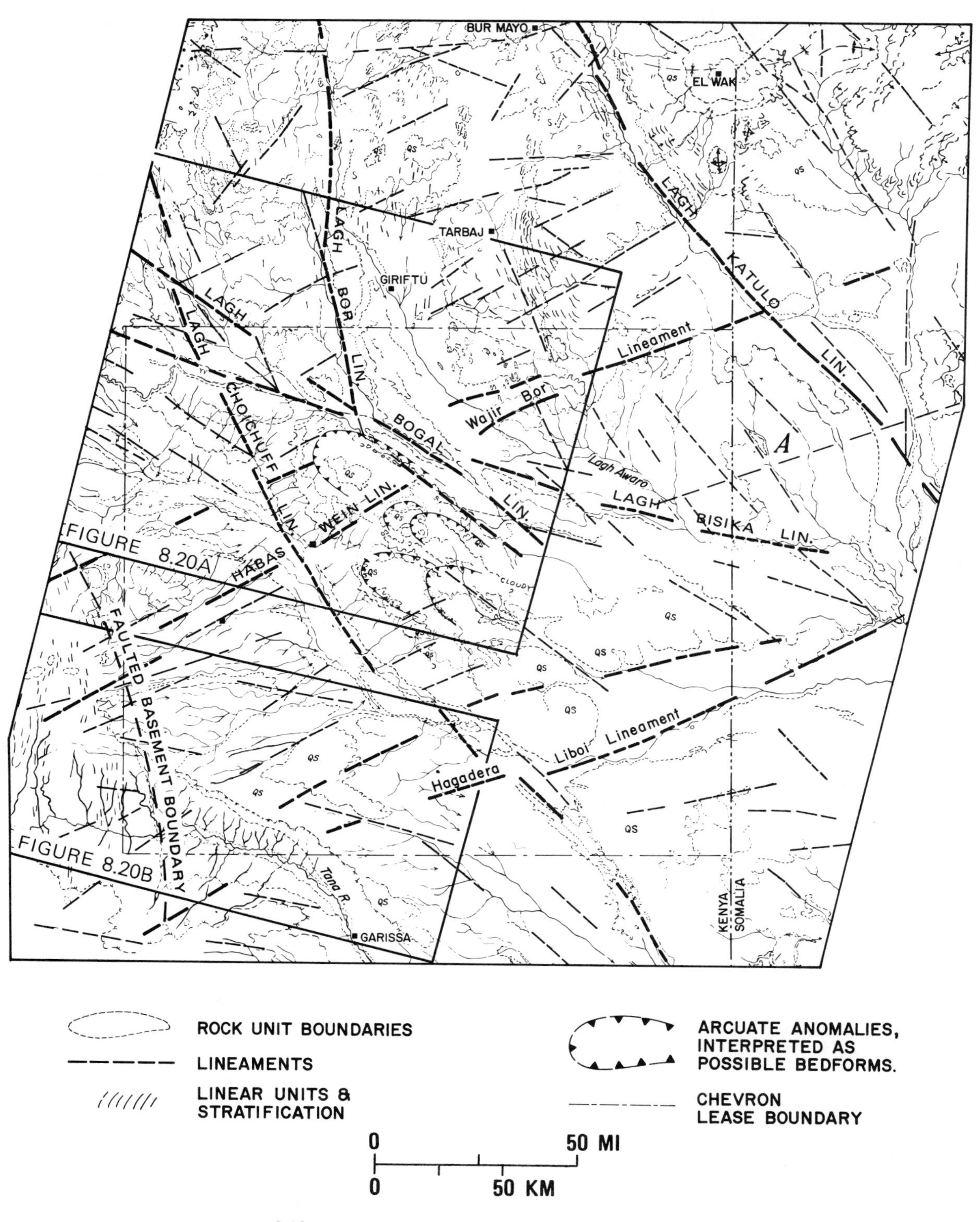

FIGURE 8.19
Interpretation of Landsat mosaic of eastern Kenya. From Miller (1975, Figure 4). Courtesy J. B. Miller, Chevron Overseas Petroleum, Inc.

A. NORTHWEST PART OF CHEVRON LICENSE AREA.
LANDSAT 1190–07054, BAND 5.

B. SOUTHWEST PART OF CHEVRON LICENSE AREA.
LANDSAT 1190–07061, BAND 5.

0 40 MI
0 40 KM

FIGURE 8.20
Landsat images of Chevron exploration license area in Kenya. See Figure 8.19 for locations. From Miller (1975). Courtesy J. B. Miller, Chevron Overseas Petroleum, Inc.

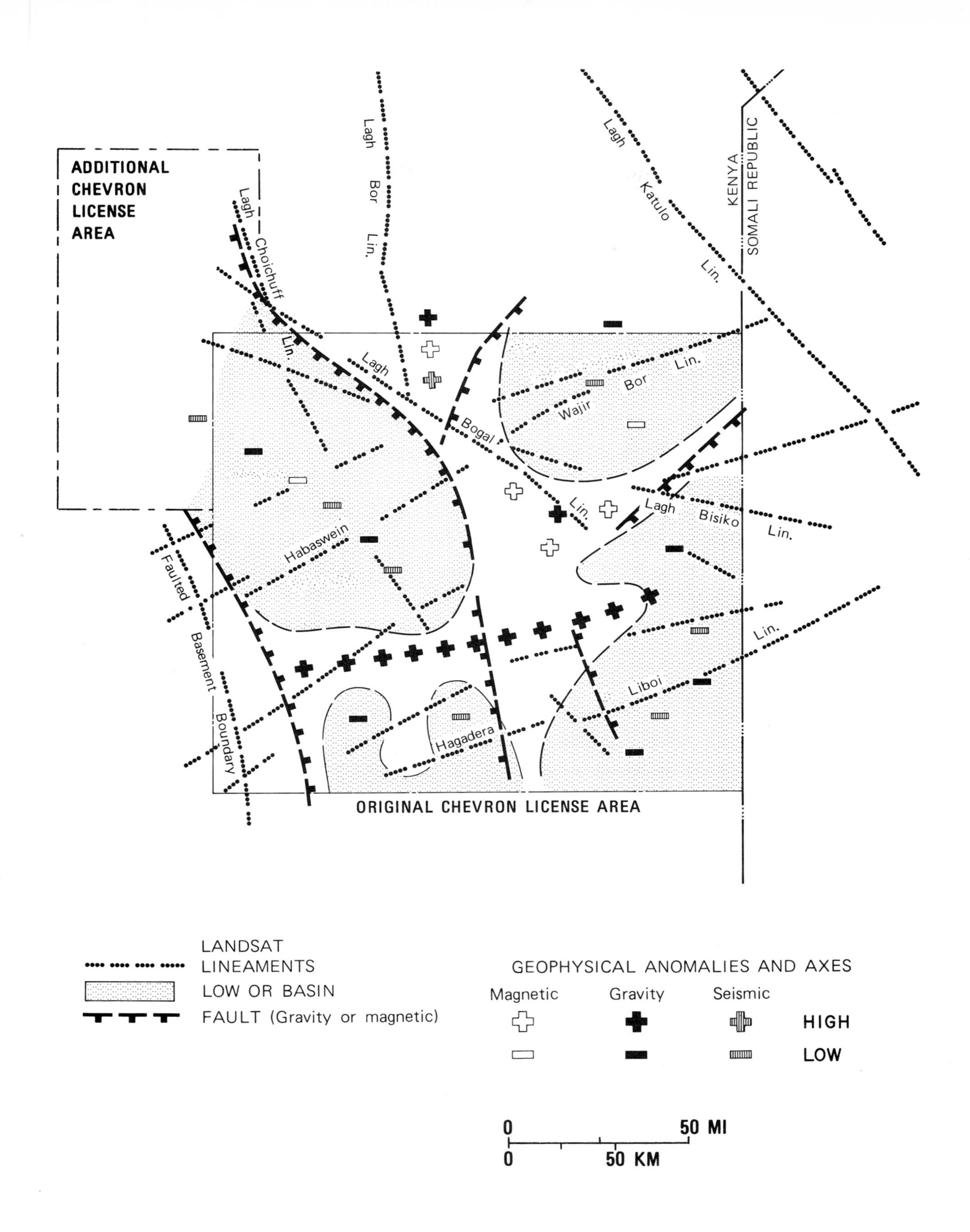

FIGURE 8.21
Comparison of Landsat features with geophysical trends. From Miller (1975, Figure 6). Courtesy J. B. Miller, Chevron Overseas Petroleum, Inc.

Egypt

Geologists of Santa Fe International Incorporated used Landsat images for reconnaissance of their East Cairo concession that covers about 4,600 km^2. The concession area, shown on the Landsat mosaic of Figure 8.22A, extends from the Great Bitter Lake on the northeast to the Damietta Branch of the Nile River on the west, where it includes some of the intensely cultivated and populated Nile Delta. Most of the area is gently rolling gravel desert bounded on the south by west-trending escarpments of Eocene limestone. Within the concession area, sedimentary rocks ranging in age from Oligocene to Recent are the predominant outcrops. Most features on published geologic maps of the area are also expressed on the Landsat images and some geologic patterns are expressed more clearly on the images than on the maps (Bentz and Gutman, 1977). The linear features interpreted from the image (Figure 8.22B) belong to a variety of structural trends:

1. The Gulf of Suez–Red Sea fracture system is represented by lineaments trending N35°W across the Landsat mosaic.
2. Linear features trending N50°W coincide with some of the major right-lateral wrench faults in the concession area. This trend and the N35°W trend have controlled the shorelines of the Gulf of Suez and the Red Sea.
3. Some west-trending lineaments are interpreted as right-lateral wrench faults, based on offsets of outcrop belts.
4. A major fold axis trending N70°E across the concession area was first discovered on Landsat images and is also apparent on geophysical maps.

As in Kenya, there is also a strong correspondence in Egypt between Landsat lineaments and geophysical trends. The gravity map (Figure 8.23A) indicates variations in density of the rocks that are caused by geologic structure or by lithologic changes. The N70°E folding trend is indicated by aligned gravity highs in the eastern part of the area. Shifts in the course of the Nile correspond to bends in the gravity contours and to Landsat linear features. The large gravity low near the juncture of the Rosetta and Damietta branches of the Nile marks an erosion channel 3,050 m deep that formed at the end of Miocene time and was filled with sediments of relatively low density. The nature of the gorge was determined from seismic work of Santa Fe International (Bentz and Gutman, 1977). Both the gorge and the present course of the Nile appear to be controlled by faults.

In the East Cairo concession area, the sedimentary sequence overlying the crystalline basement is not cut by intrusive rocks, nor are extrusive rocks present. The trends on the magnetic map (Figure

8.23B) therefore indicate the configuration of the basement rocks. High magnetic values generally indicate rocks with higher magnetic susceptibilities. Steep gradients and flattening of the magnetic contours may indicate faults in the basement. The N70°E fold trend observed on Landsat images is expressed by the subparallel alignment of magnetic highs and lows. This trend is interrupted in the south-central part of the concession area by a N10°W trend on the magnetic map that correlates with a major left-lateral strike-slip fault observed on the Landsat images. The gorge shown on the gravity map was eroded into the sedimentary section and did not involve the basement; therefore, the gorge is not expressed on the magnetic map. The Egyptian project illustrates the reconnaissance use of Landsat images, followed by geophysical surveys to locate prospect areas for drilling.

Northwest Colorado

In contrast to Kenya and Egypt, northwest Colorado is a mature exploration area where the surface structures have been drilled and the relationship between Landsat features and oil and gas fields can be directly evaluated. The winter image of Figure 8.24 demonstrates the advantages of low sun elevation and light snow cover for structural mapping. The major geologic features and the local structures associated with oil and gas fields are shown in Figure 8.25. The White River and Uinta Mountain uplifts and the Piceance Creek and Sand Wash Basins are well expressed. The Piceance Creek Basin contains major oil-shale reserves of the Green River Formation (Eocene age), which crops out around the basin margins. The Grand Hogback monocline separating the White River uplift from the Piceance Creek basin is especially prominent on the image.

Rangely anticline in the west-central part of Figure 8.24 is a major oil field that is outlined by strike ridges of resistant Cretaceous sandstones surrounding the eroded Mancos Shale outcrops in the core. The asymmetry of the anticline is indicated on the image by the gentle dip-slopes on the north flank and the steeper slopes on the south. The Blue Mountain anticline to the north is equally well expressed, but is nonproductive. Moffat, Iles, and Thornburgh are small oil fields trapped at anticlinal closures that are outlined by resistant sandstone ridges on the image. Danforth Hills and Wilson Creek oil fields are also anticlinal structures, but are marked on the image by a change to very fine texture. The Piceance Creek gas field (Figure 8.25) is a combination structural and stratigraphic trap. The anticlinal structure is clearly indicated by the radial drainage pattern and by the streams that "wrap around" the structure.

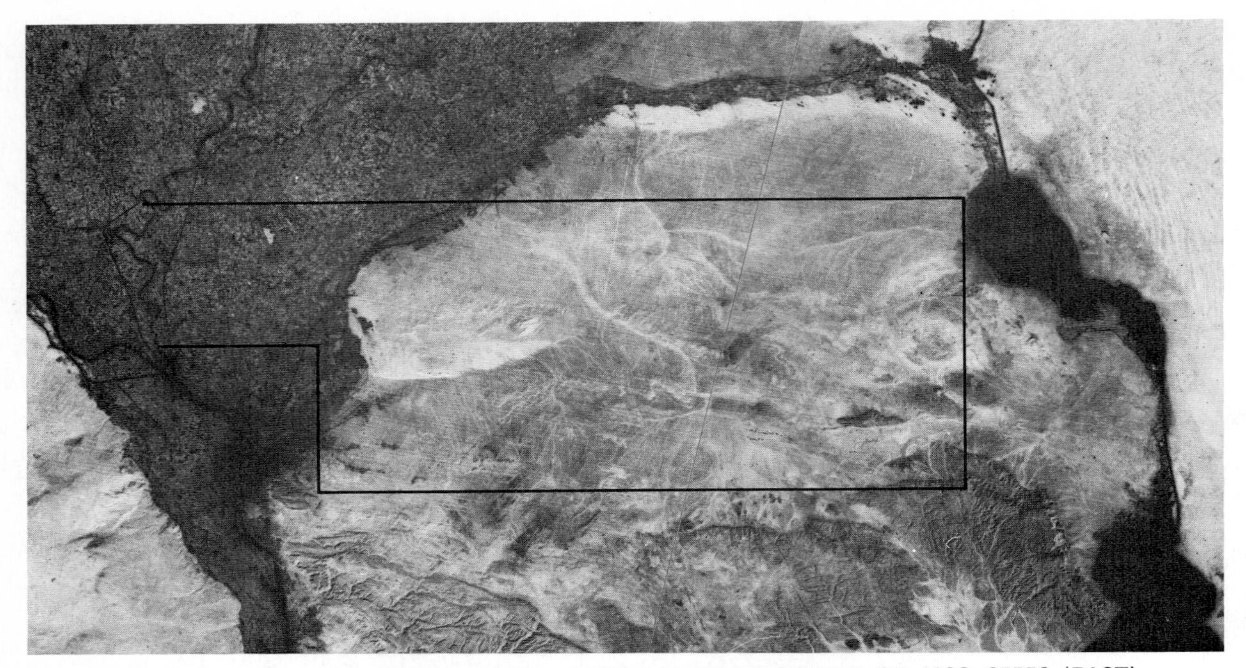

A. LANDSAT MOSAIC OF EASTERN EGYPT. COMPILED FROM BAND 5 OF 1236–07552 (EAST) AND 1165–08002 (WEST).

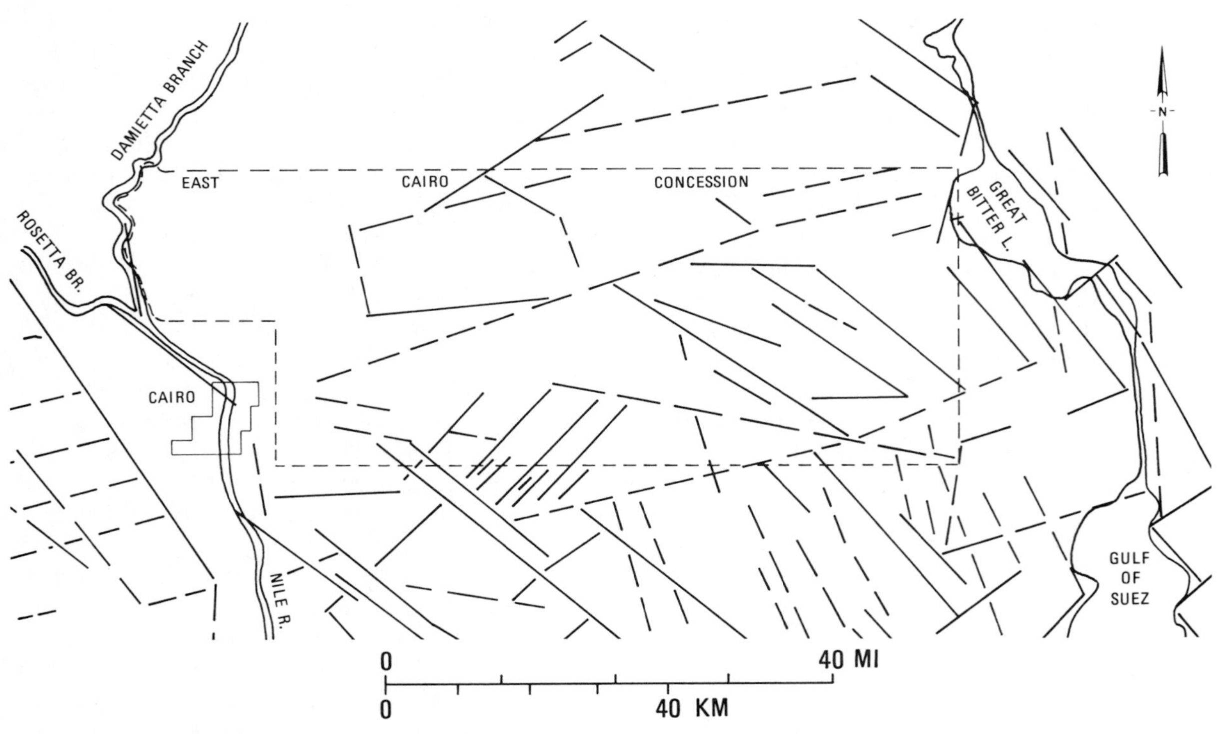

B. INTERPRETATION OF MAJOR LINEAR FEATURES ON LANDSAT MOSAIC.

FIGURE 8.22
Landsat reconnaissance mapping of oil exploration concession in eastern Egypt. From Bentz and Gutman (1977). Courtesy S. I. Gutman, Santa Fe International, Inc.

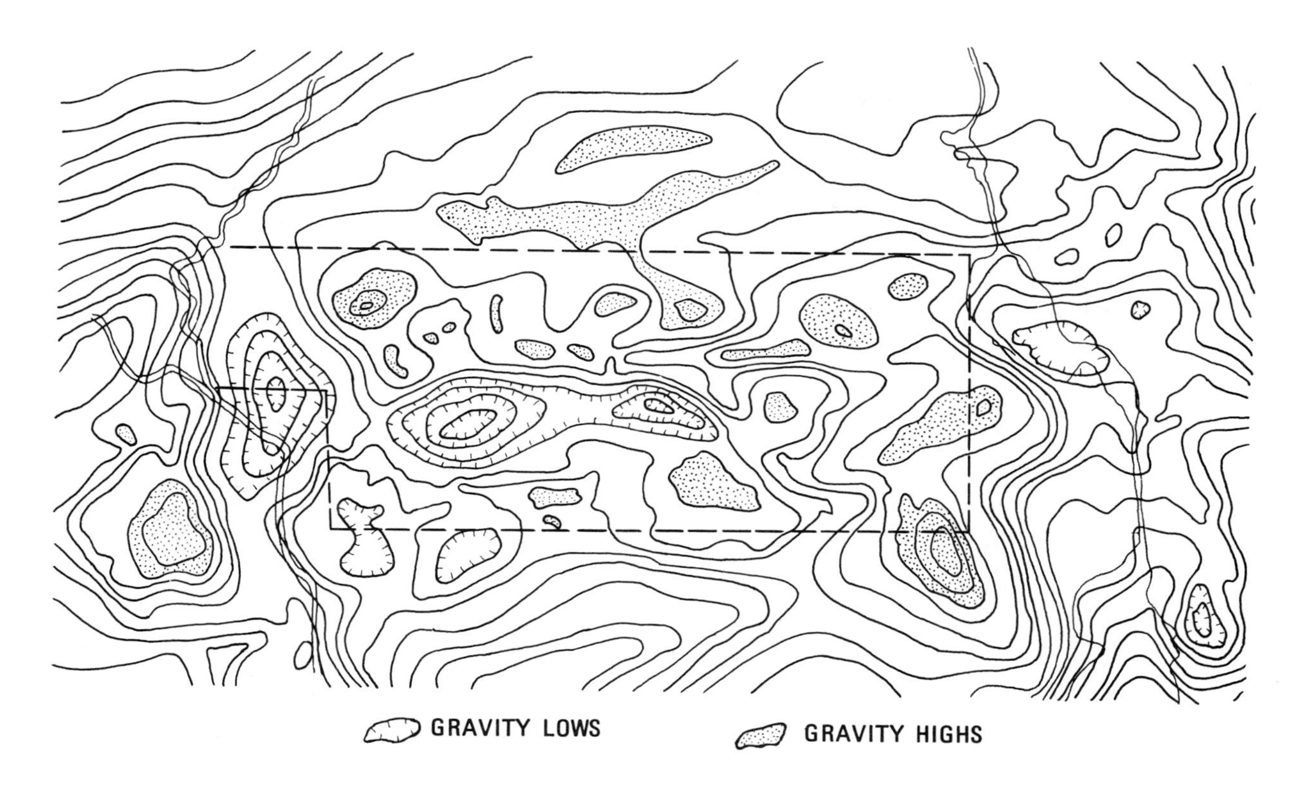

A. BOUGUER GRAVITY MAP. CONTOUR INTERVAL 2.5 MILLIGALS.

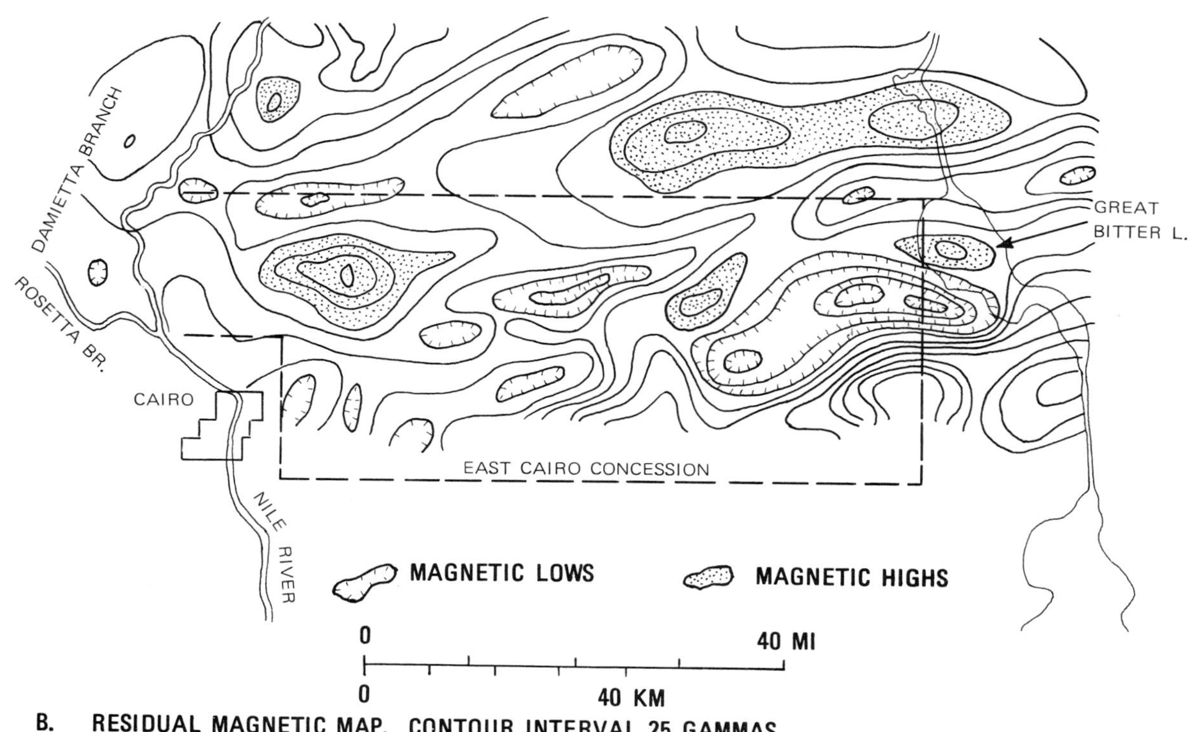

B. RESIDUAL MAGNETIC MAP. CONTOUR INTERVAL 25 GAMMAS.

FIGURE 8.23
Geophysical maps of eastern Egypt. From Bentz and Gutman (1977). Courtesy S. I. Gutman, Santa Fe International, Inc.

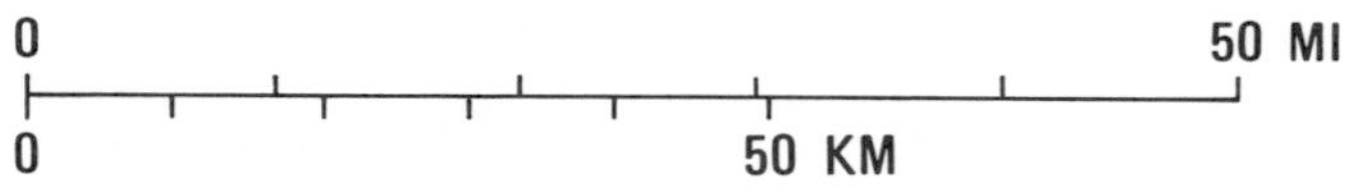

FIGURE 8.24
Winter image of northwest Colorado. Landsat 1156-17253 band 7 acquired December 26, 1973 at sun elevation of 21°.

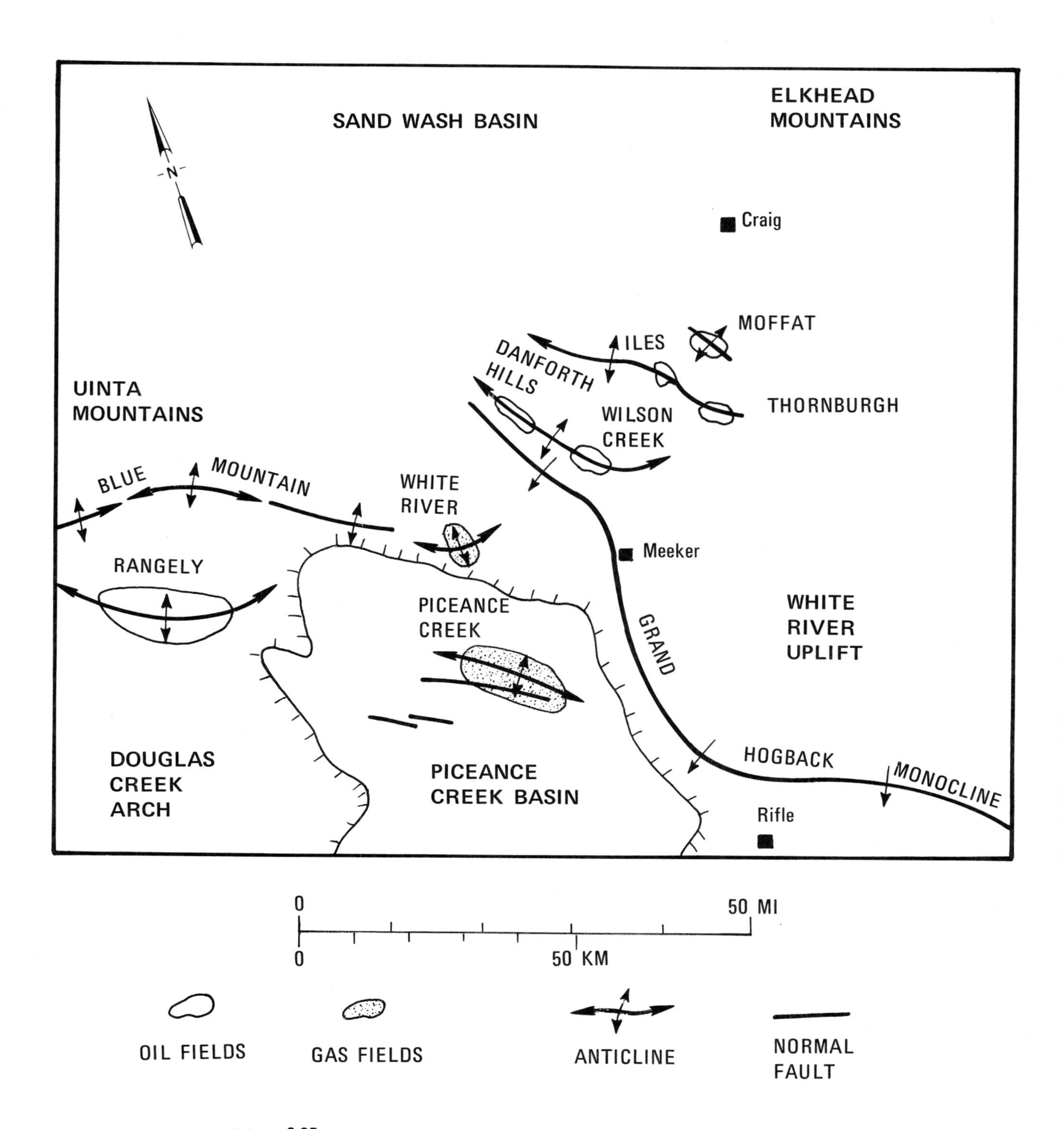

FIGURE 8.25
Location map of northwest Colorado showing oil and gas fields and associated structural features.

Applications of SLAR Images

The essentially all-weather capability and the ability to enhance geologic structure in forested terrain have made SLAR useful for oil exploration, especially in tropical regions. Wing and Mueller (1975) of Continental Oil Company described their structural reconnaissance mapping in Irian Jaya, Indonesia using SLAR images. Magnier, Oki, and Kaartidiputra (1975) published a SLAR mosaic of the Mahakam delta on the east coast of Kalimantan, Indonesia. The anticlinal trends of the onshore oil fields are clearly visible on the mosaic, despite the heavy vegetation cover.

Direct Indications of Petroleum

The Landsat and radar images may be interpreted for surface evidence of geologic structures, such as folds and faults, that may form petroleum traps at depth. This is analogous to the mapping of lineaments and fractures to locate areas favorable for ore deposition. Mineral exploration also searches for gossans and surface alteration zones that are direct indications of ore deposits. Some oil fields are marked at the surface by direct indications of the underlying hydrocarbons. Surface seeps of oil and gas that have leaked from subsurface traps are well-known examples. The original Drake well in Pennsylvania was located on the basis of oil seeps. As recently as the early 1900s, oil fields were located in California on the basis of oil seeps.

Another direct indication of petroleum is the alteration caused by the interaction between oil and gas seepage and the surface rocks overlying a hydrocarbon deposit. A classic example occurs in the surface rocks above the Cement oil field in southern Oklahoma. The sandstone outcrops are typically red, but over the Cement field they are tan and gray. Gypsum ($CaSO_4 \cdot nH_2O$) is locally replaced by calcite ($CaCO_3$) over the field. The change from red to gray has long been attributed to escaping hydrocarbons that chemically reduced the red iron oxide in the sandstone to a nonred iron compound. Another surface alteration effect was recognized by Donovan (1974), who found that the secondary calcite and dolomite ($Ca_{1/2}Mg_{1/2}CO_3$) in the surface rocks have unusual carbon isotopic values. These values indicate that hydrocarbons leaking from the reservoir were oxidized and the carbon incorporated into the secondary carbonate minerals of the surface rocks. Similar carbon isotopic values occur over the Davenport oil field in central Oklahoma (Donovan, Friedman, and Gleason, 1974).

Several unsuccessful attempts have been made to identify on Landsat images the signatures of the color and mineralogic alteration patterns at the Cement field. Such signatures could than be used to explore for other fields. Various digital processes have been applied to Landsat data of the Cement field, but no successes have been reported. Field investigations by T. J. Donovan and L. C. Rowan (personal communications) found that the color changes at Cement field are only visible at exposures in road cuts and stream beds. Most of the area is covered with soil and agriculture that has obscured the alteration effects. The ground resolution cell of Landsat is too large to detect such small and obscure targets.

Everett and Petzel (1973) interpreted Landsat images of the Anadarko Basin in the Texas Panhandle and western Oklahoma. Over a number of oil and gas fields they reported "hazy" anomalies on the images that are said to appear as if image detail had been smudged or erased. However, the anomalies are not artifacts of the image reproduction process. The hazy anomalies are only recognizable on certain Landsat images and are not visible on aerial photographs. No conclusive explanation has been given for the hazy anomalies, but it has been suggested that they are caused by man's activities. These anomalies are of interest to other investigators but no one has reported similar features. The background and status of the Anadarko basin investigation were summarized by Short (1975) who noted the lack of agreement about the cause of the hazy anomalies.

These and other efforts at remote detection of direct indications of petroleum have been generally unsuccessful, but this should not prevent further research. Some direct surface indications of petroleum do exist and may yet prove to be detectable by remote-sensing methods.

GEOTHERMAL ENERGY

Geothermal energy can be produced from subsurface reservoirs of steam or hot water that are shallow enough to be drilled and exploited economically. The following conditions are necessary for a geothermal reservoir:

1. A large, high-temperature heat source must be present at relatively shallow depth. Intrusive masses of young igneous rock are the usual heat source, and most geothermal areas are associated with surface or subsurface igneous rocks of Cenozoic age.
2. Porous and permeable reservoir rocks to hold the steam or hot water must occur near the heat source. A variety of rocks can serve as geothermal reservoirs. In the Imperial Valley of California poorly consolidated sandstones are the reservoir rocks. At the Geysers, in northern California, the reservoir occurs in sedimentary and volcanic rocks with little porosity and permeability; however, fracturing and faulting have caused the necessary porosity and permeability.
3. There must be a natural recharge system to replenish the water as steam or hot water is produced.
4. An impermeable zone above the reservoir is necessary to prevent the escape of steam and hot water. Convective flow to the surface would dissipate the heat of an unconfined reservoir. Heat losses due to thermal conduction through the rocks are relatively minor because of the low thermal conductivity of rocks.

Some geothermal reservoirs have no visible or thermal expression at the surface and are not detectable through the use of remote-sensing methods. Many geothermal reservoirs, however, have surface thermal expressions ranging in intensity from a minor elevation of ground temperature to the presence of hot springs and geysers. Hot springs and geysers commonly occur along faults and fractures that allow hot water to escape from the reservoir. Thermal springs and zones of hydrothermal alteration associated with a geothermal area in Mexico have been interpreted from thermal IR images (Valle and others, 1970). At the Geysers area, hot springs and fumaroles were also detected on thermal IR images and there is some local evidence of higher ground temperatures (Moxham, 1969). However, there is little evidence of a regional surface temperature anomaly on the imagery of the Geysers area.

Iceland Geothermal Reconnaissance

Iceland is located on the Reykjanes ridge, which is the boundary between the diverging North American and Eurasian plates. Because of this geologic setting, Iceland is the site of frequent volcanic eruptions, including six since 1946, and associated geothermal activity. Space heating and hot water for the capital city of Reykjavik have long been supplied from geothermal sources. The high-temperature geothermal areas are concentrated along the zones of active rifting and volcanism, as shown in Figure 8.26. Vatnajökull is an ice cap approximately 100 km in diameter that overlies part of the eastern zone of volcanism and covers two known high-temperature geothermal areas. These subglacial heat sources cause melting that often results in ice-cauldron subsidence features and periodic floods of melt water (Williams and others, 1974).

The aerial photograph and thermal IR image of Figure 8.27 show the Kverkfjöll geothermal area, which is located at the northern edge of the Vatnajökull ice cap (Figure 8.26). As shown on the interpretation map of Figure 8.28, the geothermal area is located between the Kverkjökull outlet glacier on the east and a re-entrant of bedrock on the west. The thermal IR image (Figure 8.27B) is not rectilinearized, which accounts for the geometric compression at the east and west margins. The ice and bedrock have relatively cool signatures (dark tones). The geothermal features, melt water, and the bedrock ridges, which confine the outlet glacier, have warm signatures. The north-trending geothermal feature is at least 2 km long and includes two separate hot areas at the northern end. At the south margin of the image, warm signatures mark concentric crevasses and an ice-cauldron subsidence feature that are caused by subsurface melting of the glacier. The warm stream emerging from the snout of the outlet glacier is melt water that has flowed along the base of the glacier from a subglacial geothermal source (R. S. Williams, Jr., personal communication). The topographic expressions of some geothermal features are detectable on the aerial photograph of Figure 8.27A, but are less pronounced than on the IR image. The U.S. Air Force Cambridge Research Laboratories and NASA, in association with the U.S. Geological Survey and Icelandic scientific organizations, have acquired thermal IR images of other geothermal areas in Iceland that have expressions similar to the Kverkfjöll geothermal area. Geothermal vents and hot springs have been detected on thermal IR images in Japan, Italy, Ethiopia, and the United States. The technique is especially useful in remote areas where the surface expression of geothermal activity has not been located by conventional means.

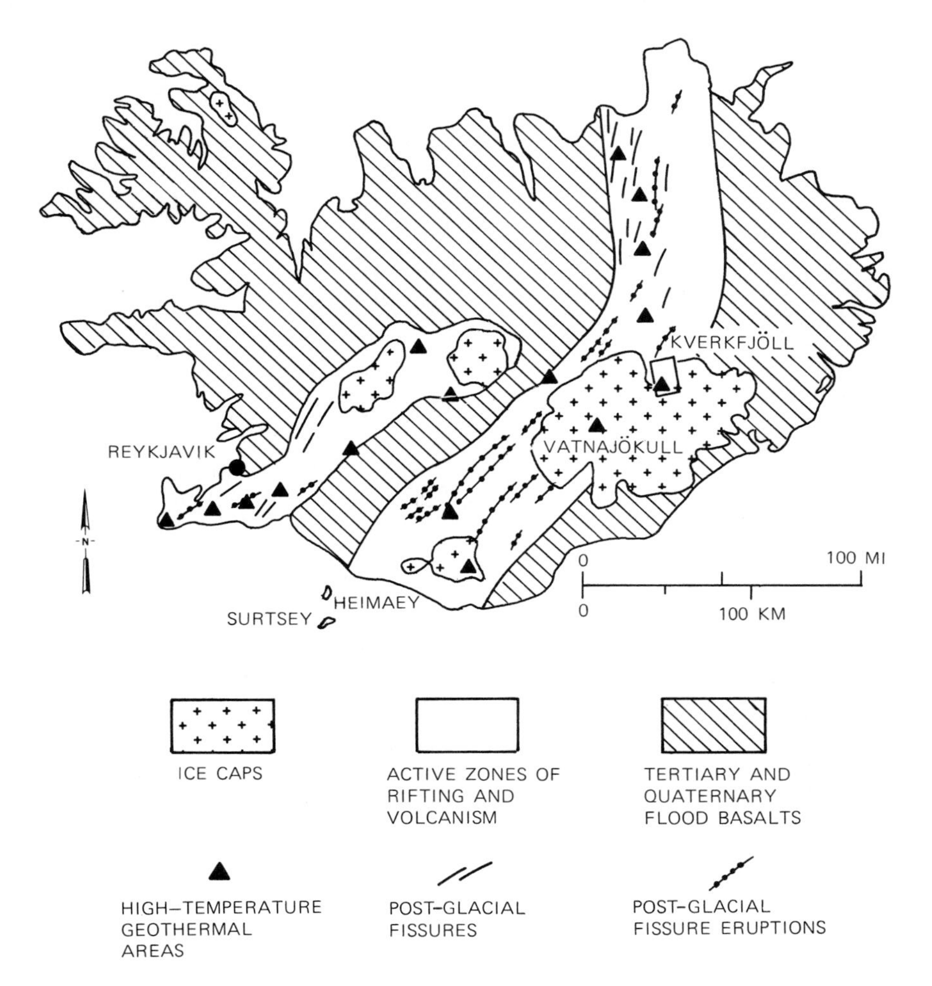

FIGURE 8.26
Generalized geologic map of Iceland showing location of Kverkfjöll geothermal area. Rectangle shows location of Figures 8.27 and 8.28. From Friedman and others (1969, Figure 4).

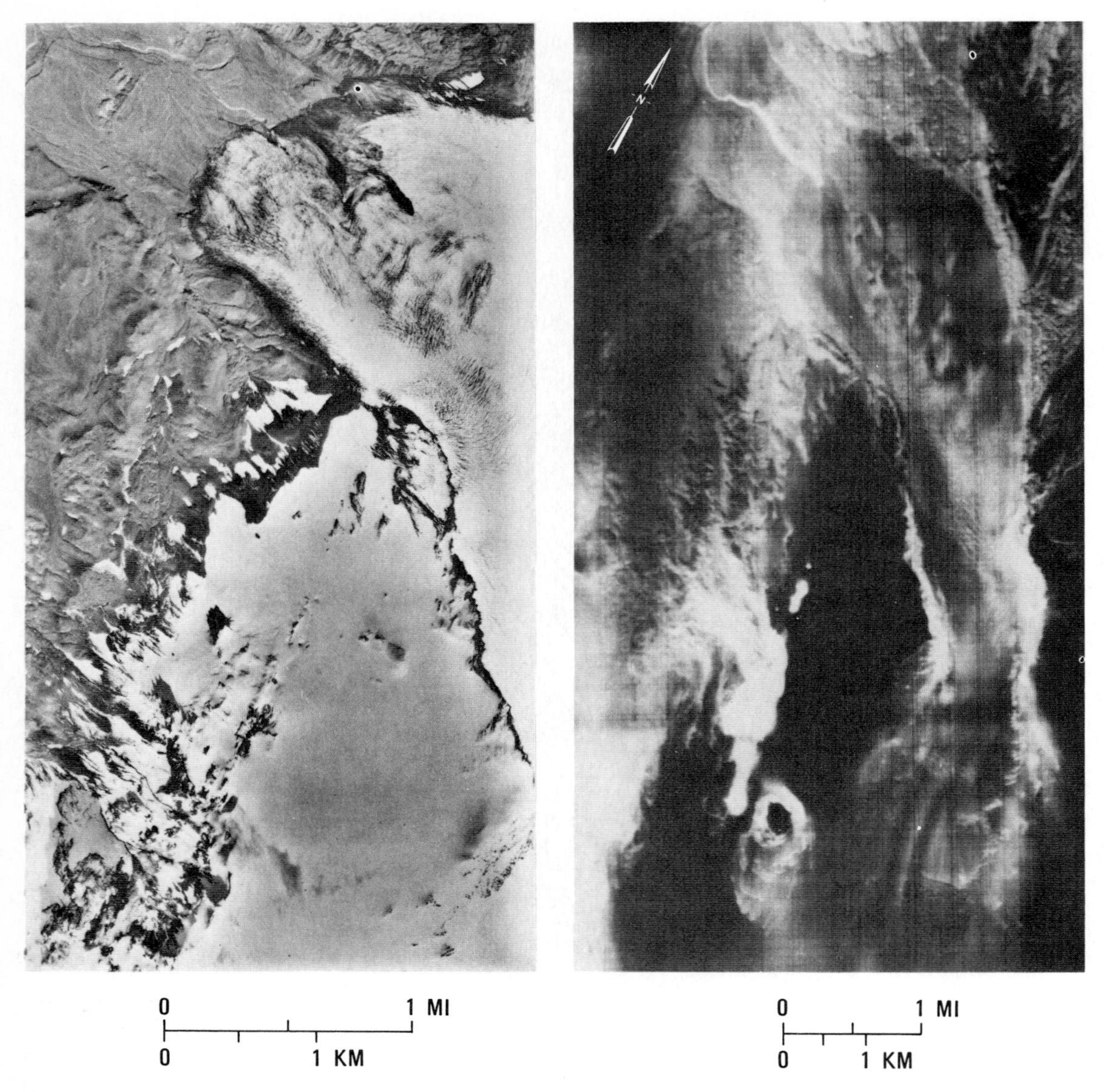

A. AERIAL PHOTOGRAPH ACQUIRED AUGUST 24, 1960 BY U.S. AIR FORCE.

B. NIGHTTIME THERMAL IR (1 TO 5.5 μm) IMAGE (NOT RECTILINEARIZED). ACQUIRED AUGUST 22, 1966 BY U.S. AIR FORCE CAMBRIDGE RESEARCH LABORATORIES.

FIGURE 8.27
Kverkfjöll geothermal area and Kverkjökull outlet glacier, Iceland. From Friedman and others (1969, Figures 10 and 11). Courtesy R. S. Williams, Jr., U.S. Geological Survey.

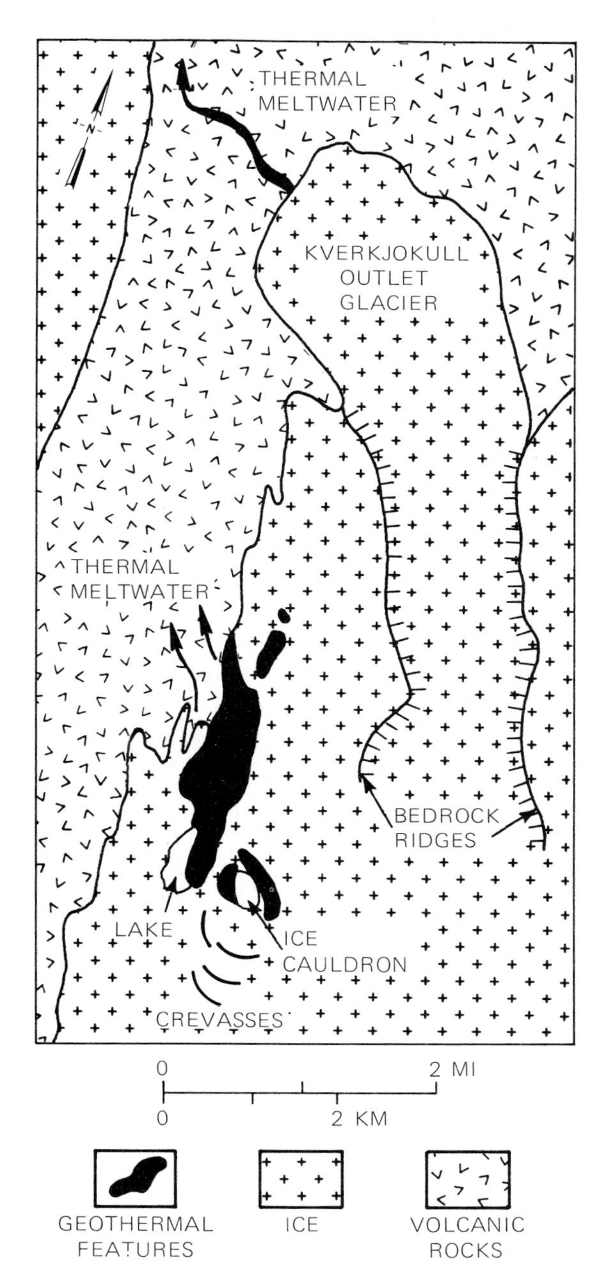

FIGURE 8.28
Interpretation map of thermal IR image of Kverkfjöll geothermal area. From Friedman and others (1969, Figure 11).

Low-Intensity Geothermal Anomalies

Thermal IR images are ideal for detecting fumaroles, steaming ground, and hot springs associated with very active geothermal areas such as Iceland and Yellowstone. Elsewhere, however, there are so-called "blind" geothermal areas that lack surface thermal activity or rock alteration. There is an additional broad category of geothermal areas in which the surface temperature is only slightly higher than the surrounding areas. These areas with low-intensity surface temperature anomalies are difficult to detect on IR images or on airborne radiometer profiles. Watson (1975) pointed out that natural variations of geology and topography can readily overwhelm surface geothermal anomalies of several hundred heat flow units (1 HFU = $1 \cdot 10^{-6} \cdot \text{cal}^{-1} \cdot \text{cm}^{-2} \cdot \text{sec}^{-1}$). Mathematical model studies were made to evaluate the relative effect of various factors on the surface radiant temperature. This analysis suggests that both thermal and reflectance images should be acquired at least three times during the diurnal cycle (Watson, 1975, p. 136). Comparison of the images may reveal subtle anomalies. A thermal IR image in the Raft River area of Idaho revealed a weak thermal anomaly that was confirmed by ground measurements.

OTHER ENERGY SOURCES

The location and distribution of large reserves of coal, oil shale, and tar sands are already known in the United States and Canada. Therefore exploration is relatively inactive and there is little application of remote sensing. Landsat images are potentially useful during mining of coal. The status of strip mining and land reclamation may be monitored by digital processing of CCT data acquired during the nine-day repetition cycle of Landsat 1 and 2 (Anderson and Schubert, 1976). In addition, hazards may be discovered by studying these images. In Indiana it was demonstrated that areas of intense fracturing on Landsat images coincided with areas of roof falls in coal mines and the fracture patterns could be used to predict the hazards (Wier and others, 1973).

FUTURE IMAGERY REQUIREMENTS

Although the Landsat multispectral scanner was not designed specifically for resource exploration, it has proven valuable for this application. Future satellite systems should produce even better results if the following improvements are made.

1. Higher spatial resolution. This is particularly needed for detecting gossans and zones of hydrothermal alteration. Gossans in a portion of the Arabian Shield are shown in Figure 8.29 with the Landsat MSS ground resolution cell and pixel at the same scale. In this area the resolution cell includes more country rock and alluvium than gossan. Reducing the dimensions of the ground resolution cell by half would improve the ability to detect the spectral signatures of gossans, but the number of pixels would be increased by a factor of four. Such a higher resolution scanner could be restricted to coverage of potential mineral areas, rather than to worldwide coverage.
2. Provision of additional spectral bands to acquire images at 1.5 to 1.8 μm and 2.0 to 2.4 μm, where distinctive spectral differences occur between altered and unaltered rocks (Figure 8.5).
3. Provision for complete stereo coverage with optimum vertical exaggeration on the stereo model. The improved capability to map folds and faults would be valuable for both mineral and oil exploration.

Radar and thermal IR images acquired from aircraft have considerable potential for mineral and energy exploration, but such images exist for only a small portion of the earth's surface. In the future, radar and thermal IR images acquired from satellites may prove to be as useful as Landsat images.

COMMENTS

Aerial photography has long been used in resource exploration and newer forms of imagery are being employed. Private industry is the largest single purchaser, in dollar value, of remote-sensing imagery from the EROS Data Center, with the principal application being for oil and mineral exploration. The extractive industries also employ contractors to acquire photography, thermal IR, and radar imagery in areas of exploration interest. These investments by cost-conscious industry are evidence for the exploration value of remote sensing. Nevertheless, critics of remote sensing frequently ask, in effect, "If remote sensing is so useful, how many mines and oil fields has it dis-

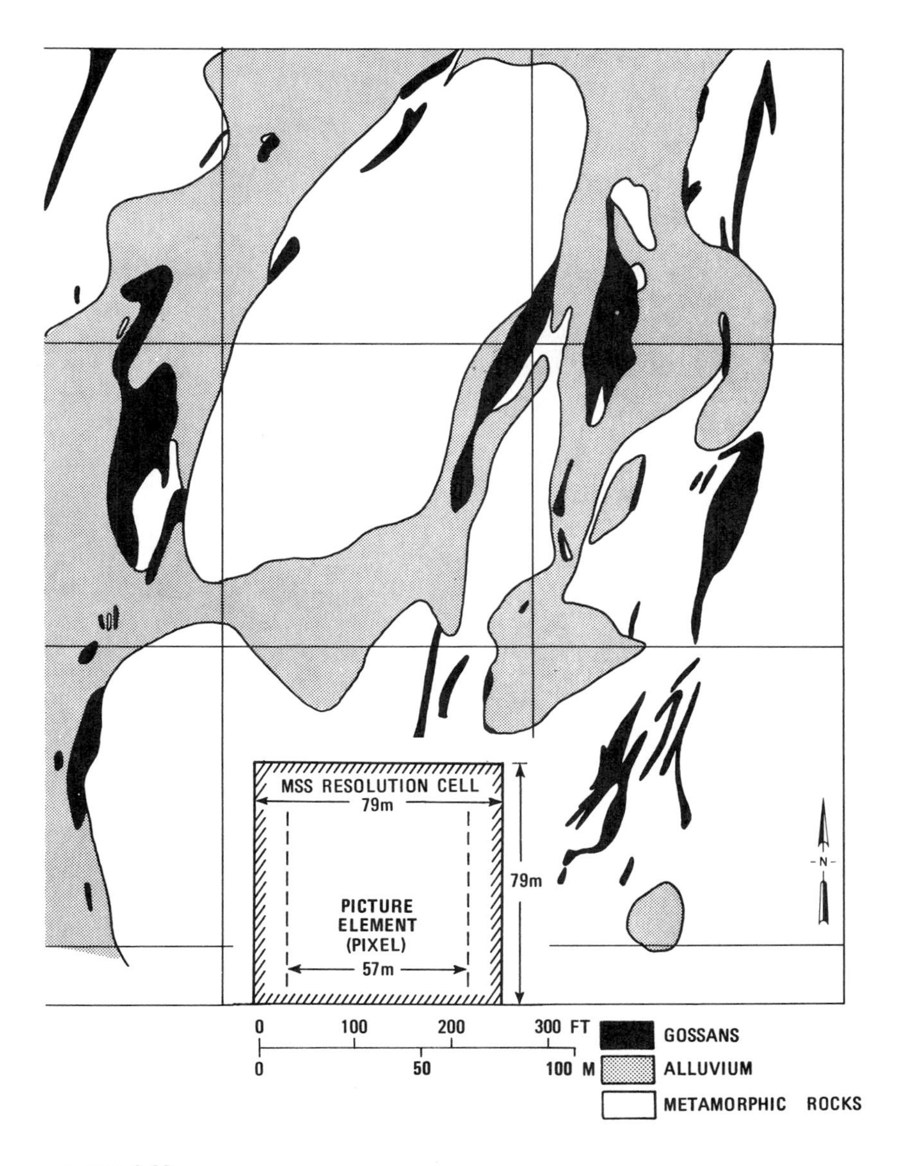

FIGURE 8.29
Size comparison of gossans with Landsat MSS ground resolution cell and pixel dimension. Locality is the Rabathan ancient mining district in the Arabian Shield (20° 25′N, 41° 25′E). Geology from Earhart and Mawad (1970, Plate 6).

covered?" Those who ask this question reveal a lack of familiarity with the exploration process. Few if any significant discoveries will ever be credited solely to remote sensing for the following reasons:

1. Remote sensing is a regional reconnaissance method that indicates target areas for follow-up surveys by more detailed and costly methods.
2. It is improbable that a mine shaft will be sunk or a wildcat well drilled strictly on the basis of remote-sensing surveys. In oil and mineral exploration the reconnaissance surveys are followed by field mapping and by geophysical and geochemical surveys that eventually define a prospect suitable for drilling.
3. An oil or mineral discovery results from a combination of exploration methods and no single method is responsible for the discovery. In practice the last survey that was conducted is commonly credited with the discovery.

The combination of Landsat MSS data and computer image processing has great potential for mineral exploration and has been demonstrated in several test cases. Improvements in spatial resolution and extended sensitivity farther into the reflected IR spectral region should improve the resource exploration capability of any future satellite systems.

REFERENCES

Abrams, M. J., R. P. Ashley, L. C. Rowan, A. F. H. Goetz, and A. B. Kahle, 1977, Use of imaging in the 0.46 to 2.36 μm spectral region for alteration mapping in the Cuprite mining district, Nevada: U.S. Geological Survey Open File Report 77-585.

Albert, N. R. D., 1975, Interpretation of Earth Resource Technology Satellite imagery of the Nabesna Quadrangle, Alaska: U.S. Geological Survey Miscellaneous Field Map MP 655J.

Anderson, A. T. and J. Schubert, 1976, ERTS-1 data applied to strip mining: Photogrammetric Engineering, v. 42, p. 211–219.

Bentz, F. P. and S. I. Gutman, 1977, Landsat data contribution to hydrocarbon exploration in foreign regions *in* Woll, T. W. and W. A. Fischer, eds., Proceedings of first annual W. T. Pecora memorial symposium, October, 1975: U.S. Geological Survey Professional Paper 1015, p. 83–92.

Canney, F. C., 1970, Remote detection of geochemical soil anomalies *in* Second Annual Aircraft Program Review: NASA Johnson Spacecraft Center, v. 1, sec. 7.

Cannon, H. L. 1971, The use of plant indicators in groundwater surveys, geologic mapping, and mineral prospecting: Taxon, v. 20, p. 227–256.

Chenoweth, W. L. and D. N. Magleby, 1971, Mine location map, Cameron uranium area, Coconino County, Arizona: U.S. Atomic Energy Commission Preliminary Map 20.

Donovan, T. J., 1974, Petroleum microseepage at Cement field, Oklahoma—evidence and mechanism: American Association of Petroleum Geologists Bulletin, v. 58, p. 429–446.

Donovan, T. J., I. Friedman, and J. D. Gleason, 1974, Recognition of petroleum-bearing traps by unusual isotopic compositions of carbonate-cemented surface rocks: Geology, v. 2, p. 351–354.
Earhart, R. L. and M. M. Mawad, 1970, Geology and mineral evaluation of the Wadi Bidah District, southern Hijaz Quadrangle, Kingdom of Saudi Arabia: U.S. Geological Survey Saudi Arabia Investigation Open File Report (IR) SA-119.
Everett, J. R. and G. Petzel, 1973, An evaluation of the suitability of ERTS data for the purposes of petroleum exploration: Third Earth Resources Technology Satellite Symposium, NASA SP-356, p. 50–61.
Friedman, J. D., R. S. Williams, G. Pálmason, and C. D. Miller, 1969, Infrared surveys in Iceland: U.S. Geological Survey Professional Paper 650-C, p. C89–C105.
Goetz, A. F. H., 1976, Remote sensing geology—Landsat and beyond: Caltech/JPL Conference on Image Processing Technology, Data Sources and Software for Commercial and Scientific Purposes, Jet Propulsion Lab SP 43-30, p. 8-1 to 8-8.
Halbouty, M. T., 1976, Application of Landsat imagery to petroleum and mineral exploration: American Association of Petroleum Geologists Bulletin, v. 60, p. 745–793.
Hunt, G. R. and J. W. Salisbury, 1970, Visible and near-infrared spectra of minerals and rocks–I silicate minerals: Modern Geology, v. 1, p. 283–300.
Hunt, G. R., J. W. Salisbury, and C. J. Lenhoff, 1971, Visible and near-infrared spectra of minerals and rocks—III oxides and hydroxides: Modern Geology, v. 2, p. 195–205.
Horton, R. C., 1964, An outline of the mining history of Nevada, 1924 –1964: Nevada Bureau Mines Report 7, pt. 2.
Kappelmeyer, O. and R. Haenel, 1974, Geothermics with special reference to application: Geoexploration Monographs, Series 1, n. 4 Gebrüder Borntraeger, Berlin.
Lowell, J. D. and J. M. Guilbert, 1970, Lateral and vertical alteration–mineralization zoning in porphyry ore deposits: Economic Geology, v. 65, p. 373–408.
Lyon, R. J. P., 1975, Mineral exploration applications of digitally processed Landsat imagery: Stanford Remote Sensing Laboratory Report 75-14, Stanford, Calif.
Magnier, P., T. Oki, and L. W. Kaartidiputra, 1975, The Mahakam Delta: Proceedings World Petroleum Congress, v. 2, p. 239–250, Tokyo.
Miller, J. B., 1975, Landsat images as applied to petroleum exploration in Kenya: NASA Earth Resources Survey Symposium, NASA TM X-58168, v. 1-B, p. 605–624.
Moxham, R. M., 1969, Aerial infrared surveys at the Geysers geothermal steam field, California: U.S. Geological Survey Professional Paper 630-C, p. C106–C122.
Nicolais, S. M., 1974, Mineral exploration with ERTS imagery: Third ERTS-1 Symposium, NASA SP-351, v. 1, p. 785–796.
Offield, T. W., 1976, Remote sensing in uranium exploration *in* Exploration of uranium ore deposits: Proceedings International Atomic Energy Agency, p. 731–744, Vienna, Austria.
Park, C. F. and R. A. MacDiarmid, 1975, Ore deposits: Third Edition, W. H. Freeman and Co., San Francisco, Calif.
Rowan, L. C., P. H. Wetlaufer, A. F. H. Goetz, F. C. Billingsley, and J. H. Stewart, 1974, Discrimination of rock types and detection of hydrothermally altered areas in south-central Nevada by the use of computer-enhanced ERTS images: U.S. Geological Survey Professional Paper 883.
Rowan, L. C., and P. H. Wetlaufer, 1975, Iron–absorption band analysis for the discrimination of iron–rich zones: U.S. Geological Survey, Type III Final Report.
Rowan, L. C., A. F. H. Goetz, and R. P. Ashley, 1977, Discrimination of hydrothermally altered and unaltered rocks in visible and near-infrared multispectral images: Geophysics, v. 42, p. 522–535.

Sabins, F. F., L. C. Rowan, N. M. Short, and R. K. Stewart, 1975, Geology *in* Summary reports: NASA Earth Resources Survey Symposium, NASA TM X-58168, v. 3, p. 21–28.

Schmidt, R., B. B. Clark, and R. Bernstein, 1975, A search for sulfide-bearing areas using Landsat-1 data and digital image-processing techniques: NASA Earth Resources Survey Symposium, NASA TM X-58168, v. 1-B, p. 1013–1027.

Short, N. M., 1975, Exploration for fossil and nuclear fuels from orbital altitudes *in* Remote Sensing energy related studies: Hemisphere Publishing Corp., p. 189–232, Washington, D.C.

Spirakis, C. S. and C. D. Condit, 1975. Preliminary report on the use of Landsat-1 (ERTS-1) reflectance data in locating alteration zones associated with uranium mineralization near Cameron, Arizona: U.S. Geological Survey Open File Report 75-416.

Valle, R. G., J. D. Friedman, S. J. Gawarecki, and C. J. Banwell, 1970, Photogeologic and thermal infrared reconnaissance surveys of the Los Negritos-Ixtlan de Los Hervores geothermal area, Michoacan, Mexico: Geothermics Special Issue no. 2, p. 381–398.

Viljoen, R. P., M. J. Viljoen, J. Grootenboer, and T. G. Longshaw, 1975, ERTS-1 imagery—an appraisal of applications in geology and mineral exploration: Minerals Science and Engineering, v. 7, p. 132–168.

Watson, K., 1975, Geologic applications of thermal infrared images: Proceedings IEEE, v. 63, p. 128–137.

Wier, C. E., F. J. Wobber, O. R. Russell, R. V. Amato, and T. V. Leshendok, 1973, Relationship of roof falls in underground coal mines to fractures mapped on ERTS-1 imagery: Third ERTS-1 Symposium NASA SP-351, p. 825–843.

Williams, R. S., Jr., and others, 1974, Environmental studies of Iceland with ERTS-1 imagery: Proceedings of Ninth International Symposium on Remote Sensing of Environment, Environmental Research Institute of Michigan, p. 31–81, Ann Arbor, Mich.

Wing, R. S. and J. C. Mueller, 1975, SLAR reconnaissance, Mimika-Eilanden Basin, southern trough of Irian Jaya; NASA Earth Resources Survey Symposium, NASA TM X-58168, v. I-B, p. 599–604.

Yost, E. and S. Wenderoth, 1971, The reflectance spectra of mineralized trees: Proceedings of Seventh International Symposium on Remote Sensing of Environment, University of Michigan, p. 269–284, Ann Arbor, Mich.

ADDITIONAL READING

Hunt, G. R., 1977, Spectral signatures of particulate minerals in the visible and near infrared: Geophysics, v. 42, p. 501–513.

Offield, T. W., E. A. Abbott, A. R. Gillespie, and S. O. Laguercio, 1977, Structural mapping on enhanced Landsat images of southern Brazil—Tectonic control of mineralization and speculations on metallogeny: Geophysics, v. 42, p. 482–500.

9

ENVIRONMENTAL AND LAND-USE APPLICATIONS

Although originally developed for other purposes, remote sensing is proving very useful for a broad range of environmental applications. All the wavelength regions, from UV through microwave, have practical applications. The synoptic regional coverage, availability of data from inaccessible areas, and repeated coverage are advantages of aircraft and satellite imagery.

MARINE ENVIRONMENT

Imagery in the photographic spectral region from aircraft and satellites is valuable for mapping bathymetry of shallow shelves and for monitoring water currents and sedimentation patterns. Thermal IR imagery from satellites and aircraft enables currents to be mapped on the basis of differences in radiant temperature of the water bodies.

Mapping the Sea Floor

Much of the world's sea floor is charted only in a general fashion or the charts are inaccurate. Accurate charts are especially needed for shallow shelf areas where submarine deposition, erosion, and growth of coral reefs can change the bottom topography within a few years after a bathymetric survey is completed. Remote sensing of the sea floor from aircraft or satellites is restricted by the fact that water absorbs or reflects most wavelengths of electromagnetic energy. Only visible wavelengths penetrate water and the depth of penetration is influenced by the turbidity of the water and the wavelength of light. As shown in Figure 9.1, transmission of light through water is essentially restricted to the wavelength region from 0.4 to 0.7 μm. Photographic IR energy is absorbed by water, as illustrated on the IR black-and-white aerial photograph in Figure 2.18B. A 10-m layer of clear ocean water transmits almost 50 percent of the incident radiation from 0.4 to 0.6 μm in wavelength, but transmits less than 10 percent of radiation from 0.6 to 0.7 μm wavelength. Coastal and bay water is more turbid than water in the open ocean. The increased turbidity results in a decrease of light transmittance and a shift in the wavelength of maximum transmittance to the 0.5 to 0.6 μm region (Figure 9.1). On Landsat MSS images, band 4 records radiation in the spectral region of maximum water penetration and band 5 records in the region of reduced penetration. These bands are useful for differentiating depths in relatively shal-

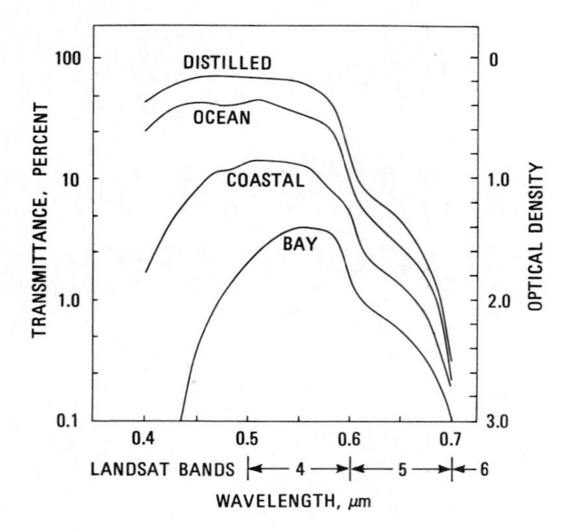

FIGURE 9.1
Spectral transmittance for 10 m of various types of water. From Specht, Needler, and Fritz (1973, Figure 1).

low water. Bands 6 and 7, in the photographic IR region, are useful for mapping the contact between land and water.

Three Landsat bands and a hydrographic chart of the Sibutu Island group in the Celebes Sea are shown in Figure 9.2. Dry land constitutes less than half of the Sibutu group with shoals, reefs, and lagoons occupying much of the area. Sibutu Island is the largest of the group with a length of 45 km, excluding the fringing reef on the south. Tumindao Island and the smaller islands to the north and south form the east rim of an elongate atoll that is bordered on the west by Tumindao Reef, which is submerged. The central lagoon has a maximum water depth of 8 fathoms and is connected with the ocean through narrow passages in the fringing reef. Meridian Reef and the reefs to the north and south form the east border of a submerged platform with water depths of several tens of fathoms. Tumindao Atoll is bordered by narrow channels with maximum depths in excess of 100 fathoms. Sibutu Island is also bordered on the east by deep water.

The dry land portion of the Sibutu group is clearly shown on the photographic IR image of band 7 (Figure 9.2C) where the very bright signature is caused by vegetation cover. The boundary between land and water is precisely shown on this image. On bands 4 and 5, however, it is impossible to distinguish land from shallow water because both have a dark signature. For example on both these images the small lagoon, shaped like a bow tie, at the south end of Tumindao Atoll could readily be mistaken for land. On band 5 the brightest signatures represent the shallowest areas of barely submerged reefs and banks of coral sand. The maximum water penetration capability of band 4 is shown by the large areas with bright signatures on this image (Figure 9.2A) that represent water depths down to about 3 fathoms. On band 4 the intermediate gray signature correlates with water down to 6 or 7 fathoms deep, such as the central lagoon of Tumindao Atoll. At greater depths no energy is reflected from the bottom which appears dark on all bands. The small lagoon at the south end of Tumindao Atoll is only 10 fathoms deep but produces the same dark signature on band 4 as does the much deeper water in the channels. The 10 to 20 fathom depths east of Meridian Reef also produce dark signatures.

The signatures on Landsat images are influenced not only by the water depth, but also by water clarity, reflectance of the bottom sediment, and by atmospheric conditions. Therefore the depth ranges associated with image signatures in the Sibutu Island group will not necessarily be the same in other areas. In Figure 9.2 a few small clouds occur over the channel between Sibutu and Tumindao Islands and over the east part of the Tumindao Lagoon. On all Landsat bands the clouds have bright signatures, which distinguishes them from dry land (dark on bands 4 and 5, bright on band 7) and from shallow reefs (bright on bands 4 and 5, dark on band 7).

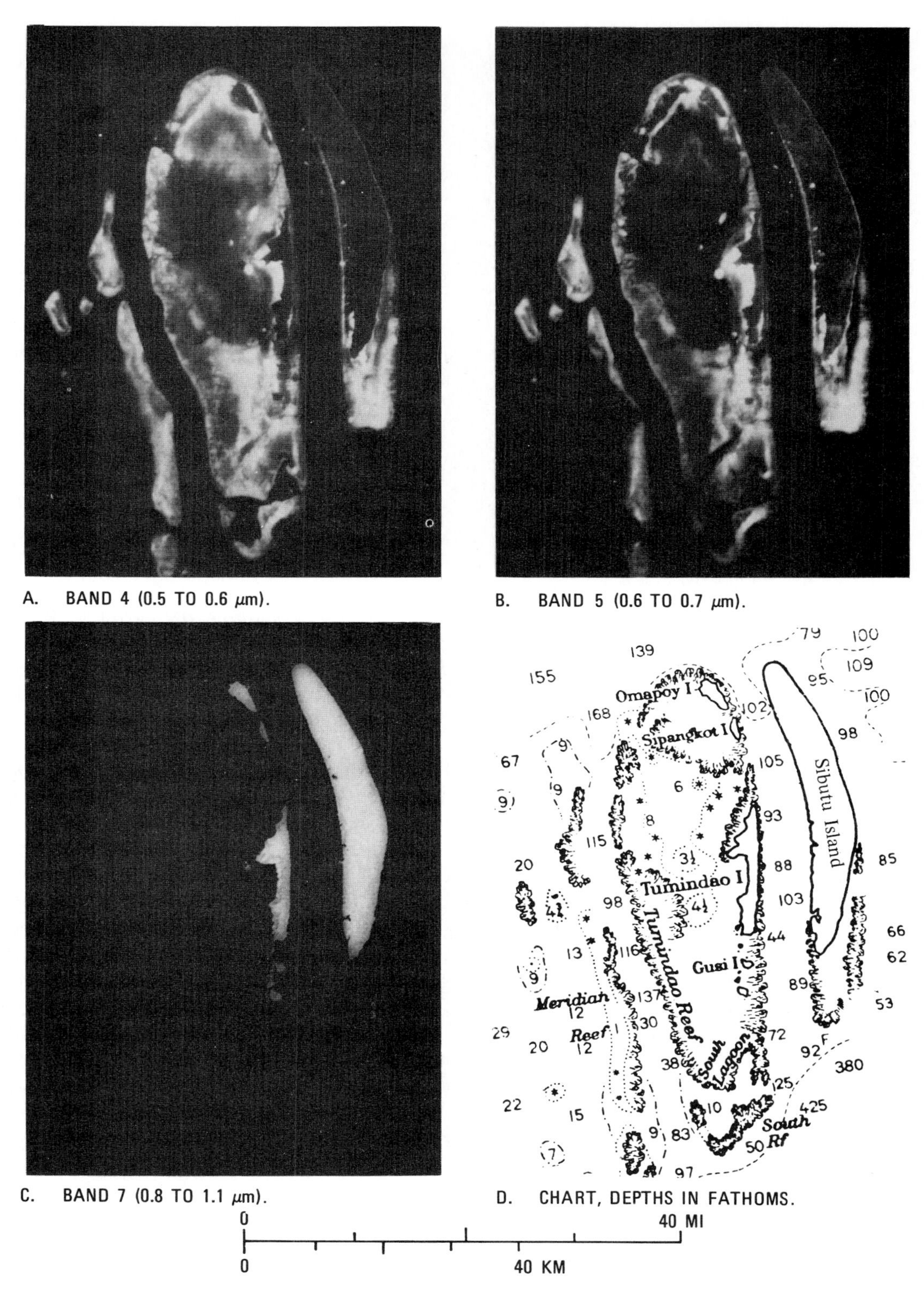

FIGURE 9.2
Landsat MSS bands and hydrographic chart of Sibutu Island group, Celebes Sea. Note differences in water penetration of the MSS bands.

The hydrographic chart of the Sibutu Island group appears to be very accurate because it closely matches the features on Landsat images. Elsewhere, however, Landsat images have revealed errors of commission or omission on hydrographic charts, many of which are based on surveys made in the 1800s. In the Chagos Archipelago of the Indian Ocean, Landsat images revealed the presence of a previously uncharted reef and a known bank that was charted 18 km east of its true position (Hammack, 1977). The U.S. Defense Mapping Agency incorporated these changes in new editions of their published hydrographic charts. Landsat images of the Georgia coast disclosed an island that was not present on existing maps or on aerial photographs acquired several years earlier. The island had formed by the accumulation of sand from longshore drift. In another offshore area marine seismic surveys were hampered by uncharted submerged coral heads that snagged cables. Digitally processed Landsat imagery was useful in locating and avoiding these hazards.

Normal color aerial photography has also been used successfully for charting shallow shelf areas. In the early 1970s the Kodak Company manufactured a two-layer color film designed for maximum water penetration (Specht, Needler, and Fritz, 1973) but this product has been discontinued.

Side-scan sonar systems acquire imagery of the sea floor by transmitting pulses of sonic energy which has maximum penetration of water. The active transmitting and receiving device is towed above the sea bottom from a cable. Narrow pulses of sonic energy are transmitted to either side of the direction of travel and reflections from the sea floor are recorded as images. The principles of operation and the geometry and appearance of the images are very similar to SLAR images. Details of the system and examples of images are given by Belderson and others (1972). All of the methods described here are applicable for mapping the floor of fresh water bodies, as well as the ocean.

Current Mapping

Conventional methods for studying ocean currents employ current meters, drift floats, and temperature measurements. In addition to the expense, these methods are hampered by the problem of simultaneously obtaining data over a broad expanse of water. These problems are largely overcome by remote-sensing systems that can provide nearly instantaneous images of current patterns over very large areas.

Landsat Imagery of Cape Mendocino, California

The small-scale and regional coverage of Landsat images emphasize major current features and reduce the confusing local details. The repetition cycle of Landsat images enables seasonal changes in circulation patterns to be monitored by mapping the plumes of suspended sediment that serve as tracers for various current systems. Several investigators report that Landsat MSS band 5 is optimum for mapping current patterns because it provides the maximum contrast between clear and turbid water (Maul and Gordon, 1975; Rouse and Coleman, 1976). The interpretation of many Landsat images of the Pacific coast, however, indicates that band 4 images are superior for this region. The contrast in reflectance between clear and turbid water is determined by the nature of the suspended sediment and other factors. An investigator should examine band 4 and 5 images of the area of interest in order to select the optimum band for interpretation of current patterns.

Seasonal changes in nearshore current patterns are shown in Figure 9.3 of the Cape Mendocino, California region. Sediment plumes on both the

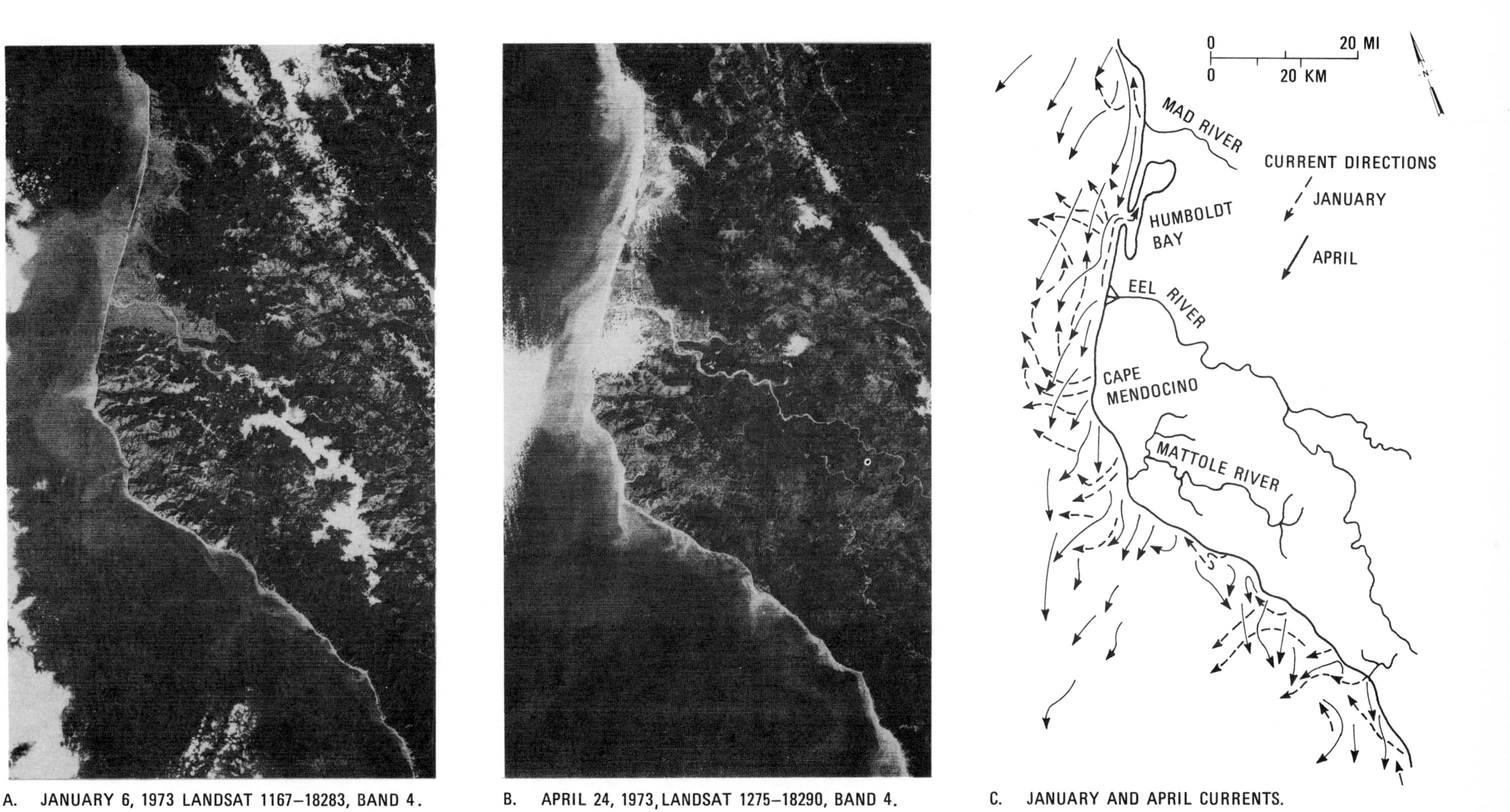

FIGURE 9.3
Seasonal changes in current and sedimentation patterns off Cape Mendocino, California, interpreted from Landsat images.

TABLE 9.1
Oceanographic conditions near Cape Mendocino, California

Landsat observation date	Wind direction	Humboldt Bay tide	Suspended sediment in Eel River
January 6, 1973	Southeast	Flood	122 $mg \cdot l^{-1}$
April 24, 1973	North	Ebb	35 $mg \cdot l^{-1}$

Source: From Carlson (1976).

January and April images originate at the mouths of the Mad, Eel, and Mattole Rivers. The California current flows slowly southward at speeds generally less than 0.25 $m \cdot sec^{-1}$ and controls offshore current patterns for the entire year. For much of the year the California current also controls the nearshore pattern, as shown by the southward movement of sediment plumes on the April image (Figure 9.3B).

The Davidson current is a deep counter current below 200 m that flows northward along the California coast. In the late fall and winter, north winds are weak or absent and the counter current appears at the surface, inshore from the main California current. The January image (Figure 9.3A) shows the northward movement of sediment plumes by the Davidson current. The effects of the current systems vary from year to year. For example on Landsat image 1527-18252 (not shown), acquired one year after the January example, the plumes are drifting southward and there is no surface expression of the Davidson current.

Oceanographic and tidal conditions in the vicinity of Humboldt Bay are given in Table 9.1 and are visible on the images. The January image was acquired during maximum flood tide and a plume of turbid water, probably from the Eel River, extends well into Humboldt Bay. The April image was acquired during ebb tide and a plume of less turbid tidal water extends offshore from the mouth of the bay. Seasonal current patterns are interpreted and compared on the map of Figure 9.3C.

NOAA-3 IR Imagery of Pacific Coast The NOAA-3 satellites, described in Chapter 5, acquire visible and thermal IR images with wider regional coverage and lower spatial resolution than Landsat. The enlarged thermal IR image of Figure 9.4 shows major upwelling along the Pacific Coast in late July, 1975. Note that signatures of NOAA-3 thermal images are the reverse of other IR images in this book. Dark signatures are relatively warm and bright signatures are cool. Upwelling along the Pacific coast is caused by strong winds from the north and northwest that move the warmer surface water offshore. It is replaced by colder bottom water (light tones in Figure 9.4) that is rich in nutrients and is biologically productive. Huge plankton blooms in the upwelling areas support a food chain that culminates in the large pelagic fishes, such as albacore. Albacore prefer temperatures of 14 to 18°C and congregate on the warmer, or seaward, side of thermal fronts where small forage fish occur. Water temperatures can be mapped at intervals of 0.5°C from the calibrated magnetic records of NOAA-3. The National Environmental Satellite Service prepares charts of the favorable temperature fronts which are used by commercial fishermen to improve their catch.

Aircraft Thermal IR Imagery of Gulf of the Farallones, California IR images acquired from aircraft have higher spatial and temperature resolution than those obtained from satellites. The trade off is the increased cost and reduction in regional coverage of aircraft images. The thermal

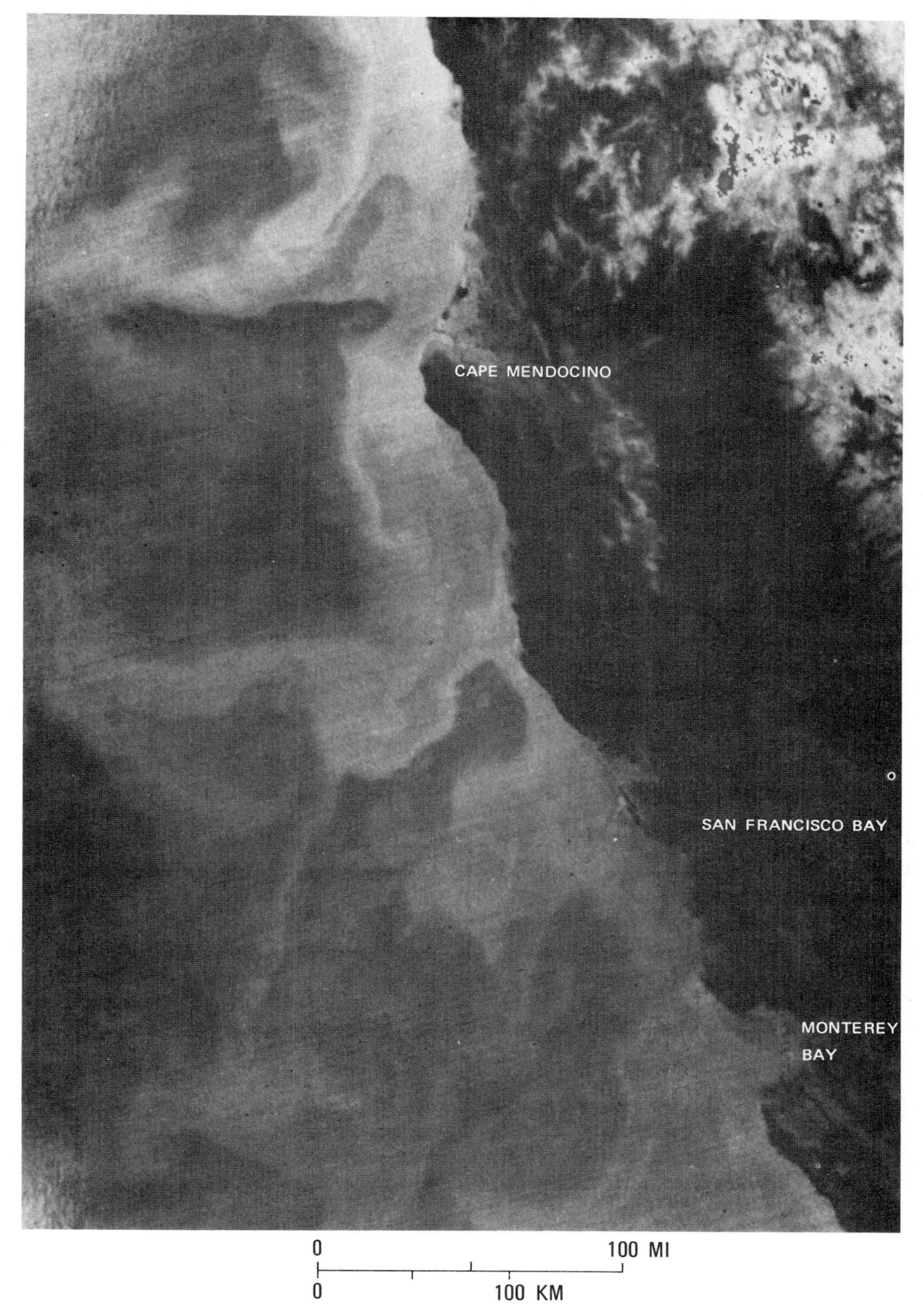

FIGURE 9.4
Enlarged NOAA-3 satellite thermal IR (10.2 to 12.5 μm) image of Pacific coast showing cold upwelling currents in late July 1975. Light tones are relatively cool radiant temperatures; dark tones are relatively warm. Courtesy L. C. Breaker, National Environmental Satellite Service.

IR image of Figure 9.5 illustrates the use of temperature differences to map the tidal flow of water from San Francisco Bay into the Pacific Ocean at the Gulf of the Farallones. The bay water is cooler than that in the Gulf and extends 19 km seaward from the Golden Gate as a cool, dark-toned plume. Thermal structure within the bay and toward the margin of the plume is clearly shown. The sharp margins of the plume are characteristic of both warm and cool plumes, whether natural or manmade. The boat wake at the west margin of the image has a cool signature because the boat disrupted the warmer surface layer and allowed the underlying cool water to come to the surface. On both satellite and aircraft IR images the radiant temperature is measured from the uppermost few micrometers of water. The temperature of the thin surface layer of water may differ somewhat from the underlying water because of the effects of wind and air temperature.

MONITORING INDUSTRIAL THERMAL PLUMES

Water is withdrawn from lakes, rivers, and the ocean to cool many industrial processes and then returned at higher temperatures to the water bodies. The heated water discharges, called *thermal plumes*, may be monitored by airborne thermal IR scanners in the same manner as natural water currents of different temperatures. Nuclear and fossil-fuel electric power plants, refineries, chemical and steel plants use large volumes of water. Aside from any chemicals or suspended matter, the heated water affects the environment in two ways: (1) Excessively high temperature may kill or inhibit growth and reproduction of organisms. In some areas, however, the heated discharge water is used for commercial cultivation of lobsters and oysters. (2) The heated water has a lower content of the dissolved oxygen that is essential for aquatic animals and for the process of oxidizing organic wastes.

Environmental legislation has been enacted to regulate thermal discharges. In California coastal waters, for example, the maximum temperature of thermal discharges shall not exceed the natural temperature of receiving waters by more than 11°C. At a distance of 300 m from the discharge, the surface temperature of the ocean shall not be increased more than 2.2°C. Temperature of discharge water may be lowered by cooling towers or by mixing with cooler water before it is discharged. Thermal IR surveys are an ideal way to monitor the temperature and pattern of the discharge outfalls.

In April, 1973 Daedalus Enterprises Incorporated acquired repeated quantitative IR images of the thermal plume discharged into Montsweag Bay, Maine from the Maine Yankee nuclear power plant. This was part of an investigation of the effect of tidal action on the shape and distribution of the thermal plume. The images in Figure 9.6 were selected from more extensive coverage to illustrate the maximum differences in the plume during a tidal cycle. The tape-recorded data were digitally processed and displayed as images with eight discrete gray-scale levels in Figure 9.6. Different gray-scale settings were used for each image in order to display the maximum thermal information. Therefore the white tone of the water in Figure 9.6A does not indicate the same temperature as the white tone in Figure 9.6C. To provide a quantitative comparison, Daedalus Enterprises prepared the thermal contour maps of Figure 9.7 using the quantitative method described in Chapter 5. The branching or merging contour lines occur where the horizontal thermal gradient is so abrupt that contour lines run together.

Comparing the images (Figure 9.6) with the corresponding digitally processed maps (Figure 9.7) demonstrates that much more thermal information can be extracted digitally than visually. The images and maps clearly show the location of the plume and the temperature distribution within the thermal plume. Note, for example, that some parts of the bay were never affected by the plume

FIGURE 9.5
Daytime thermal IR image (8 to 14 μm) of San Francisco Bay and Gulf of The Farallones at ebb tide. Image acquired January 1973 by NASA aircraft. From Pirie and Murphy (1975, Figure 6-1). Courtesy M. J. Murphy, U.S. Army Corps of Engineers.

A. FLOOD TIDE, 8:30 A.M.

B. EBB TIDE, 3:07 P.M.

C. LOW TIDE, 6:15 P.M.

FIGURE 9.6
Thermal IR images (8 to 14 μm) of thermal plume from nuclear power plant on Montsweag Bay, Maine, showing tidal influence. Gray scale calibration levels are different for each image. These digitally level-sliced images were acquired and processed by Daedalus Enterprises, Inc. Courtesy Maine Yankee Atomic Power Company and T. R. Ory, Daedalus Enterprises, Inc.

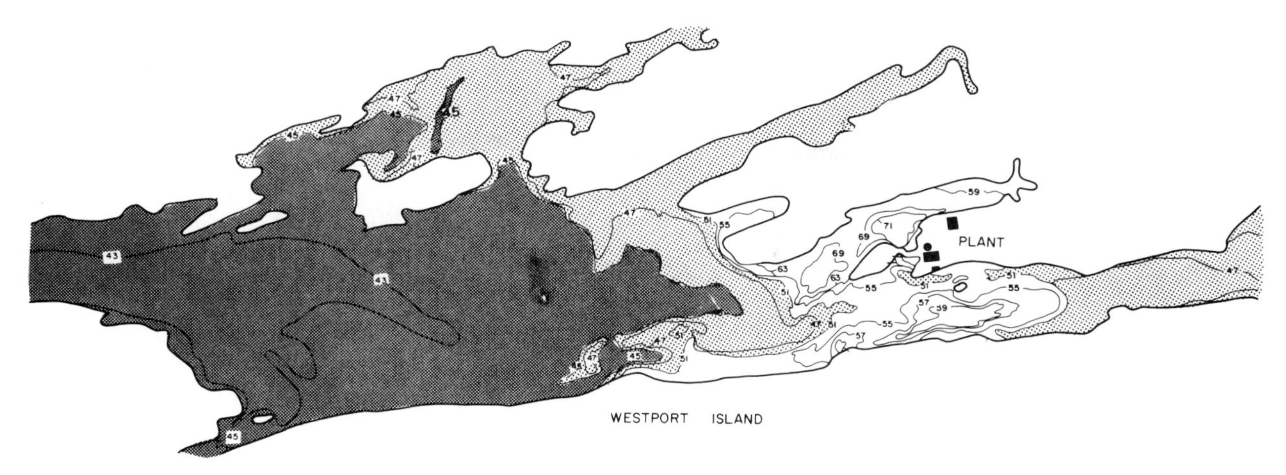

A. FLOOD TIDE, 8:30 A.M.

B. EBB TIDE, 3:07 P.M.

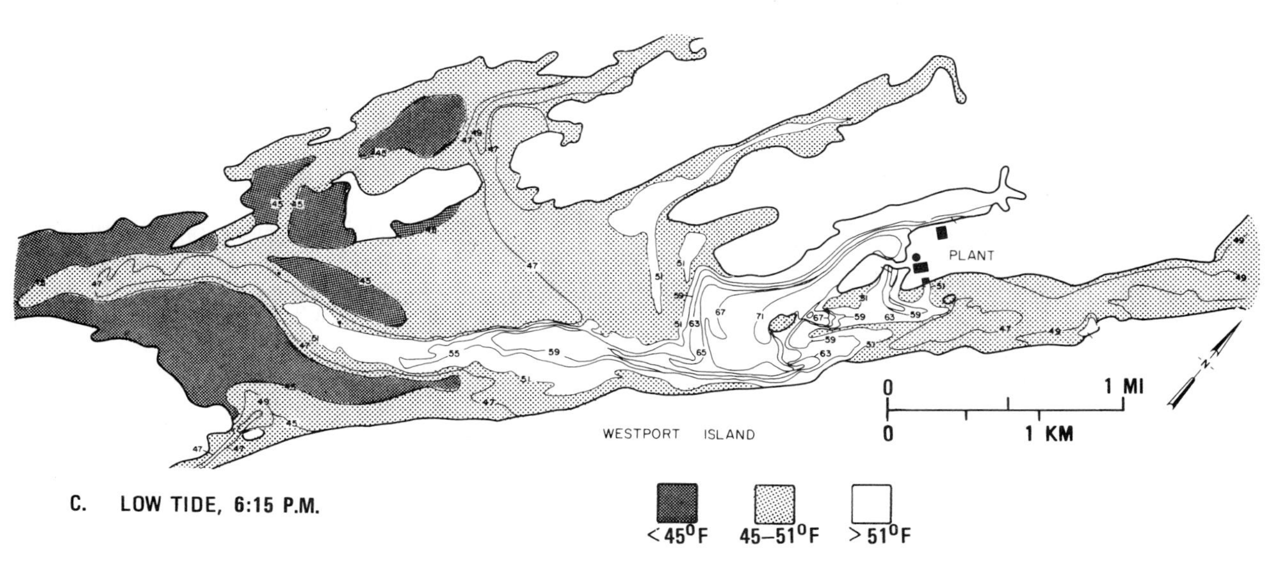

C. LOW TIDE, 6:15 P.M.

FIGURE 9.7
Temperature contours of thermal plume. Plotted from Digicolor® version of images in Figure 9.6 by Daedalus Enterprises, Inc. Courtesy Maine Yankee Atomic Power Company and T. R. Ory, Daedalus Enterprises, Inc.

TABLE 9.2
Annual input of petroleum hydrocarbons into the oceans

Source of oil	Million barrels per year	Percent of annual input
Transportation		
Tankers, dry docking, terminal operations, bilges, accidents	14.5	34.5
Coastal refineries, municipal and industrial waste	5.5	13.2
Offshore oil production		
Accidents and oil discharged with produced water	0.6	1.3
River and urban runoff	13.1	31.2
Atmospheric fallout	4.1	9.9
Natural seeps	4.1	9.9
Totals	41.9	100.0

Source: From National Academy of Sciences (1975, Table 1-5).

during this tidal cycle. The upstream and downstream extent and thermal structure of the plume are precisely shown.

To appreciate the practical value of monitoring thermal plumes by IR surveys, the reader might undertake the following exercise. Design a system using conventional surface thermometers that will produce thermal maps, with the precision and detail of those in Figure 9.7, throughout a tidal cycle. The several hundred surface thermometers must be precisely positioned and located; they must be calibrated and recorded to an accuracy of 0.5°C; they must all be read at the same times to provide simultaneous data for contouring. The thermometers must be deployed and retrieved under the influence of strong tidal currents. This survey would be impractical to conduct, but is readily accomplished with airborne IR scanning.

DETECTION AND MONITORING OF OIL FILMS

A study by the National Academy of Sciences (1975) of oil entering the world oceans annually from various sources is summarized in Table 9.2. Note that production and transportation account for 36 percent of the oil entering the oceans each year. Also note that natural seeps contribute seven times more oil than offshore production operations.

The two major applications of remote sensing of oil films are: (1) law enforcement surveillance of coastal and inland waterways for violations of pollution regulations; (2) monitoring of accidental spills to aid cleanup operations. The three aspects of oil-spill monitoring are:

1. Detection of oil spills.
2. Estimation of thickness and volume of spills. Multiple sensor techniques are giving encouraging results, but thickness estimation is still in the experimental stage.
3. Identification of the type of crude or refined oil in a spill. Several prototype systems to deal with this difficult problem are under development, but no systems are operational. Sampling and laboratory analysis are currently used to identify the type of oil in a spill.

The U.S. Environmental Protection Agency (EPA) uses the following nonquantitative classification of oil spills, given in the order of decreasing thickness:

Slick	A definite brown or black color
Sheen	Silvery sheen on water surface with no black or brown color
Rainbow	Iridescent bands of color on the water surface

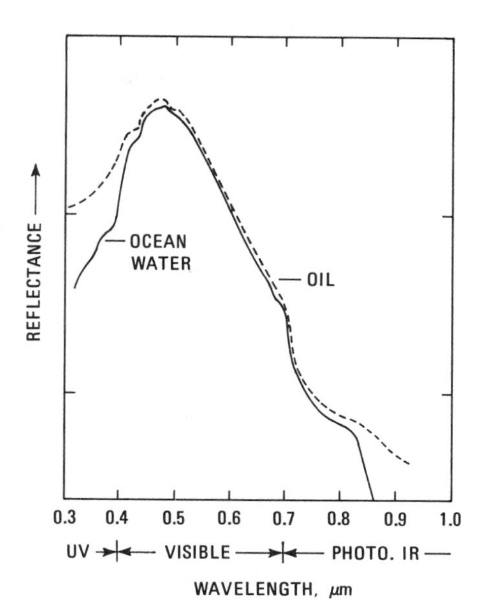

FIGURE 9.8
Spectral reflectance of ocean water and of a thin layer of crude oil. From Vizy (1974, Figure 5).

In a typical oil spill 90 percent of the volume is concentrated in 10 percent of the area, largely in the form of slicks. For remote-sensing interpretation the two thinner categories are commonly lumped as *rainbow/sheen* because they are difficult to distinguish. In this book the term *film* is used for any detectable oil floating on water with no implication about the thickness. This discussion deals with oil on water; oil that has washed ashore is more difficult to detect because of the nonuniform background. Normal color photography is the most practical system for detecting oil on the land. Crude oils and refined products are complex organic compounds and there are hundreds of varieties of each. This complexity and diversity means that there are some exceptions to any generalizations about the remote sensing signatures of oils.

The natural submarine oil seeps off Santa Barbara, California are excellent test targets because their locations are known and weather is generally good for flying. The U.S. Coast Guard and Aerojet Electrosystems Company have used this area to evaluate an airborne multiple-sensor system that is equipped with UV, photographic, thermal IR, radar, and passive microwave sensors. Images from this system are especially instructive because they are acquired nearly simultaneously and can be readily compared.

Photographic Remote Sensing

Images are acquired in the photographic spectral region with both photographic film and scanning devices. For oil to be detected there must be a difference in reflectance between the oil film and the adjacent clean water. The brown and black colors of slicks are readily detected on color photographs. The thin films that produce sheen and rainbow reflectance signatures are more difficult to detect. In Figure 9.8 the spectral reflectance of a thin layer of crude oil is compared with the reflectance of ocean water. The oil film has a higher reflectance than water in the UV, blue, and photographic IR bands. The intensity of the IR reflectance is very low, however, and may not be detectable. The photographic UV band and the blue band are the optimum portions of the photographic spectral region for detecting thin oil films.

UV Images As described in Chapter 2, UV photographs may be acquired with suitable film and filter combinations, but optical-mechanical scanners using a UV filter and detector produce better images because:

1. UV detectors are more sensitive than film and filter combinations.
2. There is no lens to attenuate the UV signal.
3. The tape-recorded data may be enhanced during playback for optimum contrast.

Images of the Santa Barbara seeps acquired with a UV scanner are illustrated in Figure 9.9A, where

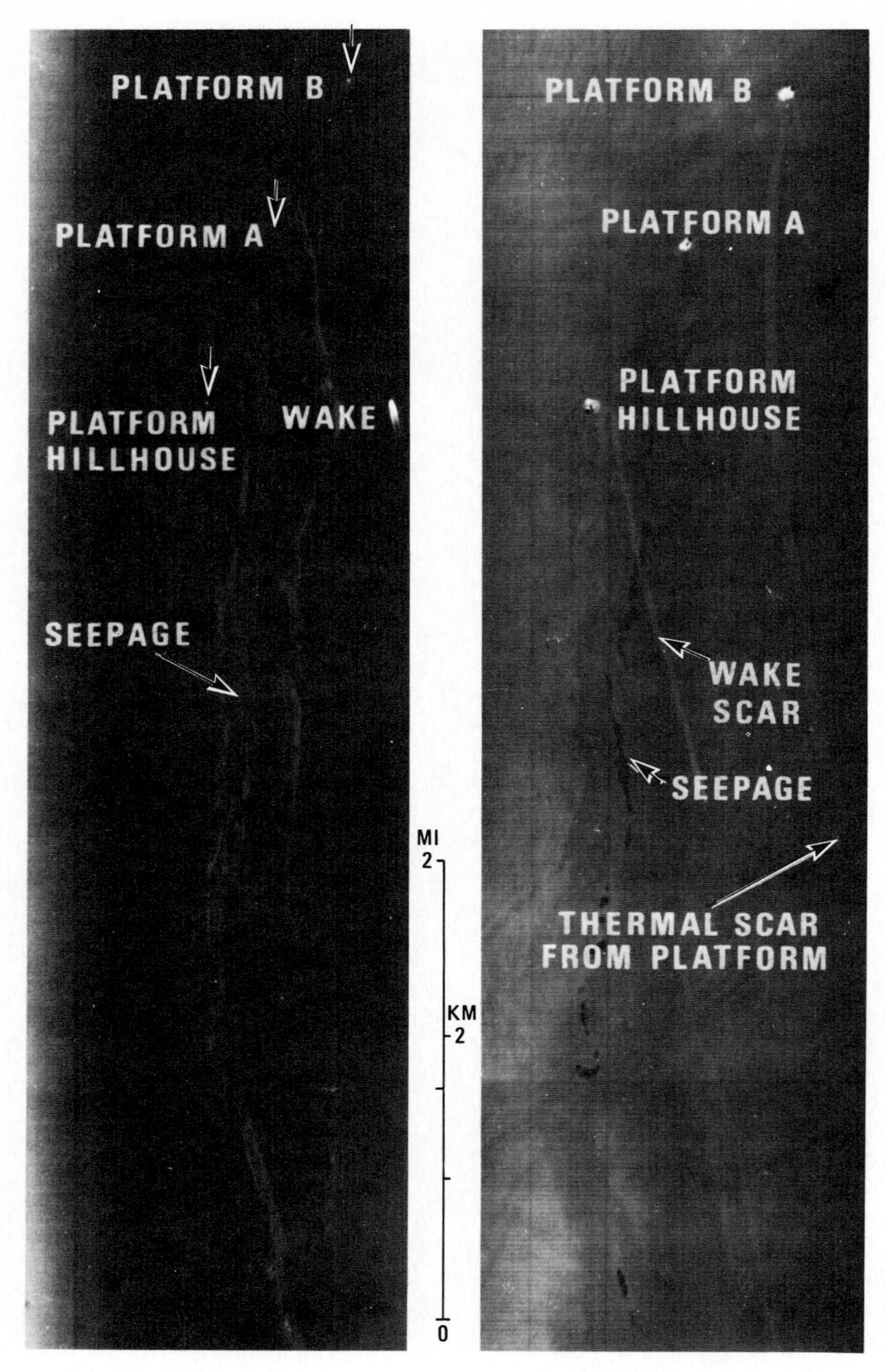

FIGURE 9.9
UV and thermal IR images of natural oil seeps along Santa Barbara coast, California. The images were acquired four minutes apart on July 31, 1974 in midafternoon with two-foot swell and wind less than six knots. From Maurer and Edgerton (1976, Figure 9). Courtesy A. T. Edgerton, Aerojet Electrosystems Company.

the oil film has a bright signature. This strong reflectance of oil in the photographic UV region is predictable from the curves of Figure 9.8. The thermal IR image (Figure 9.9B) is interpreted later in this chapter. Oblique black-and-white photographs in the visible region acquired the same afternoon as the UV and IR images are shown in Figure 9.10A. The low-angle oblique photographs indicate the bright, streaky appearance of portions of the seep in the visible region, but on the UV image the oil film is shown to be much more extensive. This reflectance difference between the UV and visible bands was also predictable from Figure 9.8.

UV imagery is the most sensitive remote method for monitoring oil on water and can detect films as thin as 0.15 μm (Maurer and Edgerton, 1976). Daylight and clear atmosphere are necessary to acquire UV images. Although UV energy is strongly scattered by the atmosphere, the effects are not too severe for images acquired at altitudes of 1,000 m or less. Floating patches of foam and seaweed have bright UV signatures that may be confused with oil. The foam and seaweed can be recognized on images in the visible band acquired simultaneously with the UV images.

Research is underway on active remote sensing methods in the photographic spectral region. The water is irradiated by a laser energy source, tuned to a specific wavelength, that stimulates fluorescence from oil films on the water. The spectral distribution of the fluorescence is detected and recorded as a trace that can be compared with traces of various oils. Under certain conditions, such a system is capable of recognizing broad classes of oils (Fantasia and Ingrao, 1974). Another active system under investigation uses a xenon light source to illuminate the terrain. These active systems enable images to be acquired both day and night.

Photographic IR and Visible Images A thin film of oil has a slightly higher reflectance than water in the blue spectral band (Figure 9.8). The aerial photographs of the Santa Barbara oil seeps (Figure 9.10A) were acquired in the visible spectral region (0.4 to 0.7 μm). The bright signatures of the oil films are caused primarily by the high reflectance in the blue spectral band. The green and red spectral bands are most useful for detecting the dark portions of heavy slicks. There is some contrast between oil and water in the photographic IR region, but this band is less useful than the UV and blue bands for oil detection.

Normal color photographs, and to a lesser extent IR color and IR black-and-white photographs, are useful for recording thick black and brown slicks. Sheens and rainbows are more difficult to photograph because of the effects of cloud shadows, glitter from small waves, and differences in sun elevation and azimuth. The optimum conditions for acquiring oblique aerial photographs of oil films are at intermediate sun elevations and with the camera aimed normal to the solar azimuth.

Landsat was not designed to detect oil films and the lack of image data in the important UV and blue spectral regions may limit the usefulness of the system. Nevertheless several confirmed and possible oil films have been detected on Landsat images. Confirmed oil films are those that have been verified by visual observation of the oil film.

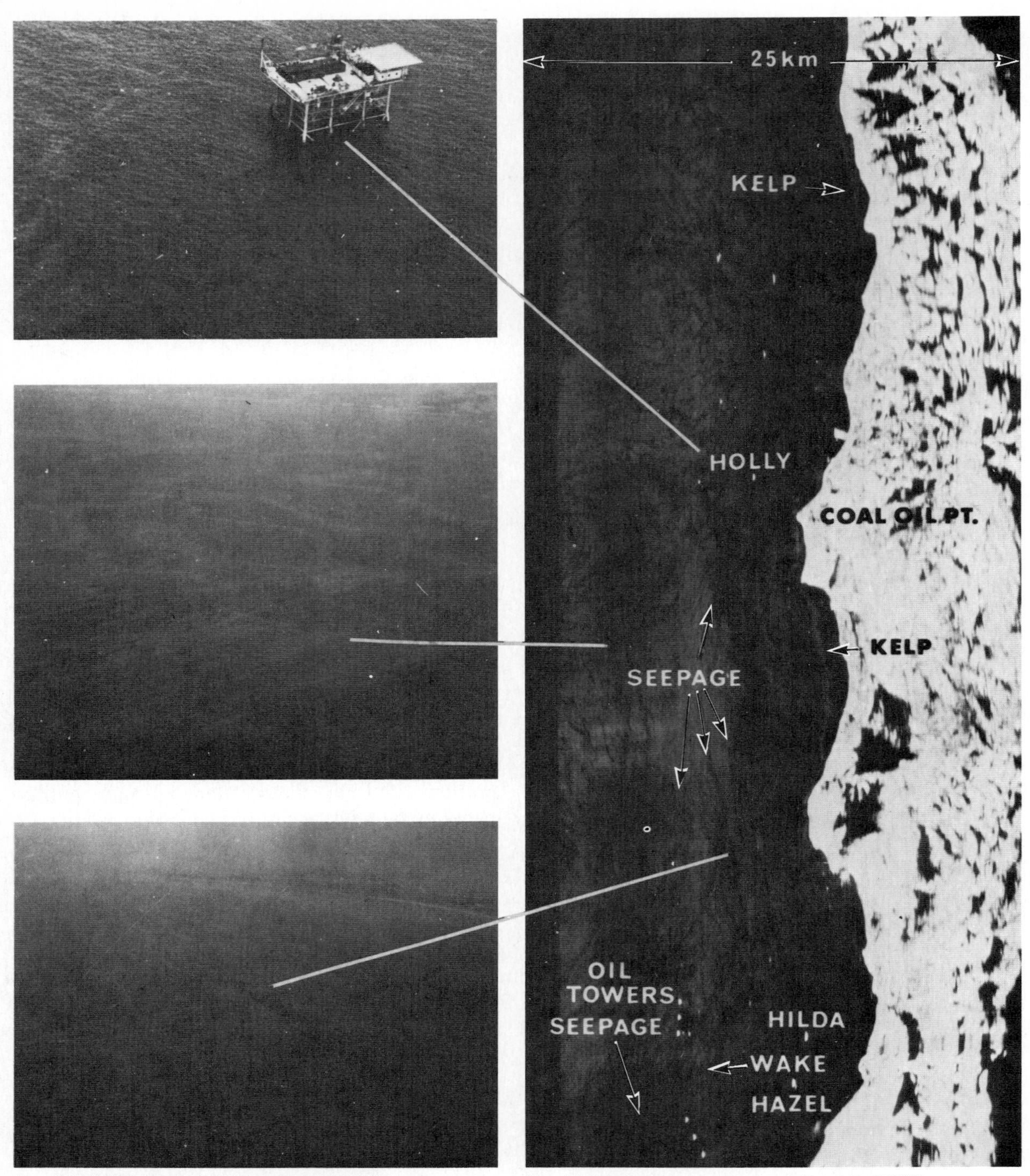

A. OBLIQUE AERIAL PHOTOGRAPHS WITH OIL SEEPS IMAGING BRIGHT.

B. X–BAND RADAR IMAGE WITH OIL SEEPS IMAGING DARK.

FIGURE 9.10
Simultaneously acquired radar image and aerial photographs of natural oil seeps along Santa Barbara coast, California. The images were acquired July 31, 1974 in midafternoon with two-foot swell and wind less than six knots. From Maurer and Edgerton (1976, Figure 7). Courtesy A. T. Edgerton, Aerojet Electrosystems Company.

Possible oil films have a signature similar to the confirmed oil films, but have not been verified by observation. There are only a few examples of confirmed oil films on Landsat images for two reasons. (1) Typically a month or more elapses between the acquisition of an image by Landsat and the time it is available for interpretation. During this time oil films have dissipated and are no longer present to be verified. (2) Aircraft and boat searches to confirm suspected oil films are difficult and expensive. Deutsch, Strong, and Rabchevsky (in press) have tabulated the confirmed and possible oil films reported from Landsat images.

Deutsch, Strong, and Estes (1977) illustrated several Landsat images of confirmed and possible oil films and noted that all reported examples were acquired at a sun elevation of 39 degrees or higher. With a few exceptions the oil films have a dark signature on all Landsat bands, which is surprising because oil films typically have a higher reflectance than clear water (Figure 9.8). Deutsch, Strong, and Estes (1977) explained the dark signatures as follows. Winds produce capillary waves on the ocean surface which cause a high reflectance (called *sunglint*) when viewed with a moderate to high sun elevation. An oil film eliminates the capillary waves and appears dark relative to the sunlight on the adjacent clean water. The elimination of capillary waves by oil films is also the basis for radar detection of oil films, described later in this chapter.

On Landsat images of the Gulf of Suez, Otterman, Ohring, and Ginsburg (1974) detected dark signatures that were interpreted as oil films. These signatures are classed as confirmed oil films because the operating crew on a nearby oil production platform reported a break in an underwater pipeline at the time the images were acquired. Dark signatures on a Landsat image of the Atlantic Ocean off the Florida Keys are probably the same oil slick reported by the U.S. Coast Guard at that time (Deutsch, Strong, and Estes, 1977). The contrast ratio between oil films and clean water is typically low on Landsat images and must be enhanced to detect the oil film. A good example is Landsat image data of the natural oil seeps off Santa Barbara that were collected at a time when viewing conditions were favorable for detecting the known oil films. Deutsch, Strong, and Estes (1977) used an interactive digital system to produce ratio images and images of individual bands with linear contrast stretch, but the signatures of the oil films were very indistinct. Finally they employed a "histogram equalization stretch" that produced the most definitive dark signatures of the oil films. This contrast stretch appears to be similar to the uniform distribution stretch illustrated in Chapter 7 (Figure 7.14C). At present the time delay in obtaining Landsat images precludes their use for real-time detection of oil spills.

Thermal IR Images

On thermal IR images oil films typically have a cool signature, as shown in Figure 9.9B, which was acquired on a separate flight line four minutes before the UV image (Figure 9.9A) was acquired. This time difference accounts for the boat wake on the UV image that is absent on the earlier IR image. The UV image aids in distinguishing oil films from currents of cold water that have similar cool signatures on the IR image. On the IR image the warm streaks extending eastward from Hillhouse and B platforms are *thermal scars* or wakes. The platform structures cause local surface turbulence in the eastward-flowing currents that mixes the cooler surface layer with the underlying warmer water. This effect is the same as the warm signature of the wakes of moving boats. It is not clear why platform A lacks a wake.

On daytime and nighttime IR images both refined and crude oils have cooler signatures than the adjacent clean water. The oil and water have the same kinetic temperature because the two liquids are in direct contact. Note from Table 5.1, however, that the emissivity of pure water is 0.993, but a thin film of petroleum reduces the emissivity to 0.972. The radiant temperature of a material can be calculated from Equation (5.8) as

$$T_{rad} = \varepsilon^{1/4} T_{kin}$$

For pure water at a kinetic temperature of 291°K (18°C) the radiant temperature is

$$T_{rad} = 0.993^{1/4} \cdot 291°K$$

$$T_{rad} = 290.5°K\ (17.5°C)$$

For an oil film at the same kinetic temperature of 291°K, the radiant temperature is

$$T_{rad} = 0.972^{1/4} \cdot 291°K$$

$$T_{rad} = 288.9°K\ (15.9°C)$$

This 1.6°C difference in radiant temperature between the oil (15.9°C) and water (17.5°C) is readily detected by thermal IR detectors that typically are sensitive to temperature differences of 0.1°C.

On IR images of some oil spills, such as the Santa Barbara platform blowout of 1969, there are narrow warm streaks interspersed within the typical cool signature of the oil film. Observers reported that these warm streaks apparently correlate with thick ropey masses of emulsified oil and water. Those thick, dark masses apparently absorb solar energy, which is later radiated as thermal energy. The day and night capability makes thermal IR imagery a valuable technique for surveillance and for monitoring cleanup activities around the clock. Rain and fog, however, prevent image acquisition. Also the interpreter must be careful to avoid confusing cold water currents, such as those in Figure 9.5, with oil slicks. An experienced interpreter using simultaneously acquired UV imagery can minimize this problem.

Radar Images

The need for night and day pollution surveillance in all weather over large ocean areas prompted experiments with SLAR images at the U.S. Naval Research Laboratory (Guinard, 1971), the U.S. Coast Guard (Maurer and Edgerton, 1976), and elsewhere. The small waves that cause backscatter on radar images of the sea are dampened by a thin film of oil. As illustrated in Figure 9.10, specular reflection from the calm water caused by an oil film results in anomalously low backscatter (dark signature) surrounded by the strong return (bright signature) from slightly rough water. Comparison with the simultaneously acquired oblique aerial photographs demonstrates the accuracy of the SLAR image. When the wind is still and the sea is flat there is no roughness contrast between clean water and oil films, but only rarely do such completely calm conditions prevail. It should be noted that the response of this radar system is calibrated to enhance the ocean backscatter, hence the rougher land surface appears extremely bright in Figure 9.10. The SLAR images of land and sea, depicted in Chapter 6, were acquired with systems calibrated for land signatures, hence the ocean backscatter is reduced on those examples.

TABLE 9.3
Minimum detectable thickness of oil films for different sensors

Sensor	Wavelength	Approximate minimum detectable thickness of oil film
UV (photographic)	0.3 to 0.4 μm	0.15 μm
Radar (X-band)	3.0 cm	<1.0 μm
Thermal IR	8.0 to 14.0 μm	10.0 μm
Passive microwave	0.8 cm	—
Calm seas	—	100.0 μm
Rough seas	—	<4.0 μm

Source: From Maurer and Edgerton (1976, Figure 1).

Passive Microwave Images

The earth radiates low levels of energy at microwave wavelengths than can be detected and displayed as imagery. The emitted and reflected microwave energy is recorded as the temperature of the surface and is a function of the surface roughness and dielectric constant of the surface (Maurer and Edgerton, 1976, p. 40). Oil films affect the temperature of ocean water, measured at microwave wavelengths, in two ways: (1) in moderately rough to very rough seas the calming effect of a thin oil film causes a lower surface temperature; (2) thick films of oil (0.1 mm and greater) emit more microwave energy than clean water because of the significantly lower dielectric constant of oil. The increase in emission from relatively thick oil films greatly exceeds the effect of decreased ocean roughness. Therefore thicker oil films have higher radiant temperatures at microwave wavelengths than clean water for calm to moderately rough seas.

The radiant temperature of the surface may be recorded in image form by passive microwave scanner systems. The system used by the U.S. Coast Guard employs a radar antenna, without the transmit capability, that detects energy at wavelengths of 0.8 cm (Maurer and Edgerton, 1976). The scanner covers a conic arc in advance of the aircraft at a constant depression angle of 45°. Spatial resolution is relatively low, typically 50 to 60 m, and the coverage of the image strip is narrow relative to SLAR. Passive microwave systems operate independently of lighting conditions and can penetrate clouds and precipitation. The ability to discriminate oil films thicker than 0.1 cm is potentially valuable during cleanup operations.

Estimation of Oil Thickness

The approximate minimum thickness of an oil film that is detectable by different remote sensors is listed in Table 9.3. The presence of an oil signature on an image indicates that the film is thicker than the minimum value required for detection on that system. By comparing simultaneously acquired images of different wavelengths, thickness of oil films may be estimated. In Figure 9.9, the oil film with a bright signature on the UV image is thicker than 0.15 μm, and the film with a dark signature on the IR image is thicker than 10 μm. By comparing the two images the portion of the oil film ranging from 0.15 to 10 μm in thickness can be mapped separately from the portion thicker than 10 μm.

Identification of Oil Types

There has been much research on airborne systems for remotely identifying the type of oil or refined product in a spill. At each wavelength region the spectral signatures of a wide variety of oils may be similar, which makes identification difficult. At present the type of oil in a spill is identified by collecting samples and analyzing them in the laboratory.

Present Monitoring Practice

The capabilities of various remote-sensing systems for monitoring oil films have been demonstrated. Monitoring of coastal and inland waterways is not routinely carried out, however, and most oil spills are reported by the party responsible for the spill (this is required by law) or by a casual observer. Normal color aerial photographs are the form of imagery most widely used during the cleanup operation for the following reasons:

1. The photographs are readily acquired, whereas mobilization of the other remote-sensing systems may require days or weeks. The typical spill will be cleaned up before the systems are available.
2. The normal color photographs may be used by cleanup crews that lack the training and experience required to interpret other forms of imagery.
3. Where oil films wash ashore, normal color photographs are the most practical form of monitoring.

MAPPING OF SEA ICE

The increase in shipping and in petroleum exploration activity in Arctic waters has increased the need for information on sea ice conditions. Global weather predictions require information about heat budgets of the polar seas, which in turn are related to ice abundance. Thus information on ice cover can aid meteorologists. Remote sensing, especially from satellites, is the only practical way to map sea ice on a regional, repetitive basis.

Terminology of Sea Ice

The most important sea ice features are listed and defined in Table 9.4, which is summarized from the more extensive nomenclature of the World Meteorological Organization. The Landsat band 7 image of Dove Bay (Figure 9.11) illustrates many of the features in Table 9.4. The prominent refrozen flaw lead separates the pack ice from the fast ice that is attached to the shore. Comparison with a band 5 image (not shown) indicates that all the leads within the pack ice are refrozen at this season (J. C. Barnes, personal communication): on band *5* images, leads covered with thinner ice are darker in tone. Open leads on other images are identified by the black tone of water on both band *5* and band *7*. Fragments of broken floes occur along the southern part of the flaw lead and between large floes of the pack ice. The image illustrates the typical wide range of sizes and shapes of individual floes. The pack ice in Figure 9.11 is classified as consolidated because the leads are refrozen and there is no open water. Open pack ice has approximately equal proportions of ice and water and the floes are not in contact. The numerous icebergs in the southwest corner of Figure 9.11 calved from coastal glaciers in the summer of 1972, but were locked in the fjord by winter fast ice before they could enter Dove Bay. Remote sensing of sea ice must deal with the following conditions and requirements:

TABLE 9.4
Sea ice terminology

Feature	Description
Fast ice	Ice which forms and remains attached to the shore. May extend seaward for a few meters to several hundred kilometers from the coast.
Floe	Any relatively flat piece of sea ice 20 m or more across. Floes are classified according to size.
Concentration	Ratio (in tenths) of sea surface covered by ice to total area of sea surface.
Pack ice	General term for any area of sea ice, other than fast ice, regardless of form or occurrence. Pack ice is classified by concentration of the floes.
Lead	Any fracture or passageway through sea ice that is navigable by surface vessels. Leads may be open or refrozen. A flaw lead separates fast ice from pack ice.
First-year ice	Sea ice of not more than one winter's growth. Thickness ranges from 30 cm to 2 m.
Second-year ice	Old ice that has survived only one summer's melt. Because it is thicker and less dense than first year ice, it stands higher out of the water.
Multi-year ice	Old ice 3 m or more thick that has survived at least two summers' melt.
Pressure ridge	Wall of broken ice forced up by pressure.
Brash ice	Accumulations of floating ice made up of fragments not more than 2 m across; the wreckage of other forms of ice.
Iceberg	A massive piece of ice extending more than 5 m above sea level that has broken away from a glacier. Icebergs are classified according to shape.

Source: From World Meteorological Organization Publication no. 259. TP145.

1. Darkness and twilight persist for several months of each year.
2. Clouds and fog persist for much of the year.
3. Broad regional coverage is needed.
4. Repeated coverage is needed to analyze ice movement during the spring breakup.

SLAR imagery most nearly meets these requirements, but is very expensive relative to Landsat imagery.

Landsat Images

The Landsat orbits provide image coverage as far north and south as 81° latitude, although illumination above 70° is insufficient to acquire images from late October to late March. An optional high-gain sensor mode may extend the imaging period. Convergence of orbits at these high latitudes provides up to three or four consecutive days of coverage of the same area during each 18-day cycle. Ice features as small as 80 to 100 m wide can be detected, as shown in Figure 9.11.

Distinguishing Ice from Clouds Clouds and fog are common over Arctic waters and have bright signatures similar to those of sea ice. A surprising amount of ice information can be interpreted, despite all but the heaviest cloud cover, by using the following criteria to distinguish between clouds and sea ice:

1. Brightness is generally more uniform for ice than for clouds. Also leads and fractures can

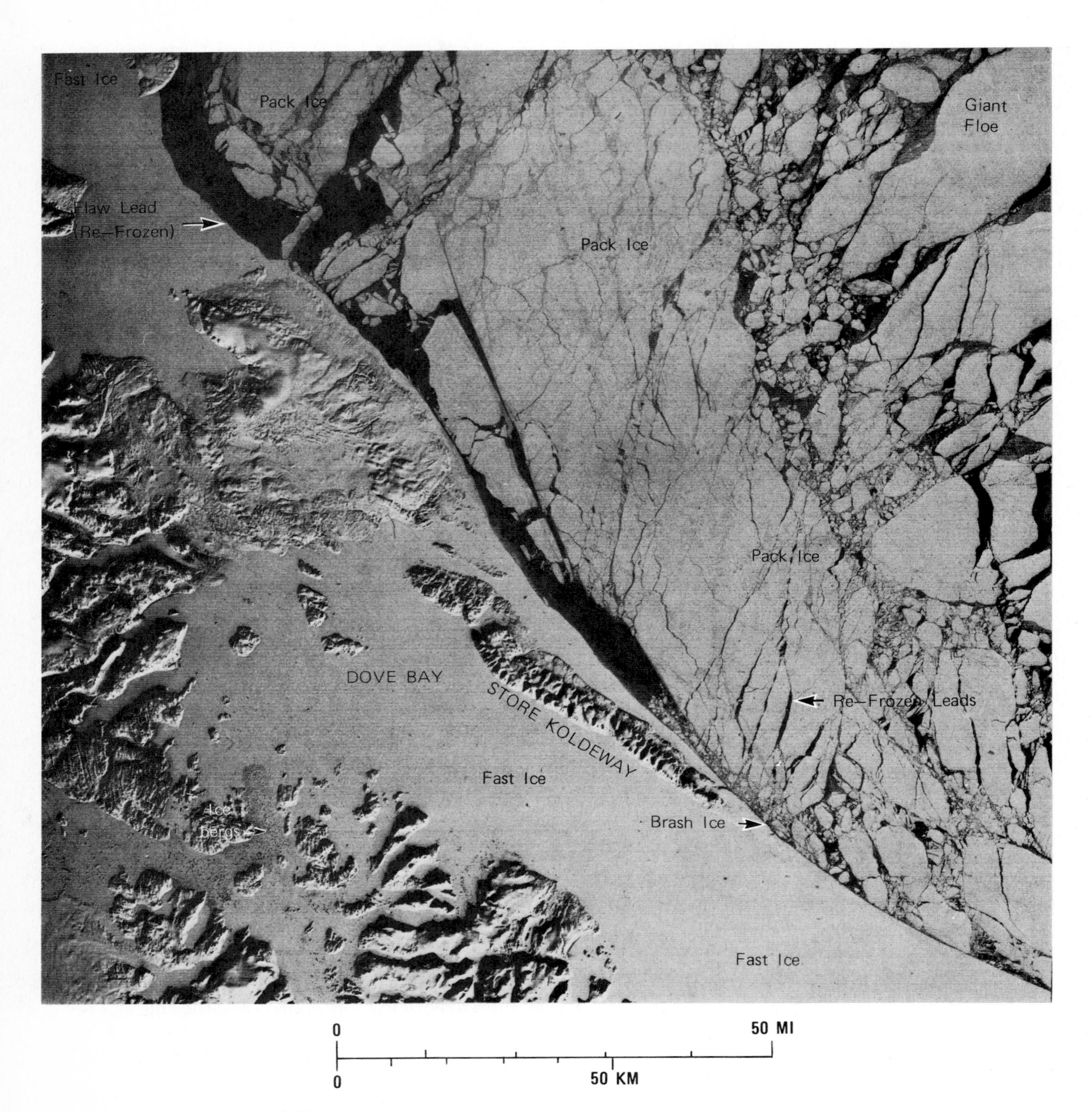

FIGURE 9.11
Landsat image of sea ice features in vicinity of Dove Bay on east coast of Greenland. Landsat 1245-13423, band 7, acquired March 25, 1973. See Figure 9.13 for details of icebergs.

be detected through a thin cloud cover and serve to identify ice.

2. Cloud shadows may be recognized on the image.
3. Clouds have gradational margins but ice floes have sharp contacts with water.
4. Textures and patterns of clouds differ from those of ice.

Terrain features are also recognizable, despite a moderate cloud cover.

Optimum MSS Bands The margins of leads and ice floes are apparent on all the MSS bands. As shown in Figure 9.12 of M'Clure Strait, it is useful to interpret both a visible band (4 or 5) and a photographic IR band (6 or 7). The band 4 image (Figure 9.12A) shows the full extent of ice cover and the surface characteristics. Meltwater on the ice surface is shown by dark patches on the band 7 image (Figure 9.12B) because a thin layer of water absorbs photographic IR radiation. The bright appearance on band 4 of these patches establishes that they are layers of water upon the ice, rather than open water. The bright lines extending across the fjords on Figure 9.12B are probably fractures along which the meltwater has drained off the ice surface. The rounded floes with bright signatures east of Eglinton Island may be second-year floes that stand above the level of the enclosing fast ice. These floes are more apparent on the band 7 image.

Icebergs Icebergs are most readily detected when they are frozen in fast ice, which provides optimum background contrast for recognizing the shadows of bergs. On the image of Kangerdluk Fjord on the west coast of Greenland (Figure 9.13A) the largest bergs are 1.5 km wide. Icebergs frozen in the fast ice of Borge Fjord on the east coast of Greenland are shown in Figure 9.13B. By measuring shadow lengths and knowing the sun elevation (16°) Barnes and Bowley (1974) calculated the height of the bergs in Borge Fjord. The largest tabular berg at the west margin of the image is 1.1 km long, 360 m wide, and 33 m high. The smaller tabular berg 5 km to the east is about 0.9 km long, 450 m wide, and 33 m high. The irregular bergs near Edvards Island range from 240 to 450 m in length and from 40 to 60 m in height. Note that the underwater mass of bergs is nine times that of the mass above water and the draft is approximately five times the height.

On summer images icebergs appear as small bright spots in the dark open water, but shadows are not detectable to aid in distinguishing bergs from small floes. In the open ocean, however, where floes have melted, any bright spots on Landsat images are almost certainly icebergs.

Ice Movement The extensive sidelap of images acquired on successive days at high latitudes enables ice movement to be measured. On March 20 and 21, 1973 nearly cloud-free images were acquired of the Davis Strait between Greenland and Baffin Island to the west. The repeated portions of two images are shown in Figure 9.14. The eastern one-quarter of both images is covered with brash ice and very small floes. Most of the area is covered by large floes measuring up to 40 km long. On March 20 (Figure 9.14A) there are few open leads, indicated by black signatures, and many of the leads are refrozen, shown by the gray signatures. The open leads are wider and more abundant on March 21 (Figure 9.14B) and some of the larger floes have broken up.

Ice movement vectors during the 24-hour period are shown by the arrows in Figure 9.14B, which have the same scale as the images. The movement is consistently southeastward at an average rate of 0.4 km per hour. This is a minimum rate, for if the ice followed an irregular course the rate of movement would be higher.

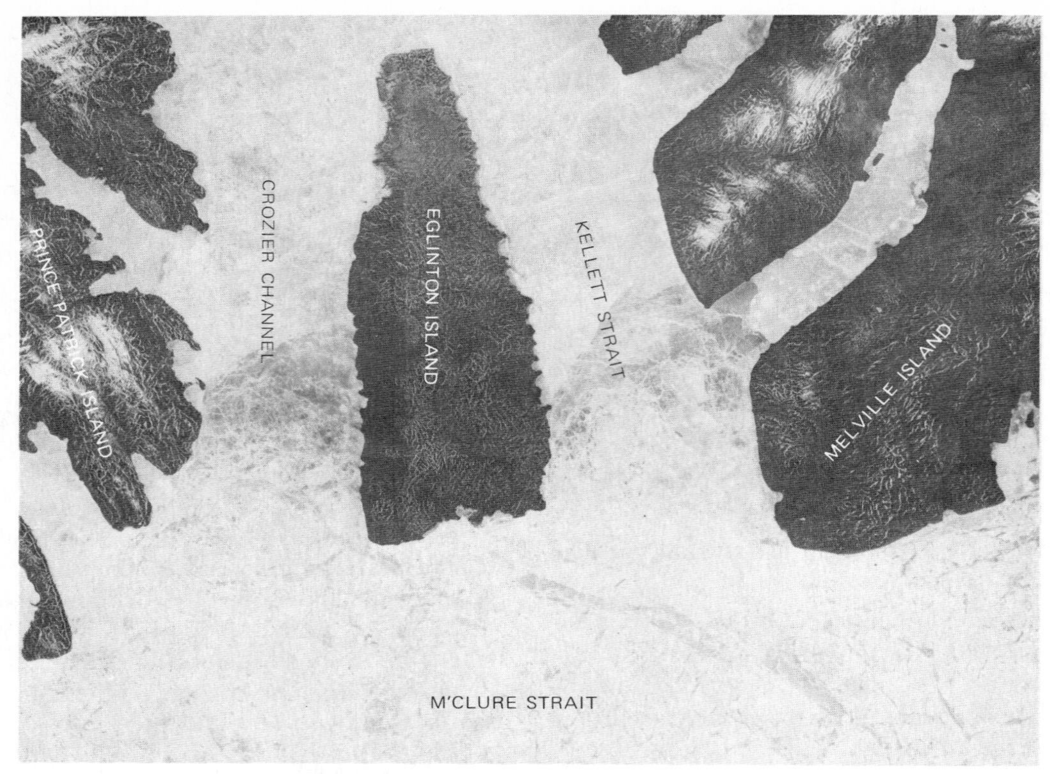

A. BAND 4 IMAGE. NOTE SHARP CONTACTS BETWEEN ICE AND LAND.

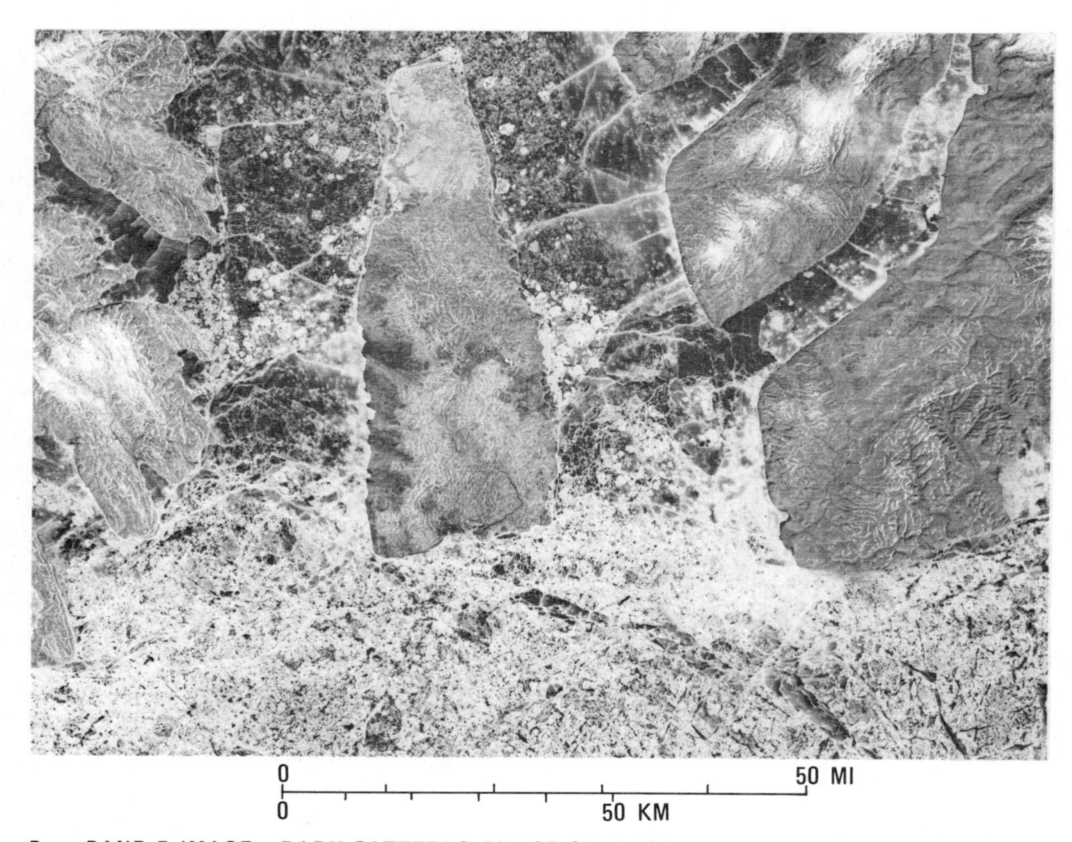

B. BAND 7 IMAGE. DARK PATTERNS ON ICE SURFACE ARE CAUSED BY ACCUMULATED MELT WATER. LINEAR DARK PATTERNS IN M'CLURE STRAIT ARE LEADS.

FIGURE 9.12
Comparison of MSS bands for Landsat 1330-20064 acquired June 18, 1973. Location is the Arctic Islands of Canada. After Barnes and Bowley (1974, Figure 4-1).

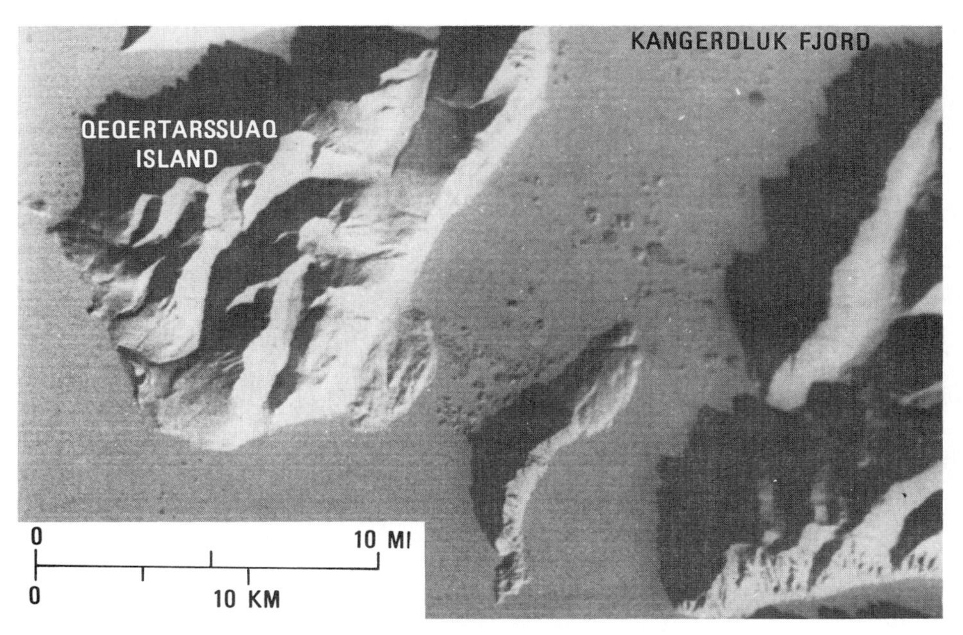

A. WEST COAST OF GREENLAND. LANDSAT 1240–14591, BAND 7, MARCH 20, 1973.

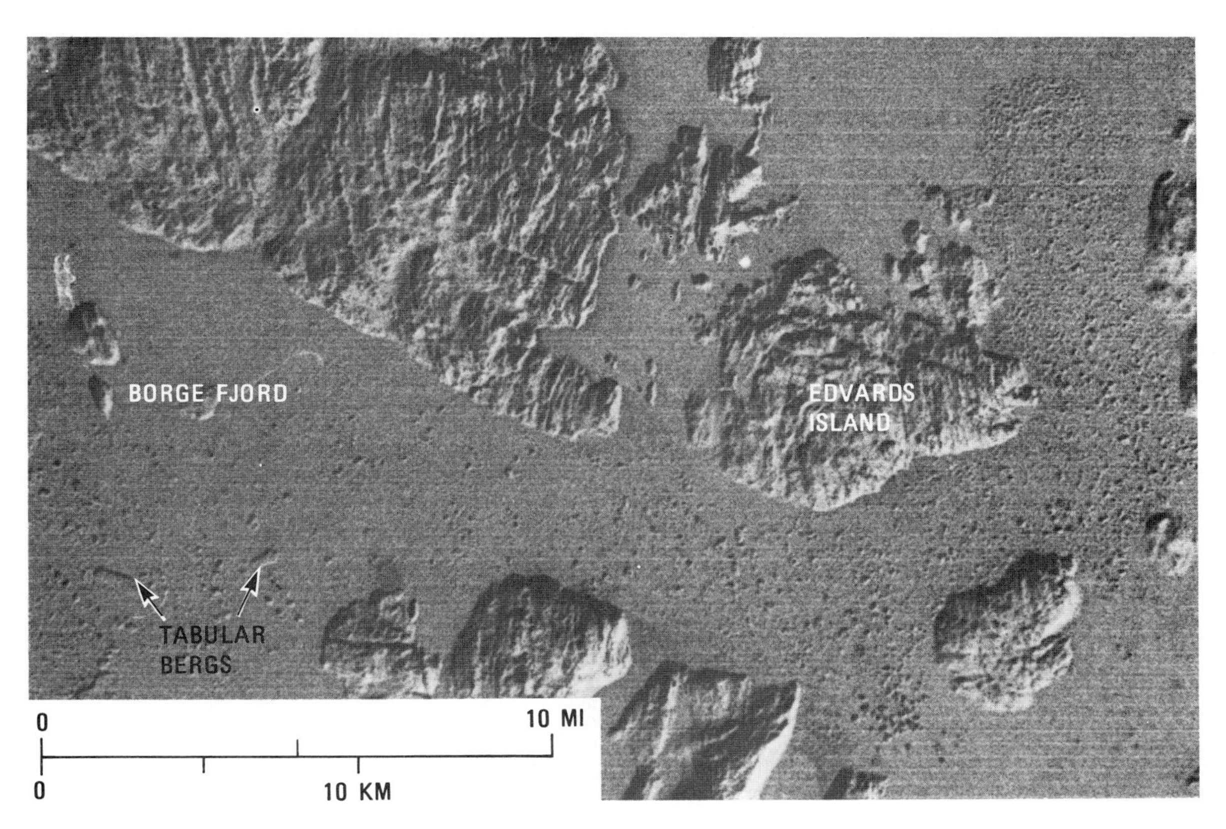

B. EAST COAST OF GREENLAND. LANDSAT 1245–13423, BAND 7, MARCH 25, 1973. ENLARGED FROM SOUTHWEST PART OF FIGURE 9.11.

FIGURE 9.13
Icebergs frozen in winter ice of Greenland.

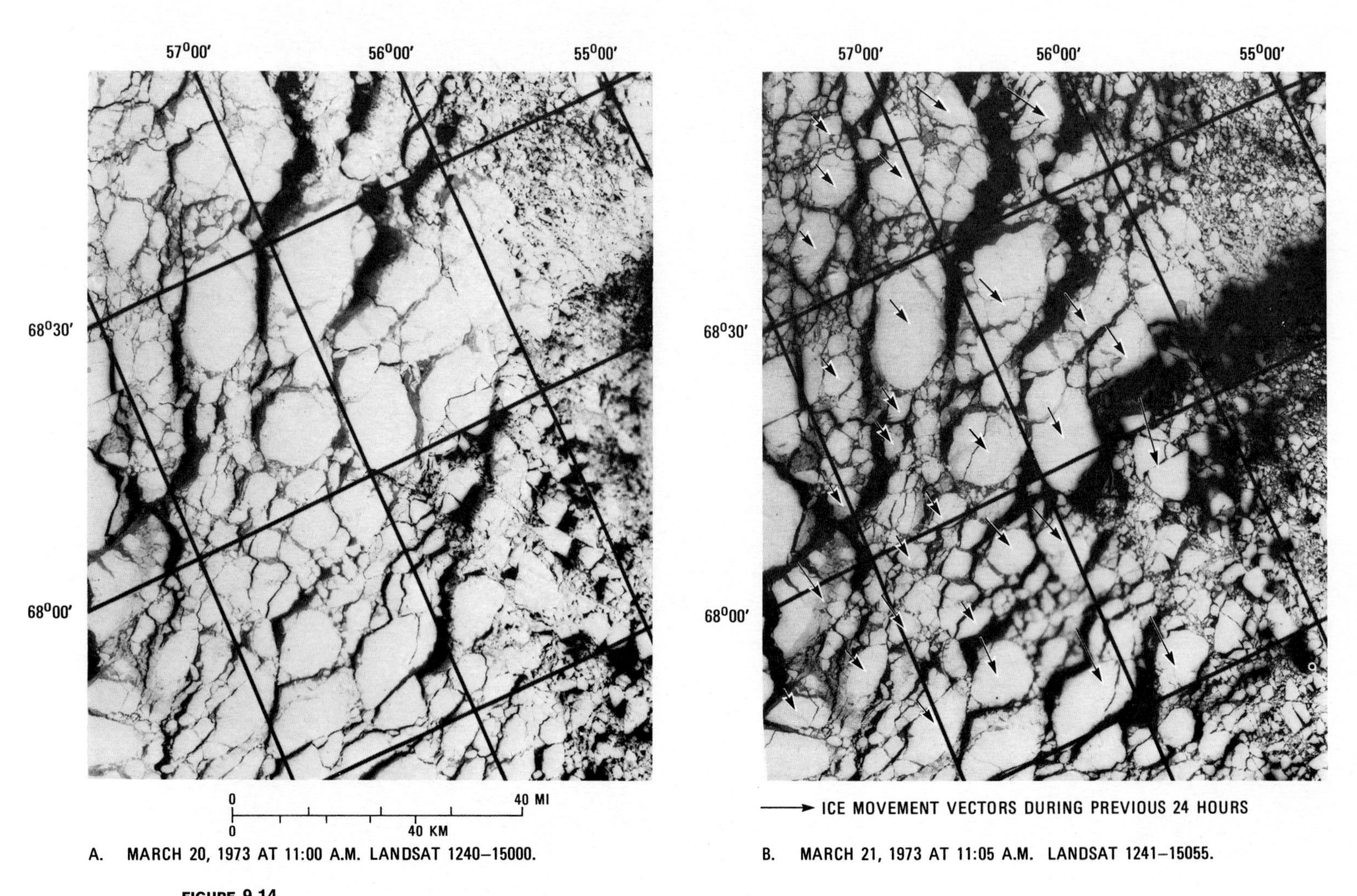

FIGURE 9.14
Movement of sea ice measured on band 7 Landsat images acquired on successive days. Davis Strait, west of Greenland.

"Quick-Look" Images The demonstrated capability of Landsat images for monitoring sea ice conditions has great potential benefit for shipping, oil exploration, and other offshore activities in the Arctic. Because of rapid changes, however, images of sea ice must be made available to the user in near-real time. Canada recognizes this requirement and their receiving station prepares and makes available on the same day a *quick-look* copy of each image. Images from the Canadian facility have been received via air mail at La Habra, California within three days of the Landsat overflight. The quick-look images are of lower quality than those illustrated here, but are adequate for interpreting and mapping major ice features. The present United States system has a minimum interval of several weeks between the time an image is acquired and the time it is available for purchase at the EROS Data Center. The improved image-processing facility planned at EDC is designed to speed up the delivery of images.

Thermal IR Images

Thermal IR images can be acquired during the periods of polar darkness, but not when heavy clouds and fog are present. The IR image and aerial photograph of Figure 9.15 were acquired simultaneously in the Northumberland Strait south of Prince Edward Island, Canada. The IR image was not corrected for scanner distortion and is compressed at the margins. The brightest tone on the IR image indicates the highest radiant temperature and corresponds to the open lead that has a dark tone on the aerial photograph. The lower surface of sea ice has a uniform temperature that is determined by the temperature of the sea water. Surface temperatures generally decrease with increasing ice thickness. However, thickness cannot be determined from IR image signatures alone because ice density, surface conditions, and snow cover also affect the surface temperature.

In the fractured reconsolidated floe on the right side of the photograph (Figure 9.15A) the light-toned angular blocks are about 50 cm thick. On the IR image (Figure 9.15B) the blocks are cooler than the refrozen lead. Surface measurements on the ice showed a radiant temperature difference of 0.5°C between the two ice types (A. O. Poulin, person communication). Icebergs (not shown) have extremely cold signatures on IR imagery. Differences in ice temperature may be transmitted through a cover of snow. Snow-covered ice with a uniform appearance on photos may show thermal signatures of concealed floes, leads, and pressure ridges on IR images.

Thermal IR images acquired from satellites provide broader coverage and are less expensive than those acquired from aircraft. The NOAA-3 satellites acquire worldwide thermal IR images on a daily basis. Their 1-km spatial resolution limits the degree of detail that can be mapped, but the concentration of ice and water can be determined. Landsat C, as previously mentioned, is planned for launch in the near future, and will include a thermal IR band with spatial resolution of approximately 240 m, which should be useful for mapping sea ice.

Radar Images

The ability to acquire radar images during darkness and bad weather makes it an excellent source

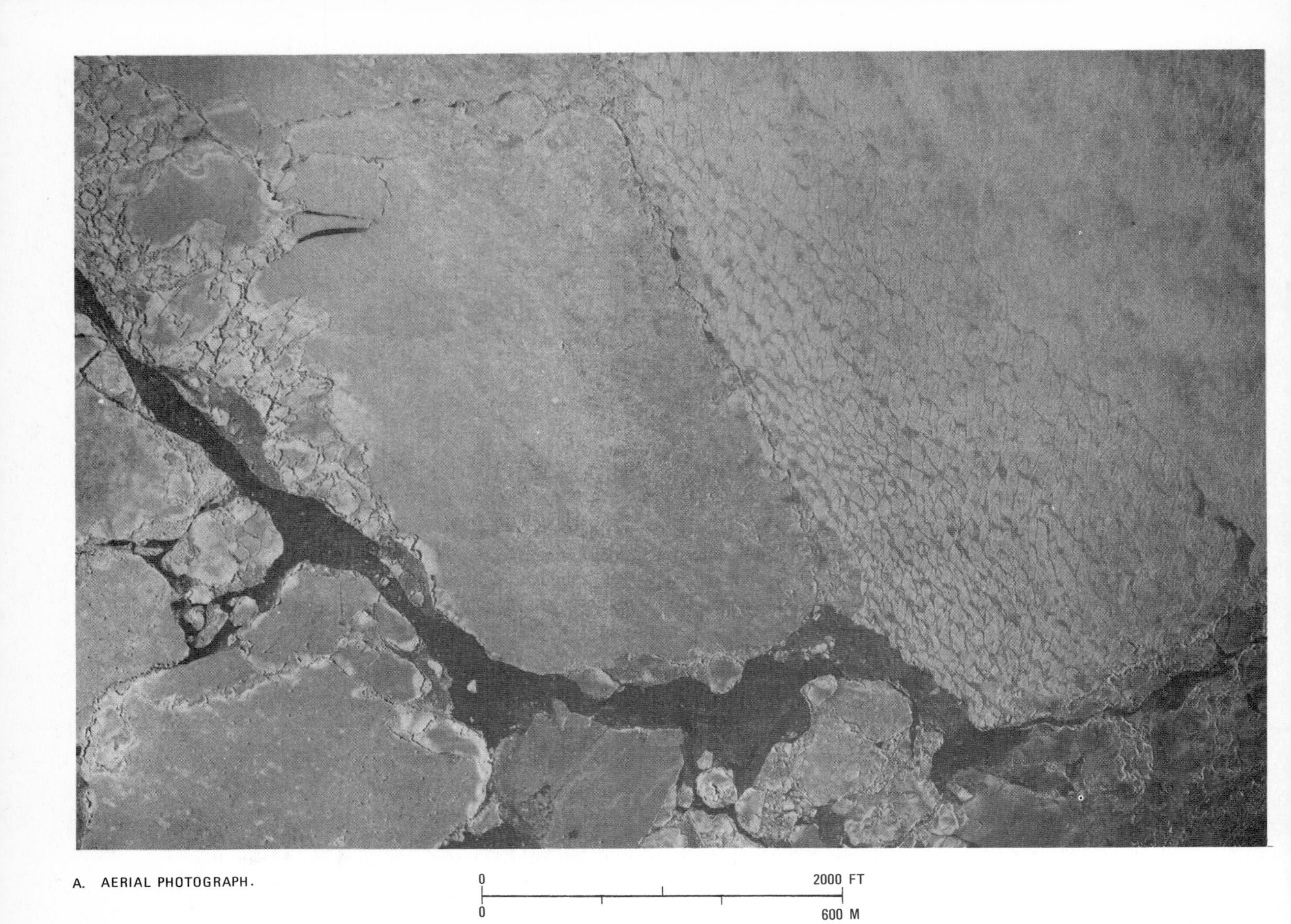

A. AERIAL PHOTOGRAPH.

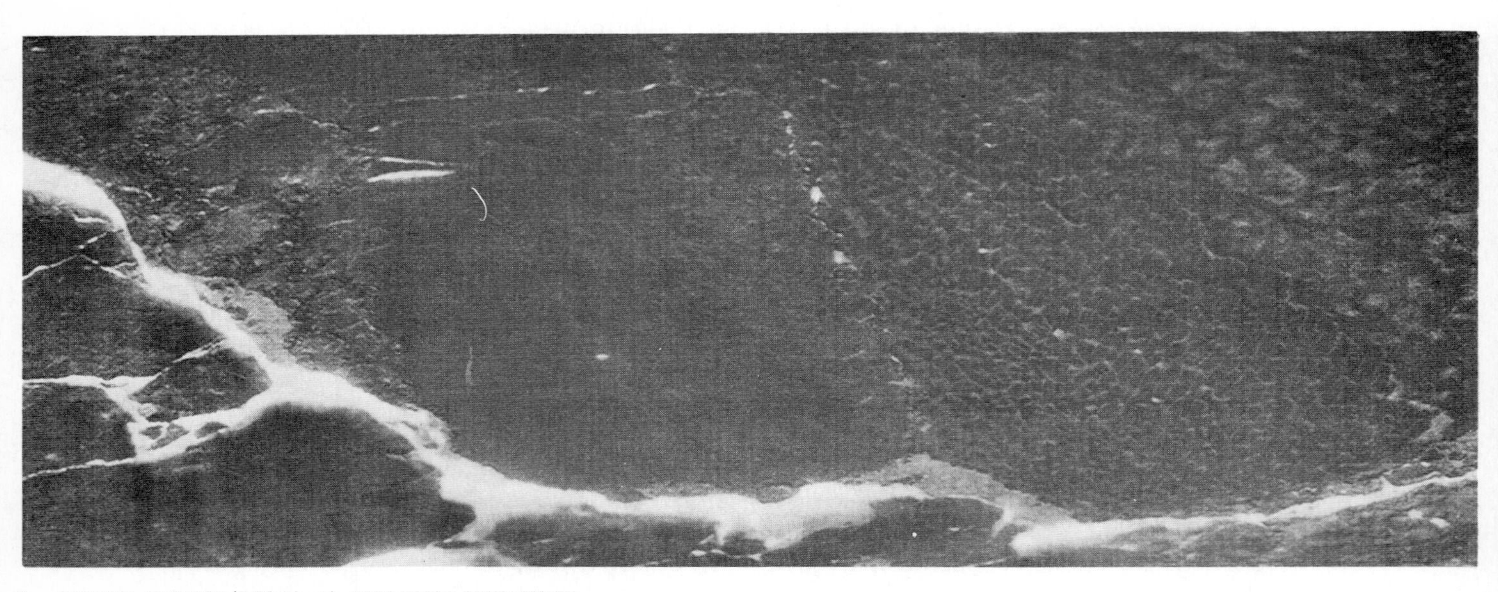

B. THERMAL IR IMAGE (8 TO 14 μm). NOTE IMAGE COMPRESSION AT UPPER AND LOWER MARGINS.

FIGURE 9.15
Aerial photograph and thermal IR image of Northumberland Strait, Gulf of St. Lawrence, acquired January 25, 1963 from altitude of 820 m by Project Bold Survey. Courtesy Ambrose Poulin, U.S. Army Engineer Topographic Laboratory.

of information for investigating sea ice. Useful images have been produced with Ka-, X-, and L-band systems. On an experimental basis, radar mosaic coverage in the Beaufort Sea was repeated to assess changes in ice conditions. Surface roughness of the ice can be estimated by the roughness criteria described in Chapter 6.

Ice Signatures Many of the ice features listed in Table 9.4 are shown on the SLAR image of the Gulf of St. Lawrence (Figure 9.16A). The interpretation (Figure 9.16B) was made without the benefit of any supporting data and therefore is somewhat generalized. The ice shown here is all first-year ice because this area thaws completely every summer. By March, when this image was acquired, the floes are rounded and marked by pressure ridges that form the bright linear features crossing the floes. The dark gray tone of the floes indicates they are relatively smooth in contrast to the bright tone of brash ice, which is known to be very rough. Numerous small floes less than 1 km wide are included with brash ice on the interpretation map. Some narrow leads filled with brash ice form bright linear features that could be confused with pressure ridges. Note the tendency for brash ice to be concentrated around the margins of larger floes.

Open water and smooth new ice are both specular surfaces and have similar dark signatures on radar images. Therefore it is difficult to distinguish between open leads and refrozen leads in Figure 9.16A. On the original larger scale image some refrozen leads are definitely recognizable by the presence of small pressure ridges and a dark gray signature that indicates some surface roughness. These criteria are also used to recognize the fast ice along the coast north and south of Cheticamp Island. The specular signature of the channel between Cheticamp Island and Cape Breton Island suggests that it is open water.

Although surface roughness explains most of the signature characteristics of sea ice, the brine content of the ice also influences the radar return. Young ice has a high content of brine that is progressively reduced with time to a minimum in multi-year ice. Brine has a higher dielectric constant than pure ice, which reduces the radar return. Thus first-year ice with a high brine content may produce a lower radar return than multi-year ice with a similar surface roughness. On radar images icebergs appear as bright spots with shadows that require a bright background to be visible.

Nonimaging Radar Methods Radar scatterometry, described in Chapter 6, provides quantitative data about radar backscatter as a function of incidence angle. The scatterometer profiles in that chapter illustrate the signatures of different types of sea ice.

Ice-sounding techniques measure ice thickness by transmitting pulses of long wavelength radar energy. Two return signals are recorded; the first from the top of the ice and the second from the base of the ice. The time delay between these signals, multiplied by the velocity of electromagnetic radiation through ice, represents the ice thickness. This technique has been used successfully to measure thickness of freshwater ice and of the Greenland icecap, both of which have sharp lower boundaries. Radar sounding has not been successful on sea ice because the lower contact with sea water is a "mushy" transitional zone of ice crystals and water that is not a sharp interface for reflecting microwave pulses.

Practical Applications In September, 1969 the tanker Manhattan successfully traversed the sea ice of the Northwest Passage to Alaska. A U.S. Coast Guard aircraft acquired radar images of the ice in advance of the tanker. The images were processed onboard the aircraft and air dropped to the ship for planning the route.

In February, 1974 NASA aircraft acquired radar images of ice on Lake Superior from which ice-condition maps were prepared. The maps were then transmitted by radio facsimile to ships. which were able to select optimum routes through the ice. This system allowed the Great Lakes shipping season to be extended for several weeks beyond the normal "freeze up" date.

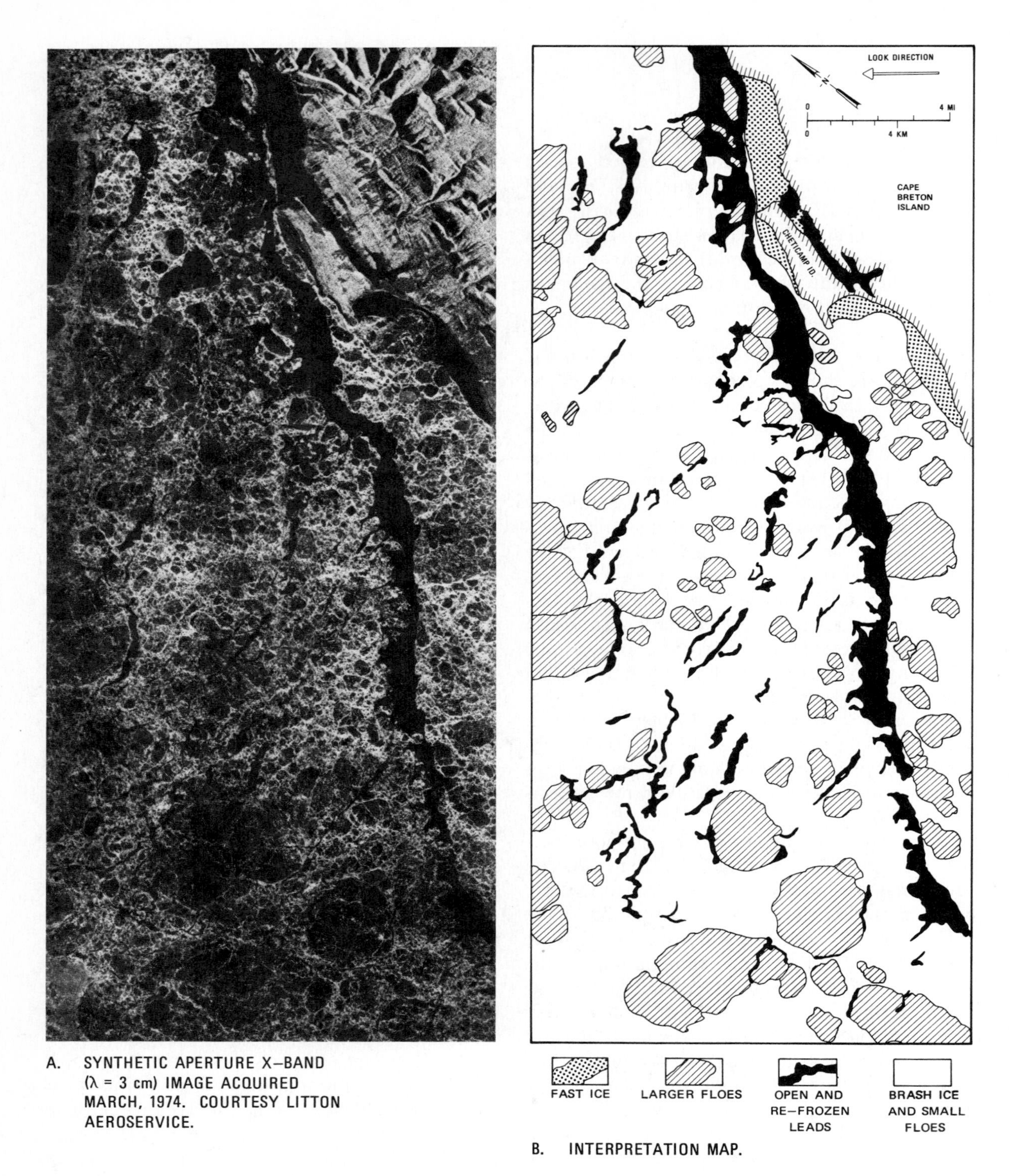

A. SYNTHETIC APERTURE X–BAND (λ = 3 cm) IMAGE ACQUIRED MARCH, 1974. COURTESY LITTON AEROSERVICE.

B. INTERPRETATION MAP.

FIGURE 9.16
Radar image of sea ice in the Gulf of St. Lawrence, Canada.

LAND-USE MAPPING

Water pollution is one aspect of man's impact on the environment that can be monitored by remote-sensing methods. Man also uses the land for habitation, agriculture, and extraction of natural resources. The impact of these activities also can be evaluated by remote sensing and image processing techniques. The mapping and classification of existing land-use patterns is essential for planning future developments. The U.S. Geological Survey (Anderson and others, 1976) developed the land-use classification shown in Table 9.5 that is designed for use with remote-sensing images. The categories listed in Level I can be identified using Landsat images at scales of 1:1,000,000 to 1:250,000 with very little supplemental information. The Level II categories can be identified on aerial photographs at scales of 1:80,000 or smaller. Definitions of the classification categories are given by Anderson and others (1976). Digital classification of Level I categories from Landsat data of the Salton Sea and Imperial Valley region, California was demonstrated in Chapter 7.

Land-Use Classification of Houston, Texas

Houston and its surroundings constitute a good test site for mapping land use by digital classification of Landsat data. A wide variety of land uses occur and correlative data are available in the form of high-altitude aircraft photographs. There are at least two significant aspects to land use in the Houston area: (1) The sustained growth rate of population and commerce in the past 30 years is among the highest in the United States. (2) Land development is not controlled by zoning regulations as it is in almost every other city in the United States.

On the enlarged Landsat image (Figure 9.17) the Level I categories of urban, rangeland, forest land, and water are recognizable. The map of Houston (Figure 9.18) emphasizes the transportation and drainage networks. Bayous are sluggish

TABLE 9.5
Classification system of land use and land cover for use with remote-sensor data

Level I	Level II
Urban or built-up land	Residential
	Commercial and services
	Industrial
	Extractive
	Transportation, communications, and utilities
	Industrial and commercial complexes
	Mixed urban or built-up land
	Other urban or built-up land
Agricultural land	Cropland and pasture
	Orchards, groves, vineyards, nurseries, and ornamental horticultural areas
	Confined feeding operations
	Other agricultural land
Rangeland	Herbaceous rangeland
	Shrub and brush rangeland
	Mixed rangeland
Forest land	Deciduous forest land
	Evergreen forest land
	Mixed forest land
Water	Streams and canals
	Lakes
	Reservoirs
	Bays and estuaries
Wetland	Forested wetland
	Nonforested wetland
Barren land	Dry salt flats
	Beaches
	Sandy areas other than beaches
	Bare exposed rock
	Strip mines, quarries, and gravel pits
	Transitional areas
	Mixed barren land
Tundra	Shrub and brush tundra
	Herbaceous tundra
	Bare-ground tundra
	Wet tundra
	Mixed tundra
Perennial snow or ice	Perennial snowfields

Source: From Anderson and others (1976, Table 2).

FIGURE 9.17
Houston, Texas, and vicinity. Landsat 1037-16244 band 5, acquired August 29, 1972. The land-use classification for this area is shown in Plate 8.

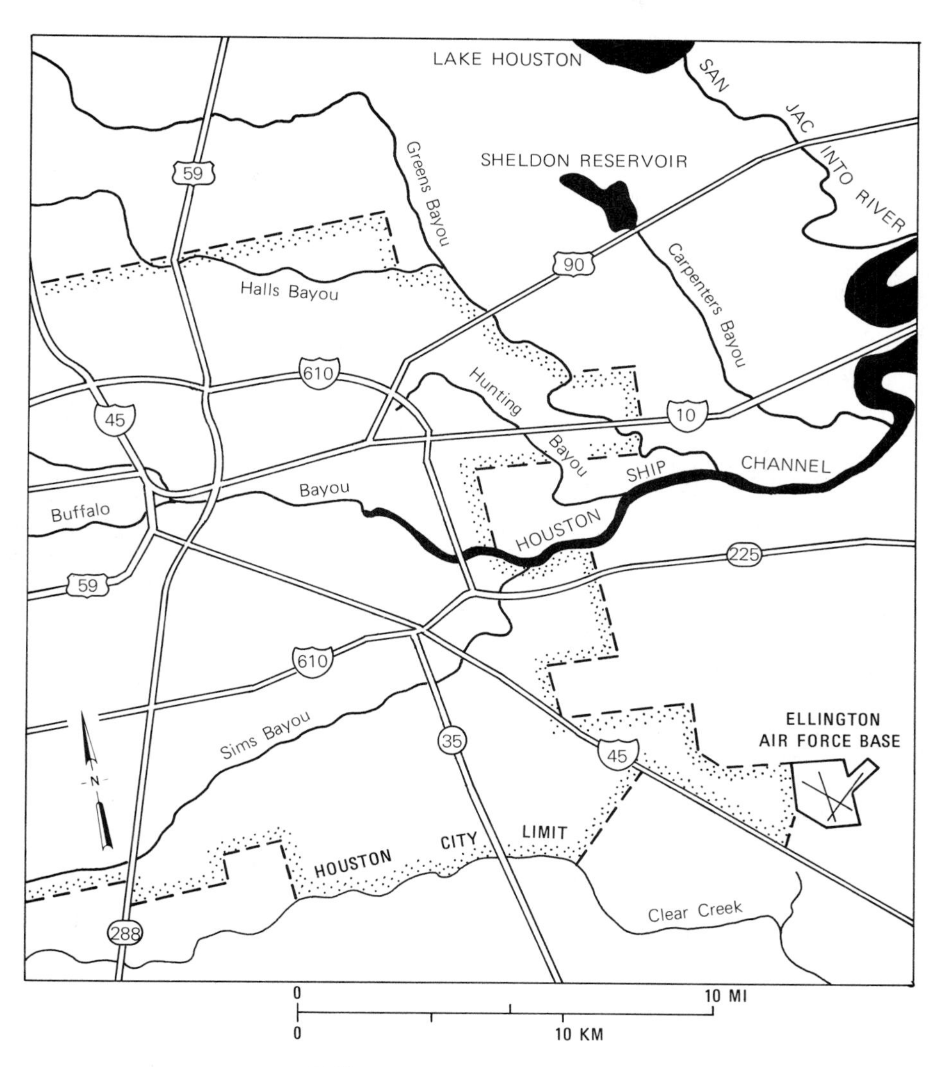

FIGURE 9.18
Location map of Houston, Texas, and vicinity.

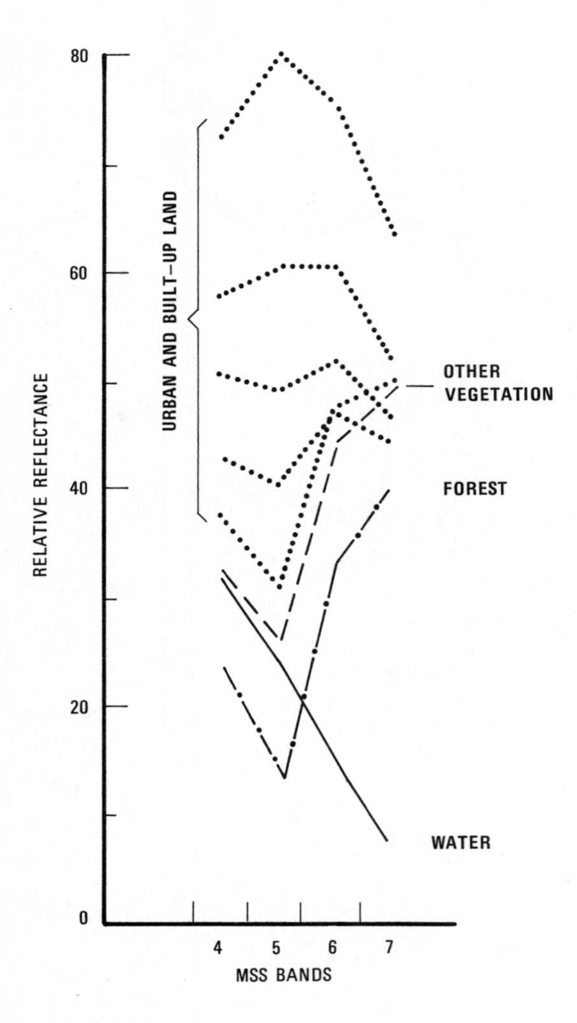

FIGURE 9.19
Spectral-reflectance curves for Level I land-use categories of Houston and vicinity. From Erb (1974A, Figure 6-3).

streams in which the water level rises and falls with the tides of the nearby Gulf of Mexico. A color composite image (not shown) aids the recognition somewhat because it combines the reflectance variations from three Landsat MSS bands. Erb (1974A, 1974B) used a variety of digital methods to classify the data on the CCTs for the scene shown in Figure 9.17. The results of one classification are shown in color in Plate 8 where the Level I categories of rangeland, forest land, and water are mapped. The urban and built-up land category is subdivided into a number of Level II categories indicated on the legend of Plate 8. The white, yellow, black, and gray signatures represent the categories of commercial, industrial, and transportation but there is no unique association between a color and an individual category. Typical spectral reflectance curves for the major categories in this scene are shown in Figure 9.19.

The Landsat classifications were compared with classifications made from high-altitude aircraft photographs. In Table 9.6 the actual land use, as determined from aerial photographs, is shown on the left, and the classifications made from Landsat data are shown along the top. For the commercial, industrial, and transportation category the Landsat analysis correctly classified 94.2 percent of the pixels, 5.5 percent were misclassified as residential, and 0.3 percent were misclassified as water. Similar high accuracies were obtained for the forest land and the water categories. There were problems with the residential, mixed urban, and rangeland categories. These categories all contain varying proportions of houses, roads, vegetation (as landscaping and open fields) and water (as swimming

TABLE 9.6
Accuracy of Landsat classification, Houston, Texas

Actual land use (from aerial photographs)	Classes of land use and land cover (from Landsat digital classification)					
	Commercial, industrial, and transportation	Residential	Mixed urban	Forest land	Rangeland	Water
Commercial, industrial, and transportation	94.2	5.5	—	—	—	0.3
Residential	2.6	66.8	23.0	4.5	3.2	—
Mixed urban	1.0	20.8	51.1	3.8	23.5	—
Forest land	—	0.7	0.2	95.1	4.0	—
Rangeland	1.1	12.1	25.7	4.8	56.2	—
Water	3.9	3.0	1.9	2.2	1.5	87.7

Source: From Erb (1974A, Table 7-3).

pools and streams). A single 79 by 79 m ground resolution cell of Landsat integrates reflectance from several (or all) of these features, which makes classification difficult for such complex categories. As advocated earlier, in Chapter 8, reducing the size of the MSS ground resolution cell should improve the accuracy of classifications.

Remote Sensing of Prehistoric Land-Use Patterns

Daytime thermal IR images and aerial photographs of northern Arizona were simultaneously acquired for geologic investigations. While interpreting an IR image (Figure 9.20A) Berlin and others (1977) noted a pattern of parallel warm (bright signature) and cool (dark) bands that is not apparent on the aerial photograph (Figure 9.20B). The dimensions of the banded area are 67 by 265 m; the 20 individual warm and cool bands are 3 to 3.5 m wide.

Field investigations established that the warm bands are low ridges of basaltic ash from eruptions of Sunset Crater in the San Francisco volcanic field; the ash is underlain by thin soil 5 to 20 cm thick. The cool bands are swales with extensive exposures of soil and minor patches of ash. Vertical relief ranges from 5.5 to 20 cm. The ridges are oriented normal to the gentle north slope; this would trap the runoff from rainfall. Berlin and others (1977) recognized the possibility that this pattern represents an ancient agricultural site. They proceeded to evaluate this possibility and determine the reason for the distinctive signature on the thermal IR image. The thermal IR signatures were investigated by inserting soil thermometers to a depth of 3 cm into the ridges and swales. In the afternoon hours, the ridges were 1.1° to 6.2°C warmer than the swales. These contact temperatures agree with the relative temperature signatures of the afternoon image. By midnight the contact temperatures of the ridges had cooled to a lower temperature than the swales and a thermal cross over had occurred. This temperature behavior indicates that the ash of the ridges has a lower thermal inertia than the soil exposed in the swales. The low albedo of the ash causes it to absorb solar energy, which contributes to the warm daytime temperature signature.

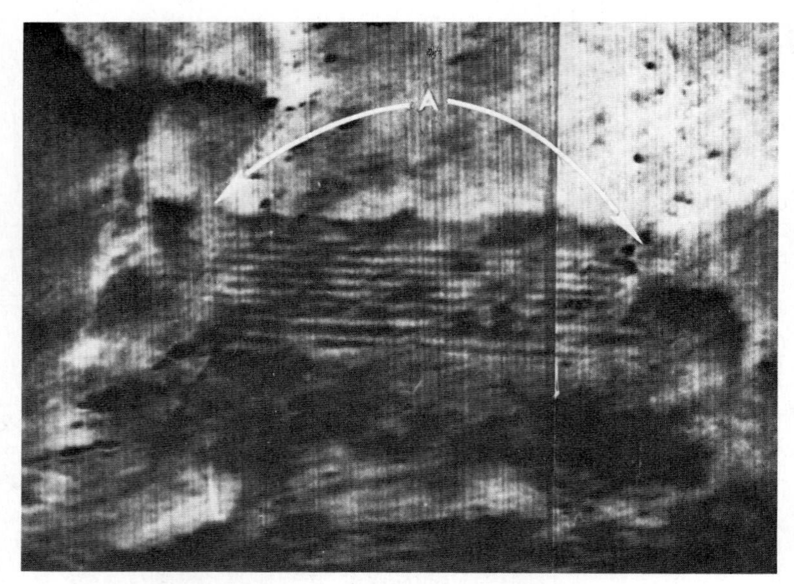

A. DAYTIME THERMAL IR IMAGE (8 TO 14 μm).

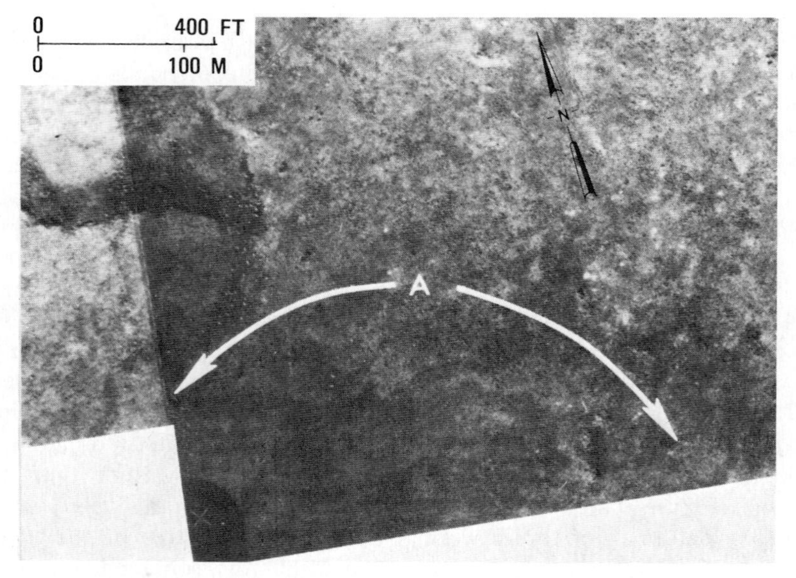

B. AERIAL PHOTOMOSAIC.

FIGURE 9.20
Simultaneously acquired IR image and aerial photomosaic of ancient corn field 40 km northeast of Flagstaff, Arizona. Reproduced by permission of the Society for American Archaeology. From *American Antiquity* 42(4):589, 1977. Courtesy G. L. Berlin, Northern Arizona University.

Analyses of pollen from soil samples showed a relatively high abundance of corn pollen in the soil beneath the ash ridges. This pattern indicates that the corn was planted in the soil of the ridges with the ash acting as a mulch to retain moisture, which is consistent with present cultivation practices of Indians in the region. The soil of the ridges is depleted in potassium, nitrogen, and phosphate relative to the soil in the swales, indicating prolonged use for agriculture. Ten ancient dwelling sites were found within a radius of 825 m from the field. The dating of pottery fragments indicates that cultivation began after the 1065 A.D. eruption of Sunset Crater that provided the cinders which were scraped up into long rows to provide mulch for corn cultivation. The depletion of soil nutrients and the change to a cooler and drier climate in the early 1200s caused the area to be abandoned, probably by 1250 A.D.

The work of Berlin and his associates is an example of the multidiscipline approach that is required to understand fully the significance of much remote sensing data. In this case the disciplines of remote sensing, chemistry, botany, and archaeology were essential to the investigation. The ancient corn field that had lain untouched for 700 years was discovered accidentally on a thermal IR image acquired for other purposes. Elsewhere, IR color photographs have been used to locate archaeological sites that are covered with natural or cultivated vegetation. The ancient ruins influence the composition and moisture content of the overlying soil, which in turn influence the reflectance of the vegetation and are detectable on the IR color aerial photographs. Remote sensing is valuable for analyzing both modern and ancient land-use patterns.

Vegetation Mapping

Aerial photographs have long been used to map vegetation. Foresters use stereo models to estimate the yield of lumber or pulp wood from timber tracts. Previsual symptoms on IR color photographs are used to determine the extent of insect infestations and to plan eradication measures. Thermal IR images of forest fires penetrate the smoke and enable the fire front and "hot spots" to be precisely located. The U.S. Department of Agriculture employs aerial photographs each year to estimate crop acreage and yield. IR color photographs have been used to monitor the spread of disease, such as corn blight in the midwest during the early 1970s.

A much more ambitious program is the Large Area Crop Inventory Experiment (LACIE) being conducted by NASA and contract organizations. The objective is to provide annual yield forecasts of the principal grain crops on a worldwide basis, beginning with wheat. Landsat CCT data are used to identify crop types and estimate the acreage. This information is then combined with rainfall data from weather satellites to estimate yield. Multispectral classification techniques are the basis for crop identification. Another valuable technique is a version of change-detection processing in which sequential Landsat images are spatially registered and seasonal changes in crop vigor are determined. The nature of these changes is related to the eventual crop yield.

COMMENTS

These examples have illustrated the many applications of remote-sensing methods for environmental and land-use monitoring. One major advantage is the flexibility of coverage that ranges from high-resolution aircraft surveys to worldwide daily coverage by NOAA satellites. The variety of available wavelength bands can provide the optimum spectral sensitivity for a particular application. Image-processing methods enable the maximum amount of information to be extracted from the data. Remote-sensing methods are typically more cost effective than surface surveys and sampling methods. Remote-sensing methods, however, are not generally utilized to the extent suggested by those potential advantages. Some problems contributing to their under-utilization are:

1. The potential applications of the methods are still not widely known.
2. There is an insufficient number of skilled and experienced interpreters.
3. Some aircraft surveys and digital-processing methods are expensive relative to available funds.

It is hoped that training courses and books such as this one will remedy the first two problems. The third can be attacked by more efficient technology to reduce costs and by more liberal funding as the potential values of remote sensing are realized.

REFERENCES

Anderson, J. R., E. E. Hardy, J. T. Roach, and R. E. Witmer, 1976, A land use and land cover classification system for use with remote sensor data: U.S. Geological Survey Professional Paper 964.

Barnes, J. C. and C. J. Bowley, 1974, The application of ERTS imagery to monitoring Arctic sea ice: Environmental Research and Technology Inc., Document 0408-F, Lexington, Mass.

Belderson, R. H., N. H. Kenyon, A. H. Stride, and A. R. Stubbs, 1972, Sonographs of the sea floor, a picture atlas: Elsevier Publishing Co., New York.

Berlin, G. L., J. R. Ambler, R. H. Hevly, and G. G. Schaber, 1977, Identification of a Sinagua agricultural field by aerial thermography, soil chemistry, pollen/plant analysis, and archaeology: American Antiquity, v. 42. p. 589.

Carlson, P. R., 1976, Mapping surface current flow in turbid nearshore waters of the northeast Pacific *in* Williams, R. S., and W. D. Carter, eds., ERTS-1, a new window on our planet: U.S. Geological Survey Professional Paper 929, p. 328–329.

Deutsch, M., A. E. Strong, and J. E. Estes, 1977, Use of Landsat data for the detection of marine oil slicks: Offshore Technology Conference, p. 311–318, Houston, Texas.

Deutsch, M., A. E. Strong, and G. Rabchevsky, in press, Detection of floating oil slicks by Landsat: Photogrammetric Engineering and Remote Sensing.

Erb, R. B., 1974A, The ERTS 1 investigation (ER 600)—ERTS 1 land use analysis of the Houston area test site: NASA TM X-58124, v. 7.

———, 1974B, The ERTS 1 investigation (ER 600)—a compendium of analysis results of the utility of ERTS 1 data for land resources management: NASA TM X-58156.

Fantasia, J. F. and H. C. Ingrao, 1974, Development of an experimental airborne laser remote sensing system for the detection and classification of oil spills: Proceedings Ninth Symposium on Remote Sensing of Environment, p. 1711–1745, Environmental Research Institute of Michigan, Ann Arbor, Mich.

Guinard, N. W., 1971, The remote sensing of oil slicks: Proceedings Seventh International Symposium of Remote Sensing of Environment, p. 1005–1025, University of Michigan, Ann Arbor, Mich.
Hammack, J. C., 1977, Landsat goes to sea: Photogrammetric Engineering and Remote Sensing, v. 43, p. 683–691.
Maul, G. A. and H. R. Gordon, 1975, On the use of Earth Resource Technology Satellite (Landsat-1) in optical technology: Remote Sensing of Environment, v. 4, p. 95–128.
Maurer, A. and A. T. Edgerton, 1976, Flight evaluation of U.S. Coast Guard airborne oil surveillance system: Marine Technology Society Journal, v. 10, p. 38–52.
National Academy of Sciences, 1975, Petroleum in the marine environment: Washington, D.C.
Otterman, J., G. Ohring, and A. Ginzburg, 1974, Results of the Israel multidisciplinary data analysis of ERTS-1 imagery: Remote Sensing of Environment, v. 3, p. 133–148.
Pirie, D. M. and M. J. Murphy, 1975, California coastal processes study: U.S. Army Engineer District, San Francisco, Aircraft SRT Project X-098, prepared for NASA-Johnson Space Center.
Rouse, L. J. and J. M. Coleman, 1976, Circulation observations in Louisiana bight using Landsat imagery: Remote Sensing of Environment, v. 5, p. 55–66.
Specht, M. R., D. Needler, and N. L. Fritz, 1973, New color film for water penetration photography: Photogrammetric Engineering, v. 40, p. 359–369.
Vizy, K. N., 1974, Detecting and monitoring oil slicks with aerial photos: Photogrammetric Engineering, v. 40, p. 697–708.

ADDITIONAL READING

Bowden, L. W., ed., 1975, Urban environments—inventory and analysis *in* Reeves, R. G., ed., Manual of remote sensing: ch. 23, p. 1815–1880, American Society of Photogrammetry, Falls Church, Va.
Estes, J. E. and L. W. Senger, eds., 1974, Remote sensing—techniques for environmental analysis: Hamilton Publishing Co., Santa Barbara, Calif.
Huebner, G. L., ed., 1975, The marine environment *in* Reeves, R. G., ed., Manual of remote sensing: ch. 20, p. 1553–1622. American Society of Photogrammetry, Falls Church, Va.
Johnson, P. L., ed., 1969, Remote sensing in ecology: University of Georgia Press, Athens, Ga.
Schanda, E., ed., 1976, Remote sensing for environmental sciences: Springer-Verlag, Berlin.

10
NATURAL HAZARDS

Earthquakes, landslides, volcanic eruptions, and floods are natural hazards that kill thousands of people and destroy billions of dollars of property each year. These losses will increase as the world population increases and more people reside in areas that are subject to these hazards. Some floods can be controlled by dams and landslide risks can be reduced by proper engineering design. Aside from these steps there is little that man can do to prevent the occurrence of natural hazards. The following steps can be taken to minimize the effects of natural hazards:

1. Analyze the risk of natural hazards. One example is to identify volcanoes that have a potential for eruption. In addition to recognizing hazards, risk analysis should delineate areas on the basis of relative susceptibility to damage.
2. Provide warning in advance of specific hazardous events. The Chinese, for example, successfully predicted a major earthquake in 1975 and were able to minimize casualties. The United States and Russia are also conducting research programs for earthquake prediction. Volcanic eruptions in Hawaii and elsewhere have been predicted on the basis of ground movements.
3. Assess the damage caused by a hazardous event. An early evaluation of damage caused by floods and earthquakes is essential for carrying out rescue, relief, and rehabilitation efforts.

Various forms of remote sensing have been used in the activities just listed. This use will undoubtedly increase as we improve our ability to interpret and understand the imagery. The data collection system on the Landsat satellites provides rapid delivery of data from flood gauges, seismometers, tiltmeters, and other devices installed at critical areas that may be inaccessible for conventional measuring methods. These data are potentially valuable for predicting and providing warning of impending hazards. As with remote-sensing imagery, we are still learning how to use these data.

EARTHQUAKES

Earthquakes are caused by the abrupt release of stress that has built up in the earth's crust. Some zones of maximum earthquake intensity and frequency occur at the boundaries between the moving plates that form the crust of the earth. Earthquakes are common along the San Andreas fault that forms part of the boundary between the North American and the Pacific plates. Major earthquakes also occur within the interior of crustal plates as demonstrated by seismic activity in China and Russia. Methods for predicting specific earthquakes are being developed, but remote sensing seems to have limited application. Remote sensing, however, is very useful for *seismic risk analysis*, which estimates the geographic distribution, frequency, and intensity of seismic activity without attempting to predict specific events. This analysis is essential for locating and designing dams, powerplants (especially nuclear), and other projects in seismically active areas.

One method of seismic risk analysis is based on the study of instrumentally recorded earthquakes and historic records of earthquakes. The instrumental records and even the historical records, which cover some 2,000 years in Japan and 3,000 years in China, do not cover enough time for valid extrapolations of future earthquakes. Another shortcoming of the historical approach was convincingly argued by Allen (1975), who reviewed the historic record in China and Turkey. In northern China the historic record showed an almost complete lack of major shocks for the period from 200 to 1000 A.D., but two major earthquakes occurred in 1556 and 1668 A.D. In Turkey historic records indicate that quiescent periods about 150 years in length separate periods of earthquake activity. This erratic temporal distribution of earthquakes is a further handicap to their analysis from historic records. In most seismic areas, however, there is clear surface evidence of major faults that moved in Quaternary time (Allen, 1975).

The second method of seismic risk analysis is based on the recognition of these *active faults*, which are defined as breaks along which movement has occurred in Holocene time (past 11,000 years). Remote-sensing analysis and field studies of active faults can provide a geologic history that overcomes many of the shortcomings of instrumental and historic records. The mapping of surface faults formed by large shallow earthquakes has not been fully utilized (Allen, 1975, p. 1041). The following examples from California, Alaska, Turkey, and China illustrate applications of remote sensing to seismic risk analysis.

Southern California

The following general relationships between faulting and earthquakes in southern California were noted by Allen (1975, p. 1043):

1. Virtually all large earthquakes (those exceeding magnitude 6.0 on the Richter scale) have occurred because of ruptures along faults that had been recognized, could have been recognized, or should have been recognized by field geologists prior to the events.
2. All of these faults have a history of earlier displacements in Quaternary and possibly Holocene time.
3. All the earthquakes are shallow, not exceeding about 20 km in depth. Most earthquakes larger than magnitude 6.0 have been accompanied by surface faulting, as have many of the smaller events.
4. The larger earthquakes have generally occurred on the longer faults, although there has been sufficiently wide variation to indicate caution in blindly applying any single formula for this relationship.
5. Generally, only a small segment of the entire length of a fault zone has broken during any single earthquake, although there are some conspicuous and significant exceptions.

Comparison of the Landsat image of southern California (Plate 2) with the map of faults having Quaternary displacements (Figure 10.1A) indicates that most of these faults are evident on the Landsat image. With few exceptions, earthquakes since 1912 have occurred in the areas of Quaternary faulting (Figure 10.1B). The smaller earth-

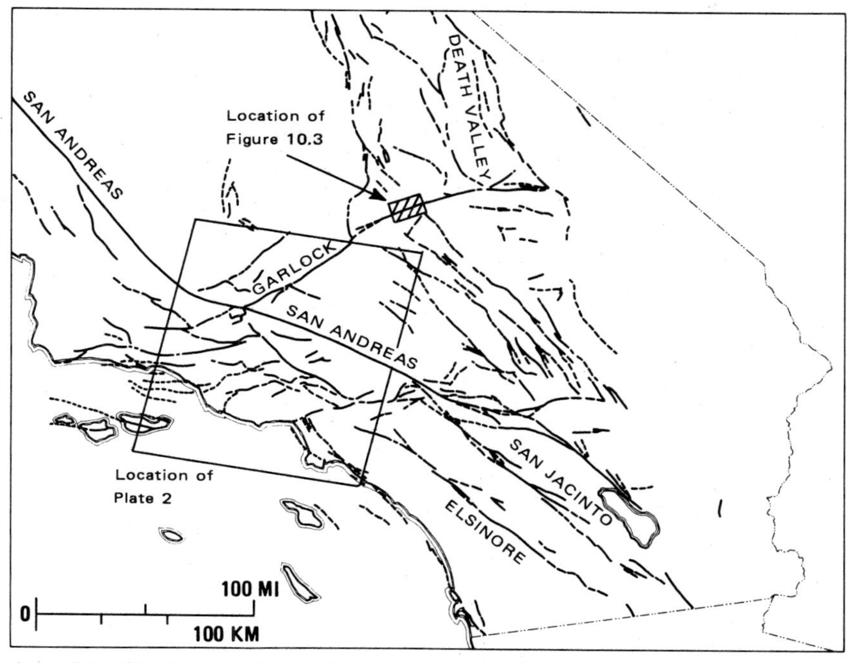

A. FAULTS WITH QUATERNARY DISPLACEMENTS.

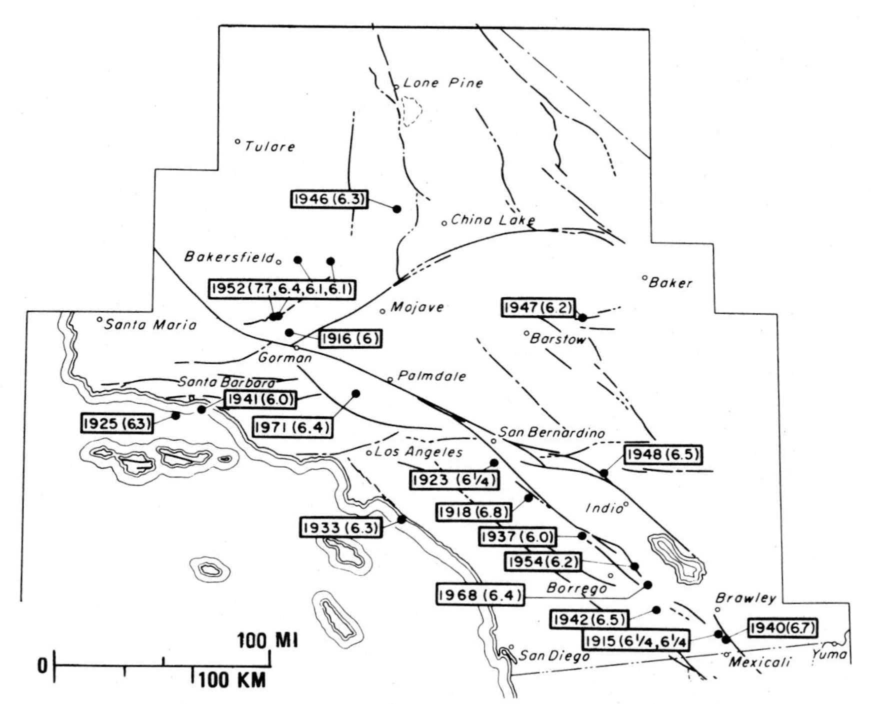

B. EARTHQUAKES OF MAGNITUDE 6.0 AND GREATER, 1912–1974.

FIGURE 10.1
Relationship of Quaternary faulting to earthquakes in southern California. From Allen (1976, Figures 3 and 4).

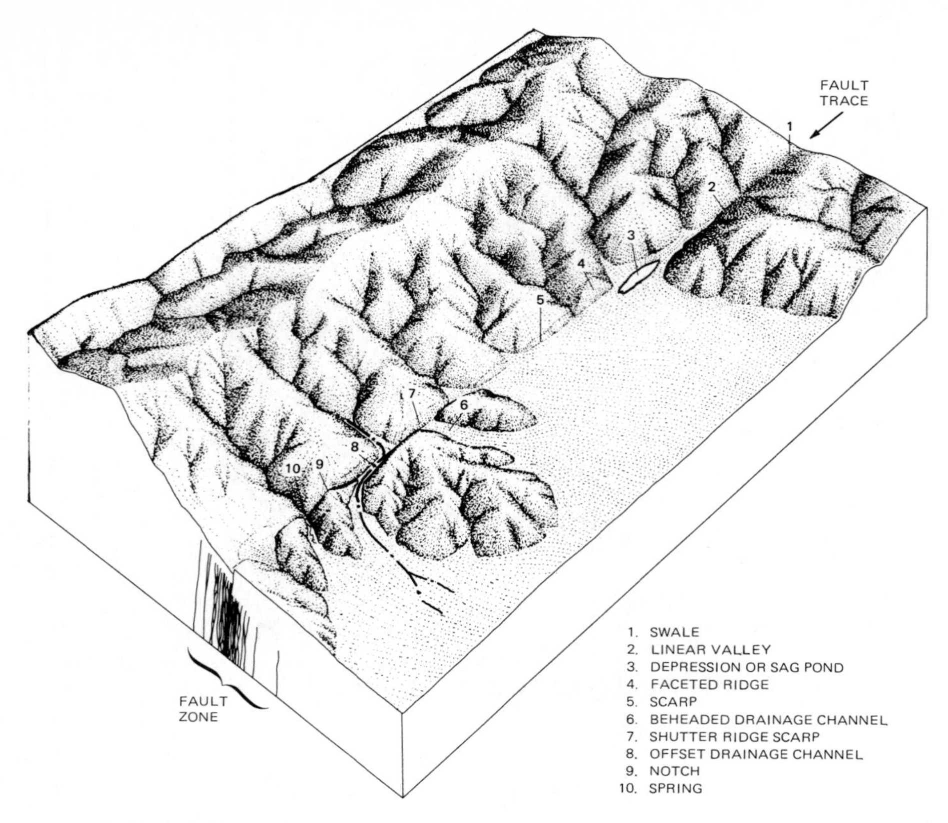

FIGURE 10.2
Generalized diagram of typical topographic features along active faults. After Sharp (1972).

quakes, not shown in Figure 10.1B, are also associated with active faults (Allen 1975, p. 1043). The lack of seismic events during the past 63 years in the Death Valley region, Garlock fault, and central segment of San Andreas fault is only a temporary pause, for these faults are marked by surface features characteristic of active faults.

As previously stated, active faults are those along which movement has occurred in Holocene time (past 11,000 years). Evidence for Holocene movement includes: (1) historic earthquakes, (2) displacement of rock units younger than 11,000 years, and (3) preservation of certain topographic features caused by faulting. Typical topographic features are illustrated in the diagram of Figure 10.2 and on the Skylab image of Figure 10.3. These features are formed by repeated horizontal and vertical displacements of a few centimeters or meters along the fault. As opposing fault blocks slide laterally along numerous individual breaks, some blocks are relatively depressed to form depressions or *sag ponds* that accumulate alluvial deposits. Other fault slivers are raised, tilted, or slid diagonally to produce *elongate ridges* and *shutter ridges. Notches, trenches,* and *troughs* may result from increased erosion of highly fractured rocks, or they may be primary fault features. *Offset drainage channels* are especially significant, because they also indicate the sense of displacement along the fault. The left-lateral, strike-slip displacement of the Garlock fault is indicated by the offset drainage channels in Figure 10.3. The preservation of these topographic features is evidence for active faulting; had the features formed in pre-Holocene time most of them would have been obliterated by erosion and deposition. As illustrated in Figure 10.2, a *fault zone* includes the entire belt of active and inactive faults that may be hundreds of kilometers in length and several kilometers in width. The *fault trace* is the surface expression of the individual active faults.

Landsat images are excellent for recognizing the

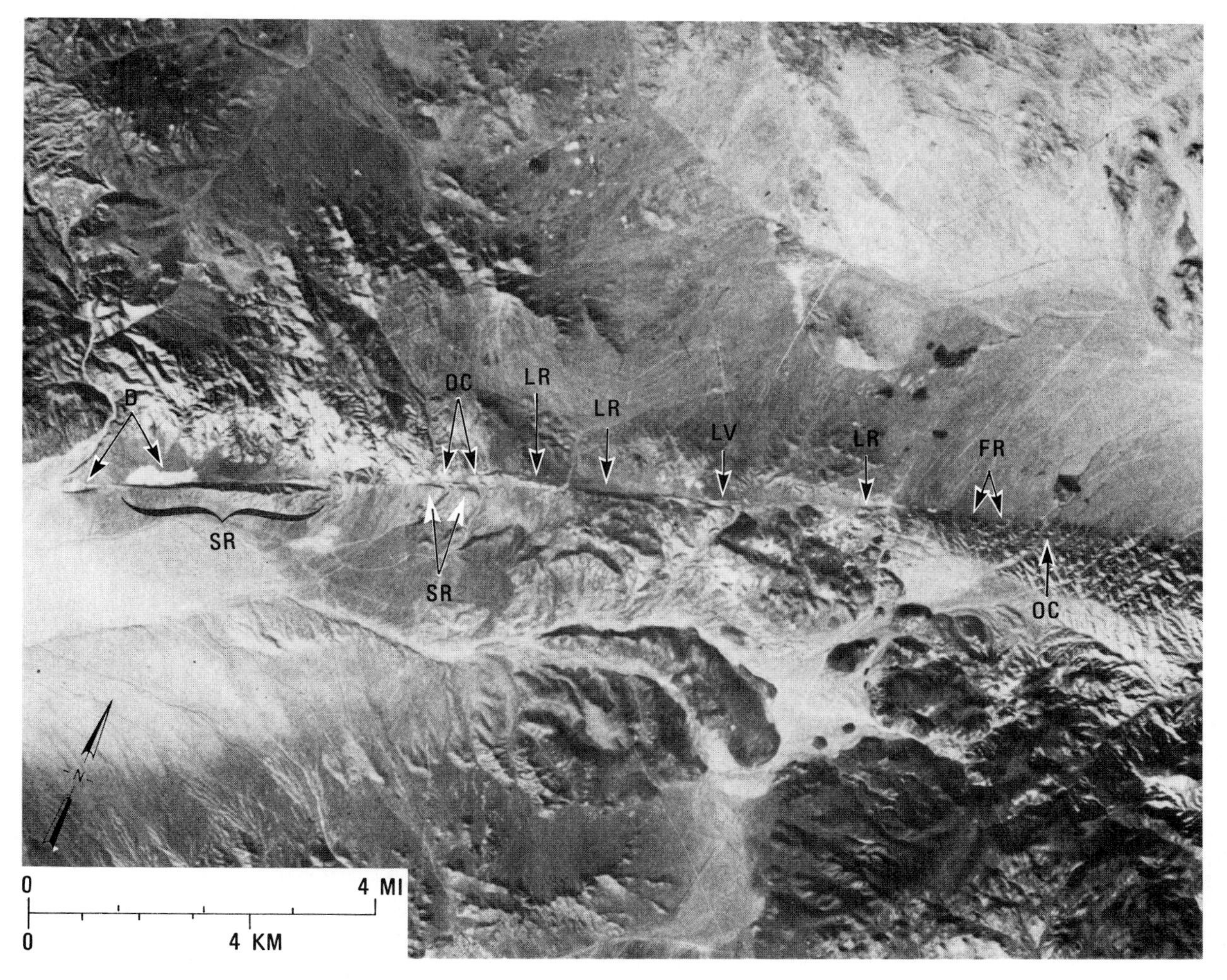

FIGURE 10.3
Skylab earth terrain camera photograph of the Garlock fault, California. The following features indicate the fault is active: D = depression; SR = shutter ridge scarp; OC = offset channel; LR = linear ridge; LV = linear valley; FR = faceted ridge. Enlarged portion of Skylab 4, S190-B, roll 92, frame 347 (original in normal color). From Merifield and Lamar (1975, Figure 2). Courtesy P. M. Merifield, California Earth Science Corporation.

continuity and regional relationships of faults, but the spatial resolution is too low to recognize any but the largest topographic features indicative of active faulting. The higher resolution of Skylab and aircraft photographs, particularly when viewed stereoscopically, is ideal for mapping these features. The use of thermal IR images for mapping the San Andreas and Superstition Hills faults, even in areas of little or no topographic expression, was described in Chapter 5. The highlighting and shadowing effects of SLAR emphasize subtle scarps and depressions along active fault zones as shown in the image that covers the San Andreas fault in the western San Gabriel Mountains (Figure 6.39).

Alaska

Known major faults are well expressed on the Landsat mosaic of the Cook Inlet region (Figure 10.4) together with numerous lineaments that previously had not been mapped (Figure 10.5). Gedney and Van Wormer (1973) plotted the epi-

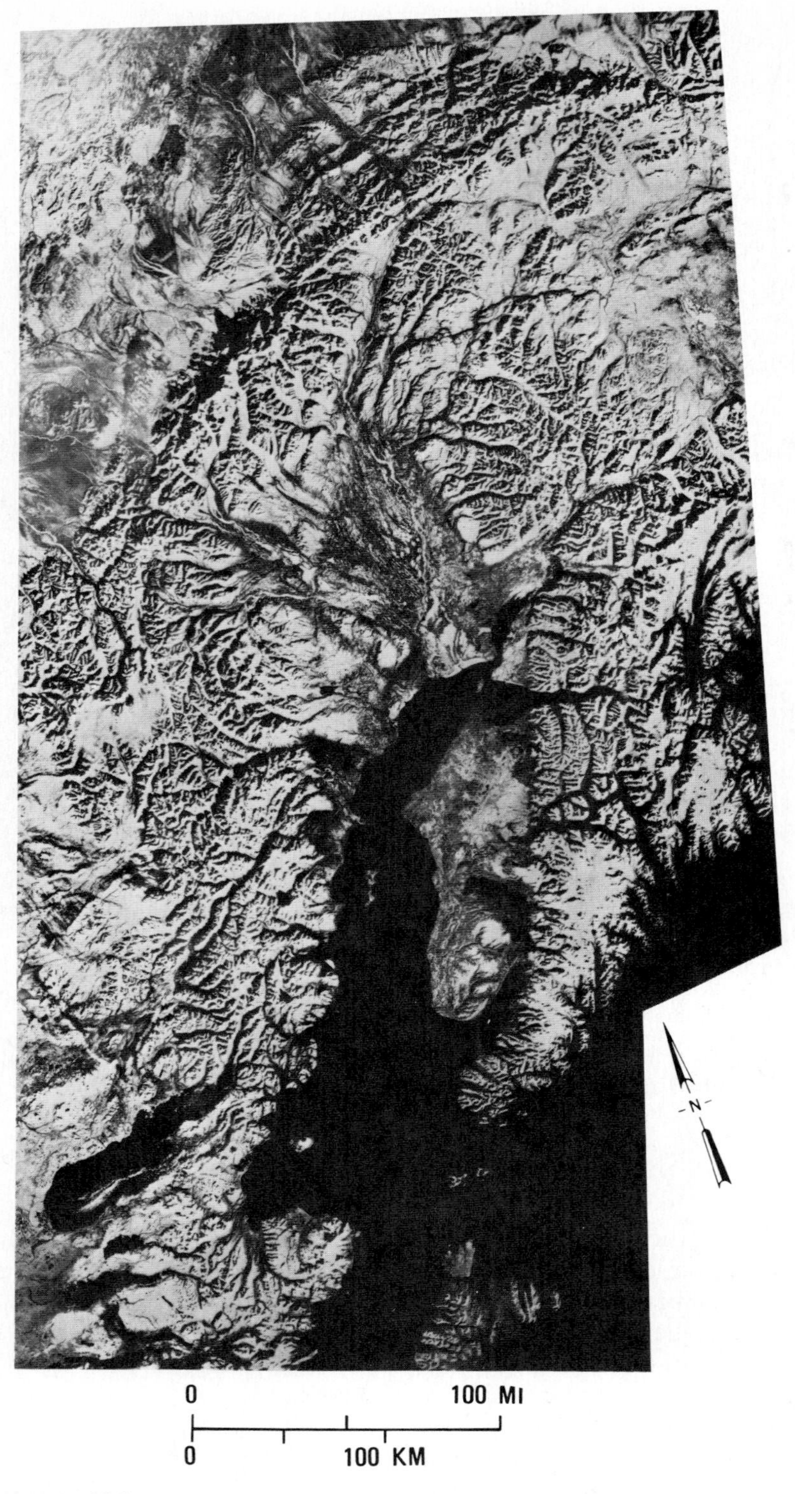

FIGURE 10.4
Landsat mosaic of south-central Alaska compiled from band 5 images acquired in early November 1972. Note shadow of Mt. McKinley in northwest part of mosaic. Mosaic by U.S. Department of Agriculture Soil Conservation Service.

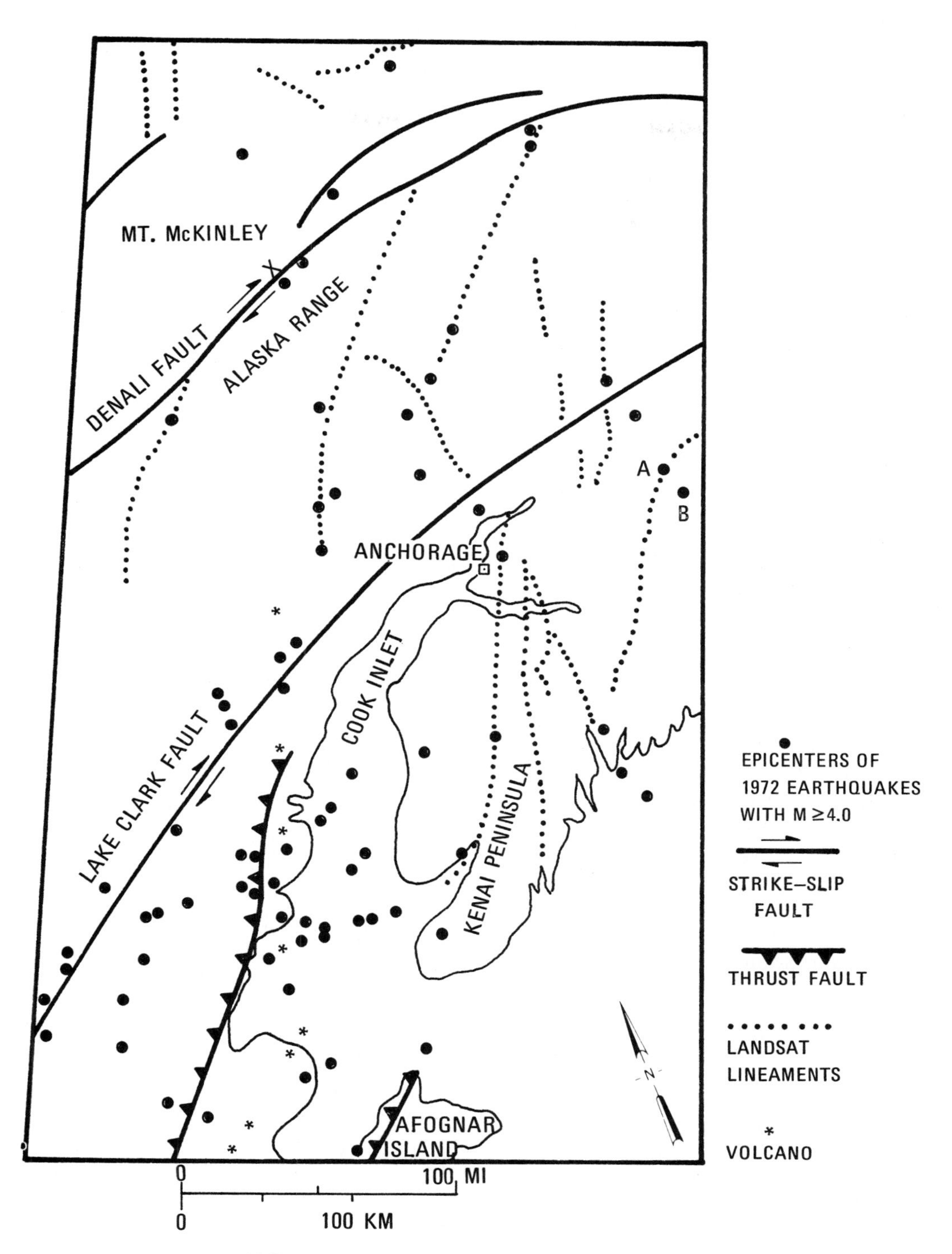

FIGURE 10.5
Interpretation map of Landsat mosaic of south-central Alaska with 1972 earthquake epicenters plotted. The major 1964 epicenter was very near A and B. From Gedney and Van Wormer (1974).

centers of earthquakes with magnitude 4.0 or greater that occurred in 1972 and noted the following relationships:

1. The epicenters in the south part of the area have little relationship to the faults and lineaments. These events were not caused by near-surface faulting, but were deep-seated earthquakes related to the subduction zone between the Pacific and North American plates.
2. A few earthquakes are associated with the Denali fault, particularly near Mt. McKinley. There is an alignment of epicenters along the Lake Clark fault.
3. Most of the previously unmapped lineaments are the site of one or more epicenters and probably should be classified as active faults. The epicenter of the 1964 Good Friday earthquake was very close to earthquake epicenters A and B along the lineament in the east-central part of Figure 10.5, although it is not clear whether this lineament played a role in the 1964 earthquake. The sharp escarpment along the west flank of the Kenai Mountains passes just east of Anchorage and was the site of three 1972 earthquakes.

In the central interior of Alaska the epicenters of the largest earthquakes occur at the intersections of lineaments and faults that are visible on Landsat images (Gedney and Van Wormer, 1973). The combination of geophysical data and remote-sensing information is a promising approach to earthquake risk analysis.

Turkey

The map of Turkey (Figure 10.6C) combines seismic data of epicenters for 65 years (large dots) with the historic earthquake record for the past 2,000 years. Figure 10.6B includes part of a southwest-trending fault that is semicontinuous between eastern Turkey and the Dead Sea rift (Allen, 1975). This is part of the wide band in eastern Turkey that experienced many historic earthquakes until about 1200 A.D. but has been relatively quiet since. The length, continuity, and evidence of Quaternary movement on the image, however, indicate a seismic potential for this fault zone. In 955 A.D. a major earthquake occurred near Palu (Figure 10.6B) that diverted rivers and changed stream courses. Farther northeast along this fault zone the 1971 Bingöl earthquake (magnitude of 6.7) was associated with 20 cm of left-lateral displacement (Allen, 1975, p. 1046).

The area of Figure 10.6A north of the Dardanelles lies along the general westward projection of the North Anatolian fault zone which is characterized by right-lateral, strike-slip displacement. This area was shaken by a major earthquake (magnitude of 7.75) in 1912. Although no clear-cut fault displacement had been mapped in association with the earthquake, the image appearance prompted Allen to field check the area and interview elderly residents. It became clear that significant surface faulting, probably strike slip, occurred in 1912 along the exact line indicated between the arrowheads in Figure 10.6A (Allen 1975, p. 1046).

China

Despite the generally restricted information and access to China, much can be learned about the great earthquakes from Landsat images. Figure 10.7 covers the southwestern end of the Shansi graben of northern China, including Sian and the Wei Ho Valley. The 1556 earthquake near Sian caused more than 820,000 deaths and could have

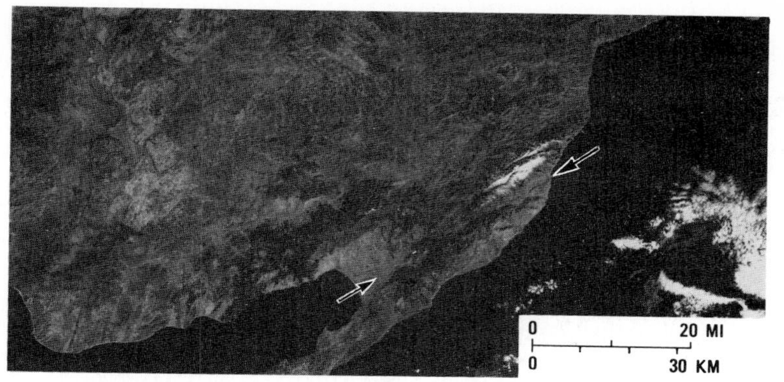

A. LANDSAT 1152–08262, BAND 5 COVERING AREA OF 1912 EARTHQUAKE (M 7.75) NORTH OF THE DARDANELLES. ARROWS INDICATE EXTENT OF SURFACE FAULTING.

B. LANDSAT 1485–07323, BAND 5 OF NORTHEAST–TRENDING FAULT ZONE (ARROWS) IN EASTERN TURKEY. A MAJOR EARTHQUAKE OCCURRED NEAR PALU IN 995 A.D.

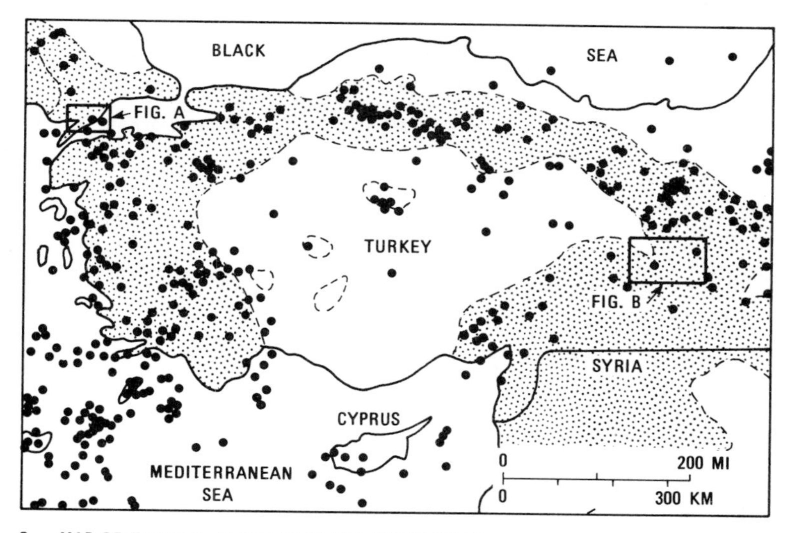

C. MAP OF TURKEY. BLACK DOTS ARE EARTHQUAKES EXCEEDING MAGNITUDE 5, 1900–1965. STIPPLED PATTERN REPRESENTS AREA OF HISTORIC SEISMICITY, FIRST THROUGH SEVENTEENTH CENTURIES.

FIGURE 10.6
Active faults on Landsat images of Turkey. From Allen (1975, Figures 10, 11, and 12).

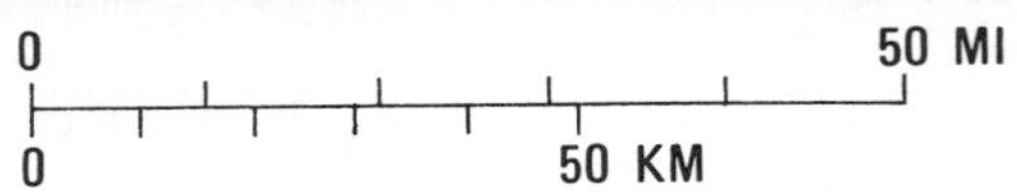

FIGURE 10.7
Sian region of Shensi Province, northern China. Dashed line outlines area of shaking greater than intensity XI during the great 1556 earthquake. Landsat 1525-02461, band 5 acquired December 30, 1973. Annotation from Allen (1975, Figure 28).

been caused by any one of several prominent Quaternary faults on the image (Allen, 1975). There is little doubt that the bounding faults of the graben structure are very active features. Chinese geologists note that the most intensive earthquake activity has occurred along the bounding faults with the greatest Cenozoic displacements. In addition to southwest-trending boundary faults, there are impressive west-trending faults in the bedrock of the southeast part of Figure 10.7. At the west boundary of the image a major fault cuts both bedrock and alluvium for 48 km and trends eastward toward the area of maximum seismic intensity in the 1556 earthquake.

Nicaragua

A major part of Managua, Nicaragua was destroyed by a severe earthquake and fires on December 28, 1972. Because this is a region with historic seismic and volcanic activity, the U.S. Geological Survey was asked to evaluate geologic hazards that might be important in the reconstruction or relocation of the city. Schmoll, Krushensky, and Dobrovolny (1975) used aerial photographs and SLAR imagery, acquired before the earthquake, to interpret faults, fractures, lineaments, and other potential geologic hazards. From these remote-sensor images they recognized that the lineaments could be grouped, on the basis of orientation, into seven sets, most of which are directly related to the regional geologic structure. During the past 10,000 years faulting has occurred along nearly all these trends; faulting and related seismic activity can be expected to continue.

The local and regional interpretation maps of Schmoll, Krushensky, and Dobrovolny (1975, Figure 5, Plate 1) indicated that no site within 15 km of Managua is significantly better than the present location, because all such sites share the geologic hazards of the present city. Based on the remote-sensing interpretation and limited field reconnaissance, two alternate localities were recognized in the interior highlands of Nicaragua as potential sites for relocating the city.

LAND SUBSIDENCE AND LANDSLIDES

Up to several meters of surface subsidence can be caused by extraction of subsurface fluids. Where subsidence occurs in populated and industrialized coastal areas, the resulting flooding can be a major problem. A classic example is the subsidence associated with the Long Beach, California oil field which has been stopped through injection of water to replace the produced oil.

In the past 30 years up to 1.5 m of land subsidence has occurred in the southeast portion of Houston, Texas and is related to the following causes (Clanton and Amsbury, 1975):

1. Active geologic processes of faulting and compaction of young sedimentary deposits.
2. Extraction of oil and gas.
3. Widespread pumping of ground water from shallow depths, which is thought to be a major cause of subsidence.

The rate of subsidence is increasing and is concentrated along active faults that have caused millions of dollars in damage to buildings, roads, and especially to underground utilities. Accurate maps of the faults are needed in order to avoid them in future construction. Clanton and Amsbury (1975) used the following criteria to map the faults on small-scale IR color photographs:

1. Fault scarps.
2. Sag ponds along fault traces.
3. Differences in drainage pattern on opposite sides of fault.
4. Linear tonal anomalies caused by higher soil moisture on the downthrown sides of faults.

These criteria should be useful for interpreting IR color photographs of other areas of subsidence and active faulting.

Landslides occur in areas of relatively steep topographic slopes underlain by unstable materials. Individual landslide events are difficult or impossible to predict by remote sensing or other methods. However, areas that are susceptible to landslides may be identified on aerial photographs by the hummocky appearance that is characteristic of unstable slopes. Old landslides may be recognized by the crescentic scarp at the head of the slide and the jumbled irregular topography at the toe of the slide. Slides are often the result of high concentrations of soil moisture that lubricate the surface material. Areas of high soil moisture should be detectable on IR color photographs by dark tones and on thermal IR images by cold signatures caused by evaporative cooling.

VOLCANOES

Most, if not all, volcanoes have been recognized and mapped because they are distinctive landforms. The active volcanoes that have erupted in historic time are also known. Considering the inactive volcanoes, it is important to distinguish those that are extinct from those that are only dormant and are likely to erupt in the future. Eruptions have occurred at a number of volcanoes that were thought to be extinct because there had been no reported activity in historic time. These include Vesuvius in 79 A.D., Lamington, New Guinea in 1951, and Arenal, Costa Rica in 1968. In addition to recognizing potentially hazardous volcanoes, there is the problem of predicting specific eruptions. A variety of physical changes and events have been reported as precursors of eruptions at different volcanoes (Bolt and others, 1975, p. 124). These include earth tremors, tilting of the earth's surface, changes in the composition of gases emanating from volcanic vents, unusual behavior of animals, and changes in temperature of fumaroles, hot springs, and crater lakes.

As part of the Landsat program a volcanic surveillance system was established to monitor seismic events and earth tilting at 15 volcanoes in Alaska, Hawaii, Washington, California, Iceland, Guatemala, El Salvador, and Nicaragua (Ward and Eaton, 1976). Each unmanned instrument installation includes an antenna for transmitting data to the Landsat data collection system from which they are relayed to the receiving stations in California and Maryland. The data are processed at Goddard Space Flight Center and relayed by teletype to the National Center for Earthquake Research in Menlo Park, California. The tiltmeters at Hawaii recorded subsidence at Kilauea Volcano during a 1973 eruption. Surface uplift was recorded at Volcán Fuego, Guatemala during the six months following a 1973 eruption. The seismic counters at Volcán Fuego showed a significant increase in activity five days before a small eruption. The seismic activity was high during the eruption and returned to a low level following the eruption (Ward and Eaton, 1976, p. 108). The data collection system and ground instrumentation provide data in near-real time from areas of difficult access. The cost of maintaining manned observatories at all these localities would be excessive.

Thermal IR Surveys

The history and status of thermal surveillance of volcanoes with remote sensing and contact temperature devices was summarized by Moxham (1971, p. 103) who concluded that ". . . results to date are more tantalizing than elucidating. At some volcanoes thermal forerunners (of eruptions) have been well documented; elsewhere, temperature measurements have been inconclusive or negative." Friedman and Williams (1968, Table 1) listed all the IR surveys of volcanoes reported to that date. More than 23 volcanoes throughout the world have been surveyed, but generally only with a single flight or, at best, several flights within a few weeks. Most of the thermal IR surveys that have been reported used older scanners with neither magnetic tape recording nor quantitative imagery. These deficiencies of equipment and coverage suggest that thermal IR methods have not been adequately evaluated for surveillance of volcanoes. The images of Mauna Loa and Kilauea, discussed in Chapter 5, illustrate the advantages of quantitative thermal data and image processing methods for mapping active volcanoes. Moxham (1971, p. 103) recommends that systematic thermal studies be made in conjunction with geophysical and geochemical investigations at sites selected to yield meaningful results about volcanic processes.

Moxham (1967) analyzed three IR surveys of Taal Volcano, Philippines, that were acquired during the nine-month quiescent period between the 1965 and 1966 eruptions. Taal is well suited for IR surveys because most of the volcano is below the water table, and there is both a crater lake and a surrounding water body. Because of the uniform and high emissivity of water, small temperature changes are readily detected and there are no masking effects of vegetation or moisture differences. The first IR survey revealed the two known fumarole fields, plus several previously unknown fumaroles on the north flank of the central cone. Both IR and ground investigations showed new hot springs along the flanks of the 1965 explosion crater, one of which enlarged prior to the 1966 eruption. Hydrothermal activity persisted around the rim of the 1965 cinder cone, which was the site of the 1966 eruption midway between the two areas of maximum discharge of thermal water.

IR surveys were made over Italian volcanoes in 1970 and 1973 to monitor any changes and to provide reference data for any future eruptions (Cassinis and others, 1974).

Mount Rainier, Washington

The last major eruption at Mount Rainier occurred about 2,000 years ago when lava and pumice were erupted and there were several mudflows. Many minor eruptions occurred in the 1800s and the last report of steam and smoke was in 1894 (Crandell and Waldron, 1976, p. 46). Mudflows are common on the flanks of active volcanoes because the steep slopes are mantled with unconsolidated, unstable volcanic ash and rock fragments. During an eruption this material mixes with melted snow and the tremors can trigger a mudflow that moves downslope at speeds of several tens of kilometers per hour. The geologic record for the past 10,000 years indicates that Mount Rainier has been the site of several mudflows that contained hundreds of million cubic meters of material and reached more than 50 km from the volcano (Crandell and Waldron, 1976, p. 47). Similar mudflows occurring today could endanger tens of thousands of residents in the region.

IR surveys of Mount Rainier were acquired for the U.S. Geological Survey in 1964, 1966, and 1969 to evaluate increased seismic and thermal activity at the summit. The aerial photograph and IR image (Figures 10.8A and 10.8B) cover the sum-

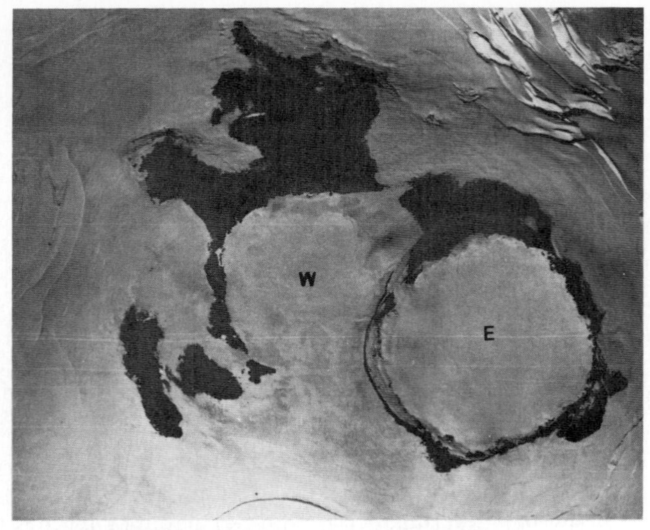

A. AERIAL PHOTOGRAPH ACQUIRED SEPTEMBER 3, 1966. W–WEST CRATER, E–EAST CRATER.

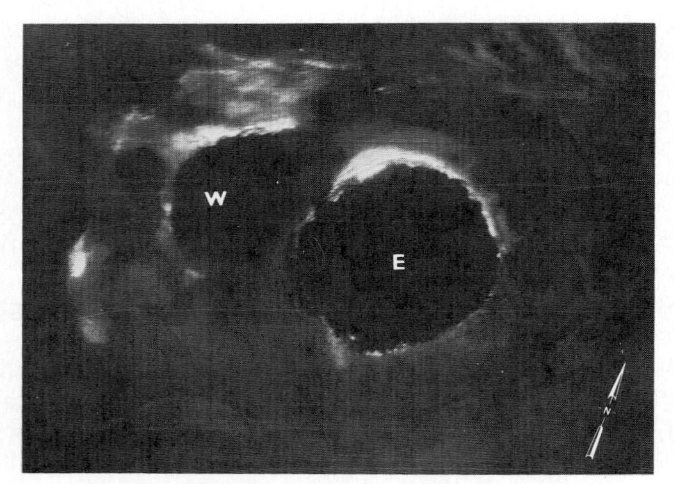

B. PRE–DAWN THERMAL IR IMAGE (8 TO 14 μm) ACQUIRED SEPTEMBER 4, 1966.

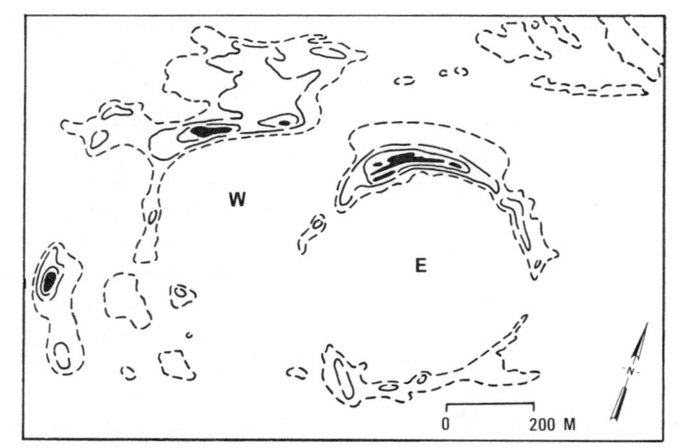

C. ISODENSITY CONTOURS OF IR IMAGE. BLACK PATTERN INDICATES AREAS OF HIGHEST APPARENT TEMPERATURE. CONTOURS INDICATE PROGRESSIVELY LOWER TEMPERATURES.

FIGURE 10.8
Summit of Mount Rainier, Washington. Part C from Moxham (1970, Figure 8). Courtesy R. M. Moxham, U.S. Geological Survey.

mit, which is a cone of andesite within a broad depression at the top of the strata volcano. The two small summit craters are nearly filled with snow and ice, but steam jets at the eastern crater have melted caverns beneath the ice. The west crater is about 300 m in diameter and is partly overlapped by the younger east crater, which is 400 m in diameter. The only rock exposed on the summit is the andesite around the crater rims which is swept bare of snow by the winds and has a dark tone on the aerial photograph. Moxham (1970) reported that fumaroles lining the rims of the two summit craters were remarkably similar in extent and apparent thermal intensity on 1964 and 1966 images. On the 1969 image several new thermal features on the southwest rim of the east crater coincide with areas of higher snow melt. The warm anomalies on the 1966 image (Figure 10.8B) occur within the andesite outcrops. The warmest areas are along the north rims of the east and west craters and an isolated spot farther west. An optical densitometer was used to contour the density of the image film, which correlates with the relative thermal radiation (Figure 10.8C). Only the areas of highest apparent temperature, shown in black on the contour map, are true anomalies.

There has been no marked increase in volcanic activity at Mount Rainier and the acquisition of thermal IR images has been discontinued. This points out one problem in accumulating a data base for predicting hazardous events. The data could be accumulated for decades or even centuries without the occurrence of an eruption. Investigators and government agencies are understandably reluctant to commit time and money on such uncertain ventures.

Satellite Imagery

The repetitive coverage and low cost of thermal IR imagery from satellites makes this an attractive alternative to aircraft for monitoring volcanoes. However, the 1 km ground resolution cell of the thermal IR imaging system on NOAA satellites is too large for mapping individual fissures, craters, and calderas. Landsat C will acquire thermal IR imagery with a ground resolution cell of 238 m. This may prove useful for monitoring volcanic activity, especially on nighttime imagery when the interfering effects of solar heating are lacking.

Although the spatial resolution of NOAA satellite systems is too large to record details of surface thermal patterns, the plume of smoke and ash from volcanoes is detectable. Less than 33 hours after the January 25, 1973 eruption of Heimaey, Iceland the plume was recorded on a NOAA 2 thermal IR image (Williams and others, 1974, Figure 11). The Tobalchik volcano in the Kamchatka peninsula erupted on July 6, 1975 and the plume was first observed on NOAA 4 thermal IR images acquired on July 9. Jayaweera, Seifert, and Wendler (1976) measured the length and azimuth of the plume on 14 NOAA images acquired from July 9 through August 17, 1975. Their most spectacular observation was on July 28 when the length of the plume was at least 960 km. From the shadow length, measured from the visible image, and from the known sun elevation, the height of the plume on July 28 was estimated at 6,500 m (Jayaweera, Seifert, and Wendler, 1976, p. 200). On the NOAA thermal IR images the plumes have a cold (bright) signature because they have cooled to the temperature of the surrounding atmosphere. On the visible images the plumes have dark signatures. The plume from the 1973 eruption of the Russian volcano Tiatia on Kunashir Island was clearly shown on Landsat 1 images. The ability to monitor size and distribution of volcanic plumes may prove useful in planning rehabilitation of areas covered by ash.

FLOODS

Images from weather satellites are used routinely for weather predictions. Data from stream gauges can be relayed by the Landsat data collection system to provide early warning of impending flood conditions. Neither of these systems is capable of providing information on the extent of flood waters that is essential for rescue and relief operations. When available, SLAR images are ideal for this purpose because of the all-weather capability. Landsat images also have proven valuable.

In the spring of 1973 record rains caused extensive flooding throughout the Mississippi River valley. Fortunately the sky over the flooded area was relatively clear on two successive days and Landsat acquired images of the entire Mississippi valley. Through special arrangements Goddard Space Flight Center processed the data and made images available very soon after acquisition. Figure 10.9 of the St. Louis area compares an image of the normal river stage with one acquired during the flood stage. Band 7 is optimum for mapping the flooded areas. The map shown in Figure 10.9C shows the extent of flooding. A more extensive analysis of Landsat images for the entire Mississippi valley was carried out by the U.S. Geological Survey (Deutsch and Ruggles, 1974).

In April, 1975 Landsat images were used to delineate flooded areas in Louisiana; these were compared with land-use classification maps to determine the extent of flood damage to urban areas, farmland, and other categories. These data were used by the state for rapid analysis of damage and to document the need for federal disaster relief funds. Landsat images have also been used for assessing flood damage in South Dakota, Iowa, Arizona, and Pakistan. For a period of several weeks after the water has receded, the maximum extent of flooding can be recognized on band 7 images by the dark tone caused by damp ground.

SUBSURFACE COAL FIRES

Underground coal mining produces large volumes of waste rock that includes combustible material such as contaminated coal, carbonaceous shale, and pyrite, which is an iron sulfide mineral. Large refuse piles containing millions of tons of this material occur at coal mining areas throughout the world. Surface fires or spontaneous combustion may ignite combustible material and the resulting fire may spread to abandoned coal mines. Such fires in refuse dumps and mines are common and some have been burning for over 50 years. Subsurface fires constitute three major hazards:

1. The large volumes of carbon monoxide and sulfur dioxide gases produced are potentially dangerous.
2. Fires in abandoned mines may cause surface subsidence problems.
3. The fires may interfere with mining operations.

Coal fires are controlled by shutting off the air supply with impervious clay coatings on the dumps or by injecting water and slurries through holes drilled into the combustion zone. For such measures to be effective, the underground fire must be accurately located. Thermal IR images have proven useful for this purpose. Figure 10.10A is a nonquantitative nighttime image of a burning coal refuse dump near Wilkes Barre, Pennsylvania. On this image the relative temperature scale is such that surface water and subsurface fires have similar bright signatures and are difficult to separate. Through processing of the magnetic tape record, the areas of maximum radiant temperature are enhanced and displayed in Figure 10.10B

A. NORMAL FALL SEASON, OCTOBER 2, 1972.
LANDSAT 1071–16104, BAND 7.

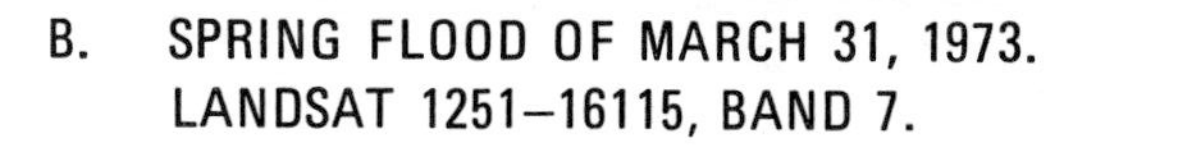

B. SPRING FLOOD OF MARCH 31, 1973.
LANDSAT 1251–16115, BAND 7.

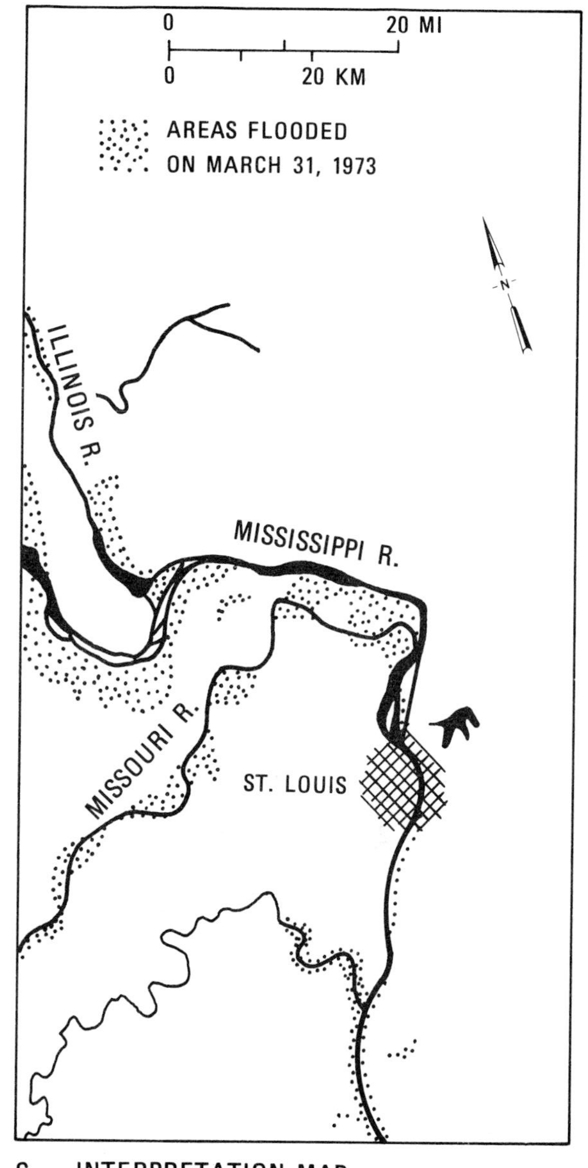

C. INTERPRETATION MAP.

FIGURE 10.9
Mapping of floodwater in upper Mississippi River valley from Landsat images.

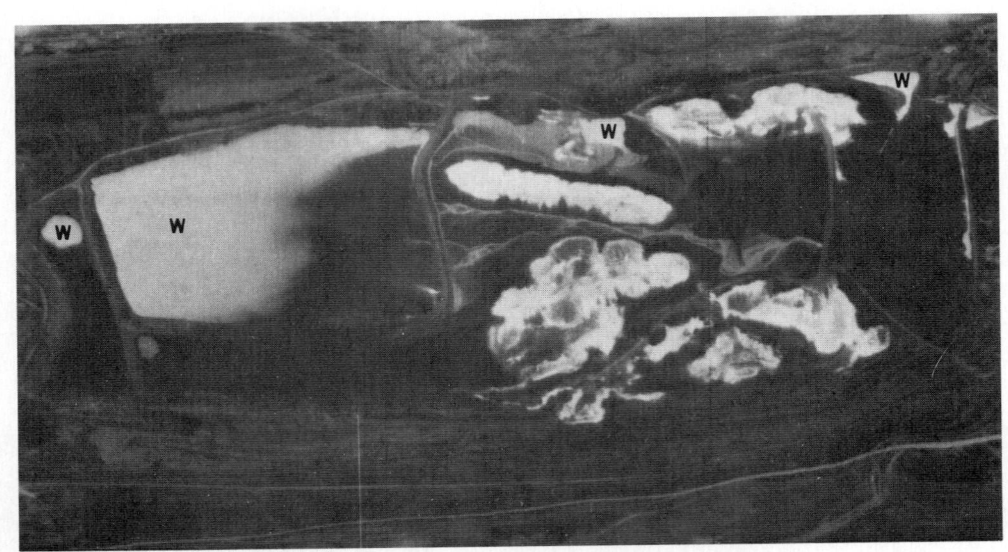

A. ORIGINAL IMAGE WITH AREAS OF OPEN WATER INDICATED BY "W."

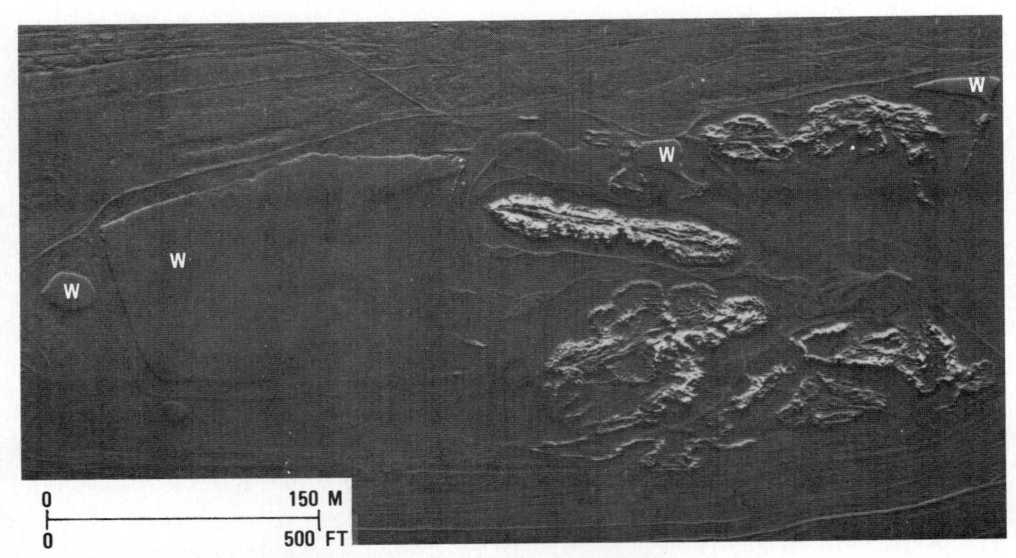

B. IMAGE AFTER PROCESSING TO ENHANCE AREAS OF MAXIMUM TEMPERATURE (WHITE TONE) THAT INDICATE FIRES WITHIN THE REFUSE BANK.

FIGURE 10.10
Original and enhanced nighttime thermal IR images (3 to 5 μm) acquired April 28, 1967 of burning coal refuse bank near Wilkes Barre, Pennsylvania. From Knuth and Fisher (1967, Figures 4 and 5). Images courtesy HRB-Singer, Inc.

where coal fires are readily distinguished from water (Knuth and Fisher, 1968). On the enhanced image the active burning fronts are emphasized and it is apparent that the narrow elongate dump is the site of the most intense burning. IR imagery should be valuable for planning fire control programs and for evaluating the effectiveness of control measures.

Greene, Moxham, and Harvey (1969) acquired thermal IR images of coal mine fires in Pennsylvania and correlated these with temperature measurements in boreholes. They reached the following conclusions:

1. Fires at shallow depths (less than 10 m) were readily detected on the images.
2. Fires at intermediate depths (10 to 30 m) were detected where heat was carried to the surface by convection in open cracks, or if the fire had been burning long enough (several years or more) for heat to reach the surface by conduction.
3. Deep fires (below 30 m) were detected only where heat reached the surface by convection in open cracks.

COMMENTS

Remote-sensing methods are useful for recognizing areas subject to natural hazards, such as earthquakes, volcanic eruptions, and floods. Imagery is also useful for assessing the damage caused by these hazardous events and for planning relief and rehabilitation efforts. Prediction of specific events is a desirable, but much more difficult, objective. The use of satellites for relaying data from seismometers, tiltmeters, and flood gauges is a valuable adjunct to the Landsat program. In the future, correlation of these data with remote-sensor imagery may contribute to developing a system for predicting hazardous events.

REFERENCES

Allen, C. R., 1975, Geological criteria for evaluating seismicity: Geological Society America Bulletin, v. 86, p. 1041–1057.

Bolt, B. A., W. L. Horn, G. A. Macdonald, and R. F. Scott, 1975, Geological hazards: Springer Verlag, New York.

Cassinis, R., C. M. Marino, and A. M. Tonelli, 1974, Remote sensing techniques applied to the study of Italian volcanic areas—the results of the repetition of the airborne survey compared to the previous data: Proceedings Ninth International Symposium on Remote Sensing of Environment, v. 3, p. 1989–2004, Environmental Research Institute of Michigan, Ann Arbor, Mich.

Clanton, U. S. and D. L. Amsbury, 1974, Active faults in southeastern Harris County, Texas: Environmental Geology, v. 1, p. 149–154.

Crandell, D. R. and H. H. Waldron, 1976, Volcanic hazards in the Cascade Range *in* Tank, R., ed., Focus on environmental geology, Second Edition, Oxford University Press, New York.

Deutsch, M. and F. Ruggles, 1974, Optical data processing and projected applications of the ERTS–1 imagery covering the 1973 Mississippi River floods: Water Resources Bulletin, v. 10, p. 1023–1039.

Friedman, J. D. and R. S. Williams, 1968, Infrared sensing of active geologic processes: Proceedings Fifth International Symposium on Remote Sensing of Environment, p. 787–815, University of Michigan, Ann Arbor, Mich.

Gedney, L. and J. Van Wormer, 1974, Earthquakes and tectonic evolution in Alaska: Third ERTS-1 Symposium, NASA SP-351 v. 1, p. 745–756.

Greene, G. W., R. M. Moxham, and A. H. Harvey, 1969, Aerial infrared surveys and borehole temperature measurements of coal mine fires in Pennsylvania: Proceedings Sixth International Symposium on Remote Sensing of Environment, p. 517–525, Environmental Research Institute Michigan, Ann Arbor, Mich.

Jayaweera, K. O. L. F., R. Seifert, and G. Wendler, 1976, Satellite observations of the eruption of Tolbachik Volcano: Transactions American Geophysical Union, v. 57, p. 196–200.

Knuth, W. M. and W. Fisher, 1968, Detection and delineation of subsurface coal fires by aerial infrared scanning *in* Abstracts for 1967 Meetings: Geological Society of America Special Paper 115, p. 67–68.

Merifield, P. M. and D. L. Lamar, 1975, Active and inactive faults in southern California viewed from Skylab: NASA Earth Resources Survey Symposium, NASA TM X-58168, v. 1, p. 779–797.

Moxham, R. M., 1967, Changes in surface temperature at Taal Volcano, Philippines 1965–1966: Bulletin Volcanologique, v. 31, p. 215–234.

———, 1970 Thermal features at volcanoes in the Cascade Range, as observed by aerial infrared surveys: Bulletin Volcanologique, v. 34, p. 77–106.

———, 1971, Thermal surveillance of volcanoes *in* The surveillance and prediction of volcanic activity: UNESCO, p. 103–124, Paris, France.

Schmoll, H. R., D. R. Krushensky, and E. Dobrovolny, 1975, Geologic considerations for redevelopment planning of Managua, Nicaragua following the 1972 earthquake: U.S. Geological Survey Professional Paper 914.

Sharp, R. V., 1972, Map showing recently active breaks along the San Jacinto fault zone between the San Bernardino area and Borrego Valley, California: U.S. Geological Survey Miscellaneous Geologic Investigations Map I-675.

Ward, P. L. and J. P. Eaton, 1976, New method for monitoring global volcanic activity *in* Williams, R. S. and W. D. Carter, eds., ERTS-1, a new window on our planet: U.S. Geological Survey Professional Paper 929, p. 106–108.

Williams, R. S. and others, 1974, Environmental studies of Iceland with ERTS-1 imagery: Proceedings Ninth International Symposium on Remote Sensing of Environment, v. 1, p. 31–81, Environmental Research Institute of Michigan, Ann Arbor, Mich.

ADDITIONAL READING

Eggenberger, A. J., D. Rowlands and P. C. Rizzo, 1975, The utilization of Landsat imagery in nuclear power plant siting: NASA Earth Resources Survey Symposium. NASA TM X-58168, v. 1, p. 799–832.

MacDonald, H. C. and R. S. Grubbs, 1975, Landsat imagery analysis—an aid for predicting landslide-prone areas for highway construction: NASA Earth Resources Survey Symposium, NASA TM X-58168, v. 1, p. 769–778.

O'Leary, D. W. and S. L. Simpson, 1977, Remote sensor applications to tectonism and seismicity in the northern part of the Mississippi embayment: Geophysics, v. 42, p. 542–548.

11

COMPARISON OF IMAGE TYPES

In the preceding chapters each type of imagery has been treated independently. Aside from comparison of radar and thermal IR images with aerial photographs, there has been little cross correlation among the different types of imagery. By comparing various types of imagery that depict the same locality, the interpreter can evaluate the response of surface materials to electromagnetic radiation ranging in wavelength from the visible through microwave spectral bands. The Pisgah Crater area in the Mojave Desert of southern California is an excellent locality for this purpose because:

1. The area has been a test site for NASA and the U.S. Geological Survey; thus the following types of images are available:
   (a) Landsat multispectral images
   (b) Skylab earth terrain camera photographs
   (c) High altitude aircraft photographs
   (d) Low altitude aircraft stereo photographs
   (e) Aircraft multispectral images
   (f) Aircraft thermal IR imagery acquired both day and night
   (g) Thermal inertia maps
   (h) Radar imagery and scatterometer data
2. Although these images were acquired at different dates there is little seasonal change in this unvegetated desert terrain; hence imagery acquired at different seasons can be compared.
3. The diverse geologic and terrain features are well mapped, and the image interpretations and correlations presented here have been checked on numerous field trips to the area.

GEOLOGIC AND GEOGRAPHIC SETTING

Pisgah Crater is actually a cinder cone, but the crater designation is so firmly established that a nomenclature change would be confusing. The crater and associated Pisgah Basalt of Recent age are located in the Mojave Desert 48 km east of Barstow, as shown on the geologic index map of Figure 11.1. Elevation of the Pisgah Basalt flow is approximately 600 m above sea level with the cinder cone rising about 78 m above the flow surface. The sparse vegetation in this arid terrain is predominantly creosote bush that occurs as scattered clumps and local concentrations in the dry washes. The broad valley in which the Pisgah flow occurs is surrounded by the low, eroded fault

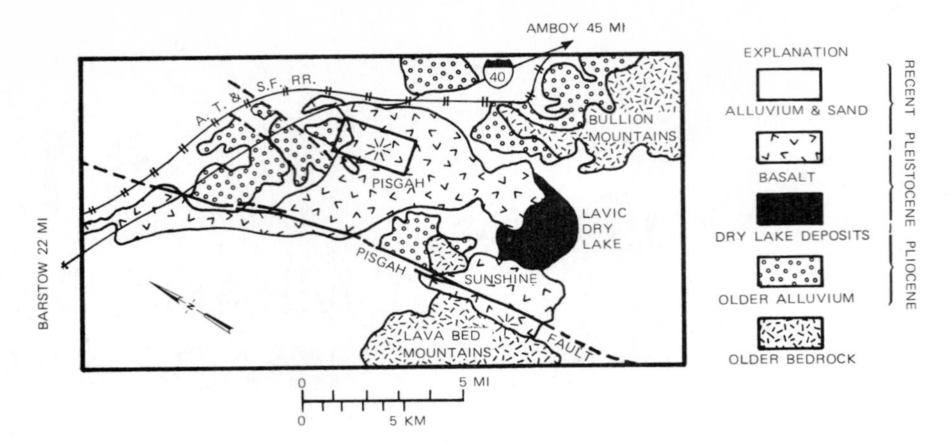

FIGURE 11.1
Geologic map of Pisgah Crater and vicinity, California. Rectangle shows location of thermal IR images and interpretation map.

blocks of the Cady, Bullion, and Lava Bed Mountains, where volcanic and intrusive rocks of Mesozoic and Tertiary age crop out.

Ground photographs of the rock units are shown at a uniform scale in Figure 11.2, together with radar scatterometer data, which are discussed later. The dry surface of Lavic Lake consists of silt and clay with well developed mud cracks (Figure 11.2A). In contrast to Cottonball Basin, there are no saline deposits exposed on Lavic Lake, which has a very smooth surface. The eolian sand consists of medium-sized grains of quartz that were deposited as thin sheets and patches by the prevailing westerly winds. The sand that overlies the basalt flows includes small fragments of basalt (Figure 11.2B). The alluvium consists of fan deposits of gravel and sand that were eroded from the older rocks. On the geologic map (Figure 11.1) the alluvium is divided into older and younger units that are distinguished by the following characteristics:

Younger alluvium
- Overlies margins of Pisgah Basalt flows
- Relatively uneroded surface
- No desert varnish
- No desert pavement

Older alluvium
- Overlapped by Pisgah Basalt flows
- Eroded and dissected surface
- Well-developed desert varnish
- Well-developed desert pavement

The desert pavement (Figure 11.2C) formed on the older alluvium through deflation by wind that removed the fine material and produced a smooth, mosaic-like surface. The fine-grained porphyritic, vesicular Pisgah Basalt consists of two distinct phases, pahoehoe and aa, that are distinguished in the visible spectral region by the medium-gray tone of the pahoehoe and the black tone of the aa. The pahoehoe phase (Figure 11.2F) is characterized by an undulating, ropey surface with flow structures. In contrast, the aa (Figure 11.2E) has a rough, scoriaceous surface that is difficult to walk across. The basalt flowed onto the north margin of Lavic Lake and also overlies the older alluvium. The Sunshine Basalt west of Lavic Lake (Figure 11.1) is similar to the Pisgah Basalt, but is older (Pliocene) and is coated with desert varnish. Pisgah Crater consists of fragments of basalt pumice that range from pebbles to cobbles in size (Figure 11.2D), with minor ash and volcanic bombs. The central depression of the cone is floored with pahoehoe basalt. A low ridge of cinders extends northward from the cone.

PHOTOGRAPHIC IMAGES

Satellite and high-altitude aircraft images in the photographic spectral region, enlarged to a uniform scale, are illustrated in Figure 11.3 with the enlargement factors indicated. The band 5 Landsat image is greatly enlarged from a raw original

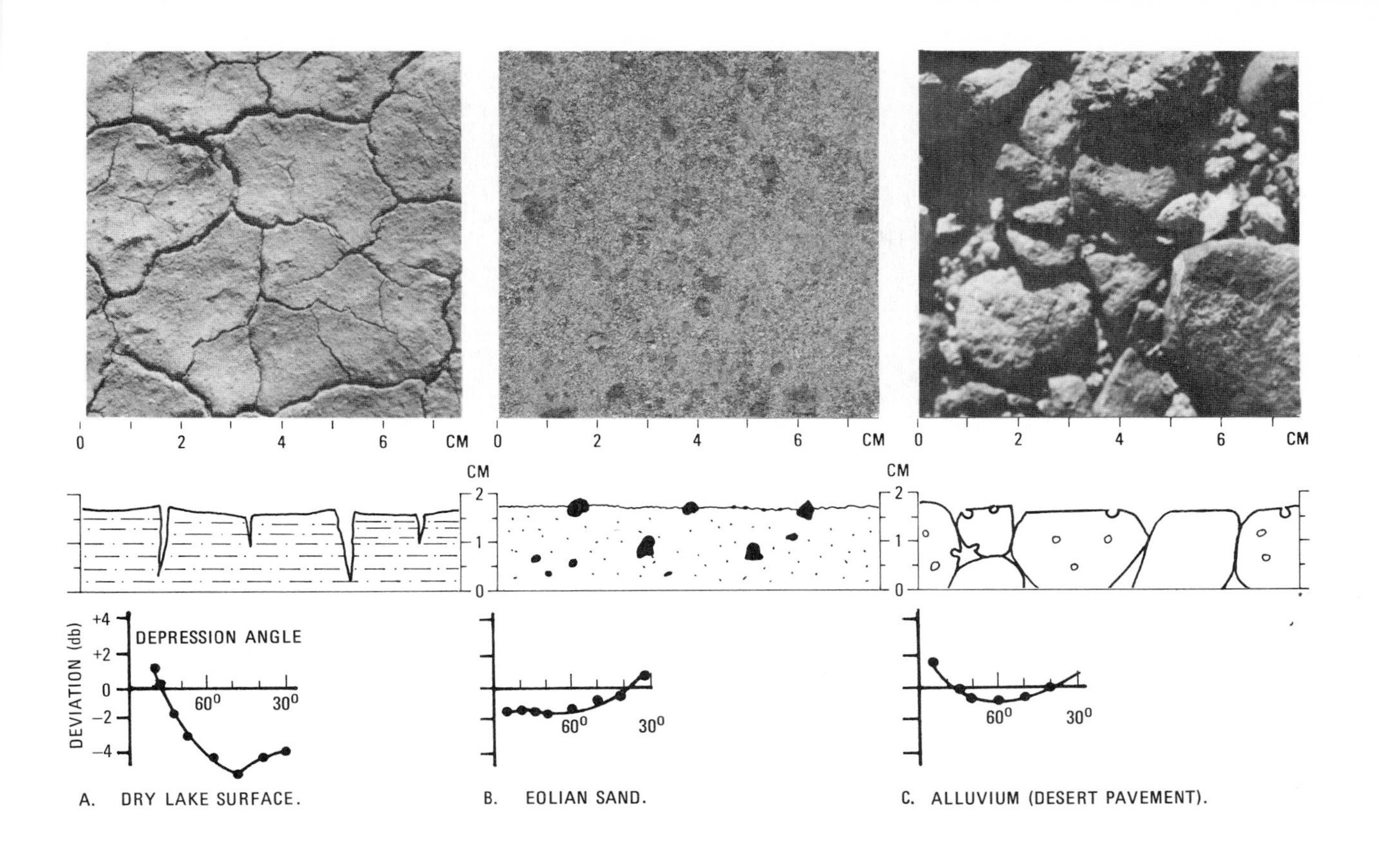

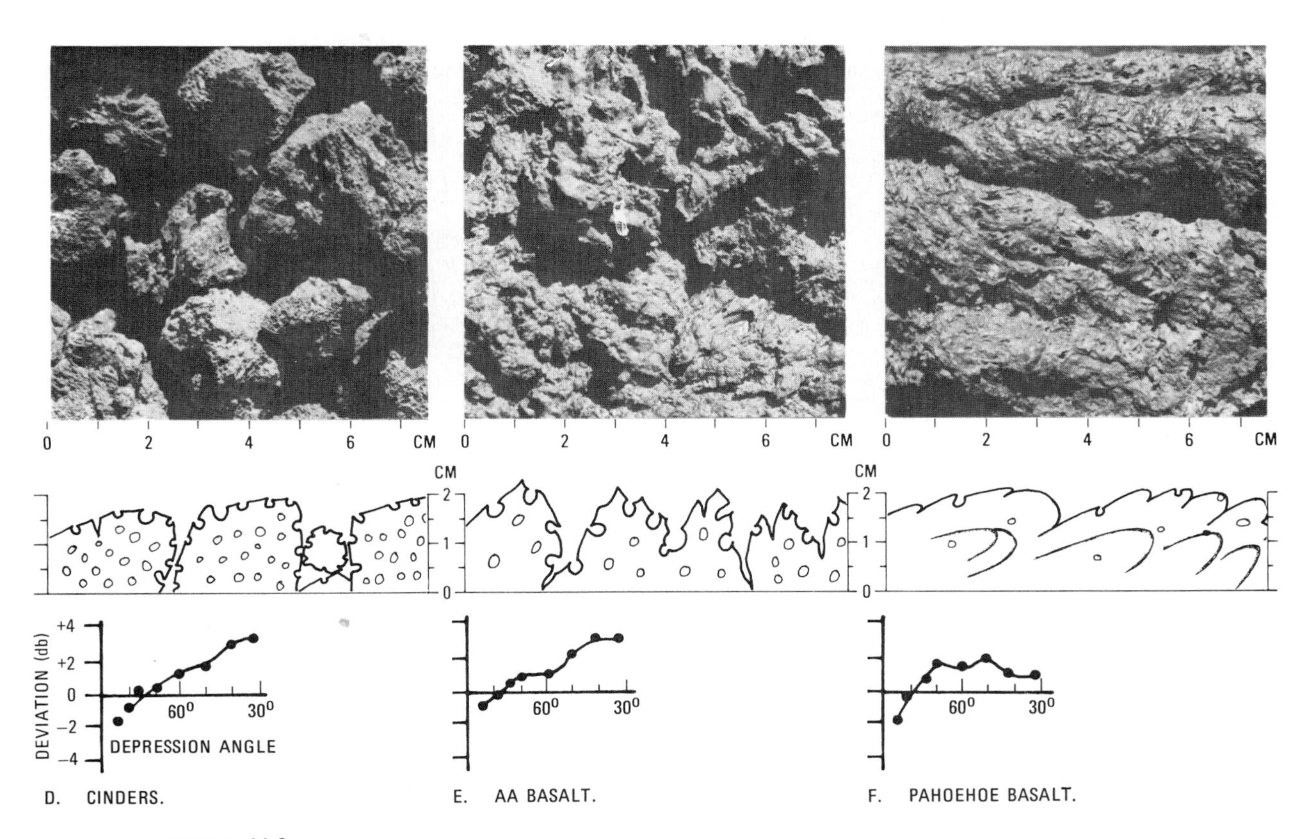

FIGURE 11.2
Ground photographs of rock types near Pisgah Crater with surface profiles and radar scatterometer deviation curves.

A. LANDSAT 1700–17422 ACQUIRED JUNE 23, 1974.
ALTITUDE 912 KM. ENLARGED 30X.

B. SKYLAB 4 EARTH TERRAIN CAMERA (S–190B) PHOTOGRAPH ACQUIRED JANUARY 26, 1974. NORMAL COLOR ORIGINAL PHOTOGRAPH. ALTITUDE 435 KM. ENLARGED 4X.

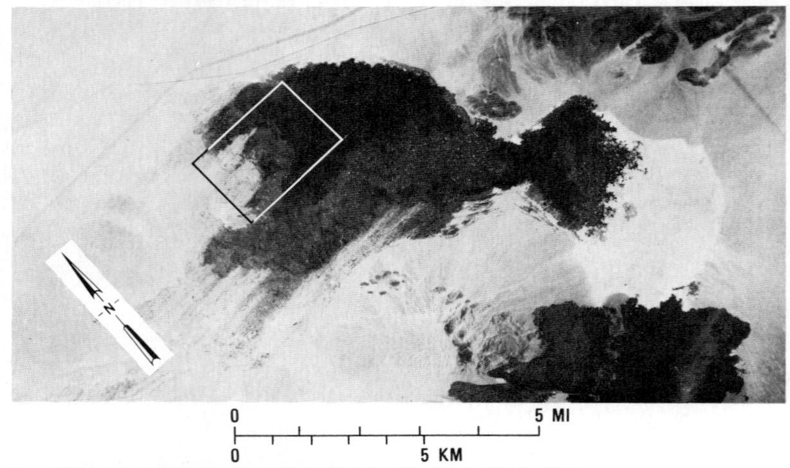

C. NASA RB–57 AIRCRAFT PHOTOGRAPH ACQUIRED APRIL 1971. IR COLOR ORIGINAL PHOTOGRAPH, ALTITUDE 18.3 KM. ENLARGED 1X. RECTANGLE SHOWS LOCATION OF STEREO PAIR OF AERIAL PHOTOGRAPHS.

FIGURE 11.3
Satellite and high-altitude aircraft photographic band images of Pisgah Crater and vicinity.

without the benefit of digital processing. Interstate Highway 40 is detectable north of the Pisgah Basalt flow and a number of small basalt outliers are detectable in Lavic Lake. On Landsat color-composite images (not shown) the aa and pahoehoe phases can be discriminated by their respective black and dark gray signatures, but these cannot be recognized on the band 5 image of Figure 11.3A. The photograph from the Skylab earth terrain camera is a black-and-white 4X enlargement from an original normal color transparency. Contrast and spatial resolution are much improved over the Landsat image. The pahoehoe and aa phases are distinguishable on the original Skylab color photograph, but not on the reproduction of Figure 11.3B. As in the Landsat image, the cinder cone at Pisgah Crater cannot be distinguished from the adjacent basalt flows. The high-altitude aircraft photograph (Figure 11.3C), which was reproduced from an original IR color transparency, is shown at approximately the original scale of 1:120,000. The aa (black tone) and pahoehoe (dark gray tone) are clearly distinguished, as is Pisgah Crater. Aside from these differences, the contrast and resolution of the aerial photograph are similar to those of the Skylab photograph, which indicates the high quality of images from the earth terrain camera.

Maximum spatial resolution is provided by the stereo pair of low-altitude aerial photographs in Figure 11.4 that covers the vicinity of Pisgah Crater and includes all the geologic materials, except for the dry lake. The geologic map (Figure 11.5) was interpreted from stereoscopic examination of the photographs, followed by field checking. As in the high-altitude aircraft photograph (Figure 11.3C) the darker aa lava can be distinguished from the lighter pahoehoe lava. Enlarged stereo viewing of the low-altitude photographs also enables the interpreter to recognize the rougher texture of the aa and smoother texture of the pahoehoe. The areas mapped as aa lava include small patches of pahoehoe that are not mapped separately. The pahoehoe areas also include patches of aa.

The higher spatial resolution and stereo viewing of the aerial photographs also enables the interpreter to recognize pressure ridges, domes, fissures, and collapsed lava tubes on the surface of the basalt flows. These features are also prominent on the thermal IR images. *Lava tubes* develop from channels of flowing lava that are covered by a crust of congealed lava. When the eruption ceases, the lava drains from the covered channelway, which remains as a cavern-like void or lava tube. The roof of the tube may cave in to form *collapsed lava tubes,* such as the prominent sinuous depression in the pahoehoe surface north of Pisgah Crater (Figures 11.4 and 11.5). Circular subsidence features are formed by local withdrawal of liquid lava from beneath the congealed crust. A prominent ridge and dome occur southeast of the crater and were formed by the pressure of lava rising from vents beneath the flows. The bulging and stretching of the surface of the dome caused several large *tensional fissures* that are up to 500 m long, 25 m wide, and 20 m deep. The walls of the fissures expose the dense basalt in the interior of the lava flows that contrasts with the vesicular rocks on the surface.

The cinders that make up Pisgah Crater have a gray tone that is intermediate between that of the surrounding aa and pahoehoe. The cinders have a much smoother texture on the aerial photographs, however, than the lava. Part of the rim of an earlier cinder cone is exposed within the west rim of the

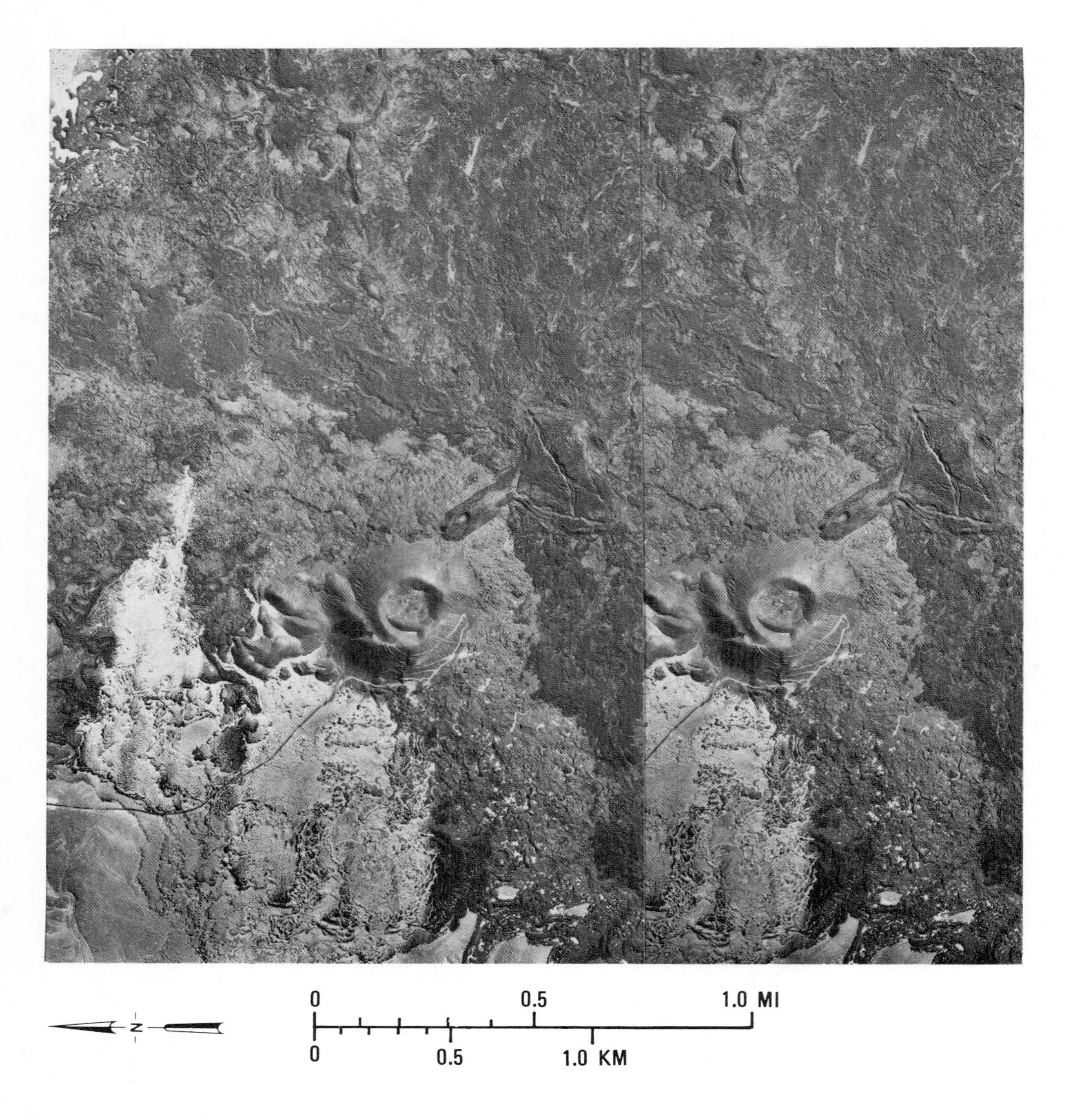

FIGURE 11.4
Stereo pair of minus-blue aerial photographs acquired February 4, 1953 at an altitude of 3,000 m with a camera lens focal length of 153 mm.

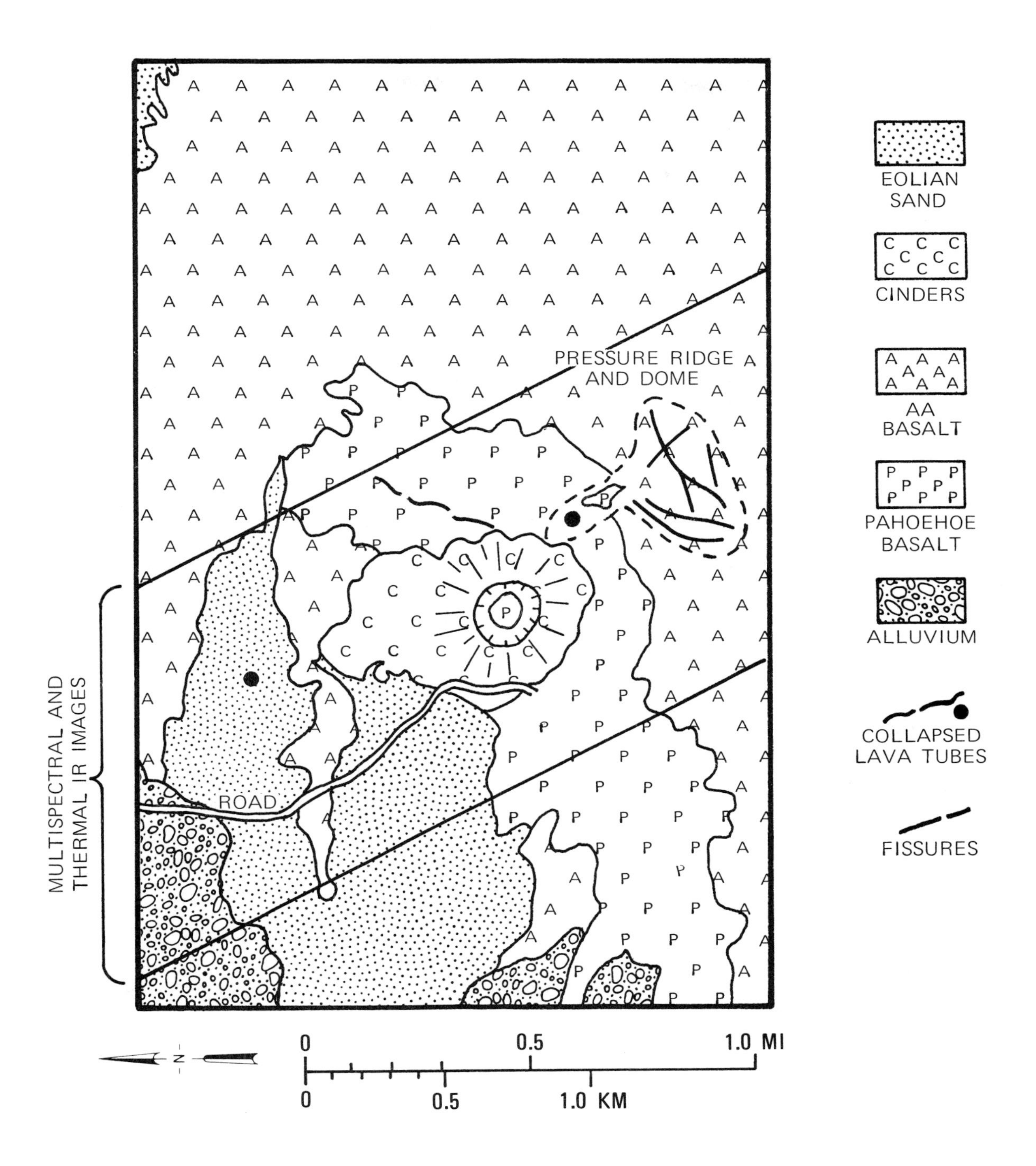

FIGURE 11.5
Geologic interpretation map of aerial photographs.

present cone. In the years following acquisition of the aerial photographs in 1953, much of the western part of the cone has been excavated for construction material.

The patches of eolian sand can be recognized on all the photographic band images because the very light tone contrasts with the surrounding lava. The sand migrates across the smooth desert pavement, but is trapped by the irregular surface of the lava flows. Thickness of the sand ranges from a few millimeters up to half a meter in the thickest patches. Older alluvium, with its smooth desert pavement surface, is exposed around the margin of the lava flow in the western part of the aerial photographs.

On the photographic images the relative reflectance of the rock units is ranked as follows:

DARKEST TONE
- Aa basalt
- Cinders
- Pahoehoe basalt
- Alluvium
- Eolian sand
- Dry lake surface

LIGHTEST TONE

AIRCRAFT MULTISPECTRAL SCANNER IMAGERY

On October 30, 1970 the Willow Run Laboratories of the University of Michigan (now the Environmental Research Institute of Michigan, or ERIM) acquired imagery of the Pisgah Crater area with an airborne multispectral scanner system. The system employs a rotating scan mirror similar to that of the thermal IR scanner described in Chapter 5. The energy radiated and reflected from the terrain is separated into 12 spectral bands that cover portions of the UV, visible, reflected IR, and thermal IR regions. The calibrated data are recorded on magnetic tape in the aircraft and played back in the laboratory to produce images. Details of the ERIM multispectral scanner system are given by Hasell and others (1975). Images acquired in the green band and photographic IR band are shown in Figures 11.6A and 11.6B. At the aircraft altitude of 900 m above terrain the ground resolution cell measures 3 m on a side. Location of the multispectral images is shown on the geologic map of Figure 11.5. On both the green and photographic IR images the basalt, alluvium, and eolian sand are readily distinguished. In these spectral bands, however, the cinders, aa, and pahoehoe have low reflectance and cannot be separated on the images. Figure 11.6C is a ratio image prepared by dividing reflectance values in the green band by values in the photographic IR band, as described in Chapter 7. The volcanic rock types are readily distinguished on the ratio image, but topographic details are suppressed, especially in the alluvium-covered areas in the north part of the ratio image. The separation of volcanic rock types in Figure 11.6C is apparently due to minor reflectance differences in the green and photographic IR bands that are emphasized on the ratio image.

THERMAL IR IMAGERY

Day and night thermal IR images acquired by the U.S. Geological Survey in February, 1965 are illustrated in Figure 11.7, together with an aerial photomosaic. Location of the images is shown in

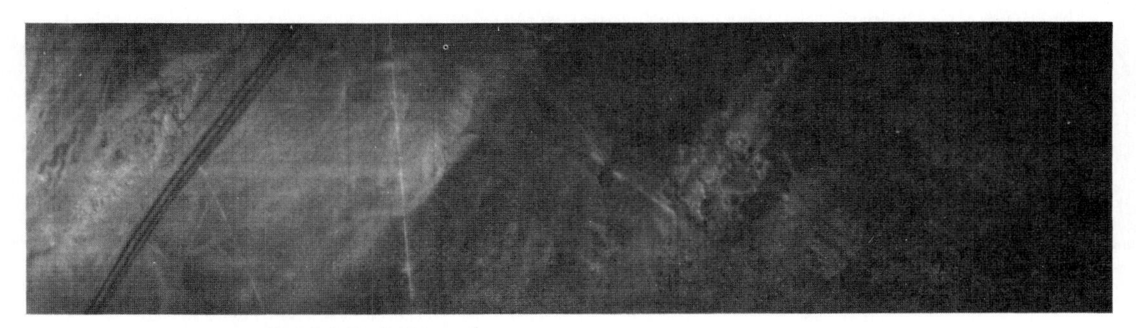

A. GREEN BAND (0.50 TO 0.52 μm).

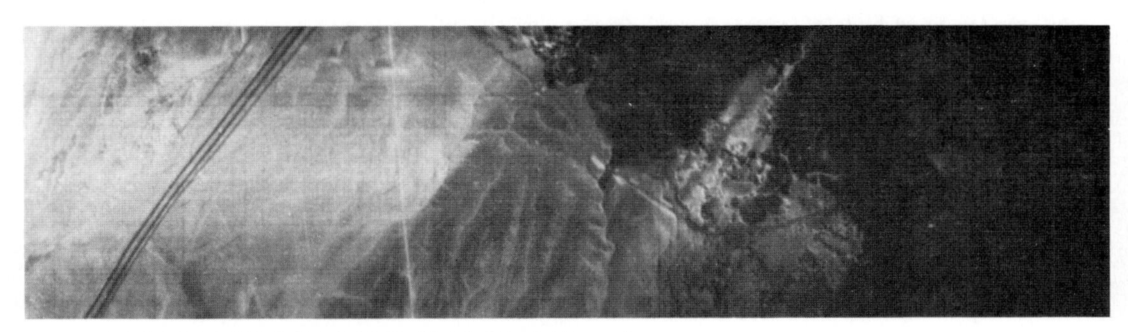

B. PHOTOGRAPHIC IR BAND (0.74 TO 0.85 μm).

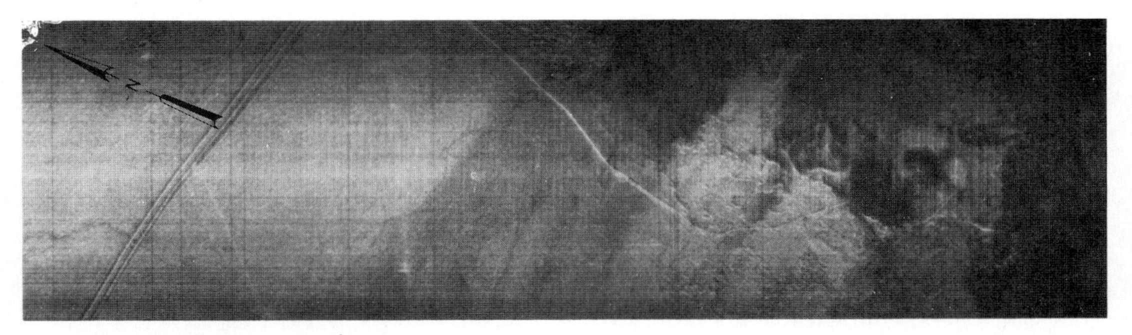

C. RATIO IMAGE OF A/B.

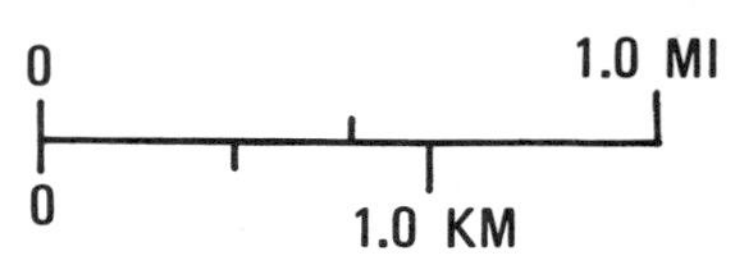

FIGURE 11.6
Aircraft multispectral scanner images acquired by Environmental Research Institute of Michigan at 8:20 A.M., October 30, 1971 from an altitude of 900 m. See Figure 11.5 for location of images. From Vincent (1972, Figure 11). Images courtesy R. K. Vincent, Geospectra Corporation.

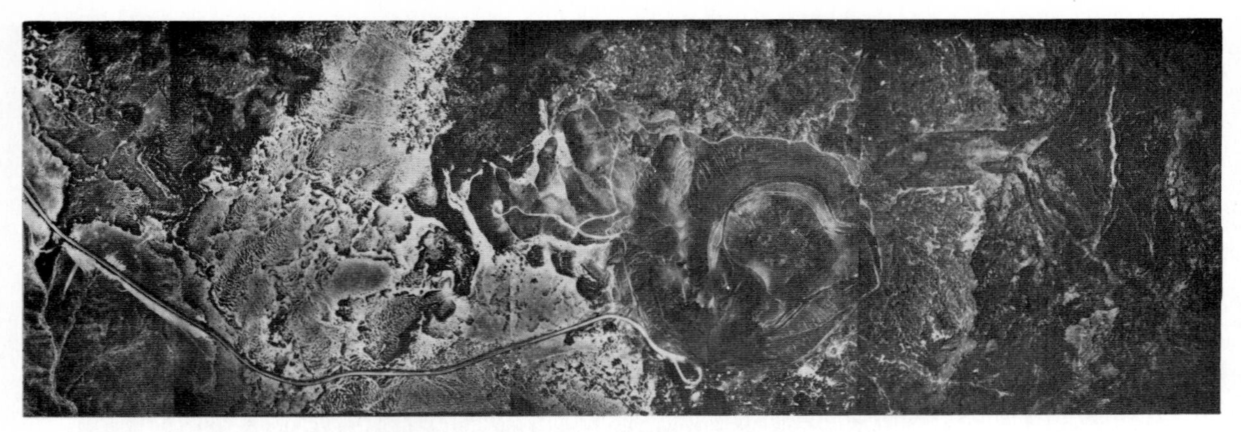

A. AERIAL PHOTOMOSAIC ACQUIRED OCTOBER 10, 1969 BY NASA.

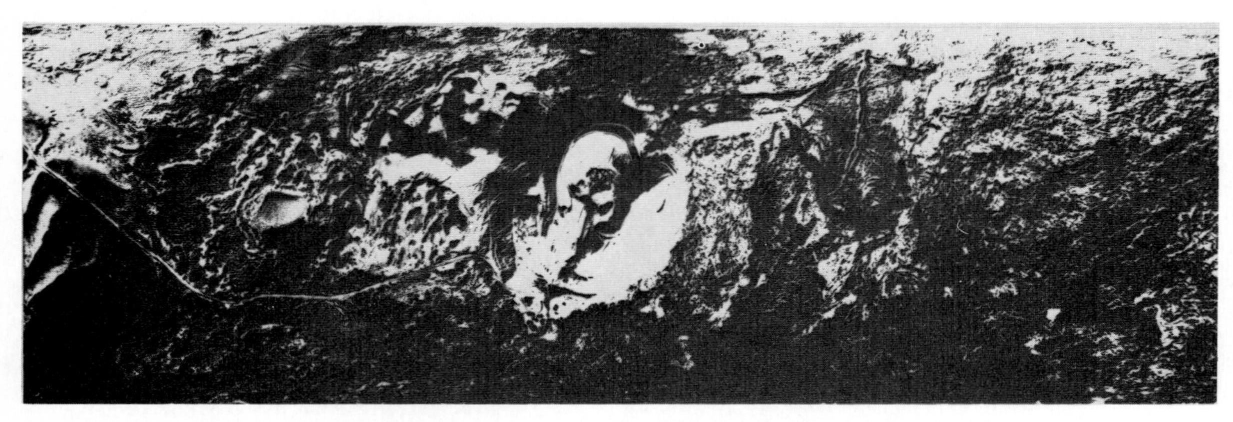

B. DAYTIME THERMAL IR IMAGE (8 TO 14 μm) ACQUIRED FEBRUARY 14, 1965 AT 2:10 P.M. BY U.S. GEOLOGICAL SURVEY.

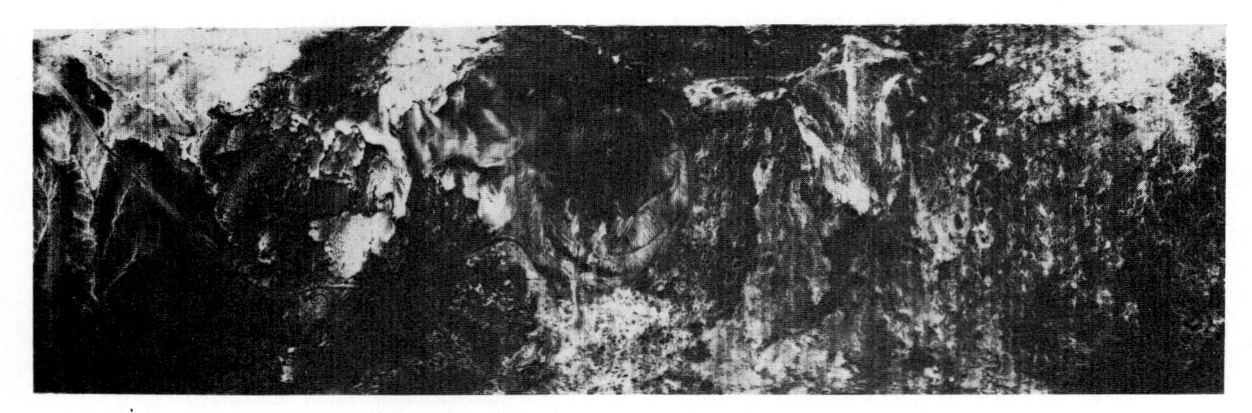

C. NIGHTTIME THERMAL IR IMAGE (8 TO 14 μm) ACQUIRED FEBRUARY 14, 1965 AT 6:29 A.M. BY U.S. GEOLOGICAL SURVEY.

0 0.5 MI
0 0.5 KM

FIGURE 11.7
Thermal IR images and aerial photomosaic of Pisgah Crater.

TABLE 11.1
Signatures of surface materials interpreted from aerial photographs and thermal IR images

Material	Visible reflectance	Radiant temperature	
		Day	Night
Aa basalt	Dark	Cool	Warm
Pahoehoe basalt	Medium	Medium to warm	Cool
Cinders	Medium	Very warm	Cool
Eolian sand	Light	Cool	Cool
Alluvium	Light to medium	Cool	Cool

Figure 11.1, and the distribution of surface materials is seen in Figure 11.5. On the daytime image warm signatures (light tones) and cool signatures (dark tones) are partly related to topographic relief, which controls heating and shadowing from the sun. The solar effect is especially apparent at Pisgah Crater, where the south-facing slopes are very warm and the north-facing slopes are very cool. The various surface materials have pronounced differences in radiant temperature on the day and night images. These differences are summarized in Table 11.1, together with the reflectance determined from the aerial photomosaic. The images of Figure 11.7 are uncalibrated. Therefore the radiant temperatures and reflectance values of Table 11.1 are relative signatures estimated by the interpreter. There is a general tendency for the aa basalt to be cool on the daytime IR image and warm on the nighttime image, whereas the pahoehoe is warm on the daytime image and cool at night. Although the radiant temperature of the cinders is influenced by topography, this material is very warm during the day and cool at night. Alluvium and eolian sand are cool on both the day and night images.

These diurnal differences in radiant temperature may be explained by the physical properties of the materials. The cinders, aa, and pahoehoe have the same mineral composition. The very low density of the cinders is the main difference between this material and the two types of basalt. The low density of the cinders contributes to their low thermal inertia, which was displayed in the thermal inertia map of Pisgah Crater (Figure 5.24B). The low thermal inertia accounts for the extreme high and low radiant temperatures of cinders on the day and night images, respectively. Although density of the aa and pahoehoe is similar, the highest thermal inertia values in Figure 5.24B are concentrated in the areas of aa lava in the northeast part of the map. This relationship agrees with the diurnal temperature variations of the two lava types shown in the 1965 images and listed in Table 11.1. The difference in surface texture between aa and pahoehoe, shown in Figures 11.2E and 11.2F, may contribute to the different patterns of diurnal temperature behavior.

The relatively low thermal inertia of sand and alluvium suggests that these materials should have a diurnal temperature pattern similar to that of cinders. This is not the case, however, for the sand and alluvium are relatively cool both day and night. This behavior is caused by their very high reflectance (albedo), which means that little solar energy is absorbed and converted into thermal energy. The silt and clay of Lavic Lake have a low thermal inertia, but on both day and night IR images (not shown) the dry lake has a cool radiant temperature. The very high reflectance, as shown in the visible band images, also accounts for this thermal pattern.

On the daytime thermal IR image the south-facing walls of the fissures on the lava dome are warm, and the north-facing walls are very cool.

TABLE 11.2
Limiting values of surface relief h for Ka-band radar (λ = 0.86 cm)

Depression angle, γ	Smooth criterion: $h < \frac{\lambda}{25 \sin \gamma}$	Rough criterion: $h > \frac{\lambda}{4.4 \sin \gamma}$
30°	0.7 mm	3.9 mm
45°	0.5 mm	2.8 mm
60°	0.4 mm	2.3 mm

TABLE 11.3
Calculated roughness categories for Ka-band radar imagery and corresponding signatures and materials

Roughness category	Vertical relief, h	Radar signature	Material
Smooth	< 0.5 mm	Dark	Dry lake surface
Intermediate	0.5 to 3.0 mm	Gray	Eolian sand, desert pavement
Rough	> 3.0 mm	Light	Cinders, aa, pahoehoe

At night, however, the entire width of each fissure is very warm relative to the surrounding surfaces of the lava flows. The collapsed lava tubes have similar diurnal temperature patterns. The sun preferentially heats the dense basalt exposed in the south-facing walls during the day. At night the absorbed thermal energy radiated from one vertical wall is absorbed and reradiated by the opposite wall. This process retains heat in the fissures and tubes, causing their warm signature at night.

RADAR IMAGERY AND SCATTEROMETER DATA

Dual-polarized Ka-band (0.86 cm) radar images of the Pisgah Crater area were acquired in 1965. A cross-polarized image with look direction to the southwest is depicted in Figure 11.8A, which corresponds to the geologic map in Figure 11.1. Despite the side-lobe banding this image is superior to the parallel-polarized image (not shown), which has lower contrast. The following features are clearly shown on the image:

1. Pisgah Crater and associated lava flows in the upper part of the image.
2. Bullion Mountains in the upper right corner.
3. Lavic Lake in the center.
4. Sunshine Crater and associated lava flows in the lower center.
5. Lava Bed Mountains at the lower margin of the image.
6. The north-trending Pisgah fault. The east-facing fault scarps are favorably oriented with respect to the radar look direction, and produce very strong returns. The west-facing scarps produce linear shadows. The combination of highlights and shadows enhances the expression of this active strike-slip fault.

Information is lacking on the depression angles at which the radar image was acquired. Based on knowledge of radar geometry, however it is reasonable to assume a depression angle of 30° for the far range of the image and 60° for the near range. The rough and smooth criteria, (from Chapter 6) may be used to calculate the limiting values of surface relief, h, at which the terrain produces smooth, intermediate, or rough signatures. The results of these calculations at representative depression angles are shown in Table 11.2. Values for the smooth criterion range from 0.4 to 0.7 mm for h across the image and a value of 0.5 mm appears representative for the entire image. The rough criterion ranges from 2.3 to 3.9 mm and a value of 3.0 mm is representative. As shown in Table 11.3 these values define the limits for the smooth, intermediate, and rough

A. REGIONAL IMAGE. COMPARE WITH GEOLOGIC MAP OF FIGURE 11.1.

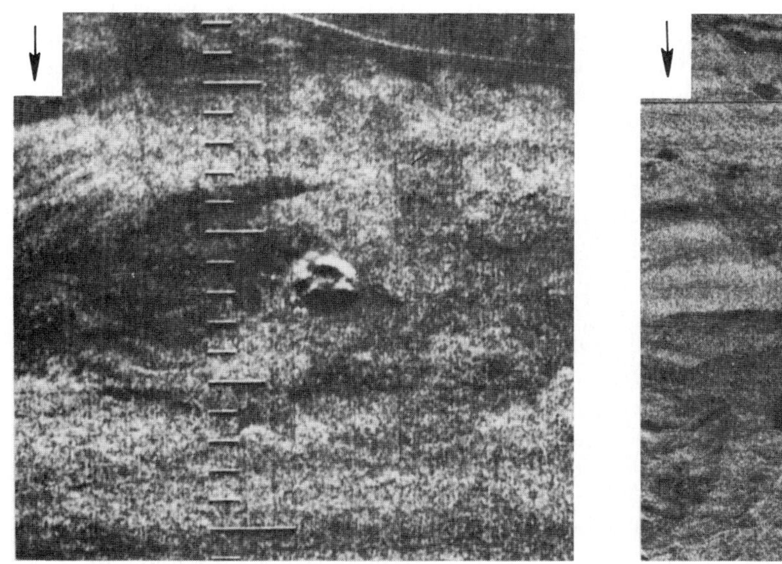

B. ENLARGEMENT OF PISGAH CRATER.

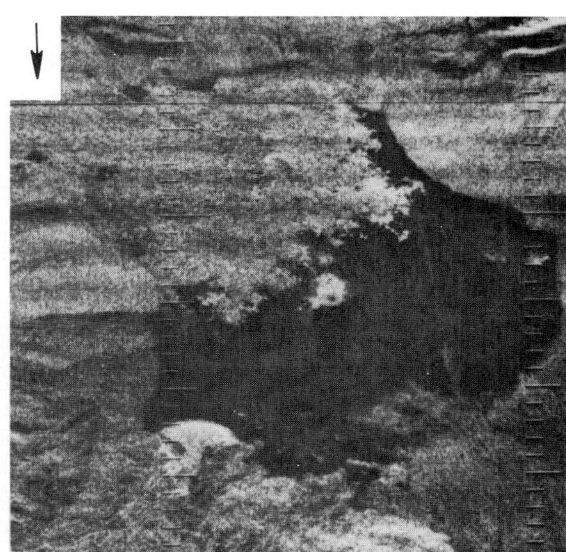

C. ENLARGEMENT OF LAVIC DRY LAKE.

FIGURE 11.8
Cross-polarized, HV, Ka-band (0.86 cm) radar images of Pisgah Crater area, acquired November 1965. Images courtesy NASA and U.S. Geological Survey.

categories that have dark, gray, and bright signatures, respectively, on the image. Table 11.3 also lists the surface materials associated with the three signatures on the images of Figure 11.8. The surface of Lavic Lake produces the only dark signature, other than shadows, on the image. As is shown by the profile in Figure 11.2A, the local surface relief is less than 0.5 mm for the dry lake, which therefore is a specular surface. The derivation of the scatterometer deviation profiles was described in Chapter 6. The scatterometer data in Figure 11.2 were acquired at a wavelength of 2.25 cm in contrast to the 0.86-cm wavelength of the radar image; nevertheless the curves provide a measure of the relative backscattering from the different materials. The portion of the deviation curve representing depression angles from 60° to 30° correlates with the near-range to far-range regions, respectively, of the radar image. The extremely low scatterometer values between these depression angles for the dry lake are consistent with the smooth surface and dark image signature of this material. The enlarged image of Lavic Lake (Figure 11.8C) illustrates the sensitivity of radar to differences in surface roughness. The narrow bright line crossing the dry lake is a narrow gravel-surfaced road with ruts and holes caused by vehicles. The extreme bright spot near the north margin of the lake is caused by a layer of basalt cobbles and pebbles that forms a dark patch on the photographic band images (Figure 11.3).

The intermediate roughness category and corresponding gray signature on the image are associated with eolian sand and desert pavement (Table 11.3). On the enlarged image of Pisgah Crater (Figure 11.8B) eolian sand causes the two dark gray signatures directly north of the crater that contrast with the bright signature of the surrounding rough lava and cinders. These patches of sand also correspond to the very bright signature on the low-altitude aerial photographs (Figure 11.4). The average diameter of the sand grains is approximately 0.5 mm, which suggests that the radar response should be that of a smooth surface. As illustrated on the aerial photographs, however, numerous patches of basalt are exposed through the sand cover to cause an average roughness in the 0.5 to 3.0 mm range. On the enlarged image of Pisgah Crater the gray signature at the center of the left edge is caused by desert pavement, and is indistinguishable from the signature of the adjacent eolian sand. Desert pavement forms a gray band parallel with the upper margin of Figure 11.8B and is readily distinguished from the bright signature of the adjacent lava flows. The bright line crossing this desert pavement is caused by the strong return from a railroad track.

On the regional radar image are large expanses of intermediate signature caused by desert pavement formed on alluvial fans adjacent to Lavic Lake and the mountains. Within the fans are local areas of bright return that are caused by patches of coarse gravel or local concentrations of desert shrubs. The intermediate nature of backscatter from the sand and desert pavement is shown by the scatterometer deviation curves that are close to the regional average value.

Very bright signatures on the radar image are associated with the cinders, aa, and pahoehoe. This is consistent with their vertical relief of several centimeters shown in Figure 11.2. Because of their limited area, these illustrations minimize the full range of surface roughness on the lava flows. In addition to the surface texture there are numerous irregular blocks and crevices, especially on the aa lava, that cause up to 1 m of local roughness. The

TABLE 11.4
Comparison of signatures on different types of images

Material	Photographic (0.4 to 0.9 μm)	Thermal IR (8 to 14 μm)		Ka band radar (0.86 cm)
		Day	Night	
Dry lake	Light	Cool	Cool	Smooth
Eolian sand	Light	Cool	Cool	Intermediate
Alluvium	Light to medium	Cool	Cool	Intermediate
Cinders	Dark	Very warm	Cool	Rough
Pahoehoe basalt	Medium	Medium to warm	Cool	Rough
Aa basalt	Dark	Cool	Warm	Rough

scatterometer deviation curve for the cinders, aa, and pahoehoe are all higher than the regional average, consistent with the surface roughness and radar signatures of these volcanic materials. The deviation curves and field observations show that the pahoehoe surface is smoother than the aa. At the short wavelength of Ka-band radar, however, both lava types have rough signatures. Radar imagery acquired at longer wavelengths should differentiate the two lava types at Pisgah. For L-band radar ($\lambda = 25$ cm) at a depression angle of 45° a vertical surface relief of 8 cm is the limiting value between intermediate and rough surfaces. Such longer wavelength imagery could distinguish aa from pahoehoe, while the shorter wavelength Ka-band imagery distinguishes between surfaces of lower vertical relief.

COMMENTS

The signatures in different spectral regions for the various surface materials near Pisgah Crater are summarized in Table 11.4. The causes for the signatures have been described. It is significant that no material has the same combination of signatures on the four major image types. However, on any single image, such as radar, several different materials may have the same signature and be indistinguishable. This is a primary argument for acquiring imagery in more than one remote-sensing spectral band. In the Pisgah Crater area, most of the geologic and geographic information can be interpreted by a careful study of the stereo aerial photographs. The Pisgah area is especially well suited for photographic band imagery acquired from satellites or aircraft because of the prevailing good weather and lack of vegetation cover. These environmental factors, plus the aridity of the area, also provide a good setting for thermal IR imagery. Any obscure or partially concealed faults and folds in the area would probably have been detected on the thermal IR images. The radar imagery was useful at Pisgah for emphasizing the linearity and scarps of the Pisgah fault. If images acquired at longer wavelengths were available, some of the materials in Table 11.4 with similar rough signatures probably would be separated into distinct signature categories. If the cloud cover, rainfall, and vegetation cover at Pisgah were increased to an extent similar to that in the tropics, radar would become the most useful form of imagery, while photography and thermal IR imagery would be of limited usefulness. The selection of remote-sensor imagery for an interpretation project should be determined not only by the objectives of the project, but also by the nature of the terrain.

REFERENCES

Hasell, P. G., 1975, Collation of earth resources data collected by ERIM airborne sensors: Environmental Research Institute of Michigan Report No. 109600–33–F, Ann Arbor, Mich.

Vincent, R. K., 1972, Rock-type discrimination from ratio images of Pisgah Crater, California: University of Michigan Willow Run Laboratories, Technical Report WRL 31650–77–T, Ann Arbor, Mich.

ADDITIONAL READING

Dibblee, T. W., 1966, Geologic map of the Lavic Quadrangle, San Bernardino County, California: U.S. Geological Survey Miscellaneous Geologic Investigations Map I-472.

Gawarecki, S. J., 1969, Infrared survey of the Pisgah Crater area, San Bernardino County, California; a geologic interpretation: U.S. Geological Survey Interagency Report NASA–99.

Lyon, R. J. P., 1972, Infrared spectral emittance in geological mapping—airborne spectrometer data from Pisgah Crater, California: Science, v. 175, p. 983–986.

12
SUMMARY

Remote sensing is a diverse field, both in the technology involved and in the applications. Chapter 1 described electromagnetic radiation and its interaction with matter. The nature of this interaction, which is specific for different wavelengths of radiation and different types of matter, is detected and recorded by remote-sensing systems. Characteristics and applications of the principal remote-sensing systems are summarized in Table 12.1. This table is designed to aid in the selection of the optimum sensor system for a particular application. At one time the concept of multiple sensor imagery was popular. It was advocated that for any interpretation problem, all possible types of imagery should be acquired and interpreted. This approach is rarely applied today, except for special problems and for research, as shown at Pisgah Crater in Chapter 11. Few of the remote-sensing systems can be used simultaneously and still acquire optimum imagery. For example, the high-altitude and side-looking geometry of radar acquisition are incompatible with other systems. The expense of multiple flights with different aircraft and imaging systems is excessive for reconnaissance surveys, which should be relatively inexpensive to be cost effective. Today the strategy is to select the remote-sensing system that provides the maximum amount of information for the project. For example, SLAR is the optimum system for mapping terrain and geologic structure in forested areas where flying conditions are poor. Thermal plumes in lakes and estuaries, on the other hand, are best mapped with thermal IR imagery acquired at relatively low altitude. Other methods can be selected by referring to Table 12.1.

LIMITATIONS AND PRECAUTIONS

We have discussed successful applications of remote sensing to a variety of activities ranging from oil and mineral exploration through environmental monitoring. These examples also have described some of the limitations and precautions associated with the various remote-sensing methods. For example, in multispectral classification for mineral prospects, clay and sand deposits in dry stream channels commonly have spectral signatures similar to those of hydrothermal alteration zones, which contain clay minerals and quartz. On thermal IR images the cold radiant temperatures of oil films may be identical with those of cold-water currents. Experienced interpreters have learned to cope with these problems. The linear

TABLE 12.1
Summary of remote-sensing systems

Image system	Wavelength region	Property detected	Operating constraints
UV imagery and photography	0.3 to 0.4 μm	Reflectance and fluorescence from solar radiation.	Daytime, good weather, low altitude to minimize atmospheric scattering.
Normal color and black-and-white photography	0.4 to 0.7 μm	Reflectance of visible solar radiation.	Daytime, good weather, any altitude.
IR color and IR black-and-white photography	0.5 to 0.9 μm	Reflectance of visible and photographic IR radiation.	Daytime, good weather, any altitude.
Landsat Multispectral Scanner	0.5 to 1.1 μm	Reflectance of visible and photographic IR radiation.	Daytime, good weather, 918 km altitude.
Thermal IR imagery	3 to 14 μm	Radiant temperature.	Day or night, good weather, any altitude.
Radar imagery	1 to 25 cm	Surface roughness and topography.	Day or night, most weather conditions; presently limited to aircraft operation.

branching pattern of stream-deposited clay and sand aids in distinguishing this material from the more localized pattern of hydrothermal alteration zones. The experienced interpreter refers to simultaneously acquired UV and visible images, when these are available, to distinguish oil films from water currents.

The so-called *principle of minimum astonishment* is a useful guideline for interpreters. When applied to remote sensing this principle means that the limitations of the imagery are sufficiently understood so that the interpreter is not astonished by the inevitable mistakes, but is able to understand and correct them on subsequent projects. Many lineaments, for example, do not correlate with mapped or unmapped faults. The interpreter should understand that these lineaments may represent zones of crustal weakness with no structural displacement, or that such lineaments may result from fortuitous alignments of drainage and other features. On the more positive side, the principle of minimum astonishment means that the successful interpretations are based on well-understood principles and can be repeated elsewhere.

The preceding limitations and precautions are technical in nature and can be dealt with objectively. Another precaution is philosophical, and concerns the excessively high success ratio that some users have been led to expect from remote-sensing technology. Prior to the launch of Landsat, for example, stories in the popular press announced that the new satellite "will locate new mines and oil fields." Today interpreters, who were not responsible for the news stories, are asked, "Where are the new mines and oil fields promised from Landsat images?" In response to this query it should first be noted that Landsat is an experimental program intended for many applications and was not specifically designed for oil and mineral exploration. The low spatial resolution, lack of adequate stereo coverage, and restricted spectral range are limitations. Despite these limitations, significant contributions have been made to oil and mineral exploration through processing and interpreting Landsat image data. No new oil or mineral discoveries have been documented as a sole result of Landsat, however, for the reasons given in Chapter 8. When new remote-sensing programs are recommended, it is imperative that the public and the user community not be misled with unrealistic promises. The resulting backlash is a handicap for future programs and investigators.

FUTURE DEVELOPMENTS

The most important future developments in remote sensing probably will occur in two fields; digital image processing and satellite imagery.

Special applications
Detecting oil films on water and fluorescent minerals.
Provides high spatial resolution and stereoscopic coverage.
See above; also used for mapping vegetation, soil moisture, land-water contacts.
Repetitive regional coverage; available in digital form.
Mapping temperature variations of land, water, and ice.
Mapping fractures and roughness; especially useful in areas with poor weather.

New software and hardware will improve the restoration, enhancement, and information extraction capabilities at reduced costs.

Table 12.2 summarizes planned United States satellite imaging programs, with the exception of meteorological systems. Most of these systems are funded, and development is well under way. The improved spatial resolution and extended spectral range of the proposed Landsat follow-on system should prove particularly valuable for mineral exploration. Stereosat is a small and relatively inexpensive system to acquire overlapping images along the orbit path with adequate base-height ratio for optimum stereo viewing. These images will be valuable for mapping regional geologic structure and terrain.

TABLE 12.2
Planned satellite remote-sensing systems

System and date	Sensors	Remarks
Landsat C (3), early 1978	Landsat MSS with addition of a thermal IR band (10.4 to 12.6 μm) with a ground resolution cell of 238 by 238 m.	Orbit pattern and coverage similar to Landsat 1 and 2.
	Return beam vidicon with a single band (0.5 to 0.75 μm) and ground resolution cell of 40 by 40 m.	The return beam vidicon system will provide higher resolution images than the MSS.
Landsat Follow-on, early 1980s	Same MSS as Landsat 3.	
	Thematic mapping scanner with ground resolution cell of 30 by 30 m. In addition to spectral bands in the 0.5 to 1.1 μm region there is a band at 0.45 to 0.55 μm and a band centered at 1.6 μm.	The improved spatial resolution and extended spectral sensitivity of the thematic mapper represent significant technologic advances.
Heat Capacity Mapping Mission, 1978	Two-band scanning system with 500 by 500 m ground resolution cell. One band in the photographic region (0.5 to 1.1 μm) and one in the thermal IR (10.5 to 12.5 μm).	Differences in radiant temperature can be determined from day and night thermal IR images. These data can be combined with albedo information from the photographic band to calculate thermal inertia of surface materials.
Seasat A, 1978	Visible and thermal IR scanning system.	Systems are designed to record sea state and radiant temperature of the oceans day and night, in all weather conditions. Some experimental imagery of land areas will be acquired.
	Synthetic aperture radar system, $\lambda = 25$ cm.	
	Passive microwave scanning radiometer.	
	Microwave scatterometer.	
Space Shuttle, 1980s	Various imaging systems including cameras, scanners, and radar.	Re-usable shuttle vehicles will carry crews and imaging systems into orbit and return. The shuttle will also serve as a stage for launching small unmanned satellites with imaging systems.
Stereosat, planning phase	Return-beam vidicon system will acquire images in a single visible band.	Acquire images with overlap along orbit path for stereo viewing.

GLOSSARY

This glossary includes terms, abbreviations, and acronyms commonly employed in remote sensing. The glossary definitions refer to the applications for which the terms are used in this book; applications outside the field of remote sensing are omitted. A more extensive glossary is provided in the *Manual of Remote Sensing,* published by the American Society of Photogrammetry in 1975. Definitions of geologic and geographic terms may be found in any standard text on those subjects or in the *Glossary of Geology and Related Sciences,* published by the American Geological Institute.

absorption band Wavelength interval at which electromagnetic radiation is absorbed by the atmosphere or by other substances. For example, there is an atmospheric absorption band at 5 to 8 μm, caused by water vapor, that absorbs thermal IR radiation of those wavelengths.

absorptivity The capacity of a material to absorb incident radiant energy.

active remote sensing Remote-sensing methods that provide their own source of electromagnetic radiation. Radar is one example.

additive primary colors The colors blue, green, and red. Filters of these colors transmit the primary color of the filter and absorb the other two colors.

air base Ground distance between principal points of successive overlapping aerial photographs.

airborne scanner A scanner designed for use on aircraft or spacecraft in which the forward motion of the vehicle provides coverage normal to the scan direction.

albedo The ratio of the amount of electromagnetic energy reflected by a surface to the amount of energy incident upon it. The symbol is A.

amplitude For waves this term represents the vertical distance from crest to trough.

analog A form of data display in which values are shown in graphic form, such as curves. Also a form of computing in which values are represented by directly measurable quantities, such as voltages or resistances. Analog computing methods contrast with digital methods in which values are treated numerically.

angle of incidence In SLAR systems this is the angle between the vertical and a line connecting the antenna and a target. The symbol is θ.

angular field of view The angle subtended by lines from a remote-sensing system to the outer margins of the strip of terrain that is viewed by the system.

anomaly An area on an image that differs from the surrounding normal area. For example, a concentration of vegetation within a desert scene constitutes an anomaly.

antenna The device that transmits and receives microwave and radio energy in SLAR systems.

aperture The opening in a remote-sensing system that admits electromagnetic radiation to the film or detector.

Apollo The United States lunar program of satellites with crews of three.

ASA index American Standards Association designation of film speed, or sensitivity to light. Higher values indicate higher sensitivity.

aspect angle Same as angle of incidence in SLAR, which is generally the preferred term.

atmosphere The layer of gases that surround some planets.

atmospheric windows Wavelength intervals at which the atmosphere transmits most electromagnetic radiation.

attitude The angular orientation of a remote-sensing system with respect to a geographic reference system.

azimuth The geographic orientation of a line given as an angle measured clockwise from north.

azimuth direction In SLAR images this refers to the direction of the aircraft ground track.

azimuth resolution In SLAR images this is the spatial resolution in the azimuth direction. In real-aperture SLAR systems azimuth resolution is determined by angular width of the transmitted beam.

background The area on an image or the terrain that surrounds an area of interest, or target.

backscatter In SLAR usage, this refers to the portion of the microwave energy scattered by the terrain surface that is directed back toward the antenna.

band A wavelength interval in the electromagnetic spectrum. For example, in Landsat the bands designate specific wavelength intervals at which images are acquired.

base-height ratio Air base divided by aircraft height. This ratio determines vertical exaggeration on stereo models.

batch processing The method of data processing in which data and programs are entered into a computer, which then carries out the entire processing operation with no further instructions.

beam A focused pulse of energy.

beam width In SLAR usage this is the angle subtended in the horizontal plane by the radar beam.

binary A numerical system using the base 2.

bit In digital computer terminology, this is a binary digit that is an exponent of the base 2.

black body A substance that radiates energy at the maximum possible rate per unit area at each wavelength for any given temperature. A black body also absorbs all the radiant energy incident upon it. No actual substance behaves as a black body although some substances, such as lampblack, approach these properties.

boresight To align the axis of one remote-sensing system with the axis of another system. The resulting images are geometrically registered to each other.

brightness Magnitude of the response produced in the eye by light.

browse files Microfilm files and viewers of Landsat images that enable one to select images. These files are maintained at U.S. Geological Survey offices and other facilities.

brute-force radar See real-aperture radar.

byte A group of eight bits of digital data.

calibration The process of comparing measurements made by an instrument with a standard.

calorie The amount of heat required to raise the temperature of 1 gm of water 1°C.

cassette A light-proof container for roll film.

cathode-ray tube A vacuum tube with a phosphorescent screen upon which images are displayed by an electron beam. The abbreviation is CRT.

CCT Computer compatible tape; the magnetic tape upon which the digital data for Landsat MSS images are recorded.

change-detection images Images prepared by digitally comparing two original images acquired at different times. The gray tones of each pixel on a change-detection image portray the amount of difference between the original images.

chlorosis Yellowing of leaves of plants resulting from an upset of the iron metabolism caused by excess concentrations of chlorine, copper, zinc, manganese, or other elements.

chopper Rotating disk of alternating blades and open spaces used in a radiometer to compare radiation from a target with that from a calibrated reference source.

classification The process of assigning individual pixels of a multispectral image to categories, generally on the basis of spectral-reflectance characteristics.

color-composite image A color image prepared by projecting individual black-and-white multispectral images in color.

color-ratio composite image A color-composite image prepared by combining, in different colors, individual spectral band ratio images for a scene.

complementary colors Two colors of light that produce white light when added together, such as red and cyan.

compositor viewer Device in which black-and-white multispectral images are registered and projected with colored lights to produce a color-composite image.

conduction The transfer of electromagnetic energy through a material by molecular interaction.

cones Receptor cells in the retina of the eye that function under daylight conditions to give color vision.

contact print A reproduction made from a photographic negative placed in direct contact with photosensitive paper.

contrast ratio The ratio between the reflectance of the brightest and darkest parts of the image. Commonly referred to as contrast.

contrast stretching Improving the contrast of images by digital processing. The original range of digital values is expanded to utilize the full contrast range of the recording film or display device.

convection The transfer of heat through the physical movement of heated matter.

corner reflector A cavity formed by three planar reflective surfaces intersecting at right angles. Electromagnetic radiation entering a corner reflector is reflected directly back toward the source of the radiation.

COSMIC Computer Software Management and Information Center, University of Georgia. This facility distributes computer programs developed by U.S. government-funded projects.

cross polarized Radar return pulse in which the polarization direction is normal to the polarization direction of the transmitted pulse. Images recorded with cross-polarized energy may be HV (horizontal transmit, vertical return) or VH (vertical transmit, horizontal return).

cycle A single oscillation of a wave.

data collection system On Landsat the system that acquires information from seismometers, flood gauges, and other measuring devices. These data are relayed to an earth receiving station. The abbreviation is DCS.

density, of images A measure of the opacity, or darkness, of an image.

density, of materials The ratio of mass to volume of a material, typically expressed as $gm \cdot cm^{-3}$. The symbol is ρ.

density slicing The process of converting the continuous gray tone of an image into a series of density intervals, or slices, each corresponding to a specific digital range.

depolarization Change in polarization of a radar pulse as a result of multiple reflections from the terrain surface.

depression angle In SLAR usage this is the angle between the horizontal plane passing through the antenna and the line connecting the antenna and the target. The symbol is γ.

detectability A measure of the smallest object that can be discerned on an image.

detector The component of a remote-sensing system that converts electromagnetic radiation into a signal that is recorded.

development The chemical processing of exposed photographic material to produce an image.

diazo film A transparent material on which image transparencies may be reproduced in specific colors.

dielectric constant Electrical property of matter that influences radar returns. Also referred to as complex dielectric constant.

diffuse reflector A surface that reflects incident radiation in a multiplicity of directions.

digital image processing Computer manipulation of the digital values for picture elements of an image.

digitization The process of converting an image recorded originally on photographic material into numerical format.

digitizer A device for scanning an image and converting it into numerical picture elements.

distortion On an image, this refers to changes in shape and position of objects with respect to their true shape and position.

diurnal Daily.

DN Digital number. The value of reflectance recorded for each pixel on Landsat CCTs.

Doppler principle Describes the change in observed frequency of electromagnetic or other waves caused by relative movement between the source of waves and the observer.

drape flying The practice of varying the altitude of a remote-sensor aircraft to maintain a constant height above the terrain, as opposed to the practice in which the aircraft flies at a fixed altitude.

EDC EROS Data Center. The U.S. Geological Survey facility at Sioux Falls, South Dakota. Here aircraft and satellite images are archived and made available for purchase. This facility also provides an indexing service for photographs and images acquired by NASA and the U.S. Geological Survey. (See also EROS.)

Ektachrome A Kodak color-positive film.

electromagnetic radiation Energy propagated in the form of an advancing interaction between electric and magnetic fields.

emission The process by which a body emits electromagnetic radiation, usually as a consequence of its kinetic temperature.

emissivity The ratio of radiant flux from a body to that from a black body at the same kinetic temperature. The symbol is ε.

emulsion A suspension of photosensitive silver halide grains in gelatin that constitutes the image-forming layer on photographic materials.

enhancement The process of altering the appearance of an image so that the interpreter can extract more information. Enhancement may be done by digital or photographic methods.

EREP Earth Resources Experiment Package on Skylab, which consisted of multispectral cameras, earth terrain camera, multispectral scanner, and non-imaging systems.

EROS Earth Resource Observation System, administered by U.S. Geological Survey.

ERTS Earth Resource Technology Satellite, now called Landsat.

exposure index A number that indicates the sensitivity of photographic film to light. Higher numbers indicate greater sensitivity.

***f* number** A representation of the speed of a lens that is determined by dividing the focal length by the diameter of the lens. Smaller numbers indicate faster lenses.

***f* stop** The focal length of a lens divided by the diameter of the adjustable diaphragm. Smaller numbers indicate larger openings that admit more light to the film.

far range Refers to the portion of a SLAR image farthest from the aircraft flight path.

film The light-sensitive photographic emulsion and its transparent base.

film speed A measure of the sensitivity of photographic film to light. Larger numbers indicate higher sensitivity.

filter, digital A mathematical procedure for removing unwanted values from numerical data.

filter, optical A material that, by absorption or reflection, selectively modifies the radiation transmitted through an optical system.

flight path The line on the ground directly beneath a remote-sensing aircraft or satellite.

fluorescence The emission of light from a substance caused by exposure to radiation from an external source.

flux See radiant flux.

focal length In cameras the distance measured along the optical axis from the optical center of the lens to the plane at which the image of a very distant object is brought into focus.

focus To adjust a remote-sensing system to produce a sharp, distinct image.

format The size and scale of an image.

fovea Small region at the center of the retina that provides maximum visual acuity.

frequency The number of wave oscillations per unit time or the number of wavelengths that pass a point per unit time. The symbol is v.

front-surface mirror Mirror with the reflective coating on the top of the glass. Used in thermal IR scanners.

GCP Ground control point. A geographic feature of known location that is recognizable on images and can be used to determine geometric corrections.

Gemini The United States program of two-man earth-orbiting satellites in 1965 and 1966.

geothermal Refers to heat from sources within the earth.

GMT Greenwich mean time. This international 24-hour system is used to designate the time at which Landsat images are acquired.

gossan Surface occurrence of iron oxide formed by weathering of metallic sulfide ore minerals.

granularity The graininess of developed photographic film that is determined by the texture of the grains of developed silver.

gray scale A calibrated sequence of gray tones ranging from black to white.

grazing angle In SLAR usage this is the angle between the terrain surface and the incident radar beam. For a horizontal surface, grazing angle and depression angle are the same.

ground range On SLAR images this is the distance from the ground track to an object.

ground-range image A SLAR image in which the scale in the range direction is constant.

ground receiving station A facility that records image data transmitted by Landsat.

ground resolution cell The area on the terrain that is covered by the instantaneous field of view of a detector. Size of the ground resolution cell is determined by the altitude of the remote-sensing system and the instantaneous field of view of the detector.

GSFC Goddard Space Flight Center. The NASA facility at Greenbelt, Maryland that is a Landsat earth receiving station. Landsat data from all United States receiving stations are converted into images at GSFC.

harmonic Refers to waves in which the component frequencies are whole number multiples of the fundamental frequency.

highlights Areas of bright tone on an image.

hue The attribute of a color that differentiates it from gray of the same brilliance and that allows it to be classed as blue, green, red, or intermediate shades of these colors.

image The representation of a scene as recorded by a remote-sensing system. Although image is a general term, it is commonly restricted to representations acquired by nonphotographic methods.

incident energy Electromagnetic radiation impinging on a surface.

index of refraction The ratio of the wavelength or velocity of electromagnetic radiation in a vacuum to that in a substance. The symbol is n.

insolation Incident solar energy.

instantaneous field of view The solid angle through which a detector is sensitive to radiation. In a scanning system this refers to the solid angle subtended by the detector when the scanning motion is stopped. Instantaneous field of view is commonly expressed in milliradians.

interactive processing The method of data processing in which the operator views preliminary results and can alter the instructions to the computer to achieve optimum results.

interpretation The extraction of information from an image.

interpretation key A characteristic or combination of characteristics that enable an object or material to be identified on an image.

IR The infrared region of the electromagnetic spectrum that includes wavelengths from 0.7 μm to 1 mm.

IR color film A color film consisting of three layers in which the red-imaging layer responds to photographic IR radiation ranging in wavelength from 0.7 to 0.9 μm. The green-imaging layer responds to red light and the blue-imaging layer responds to green light.

isotherm A line connecting points of equal temperature. Isotherm maps are often used to portray distribution of surface temperature of water bodies.

JPL Jet Propulsion Laboratory. A NASA facility at Pasadena, California operated by the California Institute of Technology.

Ka band Radar wavelength region from 0.8 to 1.1 cm.

kinetic energy Kinetic energy is the ability of a moving body to do work by virture of its motion. The molecular motion of matter is a form of kinetic energy.

kinetic temperature The internal temperature of an object, which is determined by the molecular motion. Kinetic temperature is measured with a contact thermometer, and differs from radiant temperature, which is a function of emissivity and internal temperature.

Kodachrome A normal color-positive film manufactured by Kodak.

L band Radar wavelength region from 15 to 30 cm.

Lambert conic conformal projection A map projection on which all geographic meridians are represented by straight lines that meet at a common point outside the limits of the map, and the geographic parallels are represented by a series of arcs having this common point as a center. Meridians and parallels intersect at right angles and angles on the earth are correctly represented on the projection.

Landsat An unmanned earth-orbiting NASA satellite that transmits multispectral images in the 0.4 to 1.1 μm region to earth receiving stations (formerly called ERTS).

latent image The invisible image produced by the photochemical effect of light on silver halide grains in the emulsion of film. Photographic development renders the latent image visible.

layover In SLAR images this is the geometric displacement of the top of objects toward the near range, relative to their base.

lens A piece, or combination of pieces, of glass or other transparent material that enables images to form by refraction of light.

light Visible radiation from 0.4 to 0.7 μm in wavelength that is detectable by the eye.

light meter A device for measuring the intensity of visible radiation and determining the appropriate exposure for acquiring photographs.

line-pair A pair of light and dark bars of equal sizes. The number of such line-pairs that can be distinguished per unit distance is used to express resolving power of imaging systems.

lineament A linear topographic or tonal feature on the terrain and on images and maps that may represent a zone of structural weakness.

linear An adjective that describes the straight-line nature of features on the terrain or on images and maps.

look direction Direction in which pulses of microwave energy are transmitted by a SLAR system. Look direction is normal to the azimuth direction. Also called range direction.

low-sun-angle photographs Aerial photographs acquired in the morning, evening, or winter when the sun is at a low elevation above the horizon.

luminance A quantitative measure of the intensity of light from a source, measured with a device called a photometer.

Mercury The United States program of one-man, earth-orbiting satellites in 1962 and 1963.

microwave The region of the electromagnetic spectrum in the wavelength range from 1 mm to beyond 1 m.

Mie scattering Multiple reflection of light waves by atmospheric particles that have the approximate dimensions of the wavelength of light.

modulate To vary the frequency, phase, or amplitude of electromagnetic waves.

mosaic An image or photograph made by piecing together individual images or photographs covering adjacent areas.

MSS Multispectral scanner system of Landsat that acquires images at four wavelength bands in the visible and reflected IR regions.

multispectral camera A system that simultaneously acquires photographs at different wavelengths of the same scene. Also called multiband camera.

multispectral scanner A scanner system that simultaneously acquires images in various wavelength regions of the same scene.

nadir The point on the ground vertically beneath the center of a remote-sensing system.

NASA National Aeronautical and Space Administration.

near range Refers to the portion of a SLAR image closest to the aircraft flight path.

negative photograph A photograph on film or paper in which the tones are reversed from the brightness of the features on the terrain.

noise Random or repetitive events that obscure or interfere with the desired information.

nonsystematic distortion Geometric irregularities on images that are not constant and cannot be predicted from the characteristics of the imaging system.

normal color film A film in which the colors are essentially true representations of the color of the terrain.

oblique photograph A photograph acquired with the camera axis intentionally directed between the horizontal and vertical orientations.

orbit The path of a satellite around a body under the influence of gravity.

overlap The extent to which adjacent images or photographs cover the same terrain, expressed in percent.

parallel polarized Radar return pulse in which the polarization is the same as the transmitted pulse. Images recorded with parallel-polarized energy may be HH (horizontal transmit, horizontal return) or VV (vertical transmit, vertical return).

pass In digital filters this refers to the spatial frequency of data transmitted by the filter. High-pass filters transmit high-frequency data; low-pass filters transmit low-frequency data.

passive remote sensing Remote sensing of energy naturally reflected or radiated from the terrain.

pattern The regular repetition of tonal variations on an image or photograph.

photodetector A device for measuring energy in the photographic band.

photogeology The mapping and interpretation of geologic features from aerial photographs.

photograph A representation of targets formed by the action of light on silver halide grains of an emulsion.

photographic IR The short wavelength portion of the IR band from 0.7 to 0.9 μm wavelength that is detectable by IR color film or IR black-and-white film.

photon The elementary quantity of radiant energy.

picture element In a digitized image this is the area on the ground represented by each digital value. Because the analog signal from the detector of a scanner may be sampled at any desired interval, the picture element may be smaller than the ground resolution cell of the detector. Commonly abbreviated as pixel.

pitch Rotation of an aircraft about the horizontal axis normal to its longitudinal axis that causes a nose-up or nose-down attitude.

pixel A contraction of picture element.

polarization The direction of vibration of the electrical field vector of electromagnetic radiation. In SLAR systems polarization is either horizontal or vertical.

positive photograph A photographic image in which the tones are proportional to the brightness of the terrain.

previsual symptom The phenomenon whereby the ability of vegetation to reflect photographic IR energy diminishes because of stress. This commonly occurs before the normal green color changes, and is recognizable on IR color film by a drop in brightness of the red hues.

primary colors The three colors, either additive or subtractive, that may be combined to produce the full range of colors.

principal point The center of an aerial photograph.

printout Display of computer data in alphanumeric format.

pulse A short burst of electromagnetic radiation transmitted by a radar antenna.

pulse length Duration of a burst of energy transmitted by a radar antenna, measured in microseconds.

radar Acronym for radio detection and ranging, an active form of remote sensing that operates at wavelengths from 1 mm to 1 m.

radar-rock units Rock units that are characterized by distinctive signatures on radar images. The signatures are determined by surface roughness of the rocks.

radar shadow A dark area of no return on a radar image that extends in the far-range direction from an object on the terrain that intercepts the radar beam.

radian The angle subtended by an arc of a circle equal in length to the radius of the circle; 57.3°.

radiant flux The electromagnetic energy radiated from a source, expressed in cal · cm^{-2} · sec^{-1}. The symbol is *F*.

radiant power peak The wavelength at which the maximum electromagnetic energy is radiated at a particular temperature.

radiant temperature Concentration of the radiant flux from a material. Radiant temperature is the product of the kinetic temperature multiplied by the emissivity to the one-fourth power.

radiation The propagation of energy in the form of electromagnetic waves.

radiometer A nonimaging device for quantitatively measuring radiant energy, especially thermal radiation.

random-line dropout A defect in scanner images caused by the loss of data from individual scan lines in a nonsystematic fashion.

range direction For radar images this is the direction in which energy is transmitted from the antenna and is normal to the azimuth direction. Also called look direction.

range resolution For radar images this is the spatial resolution in the range direction and is determined by the length of the transmitted pulse of microwave energy.

raster lines The individual lines swept by an electron beam across the face of a CRT that constitute the image display.

ratio image An image prepared by processing digital multispectral data. For each pixel the value for one band is divided by that of another. The resulting digital values are displayed as an image.

Rayleigh criterion For radar images, this is the relationship between surface roughness, depression angle, and wavelength that determines whether a surface will respond in rough or smooth fashion to the radar pulse.

Rayleigh scattering Selective scattering of light by particles in the atmosphere that are small relative to the wavelength of light. The scattering is inversely proportional to the fourth power of the wavelength.

RBV Return-beam vidicon.

real-aperture radar SLAR system in which azimuth resolution is determined by the physical length of the antenna and by the wavelength. The radar returns are recorded directly to produce images. Also called brute-force radar.

real time To make images or data available for inspection simultaneously with their acquisition.

recognizability The ability to identify an object on an image.

rectilinear Refers to images with no geometric distortion in which the scales in the X and Y directions are the same.

reference source In thermal IR scanners and radiometers these are electrically heated cavities maintained at a known radiant temperature and used for calibrating the radiant temperature detected from the target.

reflectance The ratio of the radiant energy reflected by a body to that incident upon it.

reflectance, spectral Reflectance measured at a specific wavelength interval.

reflected IR Wavelengths from 0.7 to about 3 μm that are primarily reflected solar radiation.

reflectivity The ability of a surface to reflect incident energy.

refraction The bending of electromagnetic rays as they pass from one medium into another.

registration The process of superimposing two or more images or photographs so that equivalent geographic points coincide. Registration may be done digitally or photographically.

relief The vertical irregularities of a surface.

relief displacement The geometric distortion on vertical aerial photographs. The tops of objects are located on the photograph radially outward from the base.

remote sensing The collection of information about an object without being in physical contact with the object. Remote sensing is restricted to methods that record the electromagnetic radiation reflected or radiated from an object, which excludes magnetic and gravity surveys that record force fields.

resolution The ability to distinguish closely spaced objects on an image or photograph. Commonly expressed as the spacing, in line-pairs per unit distance, of the most closely spaced lines that can be distinguished.

resolution target Regularly spaced pairs of light and dark bars that are used to evaluate the resolution of images or photographs.

resolving power A measure of the ability of individual components, and of remote-sensing systems, to define closely spaced targets.

retina The lining of the eyeball that includes the light-sensing rods and cones.

return In SLAR systems this is a pulse of microwave energy reflected by the terrain and received at the radar antenna. The strength of a return is referred to as return intensity.

return-beam vidicon A little-used imaging system on Landsat that consists of three cameras operating in the green, red, and photographic IR spectral regions. Instead of using film, the images are formed on the photosensitive surface of a vacuum tube. The image is scanned with an electron beam and transmitted to earth receiving stations.

reversal film A photographic film in which the negative image is converted to a positive image during the developing process.

rods Receptor cells in the retina of the eye that function under low levels of illumination and provide vision in tones of gray.

roll Rotation of an aircraft about the longitudinal axis to cause a wing-up or wing-down attitude.

roll-compensation system The part of a scanner system that measures and records roll of the aircraft. These data are used to correct the imagery for this source of distortion.

rough criterion For radar images this is the relationship between surface roughness, depression angle, and wavelength that determines whether a surface will scatter the incident radar pulse in rough or intermediate fashion.

roughness For radar images this term describes the average vertical relief of small-scale irregularities of the terrain surface.

satellite An object in orbit around a celestial body.

scale The ratio of distance on an image to the equivalent distance on the ground.

scan line The narrow strip on the ground that is swept by the instantaneous field of view of a detector in a scanner system.

scan skew Distortion of scanner images caused by forward motion of the aircraft or satellite during the time required to complete a scan.

scanner An optical-mechanical imaging system in which a rotating or oscillating mirror sweeps the instantaneous field of view of the detector across the terrain. The two basic types of scanners are airborne and stationary.

scanner distortion The geometric distortion that is characteristic of scanner images. The scale of the image is constant in the direction parallel with the aircraft or spacecraft flight direction. At right angles to this direction, however, the image scale becomes progressively smaller from the nadir line outward toward either margin of the image. Linear features, such as roads, that trend diagonally across a scanner image are distorted into S-shaped curves. Distortion is imperceptible for scanners with a narrow angular field of view and becomes more pronounced with a larger angular field of view. Also called panoramic distortion.

scattering Multiple reflection of electromagnetic waves by gases or particles in the atmosphere.

scattering coefficient curves A display of scatterometer data in which relative backscatter from the terrain is shown as a function of incidence angle.

scattering deviation curves A display of scatterometer data in which deviations of backscatter intensity from the regional average are shown as a function of incidence angle.

scatterometer A nonimaging radar device that records backscatter of terrain as a function of incidence angle.

scene The area on the ground that is covered by an image or photograph.

sensitivity The degree to which a detector responds to electromagnetic energy incident upon it.

sensor A device that receives electromagnetic radiation and converts it into a signal that can be recorded and displayed as numerical data or as an image.

side-lobe banding A defect on radar images consisting of light-toned bands parallel with the flight direction and concentrated in the near-range portion of the image.

sidelap The extent of lateral overlap between images acquired on adjacent flight lines.

signature A characteristic, or combination of characteristics, by which a material or an object may be identified on an image or photograph.

silver halide Silver salts that are sensitive to electromagnetic radiation and convert to metallic silver when developed.

simulated normal color image A Landsat color-composite image in which an image for the blue spectral band has been computed. This image and the green and red images are used to produce a normal color composite image.

sixth-line banding A defect on Landsat MSS images in which every sixth scan line is brighter or darker than the others. Caused by the sensitivity of one detector being higher or lower than the others.

sixth-line dropout A defect on Landsat MSS images in which no data are recorded for every sixth scan line, which is black on the image.

Skylab The United States' earth-orbiting workshop that housed three crews of three men in 1973 and 1974.

skylight The component of light that is scattered by the atmosphere and consists predominantly of shorter wavelengths of light.

slant range For radar images this term represents the distance measured along a line between the antenna and the target.

slant-range image For radar images this term represents an image in which objects are located at positions corresponding to their slant-range distances from the aircraft flight path. On slant-range images the scale in the range direction is compressed in the near-range region.

SLAR Acronym of side-looking airborne radar.

smooth criterion For radar images this term represents the relationship between surface roughness, depression angle, and wavelength that determines whether a surface will scatter the incident radar pulse in smooth or intermediate fashion.

software The programs that control computer operations.

specific heat The ratio between thermal capacity of a substance and thermal capacity of water.

spectral reflectance The reflectance of electromagnetic energy at specified wavelength intervals.

spectral sensitivity The response, or sensitivity, of a film or detector to radiation in different spectral regions.

spectrometer Device for measuring intensity of radiation absorbed or reflected by a material as a function of wavelength.

spectrum A continuous sequence of energy arranged according to wavelength or frequency.

specular Refers to a surface that is smooth with respect to the wavelength of incident radiation.

stationary scanner A scanner designed for use from a fixed location. Coverage in the horizontal direction is provided by a faceted mirror that rotates about a vertical axis. Coverage in the vertical direction is provided by a planar mirror that tilts about a horizontal axis. Stationary scanner images are individual frames, rather than the continuous strips acquired by airborne scanners.

Stefan-Boltzmann constant $5.68 \cdot 10^{-12}$ W $\cdot$ cm^{-2} $\cdot$ °K^{-4}.

Stefan-Boltzmann law States that radiant flux of a black body is equal to the temperature to the fourth power times the Stefan-Boltzmann constant.

stereo model A three-dimensional mental impression produced by viewing the left and right images of an overlapping pair with the left and right eye, respectively.

stereo pair Two overlapping images or photographs that may be viewed stereoscopically.

stereoscope A binocular optical device for viewing overlapping images or diagrams to obtain the mental impression of a three-dimensional model.

subtractive primary colors Yellow, cyan, and magenta. When used as filters for white light these colors remove blue, red, and green, respectively.

sun synchronous An earth satellite orbit in which the orbit plane is near polar and the altitude such that the satellite passes over all places on earth having the same latitude twice daily at the same local sun time.

surface phenomenon Interaction between electromagnetic radiation and the surface of a material.

synthetic-aperture radar SLAR system in which high resolution in the azimuth direction is achieved by utilizing the Doppler principle to give the effect of a very long antenna.

synthetic stereo images A stereo model made by digital processing of a single image. Topographic data are used in calculating the geometric distortion.

systematic distortion Geometric irregularities on images that are caused by the characteristics of the imaging system and are predictable.

target An object on the terrain of specific interest in a remote-sensing investigation.

telemeter To transmit data by radio or microwave links.

terrain The surface of the earth.

texture The frequency of change and arrangement of tones on an image.

thermal capacity The ability of a material to store heat, expressed in cal $\cdot$ g^{-1} $\cdot$ °C^{-1}. The symbol is c.

thermal conductivity The measure of the rate at which heat passes through a material, expressed in cal $\cdot$ cm^{-1} $\cdot$ sec^{-1} $\cdot$ °C^{-1}. The symbol is K.

thermal cross over On a plot of radiant temperature versus time, this refers to the point at which the temperature curves for two different materials intersect.

thermal diffusivity Governs the rate at which temperature changes within a substance, expressed in cm^{2} $\cdot$ sec^{-1}. The symbol is k.

thermal inertia A measure of the response of a material to temperature changes, expressed in cal $\cdot$ cm^{-2} $\cdot$ sec$^{-1/2}$. The symbol is P.

thermal IR The portion of the IR region from approximately 3 to 14 μm that corresponds to heat radiation. This spectral region spans the radiant power peak of the earth.

thermal IR image An image acquired by a scanner that records radiation within the electromagnetic band that ranges from approximately 3 to 14 μm in wavelength.

thermal model A mathematical expression that relates thermal and other physical properties of a material to its temperature. Models may be used to predict temperature for given properties and conditions. Thermal properties may be estimated on the basis of observed variations in temperature.

thermography The medical applications of thermal IR images. Images of the body, called thermograms, have been used to detect tumors and monitor blood circulation.

tone Each distinguishable shade of gray from white to black on an image.

trade off The compensating change that occurs in a remote-sensing system as a result of changing one factor elsewhere in that system. For example, an increase in altitude increases the lateral coverage of an imaging system but the trade off is a decrease in scale and resolution of the image.

transmissivity The property of a material that determines the amount of energy that can pass through the material.

transparency A positive or negative image on a transparent photographic material. The capability of a material to transmit light.

transpiration The production of water vapor and oxygen by vegetation.

travel time In radar systems this term refers to the time interval between transmission of a pulse of microwave energy and its return from the terrain.

UV The ultraviolet region consisting of wavelengths from 0.01 to 0.4 μm.

vegetation anomaly A deviation from the normal distribution pattern or properties of vegetation. May be caused by variations in moisture, trace elements, or other factors in the soil.

vertical exaggeration In a stereo model this is the extent to which the vertical scale is larger than the horizontal scale.

VICAR Video Image Communication and Retrieval. A software system for digital processing of Landsat and other image data.

visible radiation Energy at wavelengths from 0.4 to 0.7 μm that is detectable by the eye.

volume phenomenon The interaction between electromagnetic radiation and the interior of a material.

watt Unit of electrical power equal to rate of work done by one ampere under a pressure of one volt. The symbol is W.

wavelength The distance between successive wave crests, or other equivalent points, in a harmonic wave. The symbol is λ.

Wien's displacement law Describes the shift of the radiant power peak to shorter wavelengths with increasing temperature.

X band Radar wavelength region from 2.4 to 3.8 cm.

yaw Rotation of an aircraft about its vertical axis to cause the longitudinal axis to deviate from the flight line.

zenith The point in the celestial sphere that is exactly overhead.

INDEXES

NAME INDEX

LOCALITY INDEX

SUBJECT INDEX

5.18
5.19
7.26
5.12
10.8
5.20
7.19
7.8
7.7
7.20
7.21
2.20
PL. 1
6.4A
9.3
4.11
4.12
4.13
8.1
8.12
8.13
8.24
8.25
9.4
8.6
8.7
8.8
PL. 6
3.7
3.8
7.12
7.13
PL. 5
8.2
8.3
5.15
9.20
8.14
PL. 7
4.3, PL. 2
4.5
7.17
7.18
4.14
4.15
3.6
6.29
9.5
6.10
7.6
3.1
3.2
6.25
6.13
6.14
6.17
6.18
6.26B
10.3
5.33
6.26A
9.10
9.9
6.39
5.23
5.24
6.36
11.1–11.8
2.7
2.8
2.15
5.26
6.31
6.20
6.30
2.10
6.24
6.26C
6.4B
5.30
7.24